Predicting, Monitoring, and Assessing Forest Fire Dangers and Risks

Nikolay Viktorovich Baranovskiy
National Research Tomsk Polytechnic University, Russia

A volume in the Advances in Environmental Engineering and Green Technologies (AEEGT) Book Series

IGI Global
DISSEMINATOR OF KNOWLEDGE

Published in the United States of America by
IGI Global
Engineering Science Reference (an imprint of IGI Global)
701 E. Chocolate Avenue
Hershey PA, USA 17033
Tel: 717-533-8845
Fax: 717-533-8661
E-mail: cust@igi-global.com
Web site: http://www.igi-global.com

Library of Congress Cataloging-in-Publication Data

Names: Baranovskiy, Nikolay Viktorovich, 1978- editor.
Title: Predicting, monitoring, and assessing forest fire dangers and risks / Nikolay Viktorovich Baranovskiy, editor.
Description: Hershey PA : Engineering Science Reference, [2020] | Includes bibliographical references and index. | Summary: "This book examines current research on predicting, monitoring, and assessing forest fire dangers and risks"-- Provided by publisher.
Identifiers: LCCN 2019034110 (print) | LCCN 2019034111 (ebook) | ISBN 9781799818670 (hardcover) | ISBN 9781799818687 (paperback) | ISBN 9781799818694 (ebook)
Subjects: LCSH: Forest fires--Prevention and control.
Classification: LCC SD421 .P64 2020 (print) | LCC SD421 (ebook) | DDC 634.9/618--dc23
LC record available at https://lccn.loc.gov/2019034110
LC ebook record available at https://lccn.loc.gov/2019034111

This book is published in the IGI Global book series Advances in Environmental Engineering and Green Technologies (AEEGT) (ISSN: 2326-9162; eISSN: 2326-9170)

British Cataloguing in Publication Data
A Cataloguing in Publication record for this book is available from the British Library.

Advances in Environmental Engineering and Green Technologies (AEEGT) Book Series

Sang-Bing Tsai
University of Electronic Science and Technology of China Zhongshan Institute, China
Ming-Lang Tseng
Lunghwa University of Science and Technology, Taiwan
Yuchi Wang
University of Electronic Science and Technology of China Zhongshan Institute, China

ISSN:2326-9162
EISSN:2326-9170

Mission

Growing awareness and an increased focus on environmental issues such as climate change, energy use, and loss of non-renewable resources have brought about a greater need for research that provides potential solutions to these problems. Research in environmental science and engineering continues to play a vital role in uncovering new opportunities for a "green" future.

The **Advances in Environmental Engineering and Green Technologies (AEEGT)** book series is a mouthpiece for research in all aspects of environmental science, earth science, and green initiatives. This series supports the ongoing research in this field through publishing books that discuss topics within environmental engineering or that deal with the interdisciplinary field of green technologies.

Coverage

- Electric Vehicles
- Cleantech
- Green Technology
- Renewable Energy
- Policies Involving Green Technologies and Environmental Engineering
- Alternative Power Sources
- Contaminated Site Remediation
- Pollution Management
- Biofilters and Biofiltration
- Waste Management

IGI Global is currently accepting manuscripts for publication within this series. To submit a proposal for a volume in this series, please contact our Acquisition Editors at Acquisitions@igi-global.com or visit: http://www.igi-global.com/publish/.

The Advances in Environmental Engineering and Green Technologies (AEEGT) Book Series (ISSN 2326-9162) is published by IGI Global, 701 E. Chocolate Avenue, Hershey, PA 17033-1240, USA, www.igi-global.com. This series is composed of titles available for purchase individually; each title is edited to be contextually exclusive from any other title within the series. For pricing and ordering information please visit http://www.igi-global.com/book-series/advances-environmental-engineering-green-technologies/73679. Postmaster: Send all address changes to above address.

Titles in this Series

For a list of additional titles in this series, please visit: https://www.igi-global.com/book-series/advances-environmental-engineering-green-technologies/73679

Nano-Phytoremediation Technologies for Groundwater Contaminates Emerging Research and Opportunities
Khushboo Chaudhary (Banasthali University, India) Suphiya Khan (Banasthali University, India) and Pankaj Kumar Saraswat (Krashi Vigyan Kendra Banasthali University, India)
Engineering Science Reference • © 2020 • 150pp • H/C (ISBN: 9781522590163) • US $165.00

Implementation and Evaluation of Green Materials in Technology Development Emerging Research and Opportunities
M.T. Mastura (Universiti Teknikal Malaysia Melaka, Malaysia) and S. M. Sapuan (Universiti Putra Malaysia, Malaysia)
Engineering Science Reference • © 2020 • 215pp • H/C (ISBN: 9781799813743) • US $175.00

Modern Techniques for Agricultural Disease Management and Crop Yield Prediction
N. Pradeep (Bapuji Institute of Engineering and Technology, India) Sandeep Kautish (Lord Buddha Education Foundation, Nepal & Asia Pacific University, Malaysia) C.R. Nirmala (Bapuji Institute of Engineering and Technology, India) Vishal Goyal (Punjabi University, Patiala, India) and Sonia Abdellatif (Seoul National University, South Korea)
Engineering Science Reference • © 2020 • 291pp • H/C (ISBN: 9781522596325) • US $195.00

Novel Advancements in Electrical Power Planning and Performance
Smita Shandilya (Sagar Institute of Research, Technology and Science, India) Shishir Kumar Shandilya (Vellore Institute of Technology, India) Tripta Thakur (Maulana Azad National Institute of Technology, India) and Atulya K. Nagar (Liverpool Hope University, UK)
Engineering Science Reference • © 2020 • 388pp • H/C (ISBN: 9781522585510) • US $225.00

Global Food Politics and Approaches to Sustainable Consumption Emerging Research and Opportunities
Luke Amadi (University of Port Harcourt, Nigeria) and Fidelis Allen (University of Port Harcourt, Nigeria)
Engineering Science Reference • © 2020 • 242pp • H/C (ISBN: 9781799801252) • US $185.00

Fuzzy Expert Systems and Applications in Agricultural Diagnosis
A.V. Senthil Kumar (Hindusthan College of Arts and Science, India) and M. Kalpana (Tamil Nadu Agricultural University, India)
Engineering Science Reference • © 2020 • 335pp • H/C (ISBN: 9781522591757) • US $245.00

Handbook of Research on Environmental and Human Health Impacts of Plastic Pollution

IGI Global
DISSEMINATOR OF KNOWLEDGE

701 East Chocolate Avenue, Hershey, PA 17033, USA
Tel: 717-533-8845 x100 • Fax: 717-533-8661
E-Mail: cust@igi-global.com • www.igi-global.com

Editorial Advisory Board

Table of Contents

Detailed Table of Contents

The chapter presents the results of experimental-analytical modeling of the surface forest fire dynamics and the process of forest fuel ignition when exposed to thermal radiation from the fire line. The regularities are established for the occurrence and spread of fires in natural ecosystems of the temperate climatic zone. Analytical solutions have been obtained that make it possible to predict the level of heat load on the soil cover of coniferous stands. The special computer program has been developed to calculate the heat load during fires. The methods of field and laboratory modeling revealed patterns of forest fuel heating and ignition depending on moisture content. A practice-oriented method is proposed to calculate the width of fire barriers that limit the spread of forest fires. The methods for creating fire barriers are proposed.

The chapter presents a mathematical model of the initiation and spread of the steppe fire. The mathematical model is based on the laws of mechanics of multiphase reacting media. The main physicochemical processes describing the drying, pyrolysis, and combustion of gaseous and condensed pyrolysis products are taken into account. As a result of the numerical solution, the distributions of the velocity, temperature, and concentration fields of the components of the gas and condensed phases were determined. The dependence of the rate of spread of the steppe fire on the main parameters of the state of vegetation cover and wind speed was studied. The mathematical model presented in the chapter can be used to predict the spread of steppe fires for various types of steppe vegetation and meteorological conditions, as well as for preventive measures.

Section 2

An anthropogenic load is the main cause of forest fires in the vicinity of settlements and various objects of transport or industrial infrastructure. It is proposed to mark out linear and point sources of anthropogenic load. For the numerical simulation, the similarity in the processes of propagation of anthropogenic load and diffusive heat transfer was used. The quantitative characteristic of the anthropogenic load is the virtual (possible) number of forest fires in the controlled forest area. A mathematical model to predict the distribution of anthropogenic load from a point source is presented. Distributions of the virtual number of forest fires for model data from a point source of anthropogenic load are obtained. Conclusions about the patterns of the distribution of anthropogenic load from a point source are formulated. The prospects for the further development of these results are described.

The annual task of forecasting forest fire danger is becoming increasingly relevant, especially in the context of global warming. The forecast of surface fires is most important, as more than 80% of all vegetation fires are surface fires. Practically all crown fires develop from surface fires. This chapter discusses the deterministic-probabilistic method for predicting the number of forest fires in a controlled forest area. This methodology is based on the assumption that the number of registered and projected forest fires is related to the probability of their occurrence. The influence of forest fire retrospective data on the predicted number of forest fires for some sites of the Timiryazevskiy forestry of the Tomsk region was studied. This chapter presents the results of a comparative analysis of forecast data and statistics.

The chapter develops a dynamic spatial-distributed model of forest fire risk. The structure of forest fire risk is represented by two interrelated groups of components describing a potential of forest fire and describing valuable objects being under fire influence. The concept of fire risk which contains the probability of forest fire occurrence, its intensity, and effect, is extended using the threat as the additional prognostic spatial-temporal component that has a predictive property and allows forecasting the possibility of losses at any time. The model of fire risk is based on three stages: potential risk, the source of which is described by fire danger; risk of threat of active forest fire which hasn't covered valuable object yet; risk of destruction when the active fire has covered valuable object. Representation of risk as a process based on three stages allows describing the dynamics of risk in real-time systems, getting insight into risk nature, as well as diagnosing the situation in real time.

the Tomsk region is considered. The causes of fire emergence is revealed. The forecast of forest fires is given and zoning of forest areas of the region as for fire danger is carried out. The research identifies the role of natural conditions in the occurrence of forest fires at various territorial levels (continents, countries, regions, areas).

The impact of smoke from forest fires in western Siberia on meteorological, atmospheric electric, and aerological variables has been analyzed. The anomalous distribution of water vapor in the atmosphere associated with the peculiarities of the evaporation regime and the absence of advective moisture transfer over the southern regions of Western Siberia during the fires. With an increase in the height of the homogeneous surface smoke layer with an unchanged aerosol optical thickness, the cooling of the earth's surface and heating of the atmosphere was weakened. The smoke plume spreads predominantly in the middle of the troposphere, creating aerosol layers elevated above the ground, the lower part of which had a negative volume charge. The effect of diurnal variations in the electrical field in the near-surface layer, differs from the known similar effects.

The chapter focuses on the role of fire as an environmental factor affecting the components of the forest ecosystem. The classification of forest fires according to the fire location in the forest ecosystem is given. Negative impact of fire on communities and people, components of forest ecosystems: a phytocoenosis (the forest under-growth, understory, ground cover), tocinos (animal world), klimatop (atmosphere and climate) and adatom (topography and soil) are discussed. The role of fire as a factor of organisms and natural ecosystems' evolution in general is studied. The problems of forest restoration on fires (fumes) are considered. Research established that the degree of damage to the ecosystem components and social consequences depend on the forest fire type, natural conditions, and forest conditions of the region.

Foreword

Forest fires have become more frequent in recent years, and their consequences are increasingly widespread in the forested territories of the states of Europe, Asia, America and Australia, despite the considerable formal efforts undertaken by the governments of most interested states to suppress forest fires. Most likely, the low efficiency of this fight is due to the lack of a general theory of forest fires as a knowledge system sufficient to significantly reduce their number and the severity of their consequences. The monograph is an important contribution necessary for the practice of suppression forest fires as a knowledge system.

It should be noted, that so far the main directions of research in this field of science have been formed: prediction of forest fire danger, prediction of characteristics of forest fire propagation processes, development of methods for suppressing the burning of forest fuels, prediction of environmental consequences of forest fires. The monograph presents the results of a large group of authors in three areas. An analysis of the content of the main results of the book shows that the main aspects of forest fire danger are considered from modern perspectives.

It can be noted, that so far, for example, in the Russian Federation there are no unambiguous conclusions and, accordingly, methods for preparing such conclusions about what is the main cause of fires in the forests. So many administrators of both the Ministry of Emergency Situations and the municipal authorities believe that there are citizens of the Russian Federation and other countries interested in burning forests that carry out these arsons. Such causes of forest fires, of course, are quite likely due to the low cost of wood remaining after the fires and then used in various areas of processing. But such a motivation is difficult to explain a number of fires, for example, in Europe or Siberia. In the Siberian taiga, forest fires often occur not only in remote, but also inaccessible areas. It is impossible to remove the wood from such territory. A large fire in California was found to be not the result of arson. Similar examples can be given from forest fire practices in France, Italy and Portugal. In this regard, studies devoted to the development of the theory of forest fire danger are certainly useful and have practical value, since they provide the basis for the analysis of objective rather than subjective causes of fires in the forest.

Similarly, it is possible to evaluate studies of the spread of forest fires. It is impossible to analyze the propagation rates over a forested area in the framework of simple mathematical models that take into account only a small group of the main significant factors. The studies and results presented in the monograph show that the velocity of the combustion front depends not only on the density of distribution of forest fuels and wind speed, but also on many other factors.

Analysis of the negative impact of forest fires on nature is also an important area of research. The monograph presents the results that are the foundation for creating effective forecast systems and, consequently, forest fire prevention. This will help minimize the damage they cause (environmental and economic).

The monograph can be useful to scientists and specialists, practitioners working in state forest services, postgraduate students, graduate and undergraduates ones, as it gives a fairly objective idea of the current level of research on forest fires.

Geniy Vladimirovich Kuznetsov
Tomsk Polytechnic University, Russia

Preface

Forest fires pose a great danger to the population and ecology of many states with forested territories. Forest fires cause environmental, economic and social damage (Baranovskiy and Kuznetsov, 2017). The problem of forest fires is multifaceted and does not have a trivial solution. The following subtasks in the indicated problem can be highlighted. First, forecasting, assessing and monitoring the occurrence of forest fires (Baranovskiy and Kuznetsov, 2017). Second, assessment and modeling of the forest fire spread (Lopes et al, 2019). Third, the assessment and forecasting of the burned-out area and the environmental consequences of fire impact (Stirling et al., 2019). Fourth, localization and extinguishing of forest fires, both by ground methods and with the help of aviation (Grishin, 1997; Szelmanowski et al., 2020). Fifth, the elimination of the forest fires consequences. On the issue of combating forest fires, approaches to forecasting the occurrence of foci of forest fires and their subsequent rapid localization and extinguishing can be highlighted (Baranovskiy and Zharikova, 2014). Or you have to extinguish a large forest fire (Fernández-Guisuraga et al, 2019). As practice shows, it is almost impossible to extinguish a forest fire when it has become large. A fire stops spreading when the rainy season begins, when the whole forested area burned out, or when meeting a natural barrier, for example, a river. Therefore, it is the methods for forecasting forest fire danger that are most promising in terms of suppression forest fires.

Currently, there are various methods and approaches to forecasting forest fire danger, for example, deterministic, deterministic-probabilistic, statistical, deterministic-statistical, probabilistic and soft calculation methods (Yatat et al., 2017; Baranovskiy, 2017; Vivalda et al., 2017; Grishin and Filkov, 2011; Gray et al., 2018; Tehrany et al., 2019). Typically, methods based on the analysis of statistical information work well if the external conditions do not change much over time. However, in the event of sudden weather or climate changes, such methods may give incorrect results. Nevertheless, one should not completely abandon statistical approaches to forecasting and assessing forest fire danger. Probably, a successful solution to the problem of forecasting and assessing forest fire danger lies in the plane of the integrated and joint application of various methods.

It should be noted that the ability to obtain initial data for soft or hard calculations of parameters and conditions for the occurrence of forest fires is important in solving the problem of forecasting forest fire danger. In this case, it is reasonable to use statistical data, the results of experimental studies, as well as data arrays from various satellite systems in the complex (Santi et al., 2017; Frazier et al., 2018). The abundance of spatial data leads to the need to use specialized geographic information systems for their analysis and visualization (Khairunnisa et al., 2019).

No matter what advanced mathematical technologies for predicting fires in the forest have been developed, but without taking into account the causes of their occurrence, the problem cannot be solved. The main causes of forest fires are thunderstorm activity and anthropogenic load (Abdollahi et al., 2019;

Rodrigues et al., 2018). This book provides readers of the world scientific community with interconnected information on various aspects of forecasting, assessing and monitoring forest fires. In forestry, there is a need to analyze the processes of occurrence of forest fires at a fairly accurate level with high spatial resolution, taking into account the complex physicochemical phenomenon of a forest fire. The book gives in many respects an answer to how it is necessary to organize forecasting, assessment and monitoring of forest fires in modern conditions.

High-performance computing equipment is now available, which allows the use of complex mathematical models to predict the occurrence of forest fires. At the same time, the development of remote sensing (both satellite platforms and specialized software) allows the use of big data to analyze the occurrence of forest fires. In addition, the book provides partial answers to a number of questions related to the environmental impact of forest fires. This is important information, because in a real situation, the forest fire-environment system is a feedback system. In this situation, forest fires affect the state of the environment, the parameters of which in turn affect the conditions of forest fires. This book meets the modern challenges of society in the fight against forest fires and is consistent with global trends in the development of this area, offering new, relevant information on modern methods for forecasting, assessing and monitoring forest fire danger.

The target audience of this book will be composed of professionals and researchers working in the field of Predicting, Monitoring and Assessing of Forest Fire Danger and Risk disciplines, e.g. forest fire danger factors, forest fire weather, forest fuel, fire statistics, software and devices to be used in related forest fire danger researches, GIS analysis and remote sensing, fire risk and different methods to predict, monitor and assess forest fire danger and risk. Moreover, the book will provide insights and support executives concerned with the management of in forestry to prevent forest fires risk. Furthermore, this book can be useful for insurance companies, computer engineers who develop new system to predict, monitor and assess forest fire danger and risk.

Section 1 is devoted to the physical and mathematical modeling of forest fires. Chapter 1 is devoted to the issues of physical modeling of forest fuel ignition by a radiant heat flux and the subsequent propagation of a surface forest fire through a layer of forest fuel. The advantage of this study is that the behavior of a forest fire in a real forest was studied. These data can be useful both for validating existing mathematical models and for developing new mathematical models.

Chapter 2 is devoted to mathematical modeling of the spread of grassland fire. It is the spread of grass fire and its subsequent entry into the forested area that often causes forest fires. This chapter discusses the mathematical model of fire propagation. As already noted, the future of forest fire forecasting is associated with deterministic methods and approaches.

Section 2 is dedicated to forest fire danger modeling. Chapter 3 offers a description of a new mathematical model of the distribution of anthropogenic load in a forested area. A point source of anthropogenic load is considered, which can be a settlement, industrial facility and camping. A new deterministic mathematical model is proposed for assessing the distribution of anthropogenic load based on the similarity of diffusion heat transfer processes and the distribution of anthropogenic load.

Chapter 4 proposes a new approach to the deterministic-probabilistic forecast of the number of forest fires in a controlled forested area. It is known, that the occurrence of a forest fire depends on current conditions, as well as the long-term retrospective of conditions affecting the occurrence of a forest fire. The chapter proposes a new mathematical model for forecasting the expected number of forest fires. The forecasting period is one day, that is, it is a short-term forecasting model. The probability of a forest fire takes into account anthropogenic load, thunderstorm activity and forest conditions. The results of a retrospective analysis of the model in the territory of the Timiryazevsky forestry of Tomsk region (Russian Federation) are presented.

Chapter 5 offers a methodology for analyzing fire risk taking into account its spatial dynamics. The methodology offers two groups of parameters describing the probability of a forest fire describing objects that may be influenced by forest fires, which will make it possible to predict losses from a forest fire in the future. Three risk categories are introduced: 1) potential fire risk or fire danger; 2) the risk of a forest fire affecting the facility; 3) the risk of destruction of the object.

Section 3 is devoted to the study of thunderstorm activity. Chapter 6 proposes the results of the assessment of the territory in accordance with the degree of thunderstorm danger. As you know, thunderstorm activity is the main natural cause of forest fires. The dynamics of thunderstorm activity in the mountains of southern Siberia (the Republic of Altai and the Republic of Buryatia) is investigated. The data from the World Wide Lightning Location Network for the period from 2011 to 2016 was used. There are daily and seasonal trends in the number of lightning discharges with a spatial resolution of 10x10 kilometers.

Chapter 7 offers the results of a cluster analysis of lightning discharges using data from the World Wide Lightning Location Network over the territory of the Tomsk Region, namely the Timiryazevskiy Forestry. The data on lightning discharges for July 2014 were used. The data included 273 lightning discharges. Cluster analysis was carried out using various algorithms, for example, complete, k-means, dbscan. The information obtained can be used to predict forest fire danger caused by lightning activity.

Section 4 is devoted to the use of geographic information systems for forecasting and assessing forest fire danger. Chapter 8 offers the results of an assessment of forest fire danger in the Republic of Buryatia using meteorological trends. The data for more than 20 recent years has been analyzed. It is noted that there is an increase in ambient temperature and climate aridity in this territory. It was noted that the main forest fires occur in the river valley, where coniferous pine forests with soils of low moisture capacity are located, and the smallest number of forest fires is observed in the mountains, where the main amount of liquid precipitation occurs.

Chapter 9 provides information on the use of geographic information systems for studying fire risks in the Jewish Autonomous Region. The description of the software package for assessing forest fire danger is presented. The structure of the software: 1) the collection and processing of data from ground-based territory research; 2) forecasting the period and place of spread of fire vegetation to the nearest forest; 3) probabilistic assessment of forest fire danger; 4) action planning of forest protection services. Software verification was carried out for the territory of the Jewish Autonomous Region.

Section 5 is devoted to remote sensing of the earth's surface. Chapter 10 offers the results of an analysis of vegetation trends and the correlation of vegetation indices and climatic factors. Climatic variables were considered such as LST, Net Radiation and TRMM. The results showed that a positive vegetative trend can be explained by an increase in vegetation growth and its productivity. The ten-year NDVI and EVI cycles in the districts of Rajket, Dwarha, Porbandar, Amreit, Jamnagar were considered. The data from the MODIS device from the Terra / Aqua platform was used.

Chapter 11 offers the results of classifying a territory according to the level of forest fire danger using geographic information systems and remote sensing. Images from the Landsat ETM + satellite platform were used. Satellite analysis was conducted using controlled and unmanaged classification. Areas of the territory covered with different vegetation were identified. As is known, coniferous forests pose the greatest danger. Mixed forests represent the average fire danger, and deciduous forests represent the low one. For data visualization, ArcGIS software was used.

Section 6 is devoted to the environmental impact of forest fires. Chapter 12 offers a study on the impact of climate on forest burns using the taiga of Western Siberia as an example. The following climatic variables are considered: thunderstorm activity, rains, wind, periods of high and low ambient temperatures. A description of the situation with forest fires in the Tomsk region using the data of the MODIS device from the Terra / Aqua platform is given. The information on meteorological data from several stations of the Tomsk region is summarized. The fire dangerous season of 2012 in the Tomsk Region is considered separately.

Chapter 13 provides a description of the conditions of the territory's predisposition to forest fires on different continents of the world. Information is provided on the types of forest fires and forest fuels. The chapter gives the dynamics of forest fires in Europe, the USA, Canada, Southeast Asia and the Russian Federation. Forest fires in the Russian Federation, in particular, in the territory of the Tomsk region, have been studied in detail. As a result of the study, the role of natural conditions in the occurrence of forest fires in territories of different scales (continents, countries, regions, individual sites) was identified.

Section 7 describes the environmental impact of forest fires. Chapter 14 offers the results of an analysis of the effects of smoke from forest fires in Western Siberia on meteorological conditions, atmospheric electricity, and aerological characteristics. The abnormal distribution of water vapor in the atmosphere of the southern part of Western Siberia is associated with evaporation regimes and the absence of advective moisture transfer during forest fires. With an increase in the height of a uniform smoke layer with a constant optical thickness of the aerosol, cooling of the earth's surface and heating of the atmosphere decrease. Plume from forest fires mainly spreads in the middle troposphere.

Chapter 15 offers an overview of the effects of fires on forest ecosystems. The negative impact of forest fires on people and ecosystem components is discussed (phytocenoses, wildlife, atmosphere and soil). The issue of reforestation is also addressed. It has been established that the degree of damage to the ecosystem depends on the type of forest fire. The effect of forest fires on trees, vegetation, animals, microorganisms, and worms is considered.

The output of this book will help strengthen the paradigm of deterministic modeling in the tasks of forecasting, assessing, and monitoring the occurrence of forest fires and their consequences. A successful solution to the problem of forecasting forest fires with the aim of preventing them and reducing the negative impact on humans and ecosystems is possible only through a comprehensive study of these problems at an interdisciplinary level with the involvement of scientists from different countries. This book contributes to the development of interdisciplinary integration of scientists from different countries on the path to solving the problems of forecasting, assessing and monitoring the occurrence of forest fires and their environmental consequences. This book can serve as a prototype for future comprehensive multidisciplinary international forest fire research. Perhaps the publication of this book will become the foundation for the formation of a new multidisciplinary international work group to solve the problem of physically sound forecasting, assessment and monitoring of forest fires and their environmental consequences.

Nikolay Viktorovich Baranovskiy
Tomsk Polytechnic University, Russia

REFERENCES

Abdollahi, M., Dewan, A., & Hassan, Q. K. (2019). Applicability of remote sensing-based vegetation water content in modeling lightning-caused forest fire occurrences. *ISPRS International Journal of Geo-Information*, *8*(3), 143. doi:10.3390/ijgi8030143

Baranovskiy, N., & Zharikova, M. (2014). A web-oriented geoinformation system application for forest fire danger prediction in typical forests of the Ukraine, Lecture Notes in Geoinformation and Cartography. *199669*, 13–22.

Baranovskiy, N. V. (2017). Forest fire hazard assessment using SPMD-model of computation for massive parallel system. *International Review on Modeling and Simulations.*, *10*(3), 193–201. doi:10.15866/iremos.v10i3.10570

Baranovskiy, N. V., & Kuznetsov, G. V. (2017). *Forest fire occurrences and ecological impact prediction: monograph*. Novosibirsk, Russia: Publishing House of Siberian Branch of Russian Academy of Science. doi:10.15372/FOREST2017BNV

Fernández-Guisuraga, J. M., Suárez-Seoane, S., & Calvo, L. (2019). Modeling Pinus pinaster forest structure after a large wildfire using remote sensing data at high spatial resolution. *Forest Ecology and Management*, *446*, 257–271. doi:10.1016/j.foreco.2019.05.028

Frazier, R. J., Coops, N. C., Wulder, M. A., Hermosilla, T., & White, J. C. (2018). Analyzing spatial and temporal variability in short-term rates of post-fire vegetation return from Landsat time series. *Remote Sensing of Environment*, *205*, 32–45. doi:10.1016/j.rse.2017.11.007

Gray, M. E., Zachmann, L. J., & Dickson, B. G. (2018). A weekly, continually updated dataset of the probability of large wildfires across western US forests and woodlands. *Earth System Science Data*, *10*(3), 1715–1727. doi:10.5194/essd-10-1715-2018

Grishin, A. M. (1997). *Mathematical modeling of forest fire and new methods of fighting them*. Tomsk, Russia: Publishing House of the Tomsk State University.

Grishin, A. M., & Filkov, A. I. (2011). A deterministic-probabilistic system for predicting forest fire hazard. *Fire Safety Journal*, *46*(1-2), 56–62. doi:10.1016/j.firesaf.2010.09.002

Khairunnisa, F., & Ratna Rosalena, I. (2019) GIS application in mapping of forest and land fires and mitigation in Banjarbaru City, South Kalimantan. In *Proceedings IOP Conference Series: Earth and Environmental Science*. Vol. 273. 012049. 10.1088/1755-1315/273/1/012049

Lopes, A. M. G., Ribeiro, L. M., Viegas, D. X., & Raposo, J. R. (2019) Simulation of forest fire spread using a two-way coupling algorithm and its application to a real wildfire, Journal of Wind Engineering and Industrial Aerodynamics, 193, 103967.

Rodrigues, M., Jiménez-Ruano, A., Peña-Angulo, D., & de la Riva, J. (2018). A comprehensive spatial-temporal analysis of driving factors of human-caused wildfires in Spain using Geographically Weighted Logistic Regression. *Journal of Environmental Management*, *225*, 177–192. doi:10.1016/j.jenvman.2018.07.098 PMID:30081279

Santi, E., Paloscia, S., Pettinato, S., Fontanelli, G., Mura, M., Zolli, C., ... Chirici, G. (2017). The potential of multifrequency SAR images for estimating forest biomass in Mediterranean areas. *Remote Sensing of Environment, 200*, 63–73. doi:10.1016/j.rse.2017.07.038

Stirling, E., Smernik, R. J., Macdonald, L. M., & Cavagnaro, T. R. (2019). Fire influences needle decomposition: Tipping point in Pinus radiata carbon chemistry and soil nitrogen transformations. *Soil Biology & Biochemistry*, *135*, 361–368. doi:10.1016/j.soilbio.2019.05.024

Szelmanowski, A., Ziej, M., Pazu, A., & Głyd, K. (2020). Studying the dynamic properties of an amplifier board execution block in terms of false tripping of an aircraft fire suppression system. *Mechanisms and Machine Science*, *70*, 121–137. doi:10.1007/978-3-030-13321-4_11

Tehrany, M. S., Jones, S., Shabani, F., Martínez-Álvarez, F., & Tien Bui, D. (2019). A novel ensemble modeling approach for the spatial prediction of tropical forest fire susceptibility using LogitBoost machine learning classifier and multi-source geospatial data. *Theoretical and Applied Climatology*, *137*(1-2), 637–653. doi:10.100700704-018-2628-9

Vivalda, C., Verda, V., Carpignano, A., Dell'Erba, C., Cagliero, D., & Guelpa, E. (2017) Forest fire risk analysis methods and simulation tools, Safety and Reliability - Theory and Applications. In *Proceedings of the 27th European Safety and Reliability Conference, ESREL 2017,* pp. 1117-1124.

Yatat, V., Couteron, P., Tewa, J. J., Bowong, S., & Dumont, Y. (2017). An impulsive modelling framework of fire occurrence in a size-structured model of tree–grass interactions for savanna ecosystems. *Journal of Mathematical Biology*, *74*(6), 1425–1482. doi:10.100700285-016-1060-y PMID:27659304

Acknowledgment

Work with book was partially supported by the Russian Foundation for Basic Researches (Grant N 17-29-05093) and by Tomsk Polytechnic University CE Program.

Section 1

Chapter 1
The Dynamics of Surface Forest Fire and Forest Fuel Ignition Under the Heat Radiation From the Fire Line

Pavel Nikolaevich Goman
University of Civil Protection of the Ministry for Emergency Situations, Belarus

ABSTRACT

The chapter presents the results of experimental-analytical modeling of the surface forest fire dynamics and the process of forest fuel ignition when exposed to thermal radiation from the fire line. The regularities are established for the occurrence and spread of fires in natural ecosystems of the temperate climatic zone. Analytical solutions have been obtained that make it possible to predict the level of heat load on the soil cover of coniferous stands. The special computer program has been developed to calculate the heat load during fires. The methods of field and laboratory modeling revealed patterns of forest fuel heating and ignition depending on moisture content. A practice-oriented method is proposed to calculate the width of fire barriers that limit the spread of forest fires. The methods for creating fire barriers are proposed.

1. INTRODUCTION

In recent years, due to global warming in a number of countries around the world, the situation with forest fires has become especially aggravated. As a result of fires, the ecology of the regions is deteriorating, natural ecosystems and settlements are being destroyed, rescuers and people are dying. In addition, layers of dust and soot from fires lead to the melting of glaciers, which affects the rise in sea level and creates the likelihood of flooding vast coastal areas around the world.

The largest forest fires of the last decade include the fires that occurred in 2009 in Australia, which destroyed about 450 thousand hectares of forest and led to the death of 173 people. In Russia in 2010, as a result of the abnormal heat, forest fires covered an area of about 2 million hectares, more than 60

DOI: 10.4018/978-1-7998-1867-0.ch001

people died, 2.5 thousand families were left homeless. 111 people became victims of forest fires in Portugal in 2017. A significant part of the victims were motorists who tried to evacuate to safe areas, but fell into a fire trap on one of the roads (Figure 1). Forest fires in California in 2018 were the largest in the entire history of the state. As a result of fires, 85 people died, about 250 people are missing. The fire destroyed about 14 thousand houses (Abduragimov, 2012; Bondur, 2011; Artsybashev, 2014; Lasuta & Goman, 2019).

Special danger is represented by fires in radiation-polluted territories, the situation with which has also aggravated in recent years. So in 2015, 2017 and 2018 large fires occurred in the Chernobyl Exclusion Zone in Ukraine(Lasuta & Goman, 2019; Bogorad et al., 2016). In these forest plantations, due to the restriction of forestry activities, a large amount of combustible material is accumulated, which increases the fire hazard and the risk of fire. Fires on this category of land lead to the re-migration of radionuclides over considerable distances, which causes deterioration of the environmental situation in several countries and is a serious international problem (Figure 2).

Figure 1. Consequences of a forest fire in Portugal 2017
https://www.bbc.com/russian/features-40321696

Figure 2. Forest fire in the Chernobyl exclusion zone in 2015
https://www.bbc.com/ukrainian/news-russian-44375930

In this situation, to increase the level of fire safety of natural ecosystems and reduce the number and size of fires, scientific research related to the study of the emergence and spread of forest fires in conditions of global climate warming is of particular importance. These studies should mainly be aimed at predicting the occurrence and early detection of fires, creating conditions for limiting the spread of fires and developing modern ways to extinguish them.

2. REVIEW OF LITERATURE

The problem of forest fires are actively engaged in specialists from around the world. On the territory of the post-Soviet space, a significant contribution to the fight against fires in natural ecosystems was made by Usenya V.V., Gusev V.G., Grishin A.M., Dorrer G.A., Artsybashev E.S., Ovsyannikov I.V., Baranovsky N.V., Zharikova M.V. and a number of other researchers (Usenya, 2002; Gusev, 2005; Grishin, 1981; Dorrer, 2008; Artsybashev, 1974; Ovsyannikov, 1978; Kuznetsov & Baranovsky, 2009; Khodakov & Zharikova, 2011). These authors obtained significant results in improving the fire safety of natural ecosystems (Usenya, 2002; Artsybashev, 1974; Ovsyannikov, 1978), predicting the occurrence of forest fires and their environmental consequences (Kuznetsov & Baranovsky, 2009), physico-mathematical modeling of fire spread (Gusev, 2005; Grishin, 1981; Dorrer, 2008), creating modern fire extinguishing and fire-retarding compounds (Usenya, 2002).

In a number of European and other countries, Ford R., Weber R., Babrauskas V., Dupuy J., Linn R., McCaw W., Albini F. and other scientists made a great contribution to the development of the theory and practice of protecting forests from fires (Ford, 1995; Babrauskas, 2003; Weber, 1991; Dupuy & Larini, 1999; McCaw, 1998; Linn, 1997; Albini, Alexander & Cruz, 2012). At the present stage, processes of the occurrence and spread of fires in natural ecosystems have been adequately studied (Ford, 1995; Babrauskas, 2003), physical, mathematical, field, and numerical models of forest fires have been developed (Weber, 1991; Dupuy & Larini, 1999; McCaw, 1998; Linn, 1997; Albini, Alexander & Cruz, 2012; Sullivan, 2009; Morvan, 2011; Mell et al., 2007; Morvan, Meradji &Accary, 2008; Morvan, 2015; Gavrilov et al., 2011; Morvan, Méradji & Accary, 2009; Morvan, 2014; Nelson, 2015; Chuah &Kushida, 2007; Battaglia et al., 2000; Lei et al., 2011; Sun et al., 2006). On the basis of the developed models, software systems (Behave, Farsite, FireLab, etc.) were created for timely forecasting the dynamics of fire development. These complexes have been introduced into the activities of specialized centers in several countries of the world - the Canadian Interdepartmental Center for Forest Fire, the Global Monitoring Center for Forest Fire in Germany, the National Integrated Fire Center in the USA, the Emergency Response Coordination Center in Greece, the Republican Emergency Management and Response Center situations in Belarus and others.

It should be noted that despite significant results in ensuring the fire safety of natural ecosystems, the aggravation of the situation with forest fires in recent years indicates a lack of effectiveness of fire protection measures used and the expediency of specifying the developed models of the occurrence and spread of fires, especially for conditions of frequent dry periods.

To assess the likelihood of forest fires, scientists from different countries simulate the processes of igniting forest fuel (hereinafter - FCM) by various ignition sources (Fateh et al., 2016; Liodakis, Katsigiannis & Kakali, 2005; Leroy-Cancellieri & Leoni, 2006; Tihay et al., 2009; Cancellieri, Innocenti & Leroy-Cancellieri, 2013; Di Blasi, 2008; Branca &Di Blasi, 2004; Mehrabian, Scharler & Obernberger, 2012; Mindykowski et al., 2011; Fernandez-Gomez et al., 2011; Weise et al., 2005; Dibble, White &

Lebow, 2007; Schemel et al., 2008; Bartoli et al., 2011; Simeoni et al., 2012; Morandini, Santoni & Balbi, 2001; Marcelli et al., 2004; Dupuy et al., 2011; Pérez et al., 2011; Miloua, 2015; Font, Conesa &Moltó, 2009; Pimont et al., 2009; Monod et al., 2009; Luche et al., 2011; Finney, 1993). A significant place in the conducted studies is given to the problems of the probability of ignition of the FCM by thermal radiation from the flame front, a burning particle, a thunderstorm discharge (Baranovsky, 2009; Grishin et al., 2002; Kuznetsov & Baranovsky, 2010; Tohid &Kaye, 2017; Suzuki &Manzello, 2017). The authors (Fateh et al., 2016; Liodakis, Katsigiannis & Kakali, 2005; Leroy-Cancellieri & Leoni, 2006; Tihay et al., 2009; Cancellieri, Innocenti & Leroy-Cancellieri, 2013; Di Blasi, 2008; Branca & Di Blasi, 2004; Mehrabian, Scharler & Obernberger, 2012; Mindykowski et al., 2011; Fernandez-Gomez et al., 2011; Weise et al., 2005; Dibble, White & Lebow, 2007; Schemel et al., 2008; Bartoli et al., 2011; Simeoni et al., 2012; Morandini, Santoni & Balbi, 2001; Marcelli et al., 2004; Dupuy et al., 2011; Pérez et al., 2011; Miloua, 2015; Font, Conesa & Moltó, 2009; Pimont et al., 2009; Monod et al., 2009; Luche et al., 2011; Finney, 1993; Baranovsky, 2009; Grishin et al., 2002; Kuznetsov & Baranovsky, 2010; Tohid & Kaye, 2017; Suzuki & Manzello, 2017) conducted a large amount of experimental research. Obtained important for the theory of forest fires data on the likelihood of ignition of FCM, which are mainly aimed at predicting fires on the territory of natural ecosystems, assessing the conditions of spread of forest fires, clarifying the parameters of firebreaks.

To assess the dynamics of forest fires, specialists from different countries study the spread of fires, study the effect of wind, terrain, temperature and air humidity on vegetation burning (Regueira, 2018; Himoto et al., 2018; Morvan et al., 2018; Frangieh et al., 2018; Pinto et al., 2017; Li & Ingason, 2017; Carlos Fernandez-Pello, 2017; Vermesi et al., 2017; Padhi, Shotorban & Mahalingam, 2017; Leonelli et al., 2017; Benkorichi et al., 2017; Chen, Aanjaneya & Atreya, 2017; Restuccia, Huang & Rein, 2017; Thomas et al., 2017; Bufacchi et al., 2016; Padhi, Shotorban &Mahalingam, 2016; Thi et al., 2016). These studies are both physical and mathematical in nature. Through conducted field and laboratory experiments, fundamental data bases are formed on the basis of which mathematical prediction of fire conditions is carried out. An important place in this direction is occupied by studies on the possibility of transfer of the flame from the forest fund to a number of localities and agricultural lands (Mell et al., 2010, Boonmee & Quintiere, 2002; Anthenien, Tse & Fernandez-Pello, 2006).

For the rapid detection and extinguishing of fires, research is being conducted on the development of modern methods for monitoring the state of natural ecosystems, creating new firefighting equipment and equipment for extinguishing forest fires, and laying protective dividing strips to limit the spread of fires (Barber, 1977; Wilson & Connarro, 1969; Perry, 1998; Richards & Bryce, 1995). Considerable attention is paid by scientists from different countries to the development of modern fire extinguishing and fire-retarding compositions designed to effectively extinguish forest and peat fires (Usenya, 2002; Konev, 1977). These compositions can be applied both in the ground-based method of eliminating fires, and in the use of aviation technology. In addition, the compositions can be applied to the ground cover of the forest to create obstacles to the spread of fire and support strips to start the oncoming fire.

Of the variety of results obtained at the present stage, it should be noted that many of them are based on the study of wood burning processes (Regueira, 2018; Li & Ingason, 2017; Vermesi et al., 2017; Chen, Aanjaneya & Atreya, 2017; Beltsova & Korolchenko, 2008), which is not entirely correct when developing effective methods and means of fire protection of natural ecosystems. After all, the majority of forest fires begin with the ignition not of wood, but of the ground-based FCM to which moss, lichen, needles, foliage, small branches, bark and other materials should be attributed. In addition, research methods used by different authors differ and are often applicable only to the climatic conditions for which they were developed. As a result, physico-mathematical models and programs developed on the basis of the pyrological properties of wood have errors, are not universal for different climatic conditions and need to be refined by taking into account the ground group's ability to predominate in a particular area of the FCM to heat, ignite and maintain the combustion.

This work is aimed at addressing the identified issues in terms of establishing patterns of occurrence of forest fires in natural ecosystems of the temperate climate zone in conditions of global climate warming. At the same time, considerable attention is paid to the processes of ignition of a ground-based FCM as a conductor of ground burning, determination of the levels of thermal loads on the FCM and substantiation of effective measures to limit the spread of fires in natural ecosystems.

3. MATERIALS AND RESEARCH METHODS

The work used a combination of experimental and analytical methods for studying the processes of the emergence and spread of forest fires. The aim of the study was to establish the laws of ignition of FCM when exposed to radiant heat flux from the flame front. The object of the study was the forest ground cover of conifer plantations in the form of moss, lichen, and needlewood. The subject of the study was the laws of ignition of the ground cover of coniferous plantations when exposed to radiant heat flux from the flame front. The study was conducted in six stages.

At the first stage, using a method of field modeling, a series of experiments was carried out to establish the conditions for the emergence and spread of a forest fire center (Goman et al., 2011). In the course of the experiments, a flame front of a cylindrical shape was created, flame temperature values were determined during the combustion of a ground-based FCM, ground burning dynamics were determined, and the levels of FCM heating were detected when exposed to radiant heat flux from the flame front.

Experimental studies were conducted at the site of the Forest Institute of the National Academy of Sciences of Belarus, as well as on the territory of the forest fund of Starodorozhsky district of the Minsk region. On the horizontal sections of the pine forest, experimental sites were created, including the central (A) and peripheral (B) zones, of which the first came out to initiate a fire. The peripheral zone was intended to identify the dynamics of the bottom fire; along its perimeter, a mineralized strip (C) 1.4 m wide was created with left zones (D) to control the temperature in the ground cover layer, separated from zone (B) by 0.2 m (Figure 3). In order to obtain stable flame combustion in the initial stage of the experiment, 1 l of a flammable liquid was evenly distributed in the FCM zone (A).

After ignition of the ground cover, the zone (A) instantly engulfed in fire initiated the formation and propagation of a cylindrical flame front of a bottom fire. Measurements of the diameter of the contour of the fire and its height were carried out at intervals of 1 min using measuring rails. After the thermal front reached the mineralized strip, the height of the flame and its temperature were measured at several points, as well as the temperature at ground cover areas in zone (D). The temperature of the flame and the ground cover was monitored by means of Sosna – 003 measuring regulators with thermoelectric converters. The measurements were carried out until complete burnout of the zone (B).

During the experiment, the complex indicator of fireiness according to the Hydrometeorological Center varied from 4050 to 4310, which corresponds to class IV fire hazard in the forest according to weather conditions, when the wind speed is in the range of 0-1 m / s, the air temperature is about 20 °C and the moisture content of the combustible material is about 20%. Ground cover consisting of moss, lichen, needles, small branches, bark, grass and leaves was used in its natural state and also formed so

Figure 3. The general scheme of the model focus forest ground fire
A - the start-up zone of fire (d = 1 m); B - FCM zone (d = 8-12 m); C - mineralized 1.4 m wide strip; D - FCM zone for temperature control in its layer

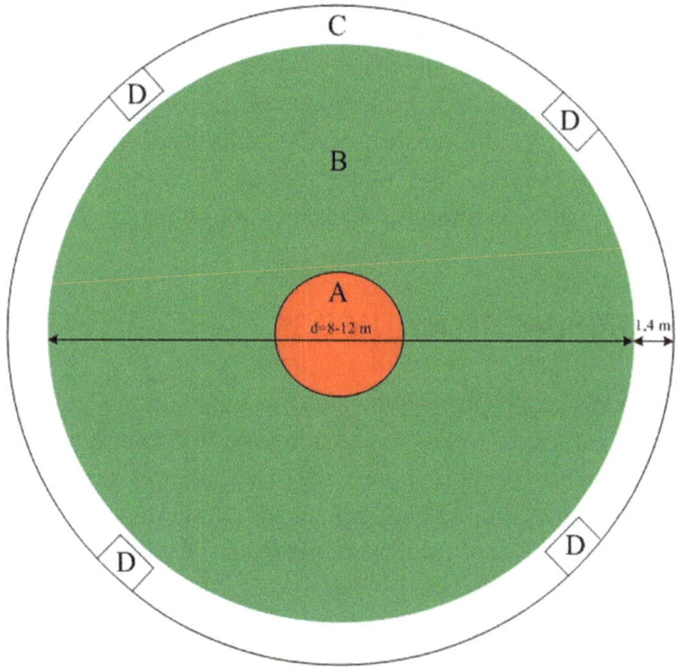

as to cover a rather wide range in thickness from 0.07 m at a density of 50 kg / m^3, up to 0.01 m at a density of 75 kg / m^3.

At the second stage, using the method of analytical modeling, the heat load from the cylindrical flame front was calculated (Sable, Goman &, Yan, 2010). Using well-known mathematical calculations, an expression is obtained that characterizes the level of heat load on the ground cover of the forest, depending on the parameters of the flame front (height, diameter, temperature). Then, using the obtained expression, the conditions of the conducted full-scale experiment were simulated, and the density values of the radiant heat flux acting on the FCM in the process of actual combustion were detected.

At the third stage, using the method of a laboratory experiment on a facility for the study of the flammability of materials (GOST 30402-96), we simulated the conditions of the effect on the FCM corresponding to the conducted full-scale experiment, the thermal load of the subcritical range (not leading to ignition). The purpose of the experiment was to establish the predisposition of the ground cover to heat when exposed to thermal radiation from a fire. The values of the heat load affecting the FCM were set in the range of values characteristic of the conducted natural experiment and calculated in the second stage. Further, the temperature in the layer of material obtained in the course of full-scale and laboratory experiments was compared. The good convergence of the obtained results made it possible to conclude that this approach is applicable. A general view of the laboratory setup is presented in Figure 4.

In accordance with the research methodology, prior to the beginning of the experiment, the FCM sample was evenly placed in a ring of asbestos located on the sample holder. Next, temperature control sensors were introduced into the sample and the installation was brought to the required temperature. When conducting experiments using a radiation panel installation on the front surface of the FCM

Figure 4. Installation diagram for impact modeling on combustible material of intensive radiation fluxes
1 - radiation panel with a heating element; 2 - protective plate with a hole in the center d = 150 mm, 3 - a ring of asbestos for the location of the sample; 4 - asbestos cement sheet; 5 - sample holder; 6 - mobile platform; 7 - lever with counterweight; 8 - supporting frame assembly

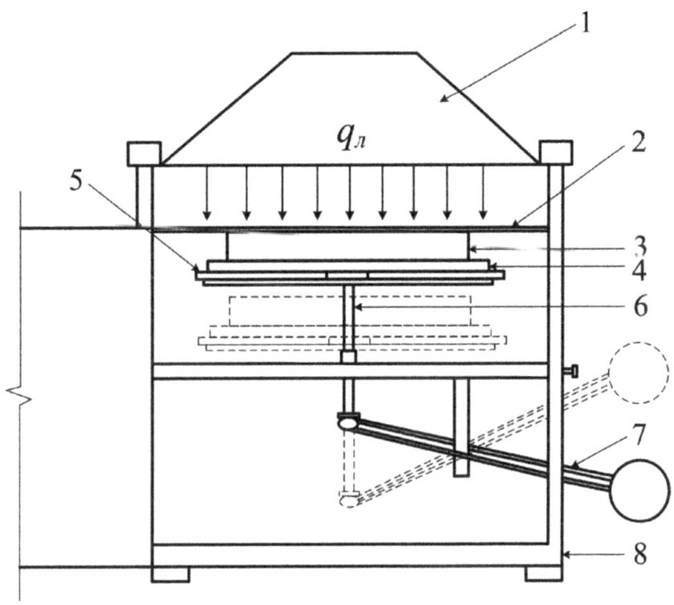

simulated the effect of heat load, characteristic of the conducted full-scale experiment. During the test, the temperature distribution over the sample height was recorded. Each subsequent sample was tested after cooling the sample holder and the asbestos ring to room temperature.

At the fourth stage, in addition to the cylindrical flame front, the method of analytical modeling carried out calculations of the thermal load affecting the FCM during the formation of flat and round flame fronts (Sable, Goman, & Yan, 2011). A comparative analysis of the results obtained for the fronts of a cylindrical, flat and round shape with the justification of the most dangerous option.

At the fifth stage, a series of full-scale studies of the ground-based FCM ignition process under the influence of heat load characteristic of forest fire conditions (Usenya, 2002; Thomas et al., 2017) were carried out at a laboratory facility for the study of the flammability of materials (GOST 30402-96). The purpose of the experiment was to obtain the dependence of the ignition time of an FCM on the heat load acting on it, as well as to determine the critical values of the density of the radiant heat flux for an FCM of different humidity. The principal difference of the fifth stage from the second is the creation of a thermal load of the critical range on the sample surface of the FCM sample (leading to ignition). During the experiment, the density of the heat flux affecting the sample was 15, 20, 25, 30, 35, 40, 45, 50 kW / m^2.

At the sixth stage, on the basis of the carried out experimental-analytical studies, a practice-oriented methodology for determining the width of fire breaks in forests with recommendations for their creation has been developed and proposed.

The text of the work provides a detailed description of the actions for the specified stages of the study.

4. THE RESULTS AND DISCUSSION

4.1. Full-Scale Modeling of Forest Ground Fire and the Process of Radiant Heat Flux on the Ground Cover of Coniferous Plantations

4.1.1. The Results of Experimental Studies of the Burning Process of the Ground Cover of Coniferous Stands

As a result of the experiments, it was established that in the initial period of fire development, when the flame transitions from zone (A) to zone (B), the speed of propagation of the flame front increases gradually, which is typical for the ignition period (Goman et al., 2011). In the future, the development of fire with the movement of its front occurs almost at a constant speed. This corresponds to an increase in the perimeter of the fire according to a law that is close to linear with the passage of time (Figure 5). The horizontality of the test sites and the relative weakness of the wind led to the realization of an almost round-shaped flame contour. However, the height of the flame along the front changed over time, which is displayed on a separate fragment (Figure 6).

The speed of movement of the flame front along the dense ground layer with a predominance of pine needles with a thickness of about 0.01 m is not great; it took about 45 minutes for the fire to reach the border (C). For layers with a predominance of moss and lichen, the rate of propagation of fire is comparable and is 20-30% higher than that set for needles, with a similarity of graphical dependence of area on time. In general, for all 10 experimental sites, the time when the flame reached the border of the mineralized band was 30-50 min, depending on the size of the model source.

The level of warming up of the forest floor under the action of the radiant heat flux and, accordingly, the efficiency of the applied standard fire barrier in the form of a mineralized strip were detected directly

Figure 5. Dependence of area (1) and perimeter (2) of fire on time burning model foci; ambient temperature of 20 °C wind speed of 1 m / s, the stock of FCM is 1.4 kg / m², its humidity is 20%

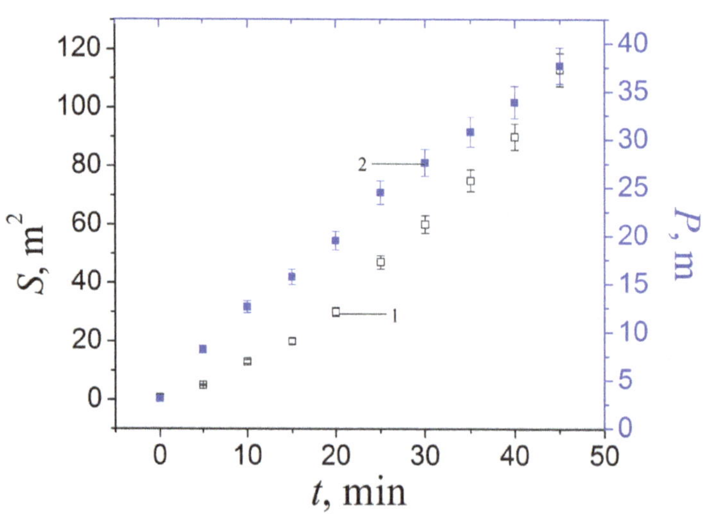

Figure 6. Combustion of FCM during a full-scale experiment

during the test. As is known (Babrauskas, 2003), heating of the ground cover is primarily determined by the level of thermal load, which depends on the height and width of the flame front, as well as its temperature.

In the literature, there are various values of the temperature of combustion of substances and materials (GOST 12.1.004-91; Cruz, 2004). However, the most reliable is the direct measurement of temperature at various points along the height of the flame in a natural experiment. According to the results of tests for the conductors of combustion in the form of moss, lichen, litter, leaves, bark, small branches, the flame temperature reaches 1070 K (Figure 7), which is lower than the flame temperature during the combustion of flammable and combustible liquids (gasoline, kerosene, etc.) - 1150 K, dry wood - 1270 K, liquefied gases - 1470 K (GOST 12.1.004-91).

Figure 7. Flame temperature over time of burning ground cover of coniferous plantations
1 - zone of contact with combustible material, 2 - center of the torch, 3 - upper region of the torch

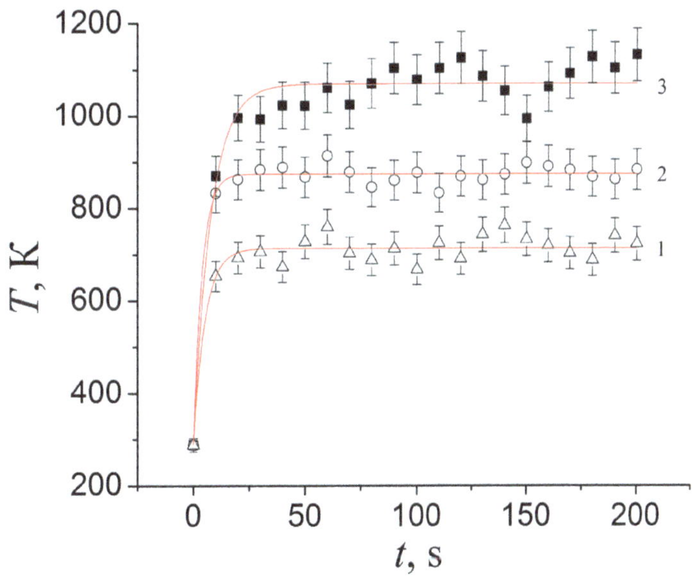

From figure 7 it follows that the formation of the flame occurs very quickly with subsequent temperature fluctuations in time associated with the movement of convection flows of combustion products, and the flow of colder air masses into the reaction zone through the lower region.

When the heat flux acts on the ground cover, it is heated, dried, pyrolized and then ignited by the pyrolysis products when the critical temperature is reached (Vermesi et al., 2017; Chen, Aanjaneya & Atreya, 2017). The speed of the required heating is largely determined by the distance of the material and the presence of obstacles in the form of barriers that can neutralize the effects of heat flow (Babrauskas, 2003). During the field experiment, it was established that the temperature values in the layer of ground cover at different distances from the flame front correspond to the hyperbolic law (Figure 8).

As can be seen in Figure 8, with increasing distance from the radiator, the temperature of the ground cover decreases. For example, when the flame height is h = 0.5 m, removal of material from the fire by 1 m leads to a decrease in temperature in its layer to 315 K, which minimizes the probability of further spread of fire. When exposed to a flame front height of 0.75 m at a distance of 0.2 m, the moss and lichen are heated to 450 K, and an increase in the distance to 1.4 m contributes to a 1.4-fold decrease in material temperature. In general, it can be concluded that the ground cover layer, at a distance of 0.2 m from the flame front, under conditions corresponding to the fourth class of fire, and a flame front height of 0.75 m, is not able to reach the critical temperature due to the action of thermal radiation.

However, for ground fires of greater intensity, safety conditions should be considered separately, taking into account the possibility of combining the results of a full-scale experiment with the data for calculating the density of radiant heat flux from extended flame fronts and the ignitability criteria for ground-based combustible material. In other words, to assess the level of irradiation of the ground cover under the action of increased thermal loads from the flame fronts, it is advisable to extrapolate the experimental results to large radiator scales, for which it makes sense to solve the problem analytically. This will allow to estimate the levels of heat load from large-scale thermal fronts and to identify the relationship between the heat flux density acting on the combustible material and the temperature

Figure 8. Temperature in the middle of the ground cover layer 0.07 m thick when exposed to thermal radiation from a cylindrical front fire flame for the height of the flame (h), m: 0.25 (1), 0.5 (2), 0.75 (3)

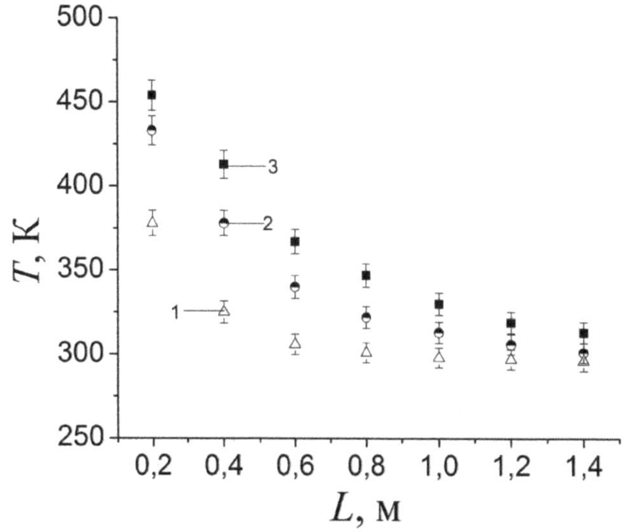

in its layer. This approach is possible due to the fact that the behavior of the experimental values of the temperature of the ground cover and the calculated values of the heat flux density for the conditions realized in the course of the experiment are close.

4.1.2. Extrapolation of Experimental Results on Large Parameters of the Thermal Front

Due to the fact that conducting large-scale full-scale experiments with the creation of extensive height and width of thermal fronts is difficult because of the danger of uncontrolled burning, to assess the level of exposure to ground cover, it is advisable to extrapolate the experimental results to large parameters of the thermal front (Bufacchi et al., 2016). Accordingly, it becomes necessary to identify patterns of formation and spatial distribution of the radiant heat flux generated by cylindrical surfaces. It is significant that among the available literature sources there is not a lot of data on the assessment of the densities of heat flux emitted by the emitters of this form (Cruz, 2004; Wang, 1979; Stepanov, Stanchits & Stankevich, 2011). Thus, in (Wang, 1979), expressions are given for the angular coefficients of irradiance between a cylinder and a rectangular surface located in a plane parallel to the axis of the cylinder, as well as for two infinitely long cylinders placed at a certain distance. In the framework of the considered geometry, the emitter and the receiving surface are positioned at the same height, which does not correspond to the situation of the conducted natural experiment and limits the possibilities of practical application of the obtained expressions for the angular irradiance coefficients. In another paper (Stepanov, Stanchits & Stankevich, 2011), an expression is given to identify the thermal damage of a target in the zone of action of a cylindrical front. The geometry presented in this work also does not fully correspond to the situation of the experiment performed and can be refined by detailed modeling of the conditions of a full-scale experiment on the effect of thermal radiation on ground cover from a long cylindrical flame front approaching the fire barrier.

Below are the results of calculating the density of radiant heat flux from a cylindrical source, which allow us to identify possible levels of heat load on ground-based combustible material located in the area of action of the heat front.

If the distance from the irradiated surface to the radiator is not large compared to its size, which is indeed the case in a significant number of practical cases, then the application of the inverse square law to determine the energy illumination is not entirely correct. The source should be represented as a large number of point emitters, followed by summing up the contributions from each (Sable, Goman, & Yan, 2010).

The general scheme of the study and the original geometric constructions are presented in Figures 9 and 10.

The source of a cylindrical shape of radius R and height h is considered under the assumption that the flow sent in all directions is the same, i.e. the radiant surface of the radiant brightness corresponds to the Lambert source. On a horizontal irradiated surface of finite size with the center *O* located in the plane of the base of the heat source, at a distance L from the radiating surface, the heat flux affects the combustible material of the ground group, which meets the conditions of the conducted field experiments. The calculation takes into account the magnitude of the visible elementary portion of the radiating front from point *O*, as well as the effective magnitude of the portion of the surface irradiated along a given direction. Then the summation of the contributions from all elementary areas along the visible surface of the heat source was made.

Figure 9. Scheme for calculating the heat load from the cylindrical source on the ground cover of the forest (side view)

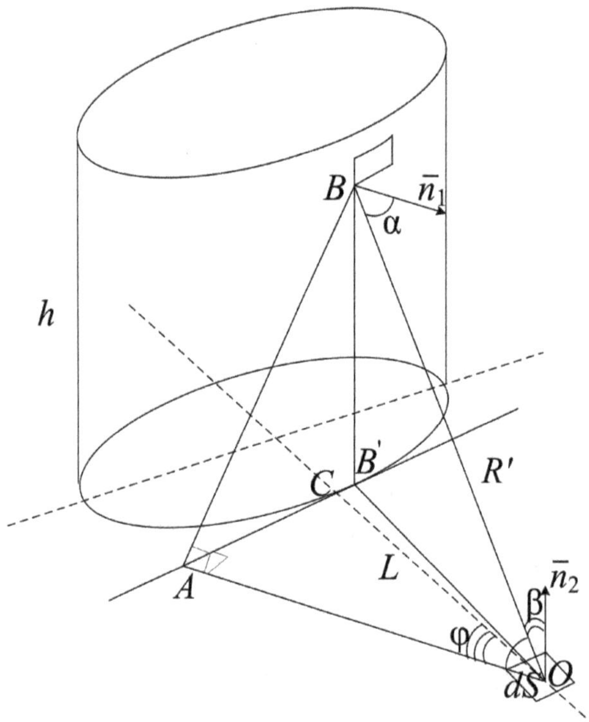

Figure 10. Scheme for calculating the heat load from the cylindrical source on the ground cover of the forest (top view)

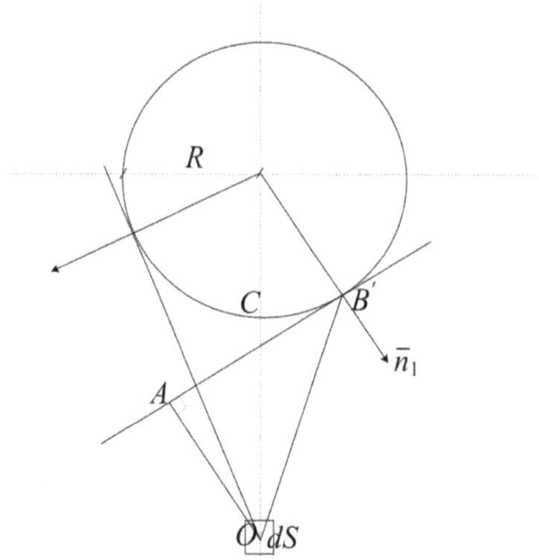

The elementary flux sent from the radiator to the irradiated surface is proportional to the portion of the radiator that is visible from point O, i.e., and the value of the solid angle, based on the irradiated surface. If the normal to the surface is an angle β with the axis of the cone, which is the boundary of the solid angle, then the required flow is proportional to the cosine of the angle between the normal and the axis of the solid angle, which is written as:

$$d\Omega = \frac{dS\cos\beta}{R'^2}, \tag{1}$$

where is the distance from the elementary part of the radiator to the irradiated surface. Thus, the falling required elementary flux to the surface along the normal to it can be expressed as:

$$dF = BdS\frac{\cos\alpha\cos\beta}{R'^2}dxdy\,, \tag{2}$$

where B is the energy brightness of the source. Accordingly, the density value of the radiant heat flux from the radiator element can be represented as:

$$dq = \frac{B}{R'^2}\cos\alpha\cos\beta dxdy\,. \tag{3}$$

After summing up all the contributions from the elementary parts of the radiator, an expression was obtained characterizing the regularities of the distribution of the density of the radiant heat flux from an extended cylindrical front, taking into account its radius, height, temperature, and the distance to the irradiated surface

$$q = \frac{\sigma T^4}{2\pi}\left(a_{L,R}(h)\left[b_{L,R}(h) - c_{L,R}(h)\right] + b_{L,R}(0) - c_{L,R}(0)\right), \tag{4}$$

$$a_{L,R}(h) = -\frac{L^2 + 2LR + h^2}{\sqrt{\left(L^2 + h^2\right)\left(\left(L + 2R\right)^2 + h^2\right)}};$$

$$b_{L,R}(h) = arctg\left(\frac{L + R}{\sqrt{\left(L^2 + h^2\right)\left(\left(L + 2R\right)^2 + h^2\right)}}\left[\frac{\left(L + R\right)^2 + R^2 + h^2}{R}\left(1 + \sqrt{1 - \frac{R^2}{\left(L + R\right)^2}}\right) - 2R\right]\right);$$

$$\varsigma_{L,R}(h) = arctg\left(\frac{L+R}{\sqrt{\left(L^2+h^2\right)\left(\left(L+2R\right)^2+h^2\right)}}\left[\frac{\left(L+R\right)^2+R^2+h^2}{R}\left(1-\sqrt{1-\frac{R^2}{\left(L+R\right)^2}}\right)-2R\right]\right);$$

q - density of radiant heat flux, W / m^2;
T is the temperature of the cylindrical radiator, K;
R is the radius of the cylindrical emitter, m;
h - height of the cylindrical radiator, m;
L is the distance from the emitter to the irradiated surface, m;
σ is the Stefan-Boltzmann constant.

Expression (4) characterizes radiation of an absolutely black body to an absolutely black surface without taking into account possible re-radiation, which meets the worst conditions in a fire and is aimed at increasing the effectiveness of preventive measures.

Modeling the conditions of the conducted natural experiment using the obtained expression shows that for a cylindrical flame front with a height of 0.25-0.75 m, a width and a temperature of 10 m and 970 K, respectively, the heat flux density at a distance of 0.2 m from the radiator is 9-17.5 kW / m^2 and further, with an increase in the distance of 7 times, it gradually decreases to 0.3-2.3 kW / m^2. With a flame front height of 0.5 m and similar parameters in width and temperature, the thermal load at a distance of 0.2 m is 15 kW / m^2.

At the same time, the revealed regularities of the distribution of radiation fluxes from a cylindrical source allow, by means of a laboratory setup for determining the flammability of materials, to simulate the conditions of a full-scale experiment (GOST 30402-96). Using this approach, under laboratory conditions, the process of influencing the ground cover of the heat load from the flame front, characteristic of the conditions of the conducted natural experiment and determined according to (4), is modeled; using thermocouples, the temperature values in the layer of the ground cover are determined and compared with the temperature measured in natural conditions. As an example, Figure 11 shows the results of comparing data obtained during laboratory and field experiments on the temperature in a layer of combustible material when exposed to thermal radiation from a flame front height of 0.5 m. Estimated estimates of the level of ground cover are also given. along the height of the fronts.

As can be seen in Figure 11, this approach allows to obtain temperature values in the layer of ground cover, which are consistent with the results of the field experiment. The discrepancy obtained in laboratory and natural conditions data on the temperature in the layer of combustible material of different types and thicknesses, depending on the parameters of thermal fronts that were observed in a natural experiment, does not exceed 10%. Accordingly, expression (4) can be used to estimate the spatial distribution of the radiant heat flux from a cylindrical flame front, characteristic of forest fires.

4.2. Estimation of the Heat Load on the Ground Cover of Coniferous Plantations From the Flame Fronts of the Flat and Round Shape

In solving practical problems of calculating the thermal stability of materials, mechanisms, protective clothing, the problem arises of estimating radiant heat fluxes generated by radiating surfaces of various

Figure 11. Dependence of temperature in the middle of the ground layer cover 0.07 m thick from the distance to the irradiated surface at a temperature and radius of the cylindrical front of 970 K and 5 m respectively, for the height of the radiator (h), m: 0.5 (1, 2), 1 (3), 1.5 (4), where 1 - the results of the field experiment; 2, 3, 4 - results laboratory experiment

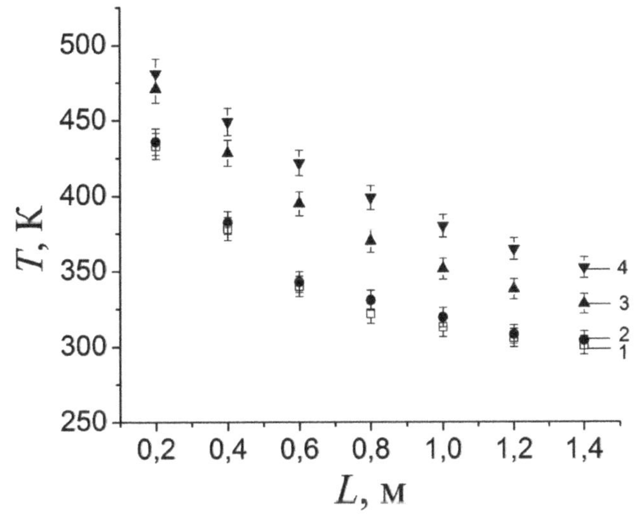

shapes, including flat sources, characteristic of forest fire conditions (Gusev, 2005; Ford, 1995; Weber, 1991). After all, when a forest burns, the front flame of a downstream fire in front of a fire barrier can reach 10 or more meters in width and 3 m in height, and to create an effective barrier it is necessary to estimate the level of heat load near it (Dorrer, 2008).

It should be noted that there is an extensive literature on the study of heat fluxes generated by flat long sources, among which a large proportion is occupied by analytical calculations, see, for example, (Mell et al., 2010, Boonmee & Quintiere, 2002; Anthenien, Tse, & Fernandez-Pello, 2006; Cruz, 2004; Wang, 1979; Stepanov, Stanchits, & Stankevich, 2011; Koshmarov & Bashkir, 1987; Gusev, 1989; Borodkin et al., 1997). However, in Ref. (Koshmarov & Bashkir, 1987), the irradiance coefficient between a flat rectangular source and a remote surface of finite dimensions with a center at a point, the projection of which coincides only with the center of the radiating surface, is presented. For observation points with projections outside the center of the source, the coefficient is not presented. The source (Wang, 1979) considered the situation when the radiation falls from one unlimited band radiator to another surface in the form of an infinite parallel strip of smaller width. Here, the irradiance coefficient is proposed to be determined by an expression operating with a combination of the ratio of the width of each of the bands to the distance between them. However, the presented ratio does not include the coordinate of the height of the observation point. Since the height of the observation point in the irradiation band is not indicated, it is not entirely correct to set the irradiance band as a geometric place for different points, because for them the coefficient is not the same. In source (Gusev, 1989), to assess the degree of human exposure to the front fire front, it is proposed to use an expression that makes sense to expand by taking into account the action of the flame front not only from the level of human growth, but also from a higher level, i.e. for the radiator of arbitrary height. The paper (Borodkin et al., 1997) presents a model of fire propagation between wooden buildings and calculates the irradiance coefficient for two opposite flat surfaces of different sizes. An expression is presented for calculating the angular irradiance coefficient depending on the height of the observation point along the vertical, in the projection of which the source center is located. However, the ratio of the calculation of the coefficient of irradiance along the width of the radiating front is not provided.

Methods for estimating the density of radiant heat flux from extended sources are also reflected in a number of current regulatory documents. Thus, the standard of the Republic of Belarus (STB 11.05.03-2010) provides an expression for determining the intensity of thermal radiation in a fire, but does not specify the shape of the flame. In addition, the presented expression for the angular coefficient is postulated as universal for all types of fronts, which can introduce significant errors. There is also no possibility of estimating the level of exposure when the irradiated surface moves along the source, which is very important in a number of practical cases. In the European standard (TCP EN 1991-1-2-2009), the formation of the radiant heat flux from the flame front is presented in more detail, expressions are given for calculating the angular coefficients when the receiving and radiating surfaces are parallel, perpendicular and angled. The given form of the radiator in the form of a plane is positioned as the most optimal, however, it should be noted that the presented estimation method is very difficult to calculate. On the one hand, the separation of the radiating surface into component fragments-rectangles is required, which leads to additional time costs in estimating the slope. On the other hand, the presented calculation algorithms contain immediate errors. As a result, there is a need to refine and coordinate the existing methods for calculating the angular coefficients and determine the most optimal way to detect the spatial distribution of radiation fluxes from an extended flat flame front.

Below are the results of the calculation of the angular irradiance coefficients for a flat, extended heat source, which makes it possible to estimate the possible thermal load on ground-based combustible material under conditions of forest fires.

The calculation is made for a flat front, located vertically with respect to the surface of the earth. At the first stage, the case of horizontal positioning of the irradiated surface is considered, which corresponds to the case of ignition of the forest ground cover.

The main elements of the geometry of the problem and the initial prerequisites for the calculation are presented in Figure 12.

The calculation was made using the previously stated generally accepted approaches, according to which the total density of the radiant heat flux on the irradiated surface can be obtained by summing all contributions from the entire radiating area:

$$q = \int_{x_1}^{x_2} \int_{0}^{h} B \frac{Lydxdy}{\left(L^2 + x^2 + y^2\right)^2}, \tag{5}$$

where L is the shortest distance from the emitter to the irradiated surface.

In this case, the position of the irradiated surface is determined by the limits of integration along the x and y coordinates, which can be set so that the projection of the irradiated surface falls inside the plane of the radiator, on its edge, outside the plane of the radiator. Essentially, expression (5) after solving the integral will set the geometry of the problem, i.e. be the product of the irradiance coefficient from the extended thermal front in a certain irradiated region and the energy brightness of the source.

The obtained expression (5) can be classified as an integral of a rational elementary function, and for its calculation the integration rules were used in two stages – according to the variables x and y in

Figure 12. Diagram of the effect of radiant heat flux from a flat source to the irradiated surface

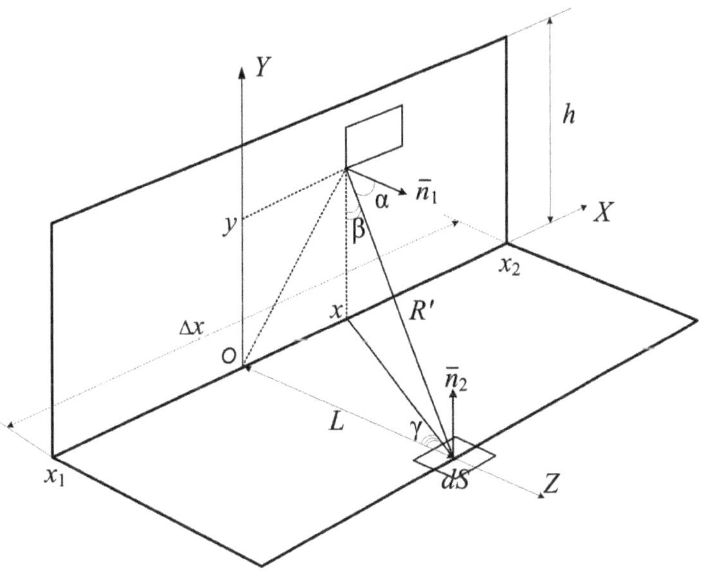

accordance with the recommendations (Gradstein & Redhead, 1962; Gusak & Gander, 1991). For the case of an isotropic flat-shaped radiator, the expression for determining the heat load acting on a horizontal surface was calculated in an approximation when the variable y changed from zero to height h, which meant the position of the irradiated surface in the plane of the radiator base, i.e. at ground level. The result is the following solution:

$$q = \frac{\sigma T^4}{2\pi}\left[arctg\left(\frac{x_2}{L}\right) - arctg\left(\frac{x_1}{L}\right) + \frac{L\left[arctg\left(\frac{x_1}{\sqrt{L^2+h^2}}\right) - arctg\left(\frac{x_2}{\sqrt{L^2+h^2}}\right)\right]}{\sqrt{L^2+h^2}}\right], \quad (6)$$

where

q is the density of the radiant heat flux, W / m^2;
T is the temperature of the radiating surface, K;
σ is the Stefan-Boltzmann constant;
x_2, x_1 - coordinates, limiting the width of the radiating surface, m;
h is the height of the radiating surface, m;
L is the distance from the radiating to the irradiated surface, m.

Obviously, the level of heat load is determined by the width of the radiator, reduced to the distance from the source to the irradiated surface x / L, as well as the position of the irradiated surface along the front. Below, the nature of the spatial distribution of heat load from a flat emitter is concretized by

Figure 13. Effect of the width of the heat front given to the distance to the irradiated surface, on the effectiveness of source for some values of the relative height of the flat front (h / L): 0.5 (1); 12); 2 (3); 3 (4); 4 (5); irradiated surface equidistant from the edges of the source

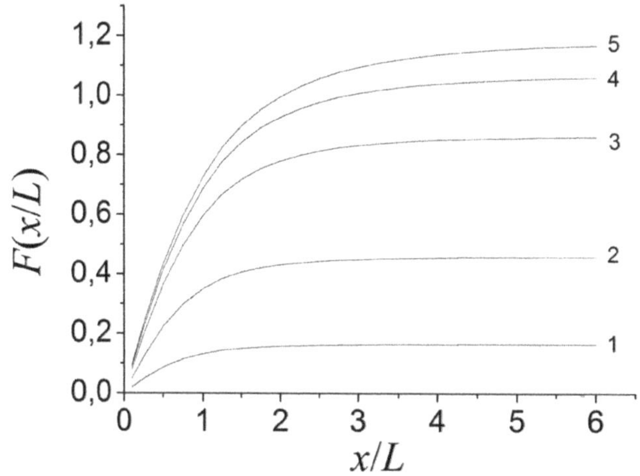

graphically representing the dependences of the density of the radiant heat flux on the variables included in the parameters of the problem under consideration.

Figure 13 shows the degree of influence of the heat front width, reduced to the distance to the irradiated surface, on the effectiveness of the source at some values of the relative height of the flat front h / L for the case when the irradiated surface is symmetric, i.e. equidistant from the edges of the source. In fact, this figure shows the coefficient of irradiance depending on the relative width of the radiating surface.

It is natural to expect that when the position of the irradiated surface changes from the center to the edge, the heat flux density should decrease. The change in heat load becomes noticeable only at the edge of the radiator, which is shown in Figure 14, which characterizes the density of the radiant heat flux depending on the displacement of the irradiated surface from the center of the flame base. This figure shows that the heat flux density slowly decreases along the radiator when the irradiated surface moves to the edge and only near the edge itself (ie, at $x \geq 4$ m), it sharply decreases, especially at small distances to the thermal front 1, $L = 1$ m).

Figure 15 shows the intensity of thermal radiation as a function of the distance to the irradiated surface and the temperature of the radiator for fixed values of the height and width of the thermal front. As can be seen in this figure, the front temperature significantly affects the level of material irradiation. So, at a height and width of the heat front of 1.5 m and 10 m, respectively, a change in the temperature of the front from 1000 K to 1500 K leads to an increase in the heat load on the irradiated surface remote from the middle of the base of the radiator by 5 m, from 1 kW / m^2 to 5 kW / m^2.

It should be noted that in conditions of forest fires, in addition to the ground cover, shrubs, grass, plants of the lower tier of the forest, tree trunks, equipment and people are effectively irradiated. In this case, it is necessary to consider the situation of the vertical location of the irradiated surface with the base at ground level. The corresponding problem was investigated in addition to the previous irradiation geometry.

Figure 14. Density of radiant heat flux depending from the removal of the irradiated surface from the center of the flame (x = 0) with the values of the distance to the heat front (L), m: 1 (1), 2 (2), 3 (3), 4 (4); front height and width 1.5 m and 10 m, respectively, emitter temperature 1300 K

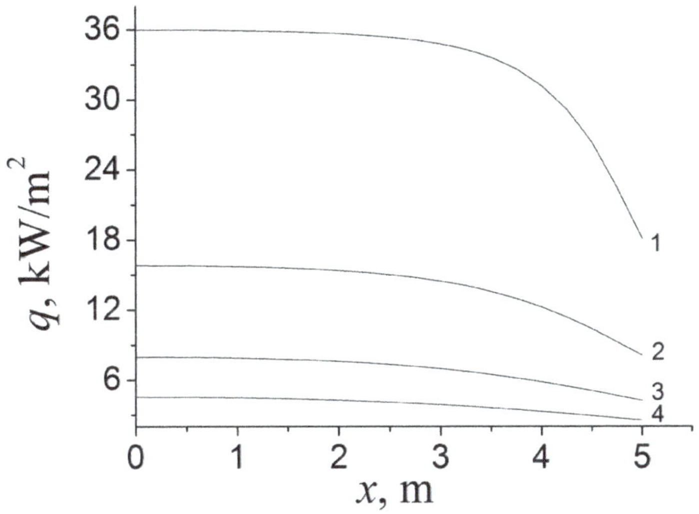

Figure 15. The dependence of the density of the radiant heat flux from the distance to the irradiated surface and the temperature of the radiator with its height and width of 1.5 m and 10 m, respectively; irradiated the surface is located opposite the middle of the base of the source

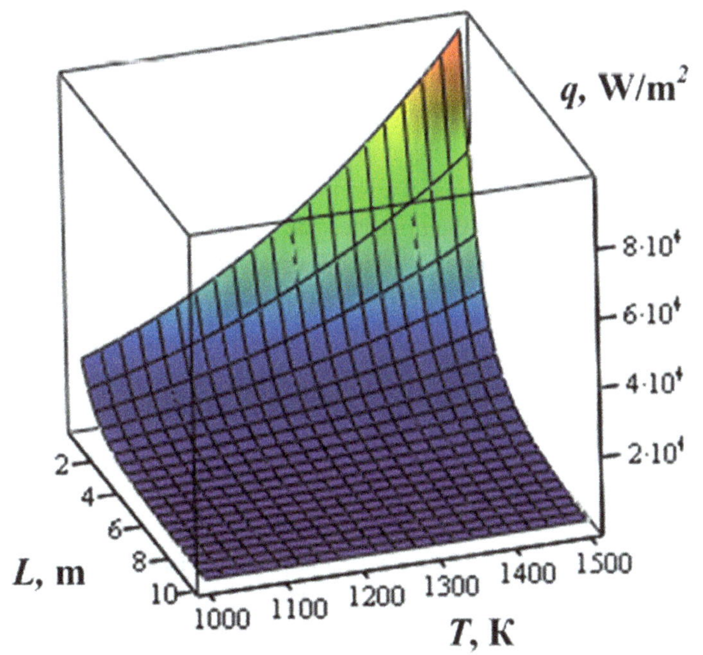

As a result, the following expression was obtained, which allows to predict the level of heat load on a vertically located irradiated surface:

$$q = \frac{\sigma T^4}{2\pi}\left(\frac{h\left[arctg\left(\frac{x_2}{\sqrt{L^2+h^2}}\right) - arctg\left(\frac{x_1}{\sqrt{L^2+h^2}}\right)\right]}{\sqrt{L^2+h^2}} - \frac{x_1 arctg\left(\frac{h}{\sqrt{L^2+x_1^2}}\right)}{\sqrt{L^2+x_1^2}} + \frac{x_2 arctg\left(\frac{h}{\sqrt{L^2+x_2^2}}\right)}{\sqrt{L^2+x_2^2}}\right), \quad (7)$$

where

q is the density of the radiant heat flux, W / m^2;
T is the temperature of the radiating surface, K;
σ is the Stefan-Boltzmann constant;
x_2, x_1 - coordinates, limiting the width of the radiating surface, m;
h is the height of the radiating surface, m;
L is the distance from the irradiated to the radiating surface, m

As an example of a solution according to (7), Figure 16 shows the simulation results of the combined effect of the distance to the irradiated surface and the height of the radiator on the level of thermal load. This figure shows that with increasing distance to the thermal front, the thermal load on combustible material gradually decreases, and with increasing height, as expected, increases. At the same time, an increase in the height of the source leads to a gradual stabilization of the heat flux density on the irradiated surface.

Indeed, a change in the distance from the center of the thermal front to the irradiated surface from 1 m to 4 m at a height of the radiator of 1 m and a temperature of 1100 K reduces the heat load from 30 kW / m^2 to 8 kW / m^2. In turn, increasing the height of the thermal front from 1 m to 4 m with a distance of 1 m to the irradiated surface and other conditions being equal increases the density of the radiant heat flux on the irradiated surface from 30 kW / m^2 to 40 kW / m^2.

It should be noted that during forest fires, not only extended flat and cylindrical fronts can be formed, but also heat sources of other forms (Ford, 1995; Albini, Alexander & Cruz, 2012; Gusev, 1989). In particular, when burning the forest canopy (tree crowns) in the absence of support for low fire, a circular flame front can form that can ignite the ground cover and lower-tier plants located behind the fire barrier (Figure 17).

In this case, the elementary flux incident on the surface is proportional to the surface area of the emitter, which is visible from the center of the irradiated surface, that is, as well as the solid angle in a given direction. As a result, the flow dF on a horizontal surface looks like:

$$dF = B\frac{RdRd\varphi \cos\alpha_1 \cos\alpha_2 dS}{R'^2}, \quad (8)$$

Figure 16. Radiant heat flux density depending from the height of the heat front and the distance from the center of the radiator to the irradiated surface. Temperature and width of the radiator 1100 K and 8 m respectively

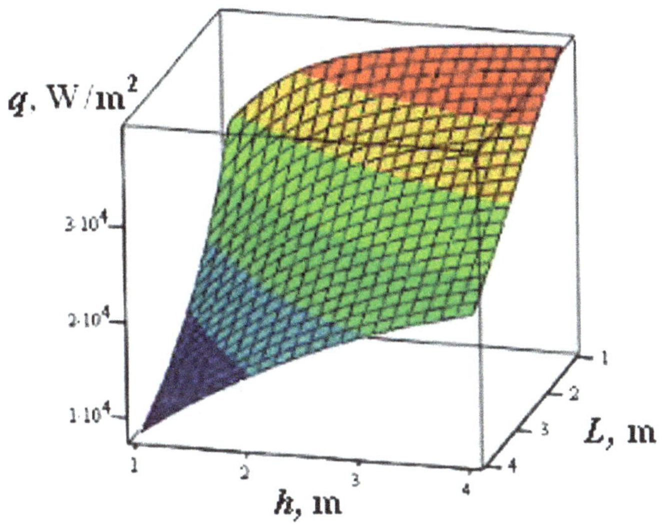

Figure 17. Diagram of the effect of radiant heat flux from a round source on the irradiated surface

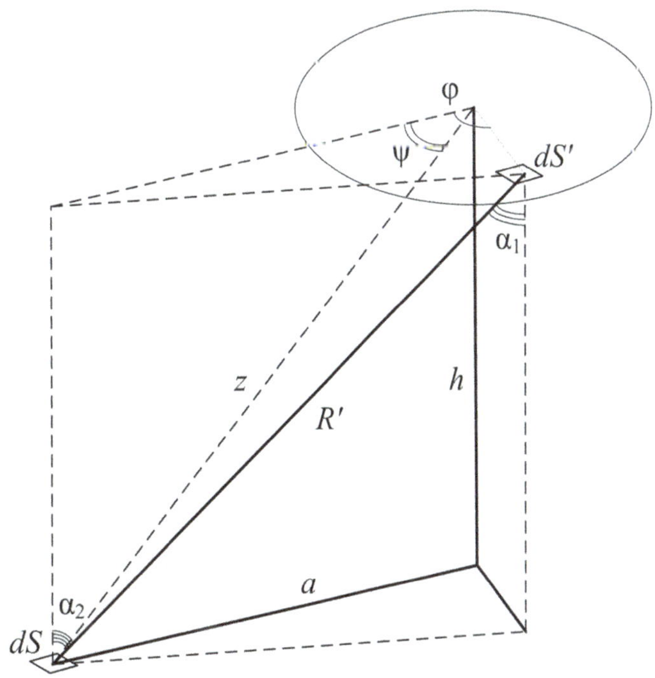

where R is the radius of the radiator. Accordingly, the density of the heat flux from the radiator element, taking into account what can be represented in the form:

$$dq = \frac{dF}{dS} = Bh^2 \frac{RdRd\varphi}{R'^4}, \tag{9}$$

where $R'$is the distance from the radiator element to the irradiated surface. Thus, the total density of the radiant heat flux on the irradiated surface is represented as the summation of all elementary contributions along the source surface:

$$q = B \int \frac{h^2}{R'^2(R,\varphi)} \frac{RdRd\varphi}{R'^2(R,\varphi)}, \tag{10}$$

As a result of solving (10) by integrating in two stages: first by the angle φ, and then by the radius of the radiator, we obtain the following expression characterizing the spatial distribution of the density of the radiant heat flux from a long circular source:

$$q = \frac{\sigma T^4 \pi}{2} \left(\frac{R^2 - a^2 - h^2}{\sqrt{\left(R^2 + h^2 - a^2\right)^2 + 4h^2a^2}} + 1 \right), \tag{11}$$

where

q is the density of the radiant heat flux, W / m^2;
T is the temperature of the radiating surface, K;
σ is the Stefan-Boltzmann constant;
a is the value specifying the displacement of the irradiated surface relative to the center of the radiator, m;
h - height of the thermal front, m;
R - emitter radius, m

As an example, Figures 18 and 19 show the dependences of the heat flux density on the input parameters of the resulting computational model. As can be seen in Figure 18, an increase in the radius of the radiator has a very significant effect on the level of irradiation of the material. Thus, when the heat front is located at a height of 15 m, the heat flux density in the radius range shown increases from 1 kW / m^2 to 26 kW / m^2. In addition, an increase in the height of the radiator from 10 m to 15 m with a radius of 5 m reduces the heat load from 52 kW / m^2 to 26 kW / m^2, i.e. 2 times.

The location of the irradiated surface relative to the center of the source also affects the level of material irradiation. So, with the height and radius of the radiator 15 m and 5 m, respectively, having a temperature of 1500 K, the displacement of the irradiated surface 10 m from the center of the radiator leads to a decrease in the thermal load on its surface from 90 kW / m^2 to 47 kW / m^2. With a decrease in the temperature of the front, a significant decrease in the level of material irradiation is observed. At the

Figure 18. The dependence of the density of the radiant heat flux on the radius and height of the radiator with a temperature of 1100 K at the location of the irradiated surface opposite the middle radiator

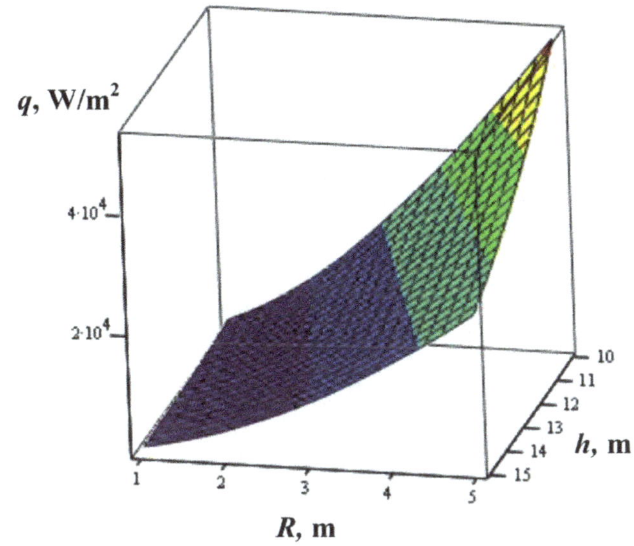

indicated values of the height and radius of the radiator, the heat flux density on the irradiated surface 10 m from the center of the radiator decreases from 47 kW / m^2 to 9 kW / m^2 as the source temperature changes from 1500 K to 1000 K (Figure 19).

Thus, the obtained expressions for calculating the density of radiant heat flux from flat, cylindrical and round sources allow us to estimate the level of heat load on ground-based combustible material located in the area of the radiating front, taking into account the size and temperature of the source, the distance to the irradiated surface and its displacement relative to center source. The calculations show that the dependence of the heat flux density on the location of the irradiated surface is quite strong in absolute terms and when developing recommendations on the design of fire barriers and thermal protection of materials, it is necessary to take into account both the characteristic dimensions of possible thermal sources and the distance to them.

4.3. Evaluation of the Flammability of the Ground Cover of Coniferous Stands When Exposed to Thermal Radiation From a Fire

Improving fire protection requirements is mainly based on the ability of combustible materials to ignite under the conditions of prolonged exposure to intense radiation fluxes (Fateh et al., 2016; Liodakis, Katsigiannis, & Kakali, 2005; Leroy-Cancellieri, & Leoni, 2006; Tihay et al., 2009; Cancellieri, Innocenti, & Leroy-Cancellieri, 2013; Di Blasi, 2008; Branca & Di Blasi, 2004; Mehrabian, Scharler, & Obernberger, 2012; Mindykowski et al., 2011; Fernandez-Gomez et al., 2011; Weise et al., 2005; Dibble, White, & Lebow, 2007; Schemel et al., 2008; Bartoli et al., 2011; Simeoni et al., 2012; Morandini, Santoni, & Balbi, 2001; Marcelli et al., 2004; Dupuy et al., 2011; Pérez et al., 2011; Miloua, 2015; Font, Conesa & Moltó, 2009; Pimont et al., 2009; Monod et al., 2009; Luche et al., 2011; Finney, 1993). This allows you to determine safe distances to the buildings covered by fire, forest areas, farmland, and take measures to localize an already occurring fire.

Figure 19. the dependence of the density of the radiant heat flux from the location of the irradiated surfaces and temperatures radiator with its height and radius 15 m and 5 m respectively

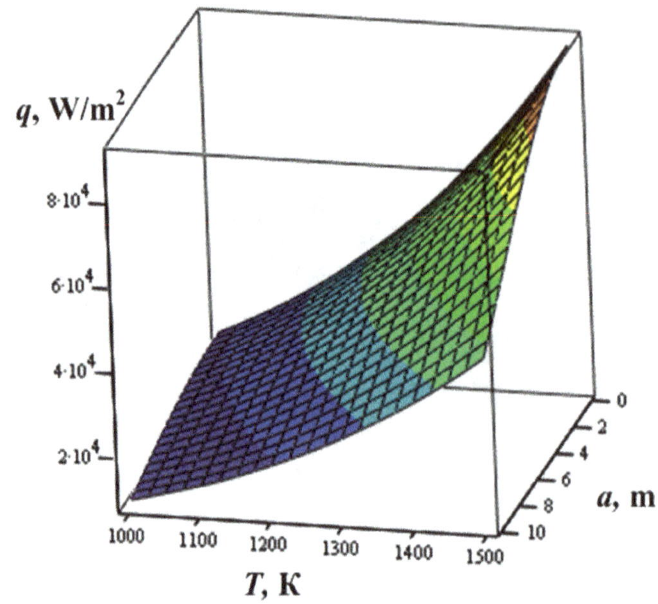

The flammability of materials is usually determined experimentally by simulating the effect of heat flux on the studied materials of small dimensions (Fateh et al., 2016; GOST 30402–96). In the future, to justify safe distances in case of fire, the method for calculating the parameters of fire breaks is used, in accordance with which the degree of material irradiation by the radiant heat flux from a given flame front is first evaluated, then the calculated heat load is compared with the material-critical value (which material is ignited with the occurrence of sustained flame combustion), obtained from an experiment (table 1) (GOST 12.1.004-91; STB 11.05.03-2010). Subsequently, it is concluded that the material is likely to ignite at a given distance to the flame front.

Thus, to obtain the required estimates for the parameters of fire breaks, it is necessary to have data on the critical thermal load for the irradiated combustible material. Sources (GOST 12.1.004-91; STB 11.05.03-2010) contain a narrow list of materials with a critical level of heat load; they lack data on the flammability of forest ground combustible material, which complicates the development of fire prevention measures in forest plantations. It should also be noted that, presented in Table 1, the data were obtained by a method designed to detect the flammability / incombustibility of solid building materials more than 40 years ago (Roitman, 1978). However, this method does not allow to explore organic materials, including vegetation. The essence of the method consists in evaluating the flammability of a sample of a solid material of strictly fixed dimensions hung on a bracket in a tube furnace of a calorimeter wrapped with a heating element, when exposed to heat flow. When conducting experiments, the heat flux from the walls of the furnace, placed in a heat-insulating medium, acts on all sides of the suspended sample, which does not correspond to the actual conditions of heat exchange during a fire. As a result, critical

Table 1. Critical values of radiant heat flux density depending on the exposure time for some materials according to (GOST 12.1.004-91; STB 11.05.03-2010)

Material	Minimal Heat Radiation, kW/m², at the Different Exposure Time, minutes		
	3	5	15
Wood (pine, moisture content 12%)	18,8	16,9	13,9
Chipboard, 417 kg/m³	13,9	11,9	8,3
Briquette peat	31,5	24,4	13,2
Lump peat	16,6	14,4	9,8
Cotton fiber	11,0	9,7	7,5
Laminate	21,6	19,1	15,4
Fiberglass	19,4	18,6	17,4
Glassine	22,0	19,8	17,4
Rubber	22,6	19,2	14,8
Coal	–	35,0	35,0

heat loads for flammable materials were often underestimated. In this regard, in determining the flammability of materials, the method described in Ref. (GOST 30402–96) is preferable.

Below are the results of studies of the ignition time of ground fuel of coniferous plantations, depending on its humidity, carried out according to the method (GOST 30402–96). Fragments of laboratory research are reflected in Figures 20 and 21. It is established (Table 2) that the moisture content of the material affects the time of its ignition more significantly than the type of ground cover, including moss, lichen, pine needle spruce or spruce. When the moisture content is up to 10%, the density of the radiant heat flux of 20 kW / m² leads to ignition of the ground layer in a time from 120 s for moss to 200 s for needles with an average material density of 75 kg / m³. When the humidity increases thirty times, the minimum value of the heat flux density, leading to ignition of the material, increases by 2.25 times, i.e. up to 45 kW / m², with exposure times of 18 s and 30 s for moss and needles, respectively. At the same time, cases of ignition of samples with a moisture content of 150% even with prolonged exposure to heat load up to 45 kW / m² were not recorded.

For ground cover moisture content of 40%, the critical value of the heat flux density for all materials considered is 40 kW / m². During the experiment, it was also established that the effect of sample density in the range of 25-125 kg / m³ on the process of their ignition is not significant and slightly more pronounced at low humidity and heat load values close to critical. So, for example, when the samples are exposed to 10% moisture of a heat flux of 40 kW / m², the time of ignition, depending on the material density, is 5-35 s versus 80-260 s with a two-fold decrease in heat load. That is, the difference in the ignition time interval is increased 6 times. In addition, the type of ground fuel and the thickness of its layer, which ranged from 1 cm for spruce to 7 cm for moss, also does not have a significant effect on its flammability, which is explained by their overall organic nature and, apparently, by reaching the critical temperature in the upper layer of the sample, which is sufficient for the occurrence of a steady flame combustion. It is also necessary to emphasize that the obtained dependences on the ignition of the tested materials are also characteristic of the cases of their mixing in various proportions.

Figure 20. Active gaseous combustion products when exposed to thermal radiation

Figure 21. Flammability of ground combustible material coniferous plantations when exposed to thermal radiation

It should be noted that in addition to the basic experimental method for estimating the heat resistance of combustible materials by the critical density of the heat flux, the conditions for their ignition can be determined analytically by the achieved critical temperature. In particular, to estimate the flammability parameters of the ground cover of high density, which is difficult to model in laboratory conditions, one can use the expression (12) given in (Lykov, 1967) and conclude that the ignition of the samples after reaching the ignition temperature, which according to (Ford, 1995) 590–620 K.

$$T = \frac{q}{\lambda}\left[\frac{\lambda t}{c\rho D} - \frac{D^2 - 3x^2}{6D} + D\sum_{n=1}^{\infty}(-1)^{n+1}\frac{2}{(n\pi)^2}\cos\left(\frac{n\pi x}{D}\right)e^{-\left(\frac{n^2\pi^2\lambda t}{c\rho D^2}\right)}\right] + T_0, \quad (12)$$

where

T is the material temperature, K;
q - density of radiant heat flux, W / m^2;
λ is the coefficient of thermal conductivity of the material, W / (m · K);
c is the specific heat capacity of the material, J / (kg · K);
ρ is the material density, kg / m^3;
t is the time of exposure to heat flow, s;
D is the thickness of the material layer, m;
T_0 is the initial temperature of the material, K;
x is the coordinate of the observation point in the material layer, m

Table 2. Time of ignition of ground fuel of coniferous plantations when exposed to thermal radiation

Heat Flux, kW/m²	Ignition Delay, s											
	Moss				Lichen				Pine Litter			
	Moisture Content, %											
	10	**40**	**150**	**300**	**10**	**40**	**150**	**300**	**10**	**40**	**150**	**300**
50	3	5	8	13	4	7	11	15	5	9	16	20
	3	6	9	15	5	7	12	17	6	10	15	23
	2	4	9	16	4	8	12	16	7	11	17	24
45	5	7	13	20	7	11	17	27	11	14	22	34
	4	8	11	18	8	13	16	25	10	16	24	32
	6	8	12	22	9	12	19	26	12	19	15	30
40	10	11	no	no	14	18	no	no	18	32	no	no
	11	13	no	no	15	24	no	no	22	30	no	no
	12	14	no	no	18	22	no	no	26	35	no	no
35	12	no	–	–	19	–	–	–	27	–	–	–
	13	no	–	–	23	–	–	–	33	–	–	–
	14	no	–	–	25	–	–	–	30	–	–	–
30	21	–	–	–	30	–	–	–	42	–	–	–
	15	–	–	–	32	–	–	–	44	–	–	–
	18	–	–	–	29	–	–	–	40	–	–	–
25	36	–	–	–	47	–	–	–	93		–	–
	34	–	–	–	50	–	–	–	107	–	–	–
	38	–	–	–	59	–	–		100	–	–	–
20	124	–	–	–	154	–	–	–	201	–	–	–
	120	–	–	–	167	–	–	–	195	–	–	–
	116	–	–	–	148	–	–	–	207	–	–	–
15	no	–	–	–	no	–	–	–	no	–	–	–
	no	–	–	–	no	–	–	–	no	–	–	–
	no	–	–	–	no	–	–	–	no	–	–	–

Moisture content can be calculated using formulae:

$$w = \frac{m_1 - m_2}{m_2} 100\%,$$

где m_1 – sample mass before drying, g;
m_2 – sample mass after drying, g.

In confirmation, Figure 22 presents the results of an experimental-analytical study of the flammability of the ground cover. The figure shows that when exposed to a heat load of 30-50 kW / m², the experimental results are fully consistent with the calculation. In particular, the radiation flux with a density of 50 kW / m² leads to the ignition of moss and needles in a time interval of 3-7 s, forming the critical temperature at the upper edge of the material. With a decrease in the density of the impact flow to 30 kW / m², as expected, before the emergence of stable combustion, the material manages to warm up

Figure 22. Flammability of FCM coniferous plantations ($D = 0.03$ m, $\rho = 75$ kg / m³, w = 10%) when exposed to thermal radiation, where 1 is moss, 2 is lichen, 3 is tree needles. Symbols - experiment, lines - area of calculation according to (12) at $\lambda = 0.06$ W / (m · K), c = 1700 J / (kg · K) for x = 0,001-0,01 m and Tverspl. = 620 K

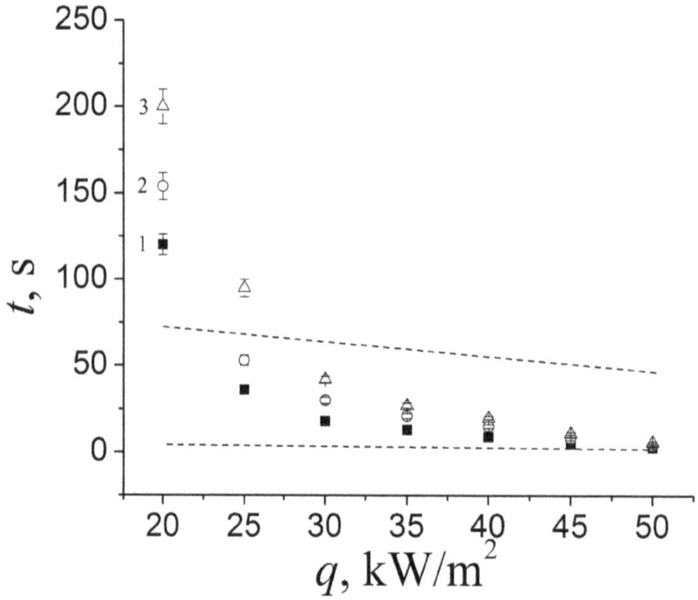

somewhat deeper until the threshold temperature is reached at a distance of 3-7 mm from its front face, depending on the type of ground fuel.

Thus, the experimental studies carried out make it possible to expand the existing list of materials with an established critical density of the radiant heat flux, which makes it possible to more adequately formulate requirements for fire fighting measures in forest plantations. At the same time, pursuing the goal of improving the reliability of preventive measures, it makes sense to determine the requirements for reducing fire loads on the ground cover to values less than 20 kW / m². This approach will determine the reliable width of the barriers, including for the worst conditions of the dry period, when the moisture content of the material is 10-20%. At the same time, the width of the obstacle to a forest fire in the dry season should be considered as universal, since it will allow to limit the spread of fire over the ground cover under all climatic conditions characteristic of the Republic of Belarus and countries with a similar climate.

4.4. Evaluation of the Influence of the Parameters of the Front of the Flame of a Forest Fire on the Level of Exposure of the Ground Cover of Coniferous Plantations. Calculation of Safe Distance in Case of Fire Based on Computer Technology

As established earlier, the process of heating, drying and subsequent ignition of ground-based combustible material significantly depends on the type of heat front, including its shape, length, height and temperature. For forest fires, the more characteristic forms of the fire front are either a plane or a cylinder (Figure 23) (Ford, 1995; Babrauskas, 2003; Dupuy & Larini, 1999; Koshmarov & Bashkir, 1987; Gusev, 1989; Borodkin et al., 1997). The determining factor affecting the geometry of a forest fire is the duration of burning, as well as the wind. For example, with complete calm, a forest fire in the initial stage of its development has the shape of a cylinder, the dimensions of which increase with time. In the future, the flame front, which propagates with the greatest speed, with increasing fire area will become more and more distinct flat shape.

At the same time, it is important to take into account the ability of forest plantations to stop the movement of air. Research Nesterov V.G. indicate that the wind speed under the forest canopy (tree crowns) decreases to 0 m / s when deepening into the forest at a distance of 100 m (Nesterov, 1949). At the same time, leaving the forest to open space, the wind again increases and reaches the initial speed at a distance of about 500 m to the forest border, i.e. up to the edge of it. At the same time, wind resistance of forest plantations is of broad national economic importance. The forest protects fields, gardens, cities, railways and highways from the harmful effects of wind. For the retention of wind, a strip of forest 10-20 m wide from 1-2 rows of trees is sometimes sufficient (Nesterov, 1949). In this regard, the modeling of a possible forest fire situation under the forest canopy without taking into account the wind load is fully

Figure 23a. Forest fire flat forms

Figure 23b. Forest fire cylindrical forms

justified. However, when studying the possible transition of fire from forest plantations to agricultural land adjacent to it or vice versa, taking into account the effect of wind on the flame front and heat exchange conditions seems important.

While ensuring fire safety in the forest and developing preventive measures, it is relevant to study the possible heat load in the area of the extended thermal fronts of the flat and cylindrical shape, or more precisely, to determine the characteristic densities of heat fluxes acting on the nearby combustible material and implemented in real conditions. Thus, according to the data of analytical calculations (4), (6) and (7), a flat front represents a greater fire hazard in comparison with a cylindrical one. At the same time, the vertically located irradiated surface is subject to a greater thermal effect as compared to the horizontal one (Figure 24).

As can be seen from Figure 24, when the heat front height is 0.5 m and its width is 10 m, which corresponds to a weak bottom fire, the probability of ground cover ignition at a distance of 1.5 m from the flame front is absent when the critical heat load is set at 20 kW / m^2. This distance corresponds to the width of currently used mineralized bands. However, with an increase in the height of the radiator to 1 m, a heat load reaches a critical value (curve 1) with a vertical arrangement of the irradiated surface. For a horizontal arrangement of the irradiated surface at small heights of the radiator for a flat and cylindrical fronts, a close level of thermal exposure is formed. However, when the height of the fire is higher than 2 m, the spread becomes more and more significant, which is due to the smaller contribution to the process of irradiation of the bent upper sections of the cylinder half-base. In this case, cylindrical heat fronts form a heat load at a distance of 1.5 m in the range of 1.5-20 kW / m^2 and are able to ignite the dried ground cover only at their height above 8 m, which meets the conditions of burning of the lower forest layer, including undergrowth and young. In addition, the achievable levels of heat load with a vertical arrangement of the irradiated surface cease to change significantly when the height of the heat

Figure 24. The dependence of the density of the radiant heat flux from the height of the radiator when the irradiated surface is removed by 1.5 m from flat (1, 2) and cylindrical heat front (3), where 1 is the calculation according to (7); 2, 3 - calculation by (6) and (4) respectively. Emitter width 10 m, temperature 1070 K

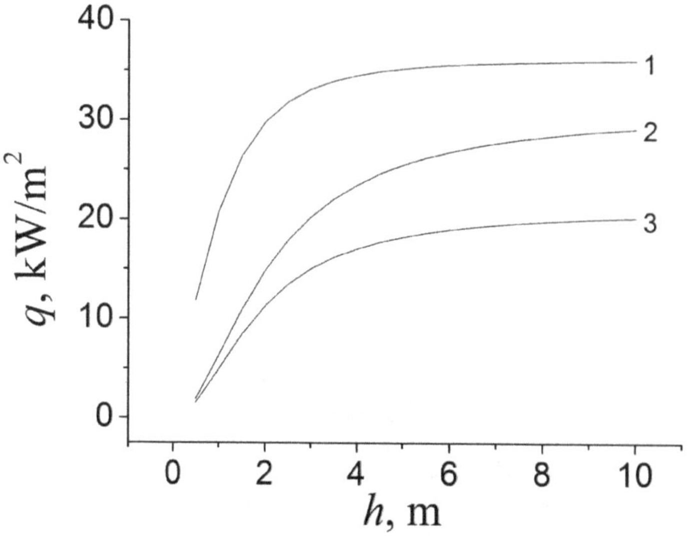

front is more than 3 m, which is due to the insignificant contribution to the resulting heat flux of remote areas in the upper part of the radiator.

Figure 25 shows the regularities of the spatial distribution of the density of the radiant heat flux at a height of the thermal front of 1.5 m, which corresponds to the conditions of an average low-level fire. It should be noted that flat and cylindrical fronts in the working range of the height of ground fires, as can be seen in this figure, are capable of forming radiation fluxes near the radiator with a density of 20-35 kW / m^2, which corresponds to the estimates given in (Fateh et al., 2016).

At the same time, as mentioned earlier, with a small height of the thermal front, the radiation from the cylindrical surface is slightly inferior to the flat one, and for the horizontal location of the irradiated surface the safe distance should be about 1 m. about 2 m.

It is significant that the achievable levels of heat load in the fire zone significantly depend on the flame temperature. Thus, under the conditions of burning of the ground cover consisting of moss, lichen, spruce needles and foliage, the working flame temperature range according to experimental data is 870-1070 K. However, in the case of burning wood materials, the temperature is slightly higher and is about 1270 K (GOST 12.1.004-91; Koshmarov & Bashkir, 1987). Figure 26 shows the effect of the temperature of the radiator on the achievable levels of heat load in the working height range and the distance to the thermal front of a flat and cylindrical shape. As you can see, the 1.5 m distance from the radiator does not provide security from thermal fronts of similar height. Under these conditions, it is possible to ignite the ground cover during the formation of an extended flat front with a temperature above 1000 K and a vertical arrangement of the irradiated surface. However, ignition of a horizontal material is possible only if there is a large amount of wood materials in the ground cover and a burning temperature of 1250 K. At that, the cylindrical heat front in the indicated temperature range of the radiator does not pose any danger from the point of view of possible critical thermal exposure.

Figure 25. Spatial distribution of radiant density heat flux at the height of the radiator 1.5 m for flat (1, 2) and cylindrical heat front (3), where 1 is the calculation by (7); 2, 3 - calculation by (6) and (4) respectively. Emitter width 10 m, temperature 1070 K

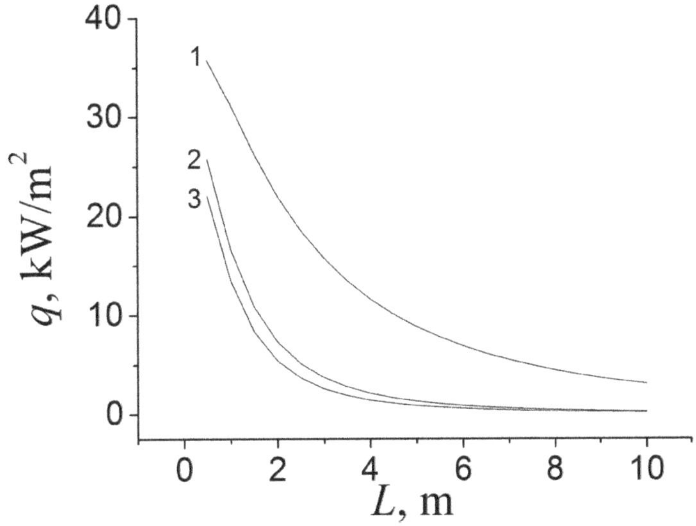

Figure 26. Influence of the temperature of the thermal front 10 m wide at achievable levels of heat load with height and distance to a 1.5 m radiator for flat (1, 2) and cylindrical heat front (3), where 1 - calculation by (7); 2, 3 - calculation by (6) and (4) respectively

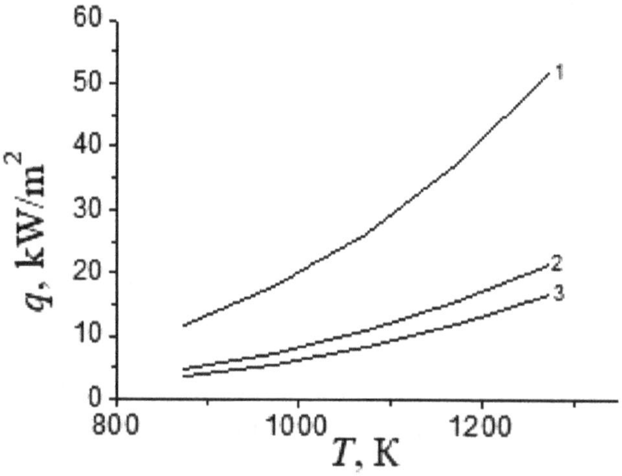

It should be noted that the width of the thermal front of 10 m chosen as an estimate is quite realizable in the conditions of forest fires. At the same time, fire fronts of large sizes can form during dry periods, but their further increase, as well as decrease, will not cause significant changes in the level of ground cover exposure due to the insignificant contribution of marginal distant sites to the resulting heat flux density. As an example, Figure 27 shows the effect of the radiator width on the heat flux density with a characteristic height of the fire front of 1.5 m for the vertical position of the irradiated surface. As can be seen in this figure, with a slight distance to the thermal front, which corresponds to the operating parameters of fire barriers arranged in the forest fund in the form of mineralized strips, the change in the width of the radiator is almost negligible. Spacing in the values of heat flux density at the distances shown in the figure to the radiator do not exceed 1.7 times.

Thus, measures to prevent forest fires in terms of limiting their spread should be envisaged on the principle of ensuring the protection of forest ground cover from the action of a flat, extended thermal front and the vertical position of the irradiated surface. This approach will provide a greater level of protection of forest plantations and adjacent agricultural land. At the same time, given that increasing the height of the heat front in excess of 3 m and width in excess of 10 m does not significantly affect the degree of exposure of the ground cover (Figure 24), these values of height and width must be considered as universal and optimal in terms of modeling safe distances, both from lower and general fires under the influence of thermal radiation.

To improve the efficiency of obtaining the necessary estimates and timely determine the level of heat load in case of fire, a computer program for estimating the safe distance from the flame front was developed using Delphi, based on the calculations of the density of radiant heat flux from a flat long front (7) and (6), which, as indicated above, allows for a greater degree of reliability of protection measures. The choice in favor of the Delphi environment is primarily determined by the speed of interface development, the possibility of increasing it, and, not unimportantly, the low requirements of the developed application for computer resources and the absence of the need to install additional service software to run the program, which also has a small exe size. file, which is convenient for its transfer (Osipov, 2006).

Figure 27. Dependence of the radiant heat flux density by (7) from the distance to the irradiated surface with the radiator width (Δx), m: 5 (1), 10 (2), 15 (3). The height and temperature of the emitter 1.5 m and 1070 K respectively

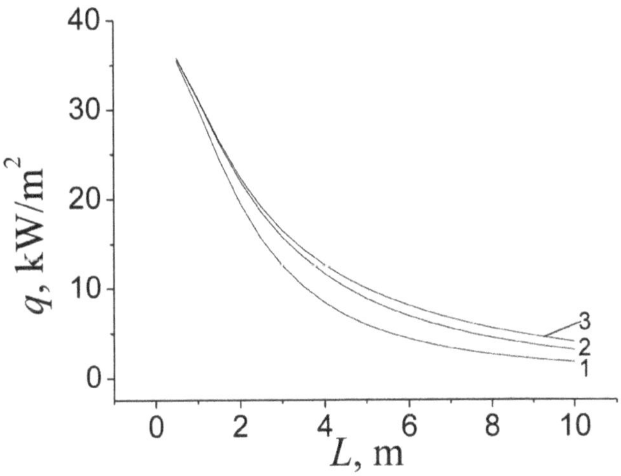

This program allows you to calculate the density of radiant heat flux Q_p (kW / m²) from the flame front, compare it with the critical value for combustible material and make a conclusion about ensuring a safe distance or finding the irradiated material or elimination forces in the danger zone with the corresponding information displayed in calculation field. It is important to note that in case of incorrect input of the values necessary for the calculation, the program automatically displays error information with explanations on the screen. The program also has the opportunity to refer to reference data on the critical values of the radiant heat flux density ($Q_{кр}$), leading to the ignition of some types of combustible material of coniferous plantations, including moss, lichen, needles, and burns of unprotected human skin (Figure 28). Additionally, you can estimate the time of safe work in special protective clothing of firefighters: heat protection suit, heat reflecting suit and means of local protection.

To construct graphical dependences of the radiant heat flux density Q_p (kW / m²) on the heat exchange parameters, a special window is provided in the right part of the interface. Moreover, the variable along the x axis can be set at the discretion of the user (L, T1, T2, r, h, e1, e2). Thereby, visualization of the dependences of the radiant heat flux density on the heat exchange parameters is achieved, which facilitates the perception of the obtained data (Figure 29). At the same time, in the right upper side with respect to the graphics side, a window appears with information about the constructed dependencies, which are displayed in different colors. In the field of plotting graphs, it is possible to set limits along the x axis, as well as the necessary step for a given variable. If you need to build a new schedule, you need to use the "Clear schedule" function and set other input data for the calculation.

The developed program can be used, first of all, by forestry workers, as well as emergency and rescue units to improve the prevention and elimination of fires, the level of protection of rescuers, the population and natural ecosystems from the dangerous factor of a forest fire in the form of heat flow. It should be noted that extended thermal fronts can be formed not only in the conditions of forest fires, but also during the burning of buildings and structures, technological equipment, storages of combustible material. To solve a problem that is not related to forest fires, it is only necessary to specify in the program the required parameters of the thermal front and the value of the critical heat load for the considered

Figure 28. General view of the interface of the program "Heat radiation in case of fire

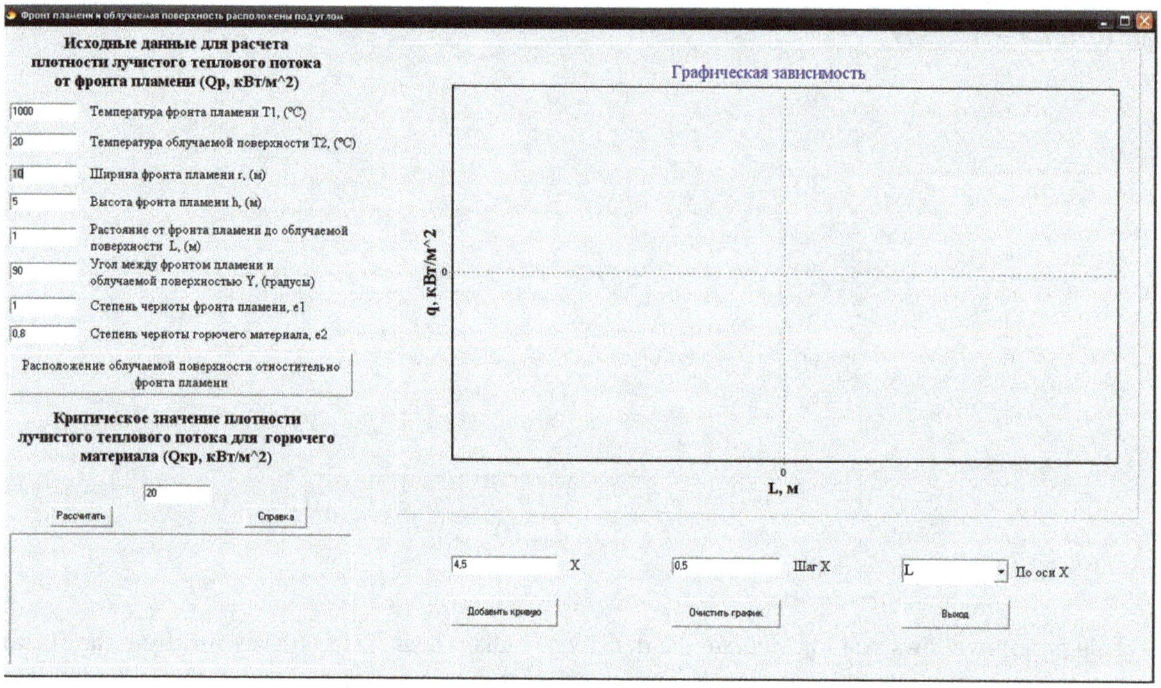

Figure 29. General view of the interface of the program "Heat radiation in case of fire." An example of building graphical dependencies

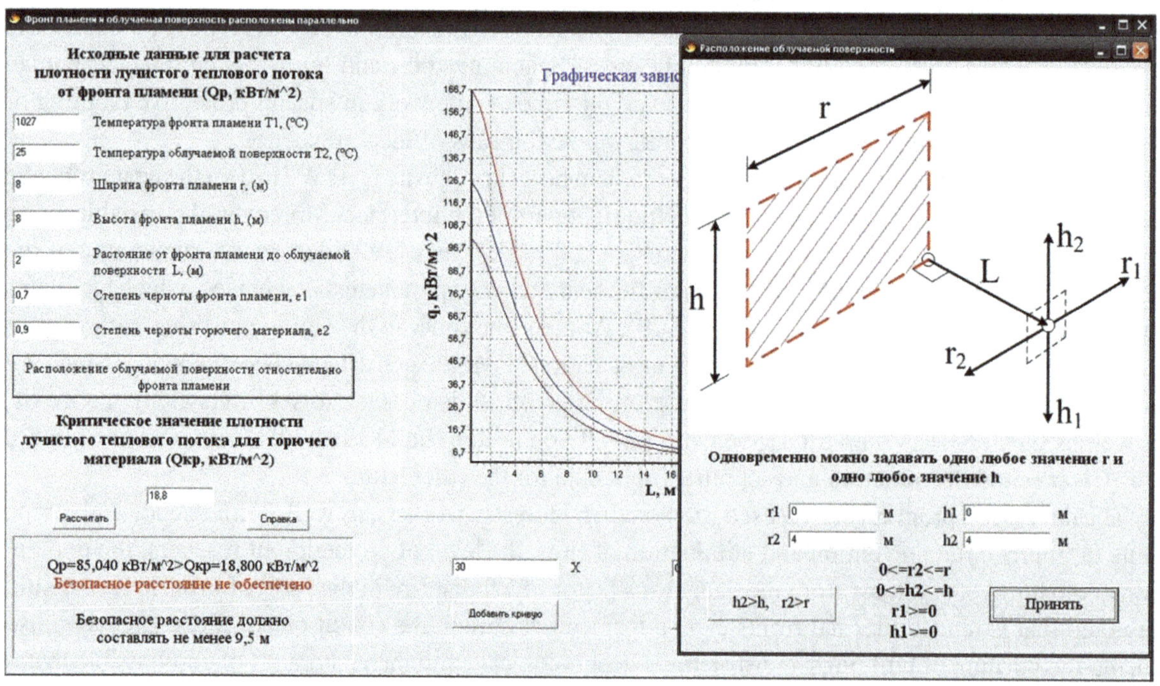

combustible material, which can be taken from reference data. The presented interface makes it possible to reduce the time of assessing the level of irradiation of combustible materials under the influence of thermal radiation from long sources and to formulate principles for the implementation of fire prevention measures.

4.5. Development of Methods for Calculating the Width of Fire Barriers, Depending on the Level of Possible Heat Load on the Ground Cover of Coniferous Plantations

The development of the methodology is based on the results of analytical studies of the laws of the spatial distribution of the density of radiant heat flux from extended flat heat sources and experimental studies of the ignitability of ground-based combustible material of coniferous plantations described in sections 4.2 and 4.3 of this work.

The calculation of the width of the fire barrier includes the following two steps:

1. Determination of the critical density of radiant heat flux for ground-based combustible material, depending on its humidity.

Approximation of experimental data on the results of a study of the flammability of ground-based fuel material of coniferous plantations in the form of moss, lichen, needles has established the following relationship between the moisture content of the material (w, %) and the critical density of radiant heat flux (q, kW / m^2) leading to its ignition:

$$q = \frac{-C_1}{1+e^{0,1w-2}} + C_2, 10 \le w \le 300, \tag{13}$$

where C_1 and C_2 are coefficients equal to 34 kW / m^2 and 45 kW / m^2, respectively.

The value of q obtained from (13) characterizes the critical (minimum) value of the heat load at which the ground-based combustible material of coniferous stands ignites. To ensure safety conditions, the heat load on the ground cover behind the fire barrier should be less than critical. Consequently, the value of the heat flux density, set by expression (13), must be reduced. It is logical to reduce the heat load value by 5 kW / m^2, as is done when determining the flammability of building materials (GOST 30402-96), and taking into account this, by (14) determine the optimal and effective width of the fire barrier. In this case, conditions are created that exclude the possibility of ground cover ignition from the heat load generated during a fire.

2. Determination of the width of the fire barrier, depending on the established critical density of the radiant heat flux and the height of the heat front for the actual parameters of the bottom ground fire.

The established regularities of the spatial distribution of the radiant heat flux density by (7) from an extended flat thermal front revealed the following relationship between the width of the fire barrier (L, m) and the height of the thermal front (h, m) at a fixed and determined according to paragraph 1 of the method, the critical heat value loads for terrestrial combustible material (q, kW / m^2):

$$L = \frac{A_1 - A_2}{1 + e^{((q-5)-B_1)/B_2}} + A_2, q \leq 40 \text{ kW/m}^2, \tag{14}$$

where A_1, A_2, B_1, B_2 - coefficients depending on the height of the thermal front (h, m), are given in table 3.

The values of the width and temperature of the flame front are taken at 10 m and 1070 K, respectively, and are argued in sections 4.1.1 and 4.4 of this work. Increasing the width of the heat front in excess of 10 m, as well as the height in excess of 3 m, ceases to significantly affect the values of the heat load on the irradiated surface.

The obtained expression (14) indicates that with an increase in the heat load required to ignite the ground cover, the required width of the fire barrier decreases. In other words, with an increase in the resistance of the material to the action of the heat flow, which is associated with an increase in its humidity, mineralized strips of smaller width are required.

The reliability of the calculation of the width of the mineralized bands by the formula (14) is confirmed by the good agreement of the calculation results with the data of field experiments. The discrepancy of these data does not exceed 10%.

If it is necessary to estimate the width of the fire barrier in a wider range of thermal front parameters (by temperature, height and width) and with different orientations of the irradiated material relative to it, it is recommended to use a computer program for calculating the heat load and safe distance in case of fire (see Section 4.4).

An example of calculation.

Conditions of the problem. It is necessary to determine the width of the mineralized strip for the conditions of prolonged drought at a soil cover moisture of about 10% and a flame front height of 3 m.

The solution of the problem.

1. In accordance with this methodology, using expression (13), we determine the critical density of the radiant heat flux (q, kW / m²) for a material with a moisture content of 10%.

$$q = \frac{-!_1}{1 + e^{0,1w-2}} + !_2 = \frac{-34}{1 + e^{0,1\cdot 10-2}} + 45 = 20 \text{ kW/m}^2.$$

Table 3. Coefficients for the expression (14)

h, m	A_1, m	A_2, m	B_1, kW/m²	B_2, kW/m²
1	51,1	0,6	-16,0	9,1
1,5	88,0	-0,1	-33,8	14,9
2	40,8	-1,3	-26,4	21,1
2,5	46,8	-4,4	-39,9	35,0
3	32,1	-13,7	-6,0	56,9

2. According to expression (14), we determine the required width of the mineralized strip, taking into account the radiative heat flux density critical for ground-based combustible material to be reduced by 5 kW / m^2.

$$L = \frac{A_1 - A_2}{1 + e^{((q-5)-B_1)/B_2}} + A_2 = \frac{32{,}1+13{,}7}{1+e^{(15+6{,}0)/56{,}9}} - 13{,}7 = 5 \text{ m}.$$

Thus, the width of the fire barrier for the accepted conditions of the task should be 5 m.

Under the condition of the formation of strong ground fires with a thermal front height of 2 m or medium with a height of 1.5 m, with other similar parameters of the problem, the width of the mineralized strip, determined by this method, should be 4 m and 3.2 m, respectively.

4.6. Recommendations for the Creation of Increased Width Fire Barriers for Global Warming Conditions

Improving the reliability of the fire protection system for forests by increasing the width of barriers to reasonable limits is reasonable and relevant, especially in the context of global warming. Considering the fact that the laying of mineralized lanes is not associated with the cutting of stands, since a vehicle with a suitable device for creating lanes can bend around existing obstacles and a 3-4 m wide section is sufficient for its passage, creating mineralized lanes seems to be the most appropriate way to limit ground fires. At the same time, the cost of creating mineralized strips of recommended width of 4-5 m, with the right choice of the method of their creation, does not significantly exceed the cost of the bands currently being arranged. Thus, regulated (TCP 193-2009), 2.8 m wide strips (Figure 30a) are laid by means of double plowing with a PKL-70 device (the width of a single plowing strip is 1.4 m). This device is the main tillage mechanism used by forestry workers in Belarus and a number of foreign countries for laying mineralized belts. To create a strip of greater width, for example, 4 m or 5 m, it is proposed the same way as in the case of a strip 2.8 m wide, double plowing, that is, the costs remain the same. In other words, it is recommended to create combined protective strips, including two standard mineralized strips 1.4 m wide and a section of combustible ground material. In this case, the only additional measure is provided - the removal of ground-based combustible material between the bands held (Figure 30b).

The most expedient in terms of efficiency and least cost are the methods of removing the ground cover by raking it (hooting) with a rake (hoes) or burning it out. Other known methods of preventing ignition of the ground cover, for example, its treatment with flame retardants, are expensive and, moreover, the treatment requires a systematic update in view of the instability of the compositions to atmospheric influences.

Thus, with the help of a rake, one person per hour can create a protective strip 0.75 m wide and 90-150 m long with a slope of terrain up to 12° and a length of 60-90 m with a slope of terrain 13-24° (Ivanov, Ivanova, & Moskalchenko, 2011). That is, for example, for a flat terrain, if necessary, to create a mineralized strip 1 km long and 5 m wide, a surface of 2.2 thousand m^2 is required (5 m - 1.4 m - 1.4 m = 2.2 m; 2.2 m × 1 thousand m = 2.2 thousand m^2). With an average capacity of 120 m / h, it is possible to treat this surface by attracting 5 people for 5 hours.

When using the second method of creating wider mineralized bands - the burning of ground-based combustible material requires the presence of special incendiary devices. The most common are the

Figure 30a. General view of the standard fire barrier 2.8 m wide, where the large width ensured by rapid removal of ground cover between two mineralized strips

Figure 30b. The proposed barrier, where the large width ensured by rapid removal of ground cover between two mineralized strips

devices AZ-1, ZA-FKT, ZA-4, ZA-1M, AZR-5.5. These devices are knapsacks, consist of a tank with a combustible mixture and a hose supplying the mixture with a burning nozzle. The cost of one unit is about 180 US dollars. These apparatuses are equipped with fire and chemical stations (PHC) at the forest enterprises of the Republic of Belarus at the rate of 1-3 apparatuses for PHS-1 and 3-5 apparatuses for PHS-2.

Productivity when creating fencing strips with incendiary devices (m / h per worker during manual work) by annealing ground fuel from the base strip is 900-1200 m / h with a terrain slope up to 12° and 600-900 m / h with a slope terrain 13-24° (Ivanov, Ivanova, & Moskalchenko, 2011). In other words, to create a 1 km mineralized strip 5 m wide by annealing a ground-based combustible material located between the 1.4-meter-wide supporting strips, 1 person's labor is required for about 1 hour. The duration of annealing is inversely proportional to the number of workers involved.

Thus, the method of annealing ground-based combustible material in comparison with its removal with a rake (hoes) is more preferable in view of the required minor costs, both physical and monetary. At the same time, the proposed measures to improve the efficiency of mineralized zones can significantly reduce the area of strong fires and the enormous annual damage from them.

It should be noted that annealing of ground-based combustible material can be used to increase the width of the barriers and increase their effectiveness in advance. However, the most expedient is to expand the barriers directly in the event of forest fires. That is, two bands 1.4 m wide are prearranged with a zone of ground-based combustible material left between them of the required width. Subsequently, in the event of strong fires with a flame front height of 2-3 m and a width of more than 10 m, operational annealing in the path of fire propagation will increase the reliability of the barriers and limit the development of the flame. This approach will reduce the area of scorched vegetation through the creation of broad lanes only in places where it is really necessary or due to the operational situation.

5. CONCLUSION

As a result of the conducted research, regularities of the occurrence and spread of fires in natural ecosystems of the temperate climatic zone under the influence of thermal radiation from the flame front were established. Analytical solutions have been obtained that make it possible to predict the level of heat load on the ground cover of coniferous plantations. For the efficiency of heat load calculations in case of fire, a special computer program has been developed. Methods of full-scale and laboratory modeling revealed patterns of heating and ignition of FCM of various humidity, which is especially important for the conditions of global warming of the climate. A practice-oriented method for calculating the width of fire barriers that limit the spread of forest fires, and ways to create them, is proposed.

The most important results of the work are the following:

- In conditions of lowland forest fires, flame fronts with a height of 1-4 m and a width of 10-20 m can form that can form heat loads reaching 40 kW / m^2, which greatly exceeds the flammability parameters of forest vegetation in droughty periods.
- The temperature of the flame of a fire that developed during the burning of the ground cover of coniferous plantations in the composition of moss, lichen, litter of needles, foliage, bark, small branches reaches 1070 K, which is slightly lower than the burning temperature of wood.
- Ignition of the ground cover of coniferous stands characteristic for the conditions of droughty periods of 10% humidity occurs when exposed to a thermal load of 20 kW / m^2 for a time not exceeding 4 minutes. The specified heat load is capable of providing shock short-term heating with the drying of the outer layer, sufficient to ignite the material and maintain a steady flame combustion.
- Moisture contained in the structure of mosses, lichens, litter, significantly affects their resistance to ignition when exposed to heat load. With an increase in the moisture content of FCM from 10% to 40%, the critical value of the density of the radiant heat flux increases at least twice, that is, to 40 kW / m^2.

- For ordinary weather conditions, corresponding to a ground cover moisture of about 40%, the effective width of fire barriers should provide the level of heat load on the ground fuel to 35 kW / m^2, which corresponds to an obstacle width of 1 m.
- For dry weather conditions leading to low groundwater moisture levels (up to 10%) to protect against low fires with a front height of about 1.5 m, it is necessary to ensure the density of radiant heat flux beyond the fire barrier to 15 kW / m^2, which can be achieved by arranging mineralized strips with a width of 3.2 m, and in case of the threat of the occurrence of strong bottom fires with a front height of 2-3 m, the indicated width of the barriers must be increased to 4-5 m.
- To create fire barriers with a width of 4-5 m, it is advisable to hold two adjacent bands 1.4 m wide with a 2.2-3.2 m wide ground cover between them set aside for its prompt removal by annealing when there is a risk of fire spreading.

REFERENCES

Abduragimov, I. M. (2012). The problem of extinguishing large forest fires and large-scale fires of solid combustible materials in buildings. *Fire and Explosion Safety*, *21*(2), 69–74.

Albini, F., Alexander, M., & Cruz, M. (2012). A mathematical model for predicting the maximum potential spotting distance from a crown fire. *International Journal of Wildland Fire*, *21*(5), 609–627. doi:10.1071/WF11020

Anthenien, R. A., Tse, S. D., & Fernandez-Pello, A. C. (2006). On the trajectories of embers initially elevated or lofted by small scale ground fire plumes in high winds. *Fire Safety Journal*, *41*(5), 349–363. doi:10.1016/j.firesaf.2006.01.005

Artsybashev, E. S. (1974). Forest fires and fight with them. Moscow, Russia: Forest industry.

Artsybashev, E. S. (2014). The impact of fires on forest biogeocenoses. *Biosphere*, *6*(1), 53–59.

Babrauskas, V. (2003). *Ignition handbook.* Issaquah, Washington: Fire Science Publishers.

Baranovsky, N. V. (2009). Estimation of probability of occurrence of forest fires taking into account meteorological conditions, anthropogenic load and thunderstorm activity. *Fire safety*, 1, 93-99.

Barber, I. R. (1977). Fire preparation as an essential factor in roadside management. *Vic. Resur*, *2*, 18–21.

Bartoli, P., Simeoni, A., Biteau, H., Torero, J. L., & Santoni, P. A. (2011). Determination of the main parameters influencing forest fuel combustion dynamics. *Fire Safety Journal*, *46*(1-2), 27–33. doi:10.1016/j.firesaf.2010.05.002

Battaglia, F., McGrattan, K., Rehm, R., & Baum, H. (2000). Simulating fire whirls. *Combustion Theory and Modelling*, *4*(2), 123–138. doi:10.1088/1364-7830/4/2/303

Beltsova, T. G. & Korolchenko, O. N. (2008). Flammability indicators of fire-protected wood. *Fire and explosion safety*, 4, 31-33.

Benkorichi, S., Fateh, T., Richard, F., Consalvi, J.-L., & Nadjai, A. (2017). Investigation of thermal degradation of pine needles using multi-step reaction mechanisms. *Fire Safety Journal*, *91*, 811–819. doi:10.1016/j.firesaf.2017.03.058

Bogorad, V. I., Litvinskaya, T. V., Shevchenko, I. A., Dybach, A. M., & Slepchenko, A. Yu. (2016). Radiation consequences of a fire in the exclusion zone of the Chernobyl NPP. *Nuclear and Radiation Safety*, *1*(69), 64–68. doi:10.32918/nrs.2016.1(69).10

Bondur, V. G. (2011). Space monitoring of wildfires in Russia in the conditions of the abnormal heat of 2010. *The study of the earth from space*, 3, 3-13.

Boonmee, N., & Quintiere, J. G. (2002). Glowing and flaming autoignition of wood. *Proceedings of the Combustion Institute*, *29*(1), 289–296. doi:10.1016/S1540-7489(02)80039-6

Borodkin, A. N. (1997). Model of the spread of fire between wooden buildings. *Fire and Explosion Safety*, *1*, 36–39.

Branca, C., & Di Blasi, C. (2004). Global intrinsic kinetics of wood oxidation. *Fuel*, *83*(1), 81–87. doi:10.1016/S0016-2361(03)00220-5

Bufacchi, P., Krieger, G. C., Mell, W., Alvarado, E., Santos, J. C., & Carvalho, J. A. Jr. (2016). Numerical simulation of surface forest fire in Brazilian Amazon. *Fire Safety Journal*, *79*, 44–56. doi:10.1016/j.firesaf.2015.11.014

Building materials. Flammability test method: GOST 30402-96. Enter 03/30/97. Minsk: Ministry of Architecture and Construction of the Republic of Belarus: FSUE TsNIISK them. V. A. Kucherenko, 1996.

Cancellieri, D., Innocenti, E., & Leroy-Cancellieri, V. (2013). WinGPYRO: A software platform for kinetic study of forest fuels. *Fire Safety Journal*, *58*, 103–111. doi:10.1016/j.firesaf.2013.01.005

Chen, Y., Aanjaneya, K., & Atreya, A. (2017). A study to investigate pyrolysis of wood particles of various shapes and sizes. *Fire Safety Journal*, *91*, 820–827. doi:10.1016/j.firesaf.2017.03.079

Chuah, K. H., & Kushida, G. (2007). The prediction of flame heights and flame shapes of small fire whirls. *Proceedings of the Combustion Institute*, *31*(2), 2599–2606. doi:10.1016/j.proci.2006.07.109

Cruz, M. G. (2004). *Ignition of crown fuels above a spreading surface fire*. (PhD Dissertation), University of Montana. Missoula.

Di Blasi, C. (2008). Modeling chemical and physical processes of wood and biomass pyrolysis. *Progress in Energy and Combustion Science*, *34*(1), 47–90. doi:10.1016/j.pecs.2006.12.001

Dibble, A. C., White, R. H., & Lebow, P. K. (2007). Combustion characteristics of northeastern USA vegetation tested in the cone calorimeter: Invasive versus noninvasive plants. *International Journal of Wildland Fire*, *16*(4), 426–443. doi:10.1071/WF05103

Dorrer, G. A. (2008). *Dynamics of forest fires*. Novosibirsk, Russia: Publishing House of the Siberian Branch of the Russian Academy of Sciences.

Dupuy, J., & Larini, M. (1999). Fire spread through a porous forest fuel bed: A radiative and convective model including fire-induced flow effects. *International Journal of Wildland Fire*, *9*(3), 155–172. doi:10.1071/WF00006

Dupuy, J. L., Maréchal, J., Portier, D., & Valette, J. C. (2011). The effects of slope and fuel bed width on laboratory fire behavior. *International Journal of Wildland Fire*, *20*(2), 272–288. doi:10.1071/WF09075

Eurocode 1. Impacts on structures. Part 1-2. General effects. Impact for fire resistance determination: TCP EN 1991-1-2-2009. Enter 01/01/10 Minsk, Belarus: Ministry of Architecture and Construction of the Republic of Belarus: Stroytekhnorm RUE, 2009.

Fateh, T., Richard, F., Batiot, B., Rogaume, T., Luche, J., & Zaida, J. (2016). Characterization of the burning behavior and gaseous emissions of pine needles in a cone calorimeter – FTIR apparatus. *Fire Safety Journal*, *82*, 91–100. doi:10.1016/j.firesaf.2016.03.008

Fernandez-Gomez, I., de Castro, A. J., Guijarro, M., Madrigal, J., Aranda, J. M., Diez, C., ... Lopez, F. (2011). Characterization of forest fuels in a Mass loss calorimeter by short open-path FTIR spectroscopy. *Journal of Quantitative Spectroscopy & Radiative Transfer*, *112*(3), 519–530. doi:10.1016/j.jqsrt.2010.10.004

Fernandez-Pello, A. C. (2017). Wildland fire spot ignition by sparks and firebrands. *Fire Safety Journal*, *91*, 2–10. doi:10.1016/j.firesaf.2017.04.040

Finney, M. A. (1993). Modeling the spread and behaviour of prescribed natural fires. *Proceedings of the 12th Conference on Fire and Forest Meteorology*. Society of American Foresters. Jekill Island, Georgia, pp. 138-143.

Font, R., Conesa, J. A., Moltó, J., & Muñoz, M. (2009). Kinetics of pyrolysis and combustion of pine needles and cones. *Journal of Analytical and Applied Pyrolysis*, *85*(1-2), 276–286. doi:10.1016/j.jaap.2008.11.015

Ford, R. (1995). *Investigation of wildfires. Bend.* Maverick Publications.

Frangieh, N., Morvan, D., Meradji, S., Accary, G., & Bessonov, O. (2018). Numerical simulation of grassland fres behavior using an implicit physical multiphase model. *Fire Safety Journal*, *102*, 37–47. doi:10.1016/j.firesaf.2018.06.004

Gavrilov, K., Accary, G., Morvan, D., Lyubimov, D., Méradji, S., & Bessonov, O. (2011). Numerical simulation of coherent structures over plant canopy. *Flow, Turbulence, and Combustion*, *86*(1), 89–111. doi:10.100710494-010-9294-z

Goman, P. N., Sobol, V. R., Barovik, D. V., & Taranchuk, V. B. (2011). Experimental and numerical simulation of the combustion process and the spread of fire in the conditions of the forest ground fire. *Technosphere Safety Technologies: Internet Journal*, *3*(37), 14. Retrieved from http://ipb.mos.ru/ttb

Gradstein, I. S., & Redhead, I. M. (1962). Tables of integrals, sums, series and products. Moscow: State. ed. Phys.-Mat. literature.

Grishin, A. M. (1981). *Mathematical models of forest fires*. Tomsk, Russia: Tomsk University Press.

Grishin, A. M. (2002). Ignition of forest combustible materials by a flow of radiant energy. *Physics of Combustion and Explosion*, *1*, 30–35.

Gusak, A. A., & Gander, G. M. (1991). Handbook of higher mathematics. Minsk, Belarus: Navuki i tehniki.

Gusev, V. G. (1989). *Assessment of firemen's exposure to firefighters in front of low and high forest fires. Forest fires and the fight against them: Sat. scientific tr. LenNIILH; Editorial* (pp. 133–137). E. S. Artsybashev. [et al.]

Gusev, V. G. (2005). *Physical and mathematical models of fire propagation and fire barriers in pine forests. SPb.* SRI LH.

Himoto, K., Shinohara, M., Sekizawa, A., Takanashi, K., & Saiki, H. (2018). A field experiment on fire spread within a group of model houses. *Fire Safety Journal*, *96*, 105–114. doi:10.1016/j.firesaf.2018.01.003

Ivanov, V. A., Ivanova, G. A., & Moskalchenko, S. A. (2011). *Handbook for extinguishing natural fires* (2nd ed.). Krasnoyarsk, Russia: Opera Printing Agency LLC.

Kasperov, G. I., & Goman, P. N. (2010). Investigation of fire hazardous properties of forest combustible materials of pine plantations. *Proceedings of BSTU. Ser. II, Forest and wood. prom.*, Vol. XVIII, 337-340.

Khodakov, V. E., & Zharikova, M. V. (2011). *Forest fires: research methods*. Kherson, Ukraine: Grin DS.

Konev, E. V. (1977). *Physical basis of burning materials*. Novosibirsk, Russia: Science.

Koshmarov, Yu. A., & Bashkir, M. P. (1987). Thermodynamics and heat transfer in fire engineering. Moscow, Russia: Vneshtorgizdat.

Kuznetsov, G. V., & Baranovsky, N. V. (2009). *Forecast of forest fires and their environmental consequences*. Novosibirsk, Russia: Publishing House of the Siberian Branch of the Russian Academy of Sciences.

Kuznetsov, G. V. & Baranovsky, N. V. (2010). Spatial formulation and numerical study of the problem of ignition of a layer of forest fuel with a particle heated to high temperatures. *Butlerov readings,* 12, 30-37.

Lasuta, G. F., & Goman, P. N. (2019). Simulation of the processes of emergence and spread of forest ground fire with an assessment of the level of heat load from the flame front. *Bulletin of the University of Civil Protection of the Ministry of Emergency Situations of Belarus*, *3*(2), 138–154.

Lei, J., Liu, N., Zhang, L., Chen, H., Shu, L., Chen, P., ... de Ris, J. L. (2011). Experimental research on combustion dynamics of medium-scale fire whirl. *Proceedings of the Combustion Institute*, *33*(2), 2407–2415. doi:10.1016/j.proci.2010.06.009

Leonelli, L., Barboni, T., Antoine Santoni, P., Quilichini, Y., & Coppalle, A. (2017). Characterization of aerosols emissions from the combustion of dead shrub twigs and leaves using a cone calorimeter. *Fire Safety Journal*, *91*, 800–810. doi:10.1016/j.firesaf.2017.03.048

Leroy-Cancellieri, V., Cancellieri, D., & Leoni, E. (2006). Thermal degradation of lingo-cellulosic fuels: DSC and TGA studies. *Thermochimica Acta*, *451*(1-2), 131–136. doi:10.1016/j.tca.2006.09.017

Li, Y. Z., & Ingason, H. (2017). Scaling of wood pallet fires. *Fire Safety Journal, 88*, 96–103. doi:10.1016/j.firesaf.2017.01.004

Linn, R. (1997). *A transport model for prediction of wildfire behaviour.* (PhD Dissertation), New Mexico State University. Las Cruces.

Liodakis, S., Katsigiannis, G., & Kakali, G. (2005). Ash properties of some dominant Greek forest species. *Thermochimica Acta, 437*(1-2), 158–167. doi:10.1016/j.tca.2005.06.041

Luche, J., Rogaume, T., Richard, F., & Guillaume, E. (2011). Characterization of thermal properties and analysis of combustion behaviour of PMMA in a cone calorimeter. *Fire Safety Journal, 46*(7), 451–461. doi:10.1016/j.firesaf.2011.07.005

Lykov, A. V. (1967). *Theory of heat conduction. M*oscow High School.

Marcelli, T., Santoni, P. A., Simeoni, A., Leoni, E., & Porterie, B. (2004). Fire spread across pine needle fuel beds: Characterization of temperature and velocity distributions within the fire plume. *International Journal of Wildland Fire, 13*(1), 37–48. doi:10.1071/WF02065

McCaw, W. (1998). *Predicting fire spread in Western Australia mallee-heath shrubland.* (PhD Thesis), W. McCaw; University of New South Wales. Canberra, Australia.

Mehrabian, R., Scharler, R., & Obernberger, I. (2012). Effects of pyrolysis conditions on the heating rate in biomass particles and applicability of TGA kinetic parameters in particle thermal conversion modeling. *Fuel, 90*, 567–575. doi:10.1016/j.fuel.2011.09.054

Mell, W., Jenkins, M. A., Gould, J., & Cheney, P. (2007). A physics-based approach to modelling grassland fires. *International Journal of Wildland Fire, 16*(1), 1–22. doi:10.1071/WF06002

Mell, W. E., Manzello, S. L., Maranghides, A., Butry, D., & Rehm, R. G. (2010). The wildland-urban interface fire problem-current approaches and research needs. *International Journal of Wildland Fire, 19*(2), 238–251. doi:10.1071/WF07131

Miloua, H. (2015). Numerical prediction of the fire spread in vegetative fuels using NISTWFDS model. *Int. J. Eng. Res. Africa, 20*, 177–192. doi:10.4028/www.scientific.net/JERA.20.177

Mindykowski, P., Fuentes, A., Consalvi, J. L., & Porterie, B. (2011). Piloted ignition of wildland fuels. *Fire Safety Journal, 46*(1-2), 34–40. doi:10.1016/j.firesaf.2010.09.003

Monod, B., Collin, A., Parent, G., & Boulet, P. (2009). Infrared radiative properties of vegetation involved in forest fires. *Fire Safety Journal, 44*(1), 88–95. doi:10.1016/j.firesaf.2008.03.009

Morandini, F., Santoni, P. A., & Balbi, J. H. (2001). The contribution of radiant heat transfer to laboratory-scale fire spread under the influences of wind and slope. *Fire Safety Journal, 36*(6), 519–543. doi:10.1016/S0379-7112(00)00064-3

Morvan, D. (2011). Physical phenomena and length scales governing the behaviour of wildfires: A case for physical modeling. *Fire Technology, 47*(2), 437–460. doi:10.100710694-010-0160-2

Morvan, D. (2014). Wind effects, unsteady behaviors, and regimes of propagation of surface fires in open field. *Combustion Science and Technology, 186*(7), 869–888. doi:10.1080/00102202.2014.885961

Morvan, D. (2015). Numerical study of the behaviour of a surface fire propagating through a firebreak built in a Mediterranean shrub layer. *Fire Safety Journal*, *71*, 34–48. doi:10.1016/j.firesaf.2014.11.012

Morvan, D., Accary, G., Meradji, S., Frangieh, N., & Bessonov, O. (2018). A 3D physical model to study the behavior of vegetation fires at laboratory scale. *Fire Safety Journal*, *101*, 39–52. doi:10.1016/j.firesaf.2018.08.011

Morvan, D., Meradji, S., & Accary, G. (2008). Wildfire behavior study in a Mediterranean pine stand using a physically based model. *Combustion Science and Technology*, *180*(2), 230–248. doi:10.1080/00102200701600978

Morvan, D., Méradji, S., & Accary, G. (2009). Physical modelling of fire spread in Grasslands. *Fire Safety Journal*, *44*(1), 50–61. doi:10.1016/j.firesaf.2008.03.004

Nelson, R. M. (2015). Re-analysis of wind and slope effects on flame characteristics of Mediterranean shrub fires. *International Journal of Wildland Fire*, *24*(7), 1001–1007. doi:10.1071/WF14155

Nesterov, V. G. (1949). *Burnability of the forest and methods for its determination. L.* Goslesbumizdat.

Occupational safety standards system. Fire safety. General requirements: GOST 12.1.004-91. Enter 07/01/92. Moscow, Russia: State. Standardization Committee: USSR Ministry of the Interior, 1991.

Osipov, D. I. (2006). *Professional programming. Moscow, Russia.* Delphi: Symbol Plus.

Ovsyannikov, I. V. (1978). Forest fire control system. Moscow, Russia: Forest industry.

Padhi, S., Shotorban, B., & Mahalingam, S. (2016). Computational investigation of flame characteristics of a nonpropagating shrub fire. *Fire Safety Journal*, *81*, 64–73. doi:10.1016/j.firesaf.2016.01.016

Padhi, S., Shotorban, B., & Mahalingam, S. (2017). A computational study of the interactions of three adjacent burning shrubs subjected to wind. *Fire Safety Journal*, *91*, 749–757. doi:10.1016/j.firesaf.2017.03.028

Pérez, Y., Pastor, E., Àgueda, A., & Planas, E. (2011). Effect of wind and slope when scaling the forest fires rate of spread of laboratory experiments. *Fire Technology*, *47*(2), 475–489. doi:10.100710694-010-0168-7

Perry, G. L. (1998). Current approaches to modelling the spread of wildland fire: A review. *Progress in Physical Geography*, *22*(2), 222–245. doi:10.1177/030913339802200204

Pimont, F., Dupuy, J. L., Caraglio, Y., & Morvan, D. (2009). Effect of vegetation heterogeneity on radiative transfer in forest fires. *International Journal of Wildland Fire*, *18*(5), 536–553. doi:10.1071/WF07115

Pinto, C., Viegas, D., Almeida, M., & Raposo, J. (2017). Fire whirls in forest fires: An experimental analysis. *Fire Safety Journal*, *87*, 37–48. doi:10.1016/j.firesaf.2016.11.004

Regueira, R., & Guaita, M. (2018). Numerical simulation of the fire behaviour of timber dovetail connections. *Fire Safety Journal*, *96*, 1–12. doi:10.1016/j.firesaf.2017.12.005

Restuccia, F., Huang, X., & Rein, G. (2017). Self-ignition of natural fuels: Can wildfires of carbon-rich soil start by self-heating? *Fire Safety Journal*, *91*, 828–834. doi:10.1016/j.firesaf.2017.03.052

Richards, G. D., & Bryce, R. W. (1995). A computer algorithm for simulating the spread of wildland fire perimeters for heterogeneous fuel and meteorological conditions. *International Journal of Wildland Fire, 5*(2), 73–80. doi:10.1071/WF9950073

Roitman, M. Y., Komissarov, E. P., & Pchelintsev, V. A. (1978). Fire prevention in construction. Moscow, Russia: stroiizdat.

Rules for fire prevention of forests in the Republic of Belarus: TCP 193-2009. Enter 11/01/09. Minsk, Belarus: Ministry of Forestry of the Republic of Belarus: Forestry Institute of the National Academy of Sciences of Belarus, 2009.

Sable, V. R., Goman, P. N., & Yan, V. I. (2010). On the influence of the form of an extended heat source on the density of the radiant energy flux. *All BDPU. Ser. III, Physics. Matematyka Infarmatika. Bialogiya. Geography (Sheffield, England), 4*, 3–9.

Sable, V. R., Goman, P. N., & Yan, V. I. (2011). On the question of the spatial distribution of the intensity of the radiant energy flux from a large area source. *Emergencies: Education and Science, 2*(6), 116-124.

Schemel, C. F., Simeoni, A., Biteau, H., Rivera, J. D., & Torero, J. L. (2008). A calorimetric study of wildland fuels. *Experimental Thermal and Fluid Science, 32*(7), 1381–1389. doi:10.1016/j.expthermflusci.2007.11.011

Simeoni, A., Thomas, J. C., Bartoli, P., Borowieck, P., Reszka, P., Colella, F., ... Torero, J. L. (2012). Flammability studies for wildland and wildland–urban interface fires applied to pine needles and solid polymers. *Fire Safety Journal, 54*, 203–217. doi:10.1016/j.firesaf.2012.08.005

Stepanov, K. L., Stanchits, L. K., & Stankevich, Yu. A. (2011). Heat radiation of an explosion: Risk of thermal injury to people and the occurrence of fires. *Vestn. Team Engineer Inst. Ministry of Emergency Situations of the Republic of Belarus, 1*(13), 31–45.

Sullivan, A. L. (2009). Wildland surface fire spread modelling, 1990-2007. 1: Physical and quasi-physical models. *International Journal of Wildland Fire, 18*(4), 349–368. doi:10.1071/WF06143

Sun, L., Zhou, X., Mahalingam, S., & Weise, D. R. (2006). Comparison of burning characteristics of live and dead chaparral fuels. *Combustion and Flame, 144*(1-2), 349–359. doi:10.1016/j.combustflame.2005.08.008

Suzuki, S., & Manzello, S. L. (2017). Experimental investigation of firebrand accumulation zones in front of obstacles. *Fire Safety Journal, 94*, 1–7. doi:10.1016/j.firesaf.2017.08.007 PMID:31092969

The system of fire safety standards. Fire safety of technological processes. Methods for assessing and analyzing fire hazard. General requirements: STB 11.05.03-2010. Enter 01/01/11. Minsk, Belarus: State. Standardization Commitee: Research Institute of Emergency Situations and Emergencies Ministry of Emergency Situations of the Republic of Belarus, 2010.

Thi, V. D., Khelifa, M., El Ganaoui, M., & Rogaume, Y. (2016). Finite element modelling of the pyrolysis of wet wood subjected to fire. *Fire Safety Journal, 81*, 85–96. doi:10.1016/j.firesaf.2016.02.001

Thomas, J. C., Mueller, E. V., Santamaria, S., Gallagher, M., El Houssami, M., Filkov, A., ... Simeoni, A. (2017). Investigation of firebrand generation from an experimental fire: Development of a reliable data collection methodology. *Fire Safety Journal*, *91*, 864–871. doi:10.1016/j.firesaf.2017.04.002

Tihay, V., Simeoni, A., Santoni, P. A., Rossi, L., Garo, J. P., & Vantelon, J. P. (2009). Experimental study of laminar flames obtained by the homogenization of three forest fuels. *International Journal of Thermal Sciences*, *48*(3), 488–501. doi:10.1016/j.ijthermalsci.2008.03.018

Tohid, A., & Kaye, N. B. (2017). Stochastic modeling of firebrand shower scenarios. *Fire Safety Journal*, *91*, 91–102. doi:10.1016/j.firesaf.2017.04.039

Usenya, V. V. (2002). *Forest fires, consequences and fight against them*. Gomel, Belarus: Forestry Institute of the National Academy of Sciences of Belarus.

Vermesi, I., DiDomizio, M. J., Richter, F., Weckman, E. J., & Rein, G. (2017). Pyrolysis and spontaneous ignition of wood under transient irradiation: Experiments and a-priori predictions. *Fire Safety Journal*, *91*, 218–225. doi:10.1016/j.firesaf.2017.03.081

Wang, H. (1979). *Basic formulas and data on heat transfer for engineers: Per. from English Handbook. Moscow, Russia: Atomizdat.*

Weber, R. (1991). Modelling fire spread through fuel beds. *Progress in Energy and Combustion Science*, *17*(1), 62–82. doi:10.1016/0360-1285(91)90003-6

Weise, D. R., White, R. H., Beall, F. C., & Etlinger, M. (2005). Use of the cone calorimeter to detect seasonal differences in selected combustion characteristics of ornamental vegetation. *International Journal of Wildland Fire*, *14*(3), 321–338. doi:10.1071/WF04035

Wilson, H. L., & Connarro, A. M. (1969). Fire breaks of many uses. *Fire Control Notes. Winter, 3*(1), 11-13.

Chapter 2
Mathematical Modeling of Steppe Fires

Valeriy Afanasievich Perminov
Tomsk Polytechnic University, Russia

ABSTRACT

The chapter presents a mathematical model of the initiation and spread of the steppe fire. The mathematical model is based on the laws of mechanics of multiphase reacting media. The main physicochemical processes describing the drying, pyrolysis, and combustion of gaseous and condensed pyrolysis products are taken into account. As a result of the numerical solution, the distributions of the velocity, temperature, and concentration fields of the components of the gas and condensed phases were determined. The dependence of the rate of spread of the steppe fire on the main parameters of the state of vegetation cover and wind speed was studied. The mathematical model presented in the chapter can be used to predict the spread of steppe fires for various types of steppe vegetation and meteorological conditions, as well as for preventive measures.

INTRODUCTION

The steppes occupy a significant territory in the Russian Federation. As a rule, it is a flat surface covered with vegetative combustible materials, which, when dried, turn into flammable materials. In the event of a fire, the fire quickly spreads under the influence of strong winds and is difficult to control. In contrast to surface forest fires, steppe fires have a significant spreading rate. Because of the change in the wind direction, a branching of the fire front often occurs, that is, a large area of simultaneously burning materials occurs. Also, there is a negative impact of the steppe fires on buildings and structures and leads to the death of people. In the forest-steppe zone of Russia, steppes and forests are distributed alternately, and after the steppe fires occur, they can cause forest fires and threaten the safety of forest areas. These reasons determine the relevance of the topic (Smelyansky et al, 2015). Very often, steppe fires, starting on the territory of neighboring states (Kazakhstan, Mongolia), cross the state border and continue on the territory of the Russian Federation. The purpose of the work is to create a mathematical model of the emergence and spread of steppe fires, the definition of the propagation contours of the front of a steppe

DOI: 10.4018/978-1-7998-1867-0.ch002

fire, as well as the dependence of the propagation velocity on the stock and moisture content of plant combustible materials. As a rule, emergency situations, and, in particular, steppe fires, begin suddenly, develop rapidly and have developmental uncertainty. In this regard, in order to successfully combat this type of fire, it is necessary to develop methods for predicting the spread of combustion, and on their basis to create methods for localizing and controlling them. The most promising ways to predict the development of emergency situations associated with wildfires are methods of mathematical modeling. For example, based on the laws of mechanics of reacting media, the approach proposed by Grishin (1997). A mathematical model of an emergency situation is called a system of relations, equations, inequalities, geometric concepts, etc., which in mathematical form display, reproduce or imitate the most important features and properties of real hazardous phenomena in order to analyze and predict their occurrence, development and consequences. Creating a mathematical model of an emergency involves several steps. The initial stage is a meaningful description of an emergency situation, which is compiled on the basis of all the knowledge about it, the results of field studies of similar situations, consultations with experts, and the study of reference and special literature. At the second stage, the formalization of a meaningful description of the model is carried out, the mathematical formulation of the problem with an indication of all the necessary input data and the desired quantities. At the third stage, the formalized scheme of emergencies should be transformed into its mathematical model. To do this, all available information must be expressed using relations, inequalities, equations, algorithms. The equations included in the model are complemented by initial and boundary conditions, as well as inequalities that determine the range of permissible values of the calculated values. In the fourth stage, the model itself is examined. By carrying out multivariate calculations, the properties of the model and its behavior under different conditions are studied. At the next stage, the model is applied to the description of real emergencies. By comparing the results of computational experiments with the available experimental data, the model parameters are identified or refined, tested, debugged, and tested for adequacy. After the adequacy of the model, i.e. its sufficient correspondence to reality has been established; the use of the model for analyzing and predicting emergencies occurring in real conditions begins.

BACKGROUND

In previous years, most scientific research was devoted to the study of forest fires. Special studies of the behavior of the steppe fires appeared only in recent years and therefore the literature review is devoted to works on the study of forest fires as the closest. And so, the results of these studies can be used in the study of steppe fires. One of the first models of the spread of natural fires during the burning of combustible plant materials was developed by Fons (1946). It was based only on the use of the heat balance equation and used as a coefficient the data obtained from the combustion of combustible substances in a wind tunnel. Also, it should be marked the model (Byram, 1970), which determined the ratio between the length of the flame, the value of consumable combustible substances and the heat of their combustion. In this model, real data were used when studying fires in pine forests. However, the disadvantage of this model was that data were used as initial data (for example, the length and height of the flame), which can only be obtained from experimental studies in a particular fire. The first model that used only those data that can be measured in advance was the model of surface fires by Rothermel (1972), which is based on the energy balance of the solid phase of a fuel. This model was one of the most successful and was tested with the help of experimental studies. Some provisions of this model are still in use. A great deal of work has been done on the theoretical problem of forest fires. The first accepted method for prediction of crown fires was given by Rothermal (1972) and Van Wagner (1977). The semi-empirical models allow to obtain a quite good data of the forest fire rate of spread as a function of fuel bulk and moisture, wind velocity and the terrain slope. But these models use data for particular cases and do not give results for general fire conditions. Also crown fires initiation and hazard have been studied and modeled in detail (eg: Alexander, 1979, Xanthopoulos, 1990, Van Wagner (1999), Cruz and et al., 2002, Albini, 1995, Scott and Reinhardt, 2001). A new mathematical model of forest fires was obtained by Grishin (1997) based on an analysis of known and original experimental data, and using concepts and methods from reactive media mechanics. The physical two-phase models used by Morvan (2004) may be considered as a development and extension of the formulation proposed by Grishin. Another approach was presented in paper using a simplified semi-empirical model equation (Karouni, A. et al., 2014). The proposed equation uses parameters of both weather and topography, where wind speed and direction, fuel moisture content and slope coefficient are introduced to obtain the expected spread rate. The significance of the proposed model is in the minimized number of parameters used and the ease of extraction. Unfortunately, in such models, the induction is reasonable in situations identical to those used in experiments. In several regions of the southern hemisphere, exotic pines have been extensively planted in native treeless areas for forestry purposes and have subsequently invaded the native environments. Planted and invading non-native plant species can alter fire regimes through changes in fuel loads and in the structure and continuity of fuels, potentially modifying the flammability of native plant communities. To estimate differences in potential fire behavior among the several selected treatments the fire behavior modeling software BehavePlus 5.0.5 (Andrews, P. L, 2013) was used (Paritsis, J., 2018). This simulation software is based also on Rothermel's mathematical model of wildfire spread (Rothermal, 1972) and has been used to characterize potential fire behavior. BehavePlus 5.0.5 was used in this study to assess an envelope of possible fire behaviors rather than as a precise predictor of fire characteristics. It was found that pine plantations and their subsequent invasion in the Patagonian shrubby steppe produced sharp changes in fine fuel amount and its vertical and horizontal continuity. These changes in fuel properties have the potential to affect fire behavior, increasing fire intensity by almost 30 times. An attempt to improve the model for modeling steppe fires as well as surface forest fires was made by Burasov and Grishin (2006). However, the authors use the same data to simulate steppe fires as for surface forest fires, and the composition and structure of plant combustible materials in these cases are significantly different. The results of experimental studies show that the rate of spread of the steppe fires far exceeds the rate of propagation of surface forest fires. In addition, in this paper a one-dimensional model is used, which does not take into account influence the finite size of front of combustion. Besides it does not allow obtaining the contour of the steppe fire distribution. A continuation of this approach to modeling steppe fires is a series of works by Asylbaev and his co-authors (2010, 2013). They also, like the previous author, use the mathematical formulation of forest fire propagation proposed by A. M. Grishin (1997). Similarly, as previous authors, Asylbaev (2010, 2013) uses thermokinetic constants, which are obtained for pine needles, to describe the drying and pyrolysis processes (A. M. Grishin et al., 2006). Thermokinetic constants for plant materials in the steppes differ from forest combustible materials A.M. Grishin (1997). Despite the three-dimensional system of equations given in the works, the calculations were carried out for horizontal and vertical coordinates. Thus, distributions are obtained in two coordinates: horizontal and vertical. Unfortunately, they do not make it possible to obtain the dynamics of the distribution of the contours of the steppe fire, as well as their dependence on various state parameters of the steppe plant material and meteorological conditions. Moreover, the contour of the steppe fire during movement is

Figure 1. The scheme of calculation domain

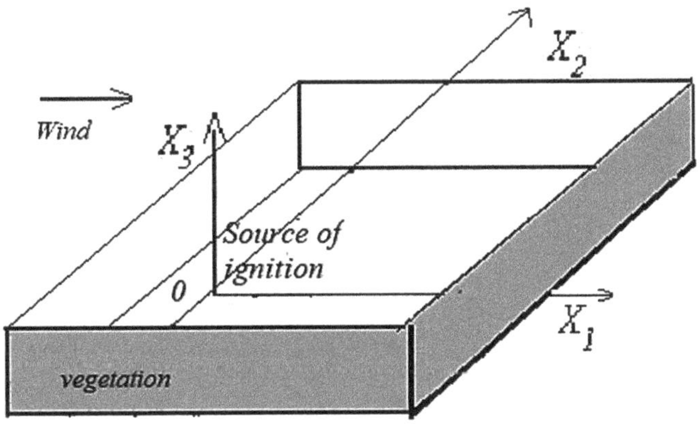

curved depending on the state of steppe combustible materials and meteorological conditions. However as following from this review, the study of wildfires has been limited mainly to cases studied of forest fires and practically no papers devoted to studying of steppe fires. The present mathematical model and results of calculation are used to study the steppe fires initiation and spread behavior at different state of vegetations and meteorological conditions.

PHYSICAL AND MATHEMATICAL MODEL

It is assumed that the steppe vegetation during a fire can be modeled as 1) a multi-phase, multistoried, spatially heterogeneous medium; 2) in the fire the vegetation is a porous-dispersed, two-temperature, single-velocity, reactive medium; 3) the steppe vegetation is supposed to be non – deformed medium (grass, plant stems and etc.), which affects only the magnitude of the force of resistance in the equation of conservation of momentum in the gas phase, i.e., the medium is assumed to be quasi-solid (almost non-deformable); 4) let there be a so-called "ventilated" vegetation massif, in which the volume of fractions of condensed steppe fuel phases, consisting of dry organic matter, water in liquid state, solid pyrolysis products, and ash, can be neglected compared to the volume fraction of gas phase (components of air and gaseous pyrolysis products); 5) the flow has a developed turbulent nature and molecular transfer is neglected; 6) gaseous phase density doesn't depend on the pressure because of the low velocities of the flow in comparison with the velocity of the sound. Let the point x_1, x_2, x_3= 0 is situated at the center of the steppe fire source at the height of the roughness level, axis 0*x* directed parallel to the Earth's surface to the right in the direction of the unperturbed wind speed, axis *Oy* directed perpendicular to *Ox* and axis 0*z* directed upward (Figure 1).

The problem formulated above is reduced to a solution of the following system of equations:

$$\frac{\partial \rho}{\partial t}+\frac{\partial}{\partial x_j}\left(\rho v_j\right)=Q, j=1,2,3; \qquad (1)$$

$$\rho\frac{dv_i}{dt} = -\frac{\partial P}{\partial x_i} + \frac{\partial}{\partial x_j}(-\rho\overline{v_i'v_j'}) - \rho s c_d v_i \,|\,\vec{v}\,|\, - \rho g_i - Qv_i; \tag{2}$$

$$\rho c_p \frac{dT}{dt} = \frac{\partial}{\partial x_j}(-\rho c_p \overline{v_j' T'}) + q_5 R_5 - \alpha_v \left(T - T_s\right); \tag{3}$$

$$\rho\frac{dc_\alpha}{dt} = \frac{\partial}{\partial x_j}\left(-\rho\overline{v_j' c_\alpha'}\right) + R_{5\alpha} - Qc_\alpha, \alpha = 1,3; \tag{4}$$

$$\frac{\partial}{\partial x_j}\left(\frac{c}{3k}\frac{\partial U_R}{\partial x_j}\right) - k\left(cU_R - 4\sigma T_S^4\right) = 0; \tag{5}$$

$$\sum_{i=1}^{4}\rho_i c_{pi}\phi_i \frac{\partial T_S}{\partial t} = q_3 R_3 - q_2 R_2 + k\left(cU_R - 4\sigma T_S^4\right) + \alpha_V\left(T - T_S\right); \tag{6}$$

$$\rho_1\frac{\partial \phi_1}{\partial t} = -R_1, \rho_2\frac{\partial \phi_2}{\partial t} = -R_2, \rho_3\frac{\partial \phi_3}{\partial t} = \alpha_C R_1 - \frac{M_C}{M_1}R_3, \rho_4\frac{\partial \phi_4}{\partial t} = 0; \tag{7}$$

$$\sum_{\alpha=1}^{3} c_\alpha = 1, P_e = \rho R T \sum_{\alpha=1}^{3}\frac{c_\alpha}{M_\alpha}, \vec{v} = (v_1, v_2, v_3), \vec{g} = (0,0,g).$$

The system of equations (1)–(9) must be solved taking into account the initial and boundary conditions:

$$t = 0: v_1 = 0, v_2 = 0, v_3 = 0, T = T_e, c_\alpha = c_{\alpha e}, T_s = T_e, \phi_i = \phi_{ie}; \tag{8}$$

$$x_1 = 0: v_1 = V, v_2 = 0, v_3 = 0, T = T_e, c_\alpha = c_{\alpha e}, -\frac{c}{3k}\frac{\partial U_R}{\partial x_1} + \frac{c}{2}U_R = 0; \tag{9}$$

$$x_1 = x_{1e}: \frac{\partial v_1}{\partial x_1} = 0, \frac{\partial v_2}{\partial x_1} = 0, \frac{\partial v_3}{\partial x_1} = 0, \frac{\partial T}{\partial x_1} = 0, \frac{\partial c_\alpha}{\partial x_1} = 0, \frac{c}{3k}\frac{\partial U_R}{\partial x_1} + \frac{c}{2}U_R = 0; \tag{10}$$

$$x_2 = -x_{2e}: \frac{\partial v_1}{\partial x_2} = 0, \frac{\partial v_2}{\partial x_2} = 0, \frac{\partial v_3}{\partial x_2} = 0, \frac{\partial T}{\partial x_2} = 0, \frac{\partial c_\alpha}{\partial x_2} = 0, -\frac{c}{3k}\frac{\partial U_R}{\partial x_2} + \frac{c}{2}U_R = 0; \tag{11}$$

$$x_2 = x_{2e} : \frac{\partial v_1}{\partial x_2} = 0, \frac{\partial v_2}{\partial x_2} = 0, \frac{\partial v_3}{\partial x_2} = 0, \frac{\partial T}{\partial x_2} = 0, \frac{\partial c_\alpha}{\partial x_2} = 0, \frac{c}{3k}\frac{\partial U_R}{\partial x_2} + \frac{c}{2}U_R = 0; \tag{12}$$

$$x_3 = 0 : v_1 = 0, v_2 = 0, \frac{\partial c_\alpha}{\partial x_3} = 0, -\frac{c}{3k}\frac{\partial U_R}{\partial x_3} + \frac{c}{2}U_R = 0,$$

$$\rho v_3 = \rho_0 \omega_0, T = T_0, \left|x_1\right| \le x_0, \left|x_2\right| \le x_0,$$

$$\rho v_3 = 0, T = T_e, \left|x_1\right| > x_0, \left|x_2\right| > x_0; \tag{13}$$

$$x_3 = x_{3e} : \frac{\partial v_1}{\partial x_3} = 0, \frac{\partial v_2}{\partial x_3} = 0, \frac{\partial v_3}{\partial x_3} = 0, \frac{\partial T}{\partial x_3} = 0, \frac{\partial c_\alpha}{\partial x_3} = 0, \frac{c}{3k}\frac{\partial U_R}{\partial x_3} + \frac{c}{2}U_R = 0. \tag{14}$$

Here and above $\frac{d}{dt}$ is the symbol of the total (substantial) derivative:

$$\frac{d}{dt} = \frac{\partial}{\partial t} + v_1 \frac{\partial}{\partial x_1} + v_2 \frac{\partial}{\partial x_2} + v_3 \frac{\partial}{\partial x_3};$$

α_v is the coefficient of phase exchange; ρ - density of gas – dispersed phase, t is time; v_i - the velocity components; T, T_S, - temperatures of gas and solid phases, U_R - density of radiation energy, k - coefficient of radiation attenuation, P - pressure; c_p – constant pressure specific heat of the gas phase, c_{pi}, ρ_i, φ_i – specific heat, density and volume of fraction of condensed phase (1 – dry organic substance, 2 – moisture, 3 – condensed pyrolysis products, 4 – mineral part of forest fuel), R_i – the mass rates of chemical reactions, q_i – thermal effects of chemical reactions; k_g, k_S - radiation absorption coefficients for gas and condensed phases; T_e - the ambient temperature; c_α - mass concentrations of α - component of gas - dispersed medium, index α=1,2,3, where 1 corresponds to the density of oxygen, 2 - to carbon monoxide CO, 3 - to carbon dioxide and inert components of air; R – universal gas constant; M_α, M_C, and M molecular mass of α -components of the gas phase, carbon and air mixture; g is the gravity acceleration; c_d is an empirical coefficient of the resistance of the vegetation, s is the specific surface of the forest fuel in the given forest stratum. To define source terms which characterize inflow (outflow of mass) in a volume unit of the gas-dispersed phase, the following formulae were used for the rate of formulation of the gas-dispersed mixture Q, outflow of oxygen R_{51}, changing carbon monoxide R_{52}.

$$Q = \left(1 - \alpha_c\right) R_1 + R_2 + \frac{M_c}{M_1} R_3, R_{51} = -R_3 - \frac{M_1}{2M_2} R_5, R_{52} = \nu_g \left(1 - \alpha_c\right) R_1 - R_5, R_{53} = 0.$$

$$R_1 = k_1\rho_1\phi_1 \exp\left(-\frac{E_1}{RT_S}\right), R_2 = k_2\rho_2\phi_2 T^{-0.5} \exp\left(-\frac{E_2}{RT_S}\right),$$

$$R_3 = k_3\rho\phi_3 S_\sigma c_1 \exp\left(-\frac{E_3}{RT_S}\right), R_5 = k_5 M_2 \left(\frac{c_1 M}{M_1}\right)^{0.5} \left(\frac{c_2 M}{M_2}\right) T^{-2.25} \exp\left(-\frac{E_5}{RT}\right).$$

The initial values for volume of fractions of condensed phases are determined using the expressions:

$$\varphi_{1e} = \frac{d(1-\nu_z)}{\rho_1}, \varphi_{2e} = \frac{Wd}{\rho_2}, \varphi_{3e} = \frac{\alpha_c \varphi_{1e} \rho_1}{\rho_3}$$

where d -bulk density, ν_z – coefficient of ashes of vegetative fuel, W – steppe fuel moisture content. It is supposed that the optical properties of a medium are independent of radiation wavelength (the assumption that the medium is "grey"), and the so-called diffusion approximation for radiation flux density were used for a mathematical description of radiation transport during steppe fires. To close the system (1)–(7), the components of the tensor of turbulent stresses, and the turbulent heat and mass fluxes are determined using the local-equilibrium model of turbulence (Grishin, 1997). The system of equations (1)–(7) contains terms associated with turbulent diffusion, thermal conduction, and convection, and needs to be closed. The components of the tensor of turbulent stresses $\rho\overline{v'w'}$, as well as the turbulent fluxes of heat and mass $\rho\overline{v_j'c_pT'}$, $\rho\overline{v_j'c_\alpha'}$ are written in terms of the gradients of the average flow properties using the formulas

$$-\rho\overline{v_i v_j} = \mu_t\left(\frac{\partial v_i}{\partial x_j} + \frac{\partial v_j}{\partial x_i}\right) - \frac{2}{3}K\delta_{ij},$$

$$-\rho\overline{\nu_j c_p T'} = \lambda_t \frac{\partial T}{\partial x_j}, -\rho\overline{\nu_j c_\alpha'} = \rho D_t \frac{\partial c_\alpha}{\partial x_j},$$

$$\lambda_T = \mu_T c_p / \Pr_T, \rho_5 D_T = \mu_T / Sc_T, \mu_T = c_\mu \rho_5 K^2 / \varepsilon,$$

where μ_t, λ_t, D_t are the coefficients of turbulent viscosity, thermal conductivity, and diffusion, respectively; $\Pr_t$, Sc_t are the turbulent Prandtl and Schmidt numbers, which were assumed to be equal to 1. In dimensional form, the coefficient of dynamic turbulent viscosity is determined using local equilibrium model of turbulence (Grishin, 1997). The length of the mixing path is determined using the formula

$$l = x_3 k_t / \left(1 + 2.5x_3\sqrt{c_d\, s / h}\right)$$

taking into account the fact that the coefficient of resistance c_d is equal to the constants $k_t = 0.4$ and h – height of the vegetation. The thermodynamic, thermophysical and structural characteristics correspond to the steppe fuels (Grishin, 1997). Because of the horizontal sizes of steppe domain more than height of vegetation – h, the system of equations of general mathematical model of forest fire (1)-(7) was integrated between the limits from height of the roughness level - 0 to h. Besides, suppose that Besides, suppose that

$$\int_0^h \phi dx_3 = \bar{\phi} h$$

$\bar{\phi}$ - average value of φ. When writing equations (15) - (21), instead of the notation for coordinates x_1, x_2 and x_3, variables *x*, *y* and *z* are used. Accordingly, instead of the velocity components v_i, the notation *u, v*, and *w* are used. The problem formulated above is reduced to a solution of the following system of equations:

$$\frac{\partial \rho}{\partial t} + \frac{\partial}{\partial x}(\rho u) + \frac{\partial}{\partial y}(\rho v) = Q - \frac{\dot{c}^- - \dot{c}^+}{h}, \tag{15}$$

$$\rho \frac{du}{dt} = -\frac{\partial P}{\partial x} + \frac{\partial}{\partial x}\left(-\rho \overline{u'^2}\right) + \frac{\partial}{\partial y}\left(-\rho \overline{u'v'}\right) - \rho s c_d u |\vec{u}| - Qu + \frac{\left(\tau_x^- - \tau_x^+\right)}{h}, \tag{16}$$

$$\rho \frac{dv}{dt} = -\frac{\partial p}{\partial y} + \frac{\partial}{\partial x}\left(-\rho \overline{u'v'}\right) + \frac{\partial}{\partial y}\left(-\rho \overline{v'^2}\right) - \rho s c_d v |\vec{u}| - Qv + \frac{\left(\tau_y^- - \tau_y^+\right)}{h}, \tag{17}$$

$$\rho \frac{dc_\alpha}{dt} = \frac{\partial}{\partial x}\left(-\rho \overline{u'c_\alpha'}\right) + \frac{\partial}{\partial y}\left(-\rho \overline{v'c_\alpha'}\right) + R_{5\alpha} + \frac{\left(J_\alpha^- - J_\alpha^+\right)}{h} - Qc_\alpha, \alpha = \overline{1,3}, \tag{18}$$

$$\rho c_p \frac{dT}{dt} = \frac{\partial}{\partial x}\left(-\rho c_p \overline{u'T'}\right) + \frac{\partial}{\partial y}\left(-\rho c_p \overline{v'T'}\right) + q_5 R_5 - \alpha_V \left(T - T_S\right) + \frac{\left(q_T^- - q_T^+\right)}{h} + k_g\left(cU_R - 4\sigma T^4\right), \tag{19}$$

$$\frac{\partial}{\partial x}\left(\frac{c}{3k}\frac{\partial U_R}{\partial x}\right) + \frac{\partial}{\partial y}\left(\frac{c}{3k}\frac{\partial U_R}{\partial y}\right) - kcU_R + 4k_g \sigma T^4 + 4k_S \sigma T_S^4 = 0,\ k = k_g + k_s, \tag{20}$$

system of equations (15)-(19) are introduced the next designations:

$$\dot{c} = \rho w, \tau_x = -\overline{\rho u' w'}, \tau_y = -\overline{\rho v' w'}, J_\alpha = -\overline{\rho w' c_\alpha'}, q_T = -\overline{\rho w' T'}, \tag{21}$$

Upper indexes "+" and "-" designate values of functions at $z=h$ and $z=0$ correspondingly. It is assumed that heat and mass exchange of fire front and boundary layer of atmosphere are governed by Newton law and written using the formulas:

$$(q_T^- - q_T^+) / h = -\alpha(T - T_e) / h,$$
$$(J_\alpha^- - J_\alpha^+) / h = -\alpha(c - c_{\alpha e}) / hc_p.$$

The system of equations (14)–(20) must be solved taking into account the initial and boundary conditions (8)-(14).

THE METHOD SOLUTION AND ANALYSIS OF RESULTS

The boundary-value problem (15)–(20) is solved numerically. A discrete analog was obtained by means of the finite volume method using the SIMPLE like algorithm (Patankar, 1981). Difference equations that arise in the course of sampling were resolved by the method of SIP. In order to efficiently solve this problem in a reactive flow the method of splitting according to physical processes was used. The basic idea of this method is based on the information that the physical timescale of the processes is great than chemical. In the first stage, the hydrodynamic pattern of flow and distribution of scalar functions was calculated. Then the system of ordinary differential equations of chemical kinetics obtained as a result of splitting was then integrated. The time step for integrating each function has to be smaller than the characteristic time of physical process to ensure the convergence of the numerical method. The time step was selected automatically. The accuracy of the program was checked by the method of inserted analytical solutions. Analytical expressions for the unknown functions were substituted in (15)–(20) and the closure of the equations were calculated. This was then treated as the source in each equation. Next, with the aid of the algorithm described above, the values of the functions used were inferred with an accuracy of not less than 1%. The effect of the dimensions of the control volumes on the solution was studied by diminishing them.

Fields of temperature, velocity, component mass fractions, and volume fractions of phases were obtained numerically. The first stage is related to increasing maximum temperature in the place of ignition with the result that a fire source appears. At this process stage over the fire source a thermal wind is formed a zone of heated vegetative fire pyrolysis products which are mixed with air penetrate into unburning grass. As a result, vegetative fuels are heated, moisture evaporates and gaseous and dispersed pyrolysis products are generated. Ignition of gaseous pyrolysis products occur at the next stage, and that of gaseous pyrolysis products in the vegetative combustible materials occurs at the last stage. As a result of heating of steppe fuel elements, moisture evaporates, and pyrolysis occurs accompanied by the release of gaseous products, which then ignite and burn away in the steppe fuel. At the moment of ignition, the gas combustible products of pyrolysis burn away, and the concentration of oxygen is rapidly reduced. The temperatures of both phases reach a maximum value at the point of ignition. The ignition process is of a gas - phase nature. As a result of numerical calculations, the temperature isotherm distributions of the gas and condensed phases, as well as the isolines of oxygen concentrations, volatile combustible products of pyrolysis and combustion, and the volume fractions of the condensed phase (dry organic matter, water, and condensed pyrolysis products) were obtained. By the distribution of these functions in the computational domain, it is possible to track the movement of the combustion front at different points in time and, consequently, the speed of propagation of the front of a steppe fire. In Figure 2 shows the temperature distribution of the gas phase at different points in time:: I-t = 5 s, II-t = 10 s, III-t = 15 s. Due to the fact that the ignition source is located at the origin, and the wind direction coincides with the direction of the x axis, the distribution of the sought functions is symmetrical about the x axis. Therefore, distributions were calculated only in the upper part of the computational domain (for positive values of y). The figure shows that the prevailing movement is carried out in the direction of the wind. However, the front of the fire expands due to the spread and in the perpendicular direction. Figures 3-5 shows the distribution of oxygen concentrations, volatile combustible pyrolysis products and temperatures in the front of the steppe fire at different points in time. From the figures it is clear that under the influence of high temperature, before the fire front, gaseous pyrolysis products are released (Figures 3-5 b), which then ignite.

This reduces the oxygen concentration in the combustion zone (Figures 3-5 a), which is consumed in the combustion process and, accordingly, gaseous combustible pyrolysis products. In this case, the combustion front moves, as can be seen from the distribution of isotherms at different points in time (Figures 3-5 c). Using the presented mathematical model, calculations were carried out to study the influence of some basic parameters characterizing the state of plant combustible materials and wind speed on the speed of propagation of a steppe fire. Variable parameters were: bulk and moisture content of vegetative combustible materials, and wind speed in an unperturbed environment. The results of the calculations are presented in Figures 6-7. Calculations showed that by increasing the wind speed, the rate of spread of the front of the steppe fire increases. With an increase in the bulk and moisture content of vegetative combustible materials, the rate of spread of fire decreases.

Figure 2. The distribution of temperature for gas phase $\bar{T}$: (1-$\bar{T}$ =5, 2 - 4, 3 - 3, 4 - 2, 5 - 1.5; $\bar{T}$ =T/Te, Te=300K)

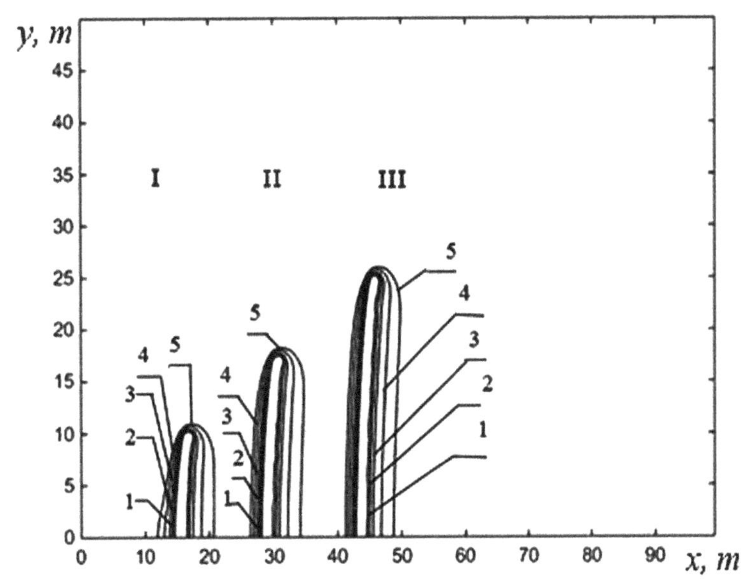

Figure 3. The distribution of concentrations $\overline{C1}$, $\overline{C2}$ *and temperature of gas phase* $\overline{T}$ *; V= 7 m/s, t= 5 s, W=0.2, d=0.8 kg/m*^{3}

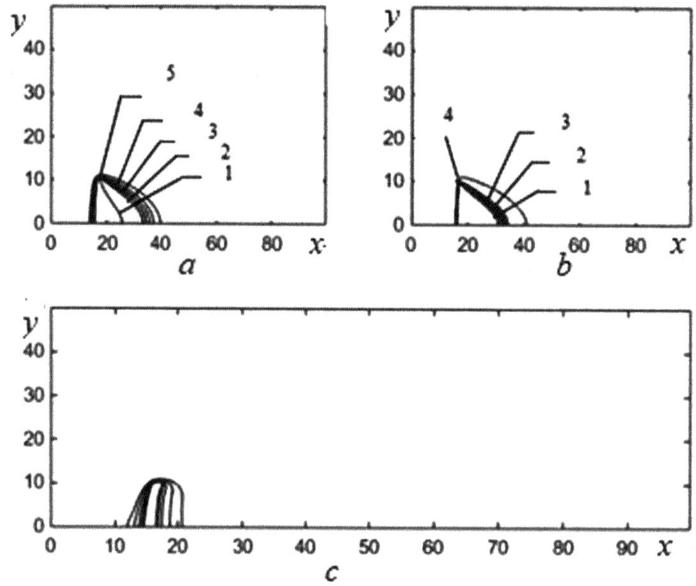

Figure 4. The distribution of concentrations $\overline{C1}$, $\overline{C2}$ *and temperature of gas phase* $\overline{T}$ *; V= 7 m/s, t= 10 s., W=0.2, d=0.8 kg/m*^{3}

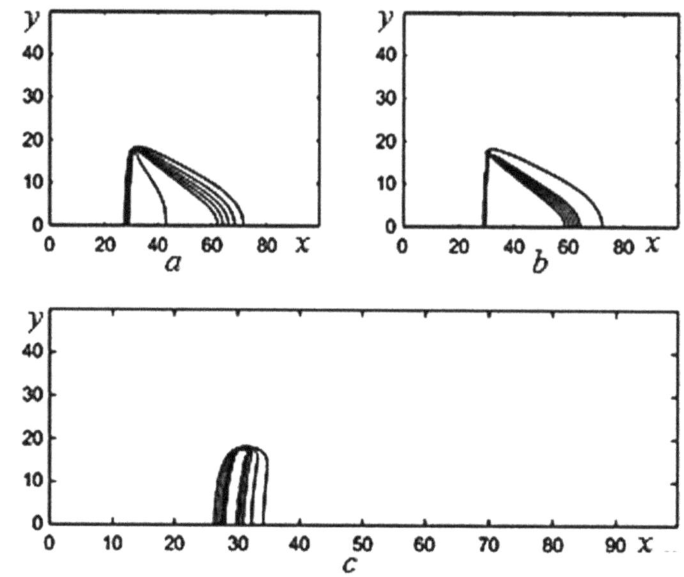

Figure 5. The distribution of concentrations $\overline{C1}$, $\overline{C2}$ *and temperature of gas phase* $\bar{T}$ *; V= 7 m/s, t= 15 s., W=0.2, d=0.8 kg/m*^{3}

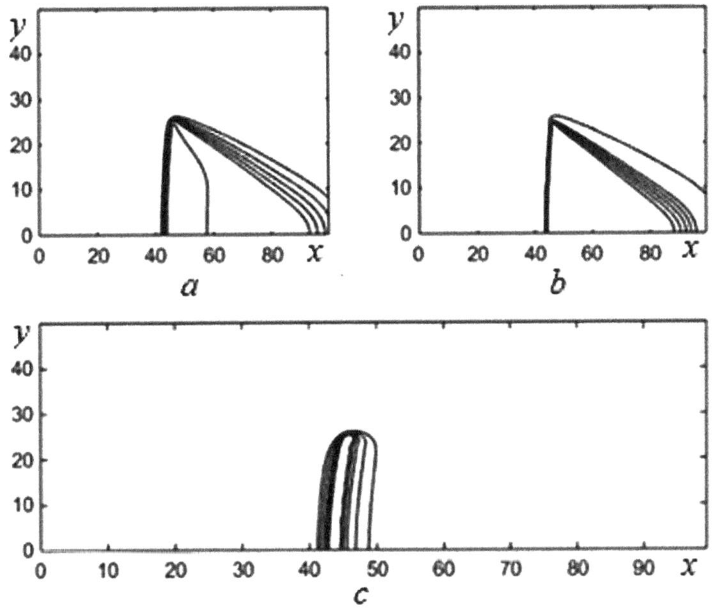

Figure 6. The influence of moisture content of the vegetative combustible materials on the rate of spread of the steppe fire: 1- W =0.2, 2 - W=0.4, 3 - W=0.6

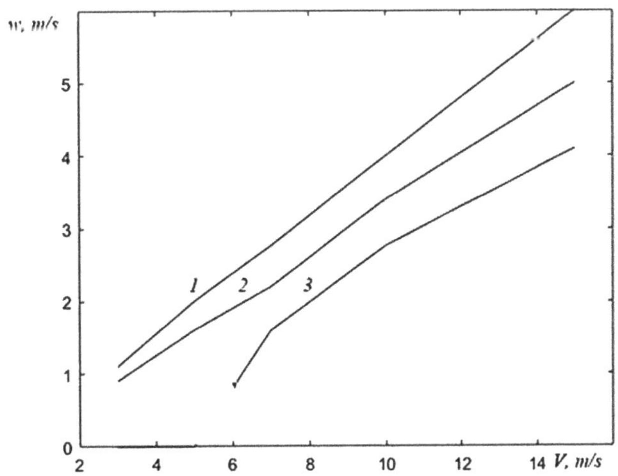

The mathematical model presented in the work can be used to predict the spread of a steppe fire if there is data on the distribution of various types of steppe vegetation, their condition (stock and moisture content), terrain and meteorological data. In addition, this mathematical model can be used to simulate the impact of a steppe fire on nearby buildings and structures. The results of such calculations will allow to estimate the safe distances from the areas covered with vegetation to various buildings. Also,

Figure 7. The influence of the bulk of the RGM on the speed of the steppe fire. 1 - d = 0.8 kg/m³, 2 - d = 1.0 kg/m³ and 3 - d = 1.2 kg/m³

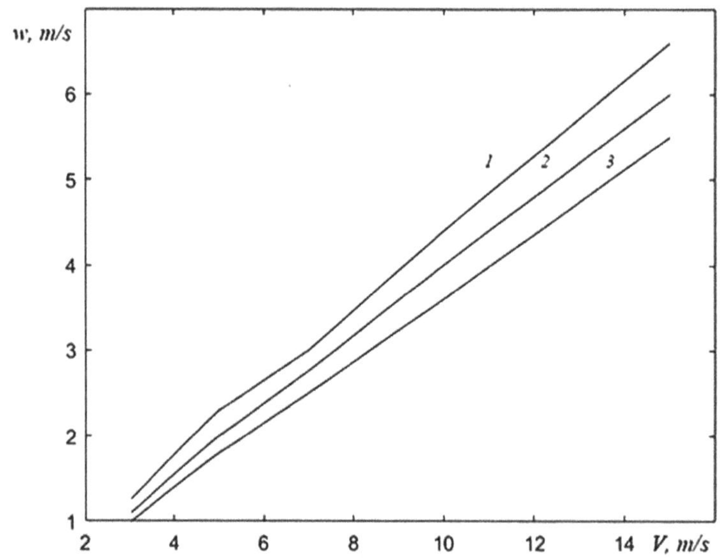

the model proposed in this paper makes it possible to assess the effect of the combustion products of a steppe fire on the surface layer of the atmosphere, calculate smoke and determine the content of hazardous components in the air, including in nearby settlements.

CONCLUSION

In this paper, a mathematical model was developed for the emergence and spread of the steppe fire. To solve the task, a computer program was created, with which the simulation of the processes of the spread of steppe fires was carried out. Distributions of temperature fields and concentrations of components of the gas and condensed phase at different points in time are obtained. For the numerical solution of the problem used the method of control volume. The effect of moisture content and bulk of vegetative combustible materials and wind speed on the speed of propagation of steppe fires has been studied.

REFERENCES

Albini, F. A., & Reinhardt, E. D. (1995). Modeling ignition and burning rate of large woody natural fuels. *International Journal of Wildland Fire*, *5*(2), 81–91. doi:10.1071/WF9950081

Alexander, M. E. (1979). Crown fire thresholds in exotic pine C.E. In *Proceedings 10th Conference on Fire and Forest Meteorology* (pp. 207-212). Ottawa, Ontario.

Andrews, P. L. (2013). Current status and future needs of the BehavePlus fire modeling system. *International Journal of Wildland Fire*, *23*(1), 21–33. doi:10.1071/WF12167

Asylbaev, N. A. (2010). Mathematical modeling of steppe fires. *Computer Research and Modeling*, *4*(4), 377–384. doi:10.20537/2076-7633-2010-2-4-377-384

Asylbaev, N. A. (2013). On the spread of a steppe fire along an inclined underlying surface. *Bulletin of Bashkir University*, *18*(4), 978–983.

Burasov, D. M., & Grishin, A. M. (2006). *Mathematical modeling of surface forest and steppe fires*. Kemerovo, Russia: Practica. (in Russian)

Byram, G. M., & Martin, R. E. (1970). The modeling of fire whirlwinds. *Forest Science*, *16*(4), 586–398.

Cruz, M. G., Alexander, M. E., & Wakimoto, R. H. (2002). Predicting crown fire behavior to support forest fire management decision-making. In *Proceedings IV International Conference on Forest Fire Research*, 11. Luso-Coimbra, Portugal: Millpress.

Fons, W. L. (1946). Analysis of fire spread in light forest fuels. *Journal of Agricultural Research*, *72*(3), 93–121.

Grishin, A. M. (1997). *Mathematical modeling forest fire and new methods fighting them*. Tomsk, Russia: Publishing House of Tomsk University.

Karouni, A., Daya, B., Bahlak, S., & Chauvet, P. (2014). A simplified mathematical model for fire spread predictions in wildland fires combining between the models of Anderson and Rothermel. *International Journal of Modeling and Optimization*, *4*(3), 197–200. doi:10.7763/IJMO.2014.V4.372

Morvan, D., & Dupuy, J. L. (2004). Modeling the propagation of wildfire through a Mediterranean shrub using a multiphase formulation. *Combustion and Flame*, *138*(3), 199–210. doi:10.1016/j.combustflame.2004.05.001

Nasanbat, E., & Lkhamjav, O. (2016). Wild fire risk map in the eastern steppe of Mongolia using spatial multi-criteria analysis. *The International Archives of the Photogrammetry, Remote Sensing, and Spatial Information Sciences*, Volume XLI-B1, 2016 XXIII ISPRS Congress, July 12–19, 2016, Prague, Czech Republic, 469-473.

Paritsis, J., Landesmann, J., Kitzberger, T., Tiribelli, F., Sasal, Y., Quintero, C., ... Nuñez, M. A. (2018). Pine plantations and invasion alter fuel structure and potential fire behavior in a Patagonian forest-steppe ecotone. *Forests*, *117*(9).

Patankar, S. V. (1981). *Numerical heat transfer and fluid flow*. New York: Hemisphere Publishing Corp.

Rothermal, R. C. (1972). *Predicting behaviour and size of crown fires in the Northern Rocky Mountains*, USDA Forest Service, Intermountain Forest and Range Experiment Station: Research Paper: INT-438. (Ogden, UT).

Rothermel, R. C. (1972). *A mathematical model for fire spread predictions in wildland fuels*. USDA Forest Service Research Paper INT-115, Ogden, 1972. (Intermountain Forest and Range Exp. Stn.).

Scott, J. H., & Reinhardt, E. D. (2001). *Assessing crown fire potential by linking models of surface and crown fire behavior*, USDA Forest Service, Rocky Mountain Research Station. Fort Collins, CO: Research Paper RMRS-RP-29.

Smelyansky, I. E., Buffalos, Yu. A., Bazhenov, Yu. A., & … . (2015). *Steppe fires and fire management in the steppe protected areas: environmental and ecological aspects. Analytical Review*. Moscow, Russia: Wildlife Conservation Center. (in Russian)

Van Wagner, C. E. (1977). Conditions for the start and spread of crown fire. *Canadian Journal of Forest Research*, *7*(1), 23–34. doi:10.1139/x77-004

Van Wagner, C. E. (1999). Prediction of crown fire behavior in two stands of jack pine. *Canadian Journal of Forest Research*, *23*, 445–449.

Xanthopoulos, G. (1990). *Development of a wildland crown fire initiation model.* (PhD thesis), University of Montana, Montana.

Section 2

Chapter 3
Mathematical Simulation of Anthropogenic Load on Forested Territories for Point Source

Nikolay Viktorovich Baranovskiy
https://orcid.org/0000-0001-9202-8171
National Research Tomsk Polytechnic University, Russia

ABSTRACT

An anthropogenic load is the main cause of forest fires in the vicinity of settlements and various objects of transport or industrial infrastructure. It is proposed to mark out linear and point sources of anthropogenic load. For the numerical simulation, the similarity in the processes of propagation of anthropogenic load and diffusive heat transfer was used. The quantitative characteristic of the anthropogenic load is the virtual (possible) number of forest fires in the controlled forest area. A mathematical model to predict the distribution of anthropogenic load from a point source is presented. Distributions of the virtual number of forest fires for model data from a point source of anthropogenic load are obtained. Conclusions about the patterns of the distribution of anthropogenic load from a point source are formulated. The prospects for the further development of these results are described.

INTRODUCTION

Forest fires occur as a result of the weather, thunderstorm activity and anthropogenic load (Baranovskiy & Kuznetsov, 2017a; Read et al., 2018; Ye et al., 2017; Benavent-Corai et al., 2007; Gorev, 2004; Perminov et al., 2014; Jiménez-Ruano et al., 2017). The processes of forest fires occurrence as a result of the thunderstorm activity have been studied quite well (Canadian Wildland Fire Information System; van Wagner, 1987; WFAS Wildland Fire Assessment System) by now compared with fires caused by the anthropogenic load (Lee et al., 2002; Ruffault & Mouillot 2017; Poulos et al., 2013; Amatulli et al., 2003). For example, a group of mathematical models was developed to determine the conditions for

DOI: 10.4018/978-1-7998-1867-0.ch003

the ignition of trees by the cloud-to-ground lightning discharge, taking into account various discharge parameters and the structure of coniferous and deciduous trees (Baranovskiy et al., 2017; Baranovskiy & Kuznetsov, 2017b; Baranovskiy & Kuznetsov, 2017c). For the mathematical description of heat and mass transfer, nonstationary partial differential equations are used (Samarskii & Vabishchevich, 1995a, 1995b). An analysis of forest fire distributions in the vicinity of settlements allowed the idea that, in the first approximation, the distribution of anthropogenic load can be described using the equations of the general theory of heat and mass transfer (Baranovskiy, 2018). That is, it is proposed to use the similarity of the processes of heat transfer and the spread of anthropogenic load in the forested area. Deterministic approach is used instead of soft and statistical computing methods (Al Janabi et al., 2017; Rawlings et al., 2015).

The aim of the research is to develop a mathematical model of the spatial-temporal dynamics of anthropogenic load from a point source based on the similarity of heat transfer processes and the spread of anthropogenic load.

BACKGROUND

Forest fires destroy the forest stands and pollute the atmosphere with the products of pyrolysis and combustion of forest fuels (Baranovskiy & Kuznetsov, 2017a). An important part of strategic measures to prevent the negative impact of forest fires on plant communities and settlements is the forecast of their occurrence.

In Canada, the forest fire risk prediction is carried out as part of the Canadian Wildland Fire Information System (CWFIS) multicomponent system (Canadian Wildland Fire Information System; Wotton, 2009; Gould et al., 2013). The Canadian Forest Fire Danger Rating System (CFFDRS) (van Wagner, 1987; Martell, 2000) is used to predict the risk of the forest fires. It produces a qualitative and quantitative prediction of the fire potential index, which is used in a wide range of management activities. The input data includes weather conditions, characteristics of forest fuel, topographic parameters that are necessary to predict forest fire maturation, the occurrence of a forest fire and a prediction of the behavior of a forest fire. CFFDRS includes the following subsystems (van Wagner, 1987):

- Canadian Forest Fire Weather Index (FWI) System; The components of the subsystem are calculated using daily data on temperature, relative humidity, wind speed and daily precipitation.
- Canadian Forest Fire Behavior Prediction (FBP) System; This subsystem provides quantitative estimates of the possible speed of fire line propagation, the consumption of forest fuel, the intensity of the fire and its description.
- Canadian Forest Fire Occurrence Prediction (FOP) System (in fact, still under development, there are several regional schemes); this system is used to predict the probability of a forest fire.
- Accessory Fuel Moisture System (under development).
- Fire Monitoring, Mapping and Modeling System (Fire M3) (Lee et al., 2002); The system was launched in 1998 as a initiative of the Canada Center for Remote Sensing and the Canadian Forest Service. System is included in Natural Resources Canada. Fire M3 goals include the use of low-resolution satellite data to identify and localize active forest fires on a daily basis, estimate the daily and annual area of burnt areas and simulate fire behavior, and biomass consumption by fire.

- The American Wildland Fire Assessment System (WFAS) has several modules to predict potential fires (WFAS Wildland Fire Assessment System):
- National Fire Danger Rating System is used to predict forest fire danger in the United States (Deeming et al, 1974; Deeming et al., 1977). There were two large-scale revisions of this system.
- The Fire Potential Index (FPI) was developed to combine satellite and field measurements into an index that correlates well with forest fires and can be used both locally and nationally with GIS.

The European Forest Fire Information System (EFFIS) has been supported by a team of experts since 1998 (European Forest Fire Information System). Currently, the group includes experts from 40 countries (European Forest Fire Information System). The forest fire danger prediction subsystem generates daily maps of the forest fire danger level for 1–10 days ahead using the predicted meteorological conditions. Previously, several fire danger indexes were tested (Viegas et al., 2000) and the Canadian Forest Fire Weather Index System was adopted.

The information system for remote monitoring of forest fires (ISDM-Rosleskhoz) (Information system for remote monitoring of forest fires ISDM-Rosleskhoz; State Standard, 1999) was established in 2003 and put into operation in 2005 (Podolskaya et al., 2011). ISDM-Rosleskhoz is designed to solve the following tasks (Podolskaya et al., 2011): 1) operational detection, registration and subsequent monitoring of areas with suspicions of a forest fire; 2) obtaining operational prediction of the parameters of existing forest fires; 3) obtaining operational predictions of the effects of forest fires.

Thus, all systems of the forest fire danger prediction, widely used in world practice, provide the calculation of a certain index showing the probability of a forest fire (Preisler et al., 2004). The most developed issue of the occurrence of forest fires from thunderstorms. The technologies for predicting forest fire danger as a result of anthropogenic load are very poorly developed. This is due to the complex nature of anthropogenic load. A probabilistic criterion to predict forest fire danger, taking into account the thunderstorm activity and anthropogenic load, was developed earlier (Bartalev et al., 2007). However, only statistical data on the anthropogenic fires that are available for individual units or quarters of a forested territory are used.

The need to use information about the presence of fire sources to predict fire danger was also noted by I.S. Melekhov (Melekhov, 1947). Increasing levels of urbanization and recreational pressure leads to changes in the forest ecosystems, including as a result of forest fires. An analysis of the socio-psychological aspects of recreational visits to the forest and the occurrence of forest fires has been carried out (Andreev & Larchenko, 1987). It has been established that with increasing distance from a populated area, the number of visits to forests and forest fires decreases (Kurbatskiy, 1964; Melluma et al., 1982) according to the Rayleigh and Poisson distributions (Telitsin, 1984).

One of the important components of the anthropogenic load is recreation in the forested areas. Predicting the magnitude of recreational loads on the forest, time and specific places to visit, will allow to solve some issues not only planning and organizing mass recreation of the population, but also preventing forest fires, increasing the efficiency of their detection and suppression (Andreev & Larchenko, 1987).

When studying population migration in geography, various mathematical models are used. Some of them were tested to describe the distribution of the forest fires in relation to settlements in rural areas (Andreev, 1986). It turned out that for this purpose, the probability density functions of a log-normal distribution and the Pareto-type function can be used.

The results of opinion polls sometimes explain some of the data on fires that have arisen in the relevant area (Andreev & Larchenko, 1987).

The process of evaluating information and deciding on a visit to a forest is dependent on many factors, among which are the following: the social and natural environment in which the recipient of a forest visit is located; the impact of psychological laws of behavior.

With the increase in the number of inhabitants, the probability of a fire near populated areas increases. The coefficient of fire dangerous attendance shows the frequency of visits (per day per 1000 inhabitants), resulting in a single fire. From here you can get an indicator of fire hazardous attendance of the settlement (Andreev & Larchenko, 1987). The distribution of fire sources over the territory and in time is similar to the distribution of forest fires and is reproduced annually with only a slight deviation from the average long-term (Telitsin, 1987).

Indicative is the GIS analysis of spatial and temporal patterns of forest fires of the 1950–1992 seasons that occurred for anthropogenic reasons in the forests of Vancouver Island (Canada) (Pew & Larsen, 2001) (Figure 1).

6329 forest fires were analyzed. The software used is Arc/Info and ArcView GIS. The study area was divided into 36050 cells. The number of burnt cells is closely related to the proximity of country roads, paved and railways and decreases with increasing distance from them (especially from country roads) (see Figure 1.a). The number of burned cells increases with distance from camps and campgrounds, reaching a maximum at a distance of 5 km from their location, but decreases with subsequent removal (see Figure 1.d). A large amount (the difference with the neighboring territories by an order of magnitude and more) burns out cells at a distance of more than 20 km from the paved and railways, municipalities of Vancouver (see Figure 1.b, c, e).

The causes of forest fires from anthropogenic load in Vancouver Island are presented in Table 1 (Pew & Larsen, 2001).

PROBABILISTIC CRITERION

The formula was obtained to predict forest fire probability for time interval *j* of a forest fire season using the basic principles of the probability theory (Baranovskiy, 2017; Baranovskiy & Zharikova, 2014; Pakshirajan, 2013):

$$P_j = [P(A)P(A_j / A)P(FF / A / A_j) + P(L)P(L_j / L)P(FF / L / L_j)]P_j(D) \quad (1)$$

It is necessary to use the definition of probabilities by means of the frequency of events and the statistical data of the specific forestry to determine all the parts in formula (5) (Baranovskiy, 2017; Baranovskiy & Zharikova, 2014):

$$P(A) \approx \frac{N_A}{N_{FS}}, P(A_j / A) \approx \frac{N_{FD}}{N_{FW}}, \quad (2)$$

Figure 1. The number of burned cells of the forested area of Vancouver Island depending on the distance to the infrastructure
(Pew & Larsen, 2001)

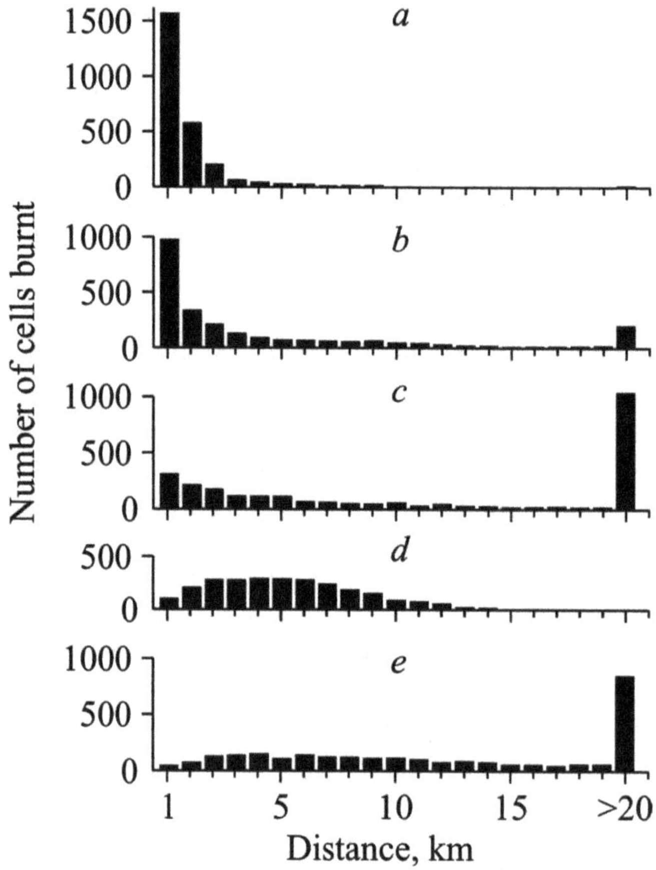

Table 1. The causes of forest fires from anthropogenic load in Vancouver Island

Fire Cause	Fire Number	Burned Area, ha	
		Total	Average (per fire)
Mixed	2567	17370	6,8
Recreation	1809	4462	2,5
Logging	1021	20927	20,5
Land clearing	512	966	1,9
Railway	189	276	1,5
Industrial objects	120	954	8,0
Motorways	104	2650	25,5
Undefined	7	8	1,2
Total	6329	47613	7,5

(Pew & Larsen, 2001)

$$P(FF / A / A_j) \approx \frac{N_{FA}}{N_{FT}}, \quad (3)$$

$$P(L) \approx \frac{N_L}{N_{FS}}, P(L_j / L) \approx \frac{N_{LH}}{N_{LD}}, \quad (4)$$

$$P(FF / L / L_j) \approx \frac{N_{FL}}{N_{FT}}, \quad (5)$$

Obviously, the more cases are considered for the specific forest, the more precise formulas (2) - (5) will be. All the parameters of a fire season have therefore to be registered every year.

P_j is the forest fire probability for interval *j* in the controlled forest area;
P(A) is the fire probability caused by the human activity;
$P(A_j/A)$ is the forest fire probability on week day *j*;
$P(FF/A,A_j)$ is the forest fire probability caused by the human activity in the forest area;
P(L) is the probability of dry thunderstorms in the forest area;
$P(L_j/L)$ is the probability of cloud-to-ground lightning discharge;
$P(FF/L,L_j)$ is the forest fire probability caused by the lightning provided that there is a dry thunderstorm in the forest area;
$P_j(D)$ is the probability that forest fuel layer will be dry;
index *j* corresponds to the day of a fire season.
N_A is the number of days during a fire season when anthropogenic load is enough for forest fuel ignition;
N_{FA} is the number of fires due to anthropogenic load;
N_{FT} is the total number of fires; N_L is the number of days when there was lightning (when dry thunderstorms occurred);
N_{FS} is the total number of days in a fire season;
N_{FL} is the number of fires due to lightning (when dry thunderstorms occurred);
N_{FD} is the number of fires on the specific day of week; N_{FW} is the total number of fires during week;
N_{LH} is the number of cloud-to-ground lightning discharges from 00:00 am;
N_{LD} is the total number of cloud-to-ground lightning discharges per day.

MATHEMATICAL MODEL OF ANTHROPOGENIC LOAD

The next stage is the spatial and temporal prediction of the fields of these parameters. The probability of the presence of sources of fire, which is determined by the first expression of formulas (2), is important. However, it is impossible to determine the number of forest fires for anthropogenic reasons during the whole week and on a specific day of the week with a spatial resolution at the level of the forest taxation unit (site). Therefore, as a parameter that will characterize the level of anthropogenic load, we will use the so-called “virtual number of forest fires”. A grid should be introduced that covers the forest area adjacent to the point source of anthropogenic load located in the forest territory. Only at certain points of the forest stand will it coincide with the actual number of forest fires (determined integrally for a sufficiently large area), and at intermediate points specific numerical values will be obtained using the compander-expander model (Baranovskiy, 2007). As a result, a parameter field will be prepared that characterizes the anthropogenic load at the initial moment of time. The next step will be the prediction of the evolution of this field at different points in time. In the literature there are facts that allow us to consider the anthropogenic load as a deterministic process. The authors in (Andreev & Larchenko, 1987) state that decision-making and behavior processes, despite the influence of various factors, are not random, since all people have common reactions to information about the structure and qualities of space, general mechanisms for perceiving this information to make decisions. People with similar fields of information show similar preferences for places of rest, that is, they behave in the geographic space almost the same. The following model of anthropogenic load was formulated:

The geometry of the solution domain for the two-dimensional case is presented in Figure 2.

The equation of the spatial-temporal dynamics of anthropogenic load:

$$\frac{\partial N^*}{\partial t} = \frac{1}{r}\frac{\partial}{\partial r}\left(r \cdot A\frac{\partial N^*}{\partial r}\right) + \frac{1}{r^2}\frac{\partial}{\partial \phi}\left(A\frac{\partial N^*}{\partial \phi}\right), \tag{6}$$

The boundary conditions will depend on the specific task. For example, the conditions:

$$r = 0,\ N^* = N^*_{r_0}, \tag{7}$$

$$r = R,\ A\frac{\partial N^*}{\partial r} = 0, \tag{8}$$

The coordinate φ is periodic boundary conditions or following ones in a symmetric setting:

$$\phi = 0,\ A\frac{\partial N^*}{\partial \phi} = 0, \tag{9}$$

$$\phi = \pi,\ A\frac{\partial N^*}{\partial \phi} = 0, \tag{10}$$

The initial conditions:

$$N^*(r,\phi)\Big|_{t=0} = N^*_0(r,\phi) \tag{11}$$

where N^* is the virtual number of forest fires, t is time, r is the radial coordinate, φ is the angular coordinate, A is the coefficient that characterizes the speed of propagation of ignition sources and has the dimension [m^2/s]. This equation is solved in a certain circle containing a forest area adjacent to a locality. At the initial time, $N^*(r,\varphi)$ has some background value. In addition, one should add boundary conditions

Figure 2. Task geometry (point source of anthropogenic load)

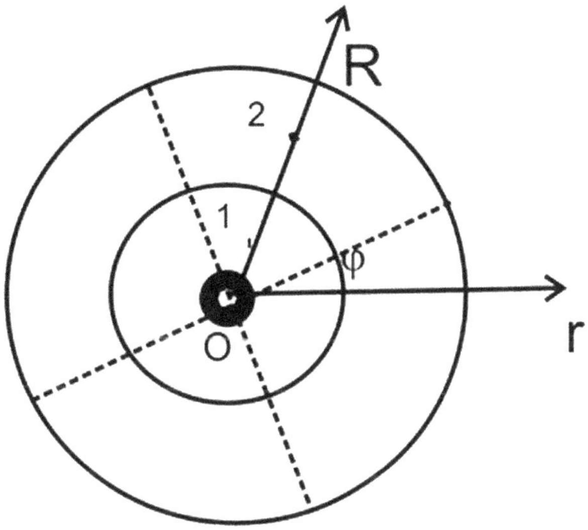

along the r coordinate. Their appearance will depend on the specific formulation of the problem. The coordinate φ should use periodic boundary conditions. For the numerical solution of this problem in the full formulation, the cyclic marching method can be used (Taflove & Hagness, 2005).

RESULTS AND DISCUSSION

To carry out computational experiments, scenario modeling should be applied (Baranovskiy, 2007). Tables 2, 3, 4 describe the main scenarios of the anthropogenic load on a model forested area from a point source.

It should be noted that the point source of anthropogenic load in the general case can be various objects (Catry et al., 2009; Ganteaume et al., 2013). In the present work, such varieties as a settlement, an industrial facility, and camping are highlighted. The settlement corresponds to the impact of the general population on the forest-covered territories. In this case, the activities of people can be associated with fishing, hunting, gathering of wild-growing plants, individual recreation, trips to the rural settlements. The scenario of an industrial facility can correspond to the activities of people in the vicinity of gas stations, forestries, energy facilities, construction organizations and potentially dangerous objects. The camping scenario summarizes the mass recreation in forested areas. For different types of point sources, different scenarios of anthropogenic impact are selected, which is expressed in different values of the virtual number of forest fires in the vicinity of such a point source. In addition, for each type of the point source of anthropogenic load, three characteristic object sizes are selected: small, medium and large.

For each such point source, a certain value of the virtual number of forest fires in the vicinity of the object is given. It should also be noted that the anthropogenic load is a significantly prolonged process over time. For definiteness, time is counted in numerical simulation from 6 am local time. Several scenarios of anthropogenic impact on forest areas are considered: 2,4,6,8,10 hours. Scenarios are selected in such a way as to cover the morning and afternoon hours. It should be noted that in the evening (for

Table 2. Description of model scenarios for the settlement

N Scenario	Source of Anthropogenic Load	VNF_L, Item	A, m^2/s	t_{end}, h
1	Settlement	40	10 000	2
2	Settlement	40	10 000	4
3	Settlement	40	10 000	6
4	Settlement	40	10 000	8
5	Settlement	40	10 000	10
6	Settlement	30	10 000	2
7	Settlement	30	10 000	4
8	Settlement	30	10 000	6
9	Settlement	30	10 000	8
10	Settlement	30	10 000	10
11	Settlement	25	10 000	2
12	Settlement	25	10 000	4
13	Settlement	25	10 000	6
14	Settlement	25	10 000	8
15	Settlement	25	10 000	10

Table 3. Description of model scenarios for an industrial facility

N Scenario	Source of Anthropogenic Load	VNF_L, Item	A, m^2/s	t_{end}, h
1	Industrial object	25	10 000	2
2	Industrial object	25	10 000	4
3	Industrial object	25	10 000	6
4	Industrial object	25	10 000	8
5	Industrial object	25	10 000	10
6	Industrial object	20	10 000	2
7	Industrial object	20	10 000	4
8	Industrial object	20	10 000	6
9	Industrial object	20	10 000	8
10	Industrial object	20	10 000	10
11	Industrial object	17.5	10 000	2
12	Industrial object	17.5	10 000	4
13	Industrial object	17.5	10 000	6
14	Industrial object	17.5	10 000	8
15	Industrial object	17.5	10 000	10

Table 4. Description of the model scenarios for camping

N Scenario	Source of Anthropogenic Load, Type	VNF_L, Item	A, m^2/s	t_{end}, h
1	Camping	15	10 000	2
2	Camping	15	10 000	4
3	Camping	15	10 000	6
4	Camping	15	10 000	8
5	Camping	15	10 000	10
6	Camping	10	10 000	2
7	Camping	10	10 000	4
8	Camping	10	10 000	6
9	Camping	10	10 000	8
10	Camping	10	10 000	10
11	Camping	7	10 000	2
12	Camping	7	10 000	4
13	Camping	7	10 000	6
14	Camping	7	10 000	8
15	Camping	7	10 000	10

definiteness, by 4 pm local time) we should expect the maximum level of anthropogenic load. In the following hours, an outflow of people from a forested area is more probable and possible. In addition, in the tables 2,3,4 there is a parameter A, which is specified in the mathematical model. It is responsible for the spread of anthropogenic load deep into forested areas. In fact, it characterizes the properties of a forested area, by analogy, as the thermal diffusivity characterizes the properties of a heat-conducting material (Kunkel et al., 2018; Li & Bai, 2019).

It should be noted that to determine the virtual number of forest fires on the border of a point source, you can use the averaged data for the adjacent forest area (based on statistics of forest fires on this territory) (Data archive on forest fires in the Timiryazevskiy forestry of the Tomsk region). In turn, zero values of the virtual number of forest fires can be specified as initial conditions. Alternatively, it is possible to set some background values of the virtual number of forest fires, which will correspond to the presence of unrecorded sources of anthropogenic load in a controlled forest area around the point source of anthropogenic load in question.

The typical distributions of the virtual number of forest fires in the vicinity of various point sources should be demonstrated. Such distributions in the geographic information system can be imposed on layers with different geographic features and layers with forest descriptions of specific forest areas. In the figures below, cold tones are responsible for the low level of anthropogenic load, and warm tones correspond to a high level of anthropogenic load.

Figures 3,4,5 show the field of the virtual number of forest fires for large, medium and small settlements for 10 hours of anthropogenic load spread. Figures 6,7,8 show the field of the virtual number of forest fires for large, medium and small industrial facilities for 10 hours of anthropogenic load spread. Figures 9,10,11 show the field of the virtual number of forest fires for large, medium and small camping for 10 hours of anthropogenic load spread.

The analysis of the presented two-dimensional distributions of the virtual number of forest fires shows significant differences in the level of anthropogenic load on the forest-covered territory from various point sources. A natural fact is a higher level of anthropogenic load from the settlement. It is clear that the population, even in a small settlement, in real situations may exceed the number of employees of any industrial facility, and even more so camping. Therefore, according to the predicted virtual number of forest fires, point sources are located in the following order: settlement, industrial facility, camping.

Different scenarios for the size of a single point source also showed that a higher level of anthropogenic load from a large point source is predicted. Then follow the medium and small source of anthropogenic load. It should be noted that in all the considered scenarios the same value of A was used, which in the simulation meant the location of all point sources on the territory with the same properties. This is more appropriate for scenario modeling. Nevertheless, in future studies, the influence of territory properties on the distribution of the virtual number of forest fires should be investigated.

For a more detailed quantitative analysis, one-dimensional distributions were constructed in the same cross section of a two-dimensional region with $\varphi=0$. Figures 12,13,14 show the comparative results in the cross section $\varphi=0$ for different point sources for 10 hours of anthropogenic exposure (settlement, industrial facility, camping). In figures 12,13,14 numbers denote point sources: 1 - large, 2 - medium, 3 - small.

Figure 3. The distribution of the virtual number of forest fires for a large settlement (10 hours of exposure)

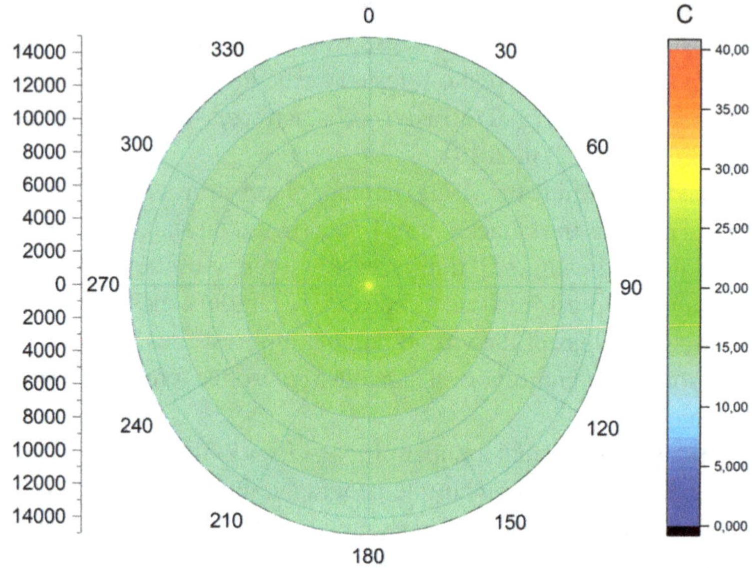

Figure 4. Distribution of the virtual number of forest fires for the medium settlement (10 hours of exposure)

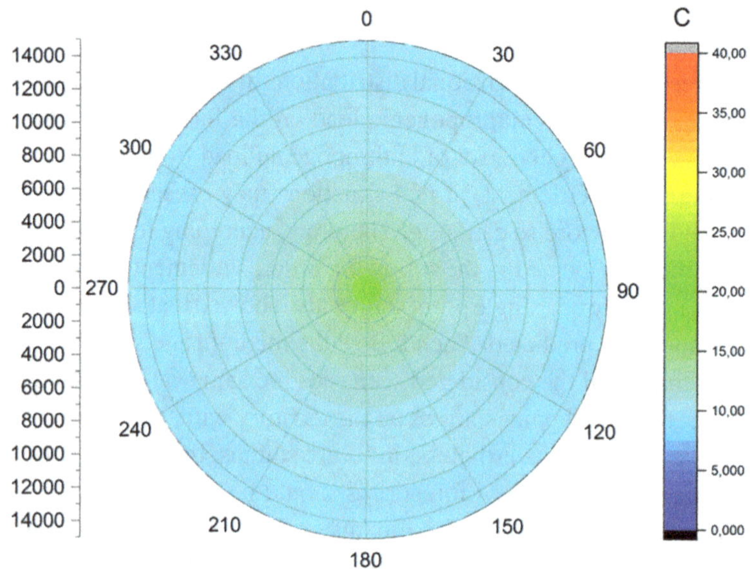

Figure 5. The distribution of the virtual number of forest fires for a small settlement (10 hours of exposure)

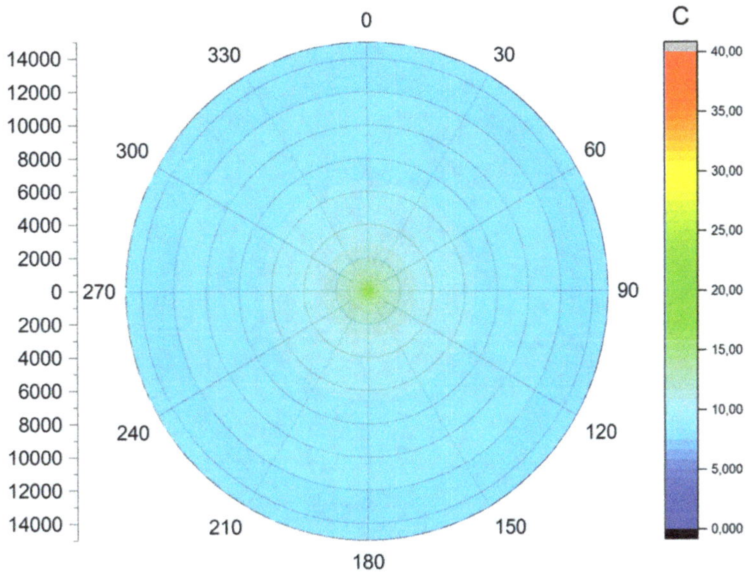

Figure 6. Distribution of the virtual number of forest fires for the large industrial facility (10 hours of exposure)

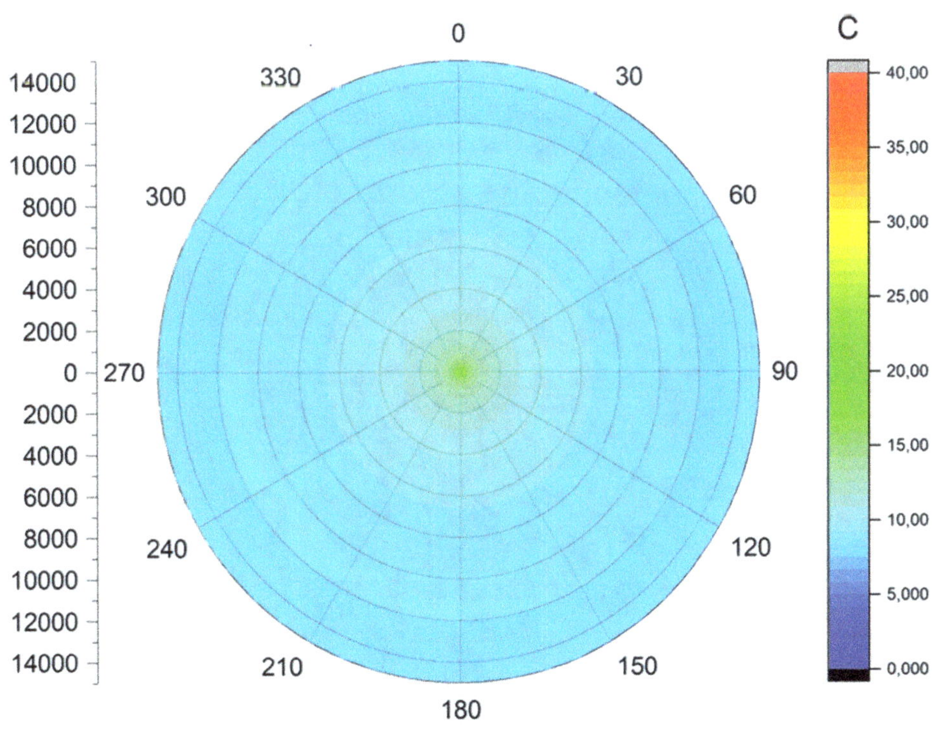

Figure 7. The distribution of the virtual number of forest fires for the medium industrial facility (10 hours of exposure)

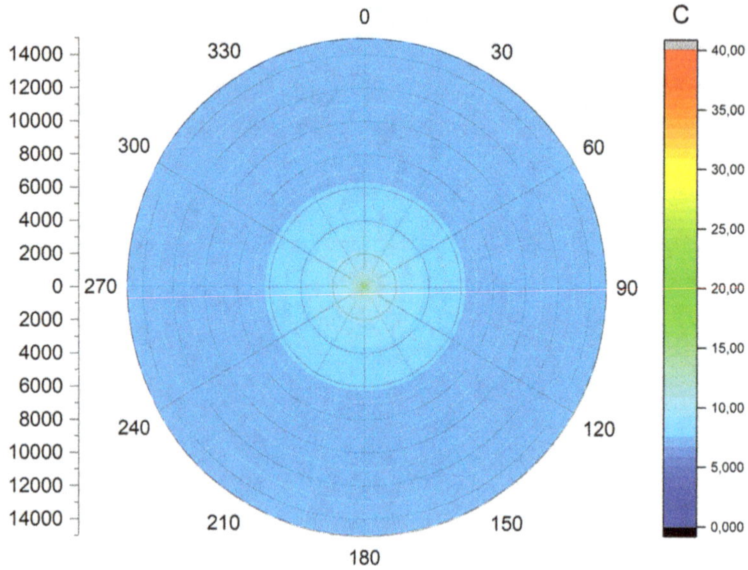

Figure 8. Distribution of the virtual number of forest fires for the small industrial facility (10 hours of exposure)

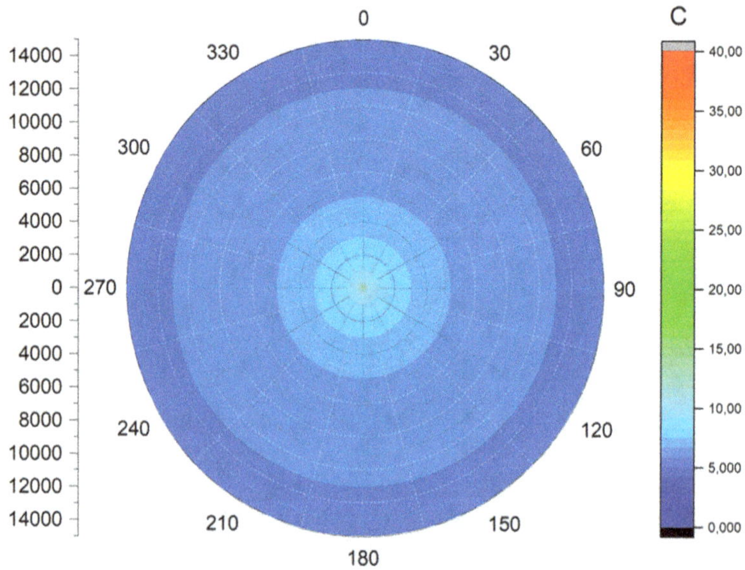

Figure 9. The distribution of the virtual number of forest fires for the large camping (10 hours of exposure)

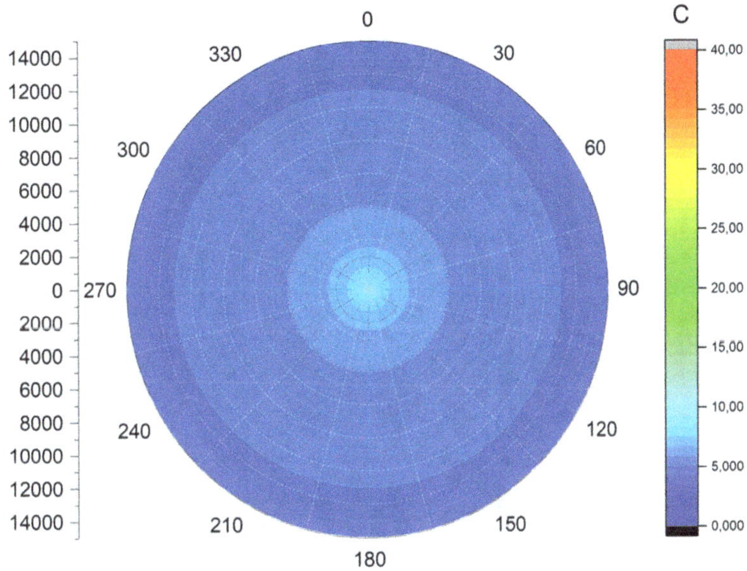

Figure 10. The distribution of the virtual number of forest fires for the medium camping (10 hours of exposure)

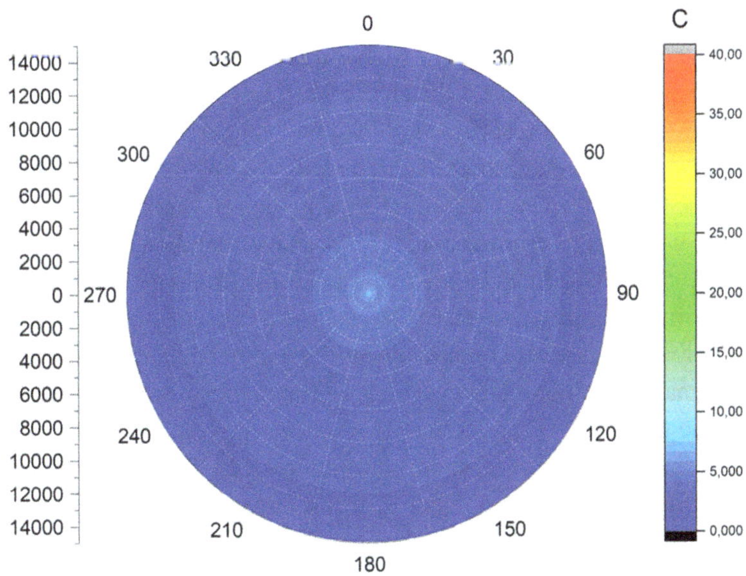

Figure 11. The distribution of the virtual number of forest fires for the small camping (10 hours of exposure)

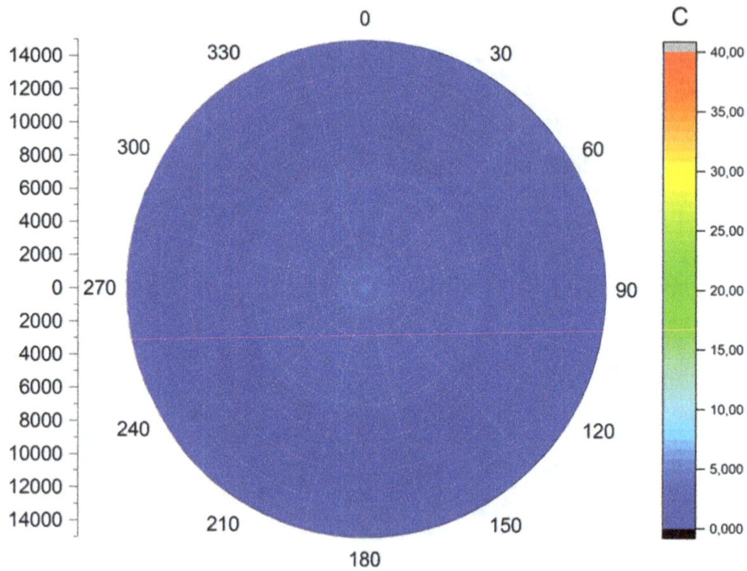

An analysis of the distributions presented in Figure 12 shows that the gradients of the virtual number of forest fires are characteristic of forest areas within 7–8 km from a point source of anthropogenic load. Further, the virtual number of forest fires is almost unchanged. Moreover, maximum gradients were found near a point source of anthropogenic load. Differences in the predicted virtual number of forest fires are more significant for large and medium settlements than for medium and small settlements. An analysis of the distributions presented in Figure 13 shows that the gradients of the virtual number of forest fires are characteristic of a forested area at a distance of 5–6 km from a point source of anthropogenic load. Further, the virtual number of forest fires is almost unchanged. For an industrial facility, the same patterns were found in large gradients near the source and the same differences for large, medium and small industrial facilities. Only the curves fall more closely to each other.

An analysis of the distributions presented in Figure 14 shows that the gradients of the virtual number of forest fires are characteristic of a forested area within 2.5 - 3 km in the vicinity of the camping site. Further, the predicted virtual number of forest fires is almost unchanged. The maximum gradients of the virtual number of forest fires are also obtained near the campsite. It should be noted that the curves for large, medium and small camping fall most closely to each other compared with other types of point sources of anthropogenic load.

Conducting a generalized analysis of these results (Figures 12 - 14), we can conclude that it is most important to predict the expected virtual number of forest fires in the vicinity of point sources of anthropogenic load. However, near such a source, forest fires can be detected in a relatively short period of time, which will contribute to its rapid localization and subsequent extinguishing. In turn, in more remote areas it is more difficult to detect forest fires and it may take more time to localize and extinguish it (Larjavaara et al., 2005).

Figure 12. Dependence of the virtual number of forest fires on the radial coordinate for the settlement (10 hours of exposure)

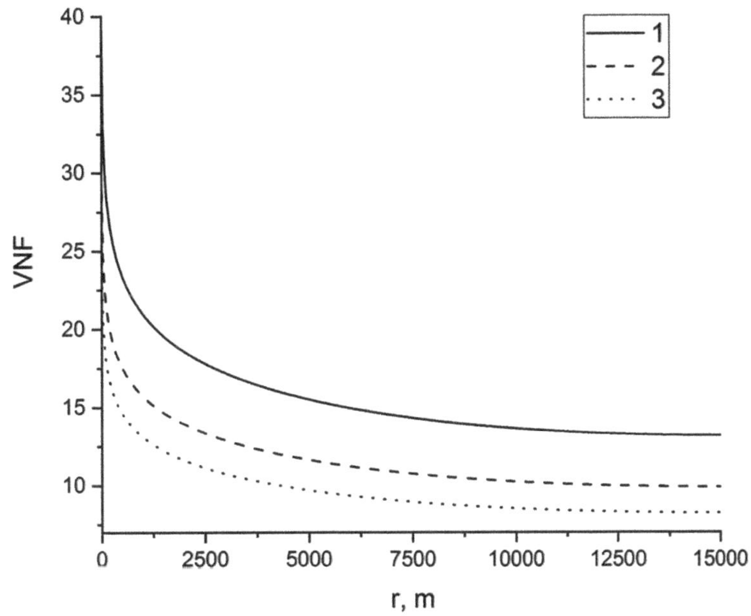

Figure 13. Dependence of the virtual number of forest fires on the radial coordinate for the industrial facility (10 hours of exposure)

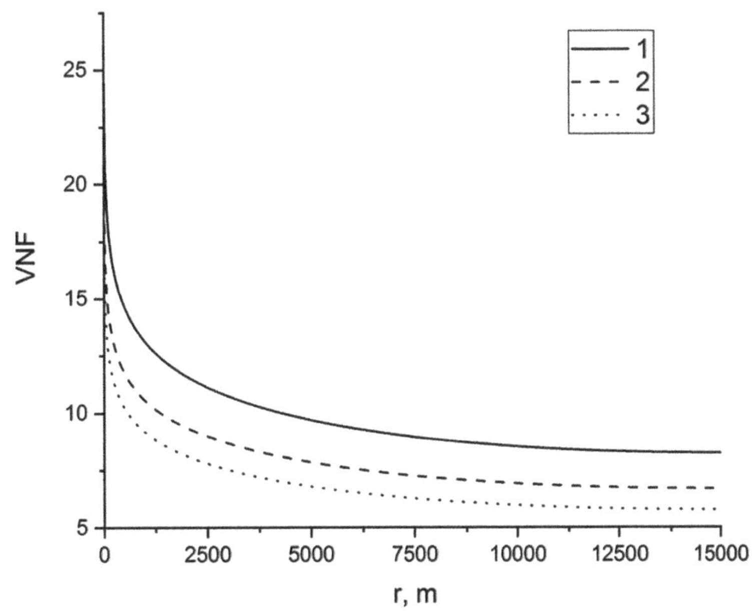

Figure 14. Dependence of the virtual number of forest fires on the radial coordinate for the camping (10 hours of exposure)

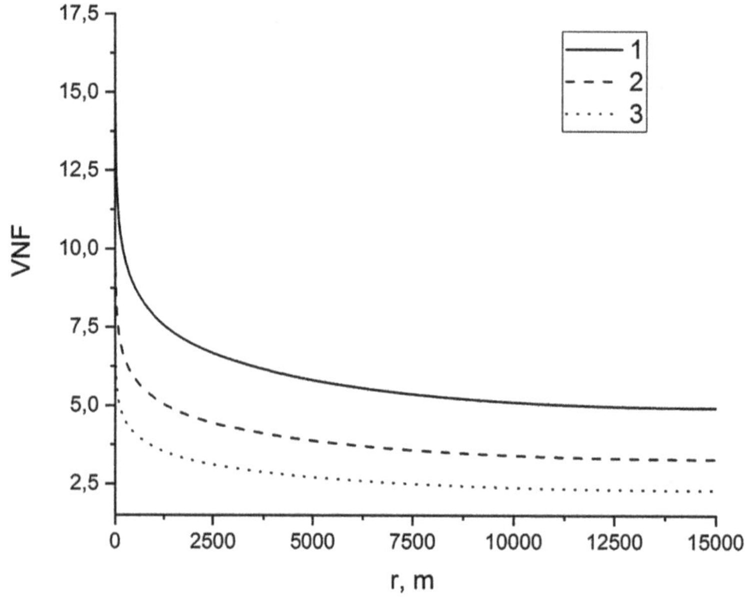

It should also be noted that in the considered mathematical model boundary conditions of the second kind were used. In future works, the impact of different boundary conditions on the distribution of the virtual number of forest fires in a controlled forest area should be assessed. In addition, in a real situation, a non-uniform distribution of the virtual number of forest fires in a forest area is possible due to different boundary conditions at the internal and external borders depending on the angular coordinate, as well as due to various properties of the forest area. That is, different values of the coefficient A for different parts of the forest area.

It is important to understand that the application of this mathematical model is most promising in conjunction with geographic information systems (Hernandez-Leal et al., 2008; Keane et al., 2001; Pummakarnchana et al., 2005). Various options for the implementation of the proposed mathematical model in the geographic information system can be proposed. In ArcGIS software, this model can be implemented as a software tool in the built-in high-level language ArcPy (a variation of the Pyton language) (Baranovskiy & Yankovich, 2015). A similar approach can be used in the GRASS GIS host system (GRASS GIS manual; Bui et al., 2017). In addition, the capabilities of the specialized toolbox in the Matlab computer system (Matlab official web-site) can be used. Alternatively, a specialized GIS system can be developed, when the calculation of anthropogenic load parameters is organized in a special interactive application written in a high level language (for example, in the rapid application development environment of Embarcadero RAD Studio in Pascal/Delphi) (RAD Studio.). In turn, the visualization of the distribution of the virtual number of forest fires can be implemented in the Origin Pro software (Origin Lab Official web-site). This option is implemented in this work. This development can be used with remote sensing, GIS and HPC technologies and with lightning monitoring systems (Diez-Delgado et al., 2004; Adab et al., 2013; You et al., 2017; Brun et al., 2017; Jacobson et al., 2006; Sakellariou et al., 2017).

CONCLUSION

Thus, this paper proposes a predictive mathematical model for the data assimilation system on the level of anthropogenic load. A parametric analysis of this model is presented. It is necessary to develop a methodology for determining the coefficient of model A. The use of this model in the system for prediction of forest fire danger will allow an increase in the level of detail, taking into account such factors as the anthropogenic load. Using this model will provide a hybrid deterministic-probabilistic prediction of the forest fires occurrence from anthropogenic load, when the distribution of anthropogenic load is modeled as a deterministic process, but ultimately the probability of a forest fire is calculated.

The overall goal of research in the field of anthropogenic load on forested areas should be the data assimilation system on the level of anthropogenic load. One should carry out an objective analysis of data, by analogy with different data assimilation systems (Carrió et al., 2019; Meng & Zhang, 2015; Srivas et al., 2017; Apte et al., 2007), can carry out the stages of analysis and predicting. The compander-expander model was developed earlier (Baranovskiy, 2007), which essentially belongs to the interpolation type of data analysis procedures (Flemming et al., 2001). Lightning monitoring systems and data assimilation systems on anthropogenic load can be used with software based on mathematical models of drying, pyrolysis and ignition of forest fuels (Grishin et al., 2001, 2006; Yashwanth et al., 2015; Howitt, 2015).

ACKNOWLEDGMENT

This work was implemented with the financial support of the Russian Foundation for Basic Research. [Project N 17-29-05093].

REFERENCES

Adab, H., Kanniah, K. D., & Solaimani, K. (2013). Modeling forest fire in the northeast of Iran using remote sensing and GIS techniques. *Natural Hazards*, *65*(3), 1723–1743. doi:10.100711069-012-0450-8

Al Janabi, S., Al Shourbaji, I., & Salman, M. A. (2017). *Assessing the suitability of soft computing approaches for forest fires prediction*. Applied Computing and Informatics; doi:10.1016/j.aci.2017.09.006

Amatulli, G., Perez-Cabello, F., & Du la Riva, J. (2007). Mapping lightning / human-caused wildfires occurrence under ignition point location uncertainty. *Ecological Modelling*, *200*(3-4), 321–333. doi:10.1016/j.ecolmodel.2006.08.001

Andreev, Yu. A. (1986). *Laws of distribution of forest fires. Methods and means of fighting forest fires*. Moscow, Russia: VNIILM. (In Russian)

Andreev, Yu. A., & Larchenko, G. F. (1987). *Socio-psychological aspects of recreational forest visits and the occurrence of fires. Forest fires and the fight against them*. Moscow, Russia: VNIILM. (In Russian)

Apte, A., Hairer, M., Stuart, A. M., & Voss, J. (2007). Sampling the posterior: An approach to non-Gaussian data assimilation. *Physica D. Nonlinear Phenomena*, *230*(1-2), 50–64. doi:10.1016/j.physd.2006.06.009

Baranovskiy, N., & Zharikova, M. (2014). A web-oriented geoinformation system application for forest fire danger prediction in typical forests of the Ukraine. *Lecture Notes in Geoinformation and Cartography*. 13-22.

Baranovskiy, N. V. (2007a). Testing of the system of data assimilation on the level of anthropogenic load for a linear source of anthropogenic load, *Environmental Safety Management System: A Collection of Proceedings of a Correspondence International Scientific and Practical Conference*, Yekaterinburg, Russia: USTU - UPI, 319-322. (In Russian).

Baranovskiy, N. V. (2007b). *Mathematical modeling of the most probably scenarios and conditions of forest fires occurrence.* (Thesis). Tomsk, Russia: Tomsk State University. (In Russian)

Baranovskiy, N. V. (2017). Forest fire danger assessment using SPMD-model of computation for massive parallel system. *International Review on Modelling and Simulations.*, *10*(3), 193–201. doi:10.15866/iremos.v10i3.10570

Baranovskiy, N. V. (2018). Mathematical simulation of anthropogenic load from a linear source using partial differential equations in the context of the forest fires occurrence. *Advances in Differential Equations and Control Processes*, *19*(3), 237–262. doi:10.17654/DE019030237

Baranovskiy, N. V., & Kuznetsov, G. V. (2017). *Forest fire occurrences and ecological impact prediction: Monograph*. Novosibirsk, Russia: Publishing House of the Siberian Branch of the Russian Academy of Science. doi:10.15372/FOREST2017BNV

Baranovskiy, N. V., & Kuznetsov, G. V. (2017). Mathematical simulation of deciduous tree ignition by cloud-to-ground lightning discharge using large vessels approximation. *JP Journal of Heat and Mass Transfer*, *14*(4), 533–545. doi:10.17654/HM014040533

Baranovskiy, N. V., & Kuznetsov, G. V. (2017). Coniferous tree ignition by cloud-to-ground lightning discharge using approximation of "ideal" crack in bark. *JP Journal of Heat and Mass Transfer*, *14*(1), 173–186. doi:10.17654/HM014010173

Baranovskiy, N. V., Kuznetsov, G. V., & Nemova, T. N. (2017). Mathematical computing of coniferous tree ignition by the cloud-to-ground lightning discharge using Joule-Lenz's law. *Iranian Journal of Electrical and Computer Engineering*, *7*(3), 1337–1346.

Baranovskiy, N. V., & Yankovich, E. P. (2015). Geoinformation monitoring of forest fire danger on the basis of remote sensing data of surface by the artificial Earth satellite. *Journal of Automation and Information Sciences*, *47*(8), 11–23. doi:10.1615/JAutomatInfScien.v47.i8.20

Bartalev, S. A., Egorov, V. A., Loupian, E. A., & Uvarov, I. A. (2007). Multi-year circumpolar assessment the area burnt in Boreal ecosystems using SPOT-vegetation. *International Journal of Remote Sensing*, *28*(6), 1397–1404. doi:10.1080/01431160600840978

Benavent-Corai, J., Rojo, C., Suarez-Torres, J., & Velasco-Garcia, L. (2007). Scaling properties in forest fire sequences: The human role in the order of nature. *Ecological Modelling*, *205*(3-4), 336–342. doi:10.1016/j.ecolmodel.2007.02.028

Brun, C., Artes, T., Cencerrado, A., Margalef, T., & Cortés, A. (2017). A high performance computing framework for continental-scale forest fire spread prediction. *Procedia Computer Science*, *108C*, 1712–1721. doi:10.1016/j.procs.2017.05.258

Bui, D. T., Bui, Q.-T., Nguyen, Q.-P., Pradhan, B., Nampak, H., & Trinh, P. T. (2017). A hybrid artificial intelligence using GIS-based neural-fuzzy inference system and particle swarm optimization for forest fire susceptibility modeling at a tropical area. *Agricultural and Forest Meteorology*, *233*, 32–44. doi:10.1016/j.agrformet.2016.11.002

Canadian Wildland Fire Information System. Official site. Retrieved from http://cwfis.cfs.nrcan.gc.ca/home

Carrió, D. S., Homar, V., & Wheatley, D. M. (2019). Potential of an EnKF storm-scale data assimilation system over sparse observation regions with complex orography. *Atmospheric Research*, *216*, 186–206. doi:10.1016/j.atmosres.2018.10.004

Catry, F. X., Rego, F. C., Bacao, F. L., & Moreira, F. (2009). Modeling and mapping wildfire ignition risk in Portugal. *International Journal of Wildland Fire*, *18*(8), 921–931. doi:10.1071/WF07123

Data archive on forest fires in the Timiryazevskiy forestry of the Tomsk region, CD. (2010). Department of Forestry of Tomsk Region. (In Russian)

Deeming, J. E., Burgan, R. E., & Cohen, J. D. (1977). *The National Fire Danger Rating System - 1978.* United States Department of Agriculture, Forest Service, General Technical Report INT-39, Intermountain Forest and Range Experiment Station, Odgen, UT.

Deeming, J. E., Lancaster, J. W., Fosberg, M. A., Furman, W. R., & Schroeder, M. J. (1974). *The National Fire Danger Rating System.* United States Department of Agriculture, Forest Service, Research Paper RM-84, Rocky Mountain Forest and Range Experiment Station, Fort Collins, CO.

Diez-Delgado, R., Lloret, F., & Pons, X. (2004). Statistical analysis of fire-frequency models for Catalonia (NE Spain, 1975-1998) based on fire scar maps from Landsat MSS data. *International Journal of Wildland Fire*, *13*(1), 89–99. doi:10.1071/WF02051

European Forest Fire Information System. Official site. Retrieved from http://effis.jrc.ec.europa.eu

Flemming, J., Reimer, E., & Stern, R. (2001). Data assimilation for CT-Modelling based on optimum interpolation. *Preprints of 25th NATO/CCMS International Technical Meeting on Air Pollution Modelling and its Application.* Belgium, Louvain-la-Neuve: Catholic University of Louvain.

Ganteaume, A., Camia, A., Jappiot, M., San-Miguel-Ayanz, J., Long-Fournel, M., & Lampin, C. (2013). A review of the main driving factors of forest fire ignition over Europe. *Environmental Management*, *9*(3), 651–662. doi:10.100700267-012-9961-z PMID:23086400

Gorev, G. V. Evaluation of the climate predisposition of the territory to the occurrence of forest fires (on the example of the Tomsk region). (Dissertation). (Tomsk, TSU, 2004) (In Russian)

Gould, J. S., Patriquin, M. N., Wang, S., McFarlane, B. L., & Wotton, B. M. (2013). Economic evaluation of research to improve the Canadian forest fire hazard rating system. *Forestry*, *86*(3), 317–329. doi:10.1093/forestry/cps082

GRASS GIS manual: i.atcorr. Retrieved from https://grass.osgeo.org/grass76/manuals/i.atcorr.html

Grishin, A. M., Golovanov, A. N., Kataeva, L. Y., & Loboda, E. L. (2001). Formulation and solution of the problem of drying of a layer of combustible forest materials. *Combustion, Explosion, and Shock Waves*, *37*(1), 57–66. doi:10.1023/A:1002868709387

Grishin, A. M., Golovanov, A. N., & Sukov, Y. V. (2006). Experimental determination of thermophysical, thermokinetic, and filtration characteristics of peat. *Journal of Engineering Physics and Thermophysics*, *79*(3), 557–562. doi:10.100710891-006-0135-9

Hernandez-Leal, P. A., Gonzalez-Calvo, A., Arbelo, M., Barreto, A., & Alonso-Benito, A. (2008). Synergy of GIS and remote sensing data in forest fire danger modeling. *IEEE Journal of Selected Topics in Applied Earth Observations and Remote Sensing*, *1*(4), 240–247. doi:10.1109/JSTARS.2008.2009043

Howitt, D. G. (2015). An assessment of hot metal fragments from heavy mechanical equipment as potential ignition source for forest litter. *Journal of Fire Sciences*, *33*(6), 427–444. doi:10.1177/0734904115611820

Information system for remote monitoring of forest fires ISDM-Rosleskhoz. Official site. Retrieved from https://nffc.aviales.ru/main_pages/index.shtml

Jacobson, A. R., Holzworth, R., Harlin, J., Dowden, R., & Lay, E. (2006). Performance assessment of the world wide lightning location network (WWLLN) using the Los Alamos Sferic Array (LASA) as ground truth. *Journal of Atmospheric and Oceanic Technology*, *23*(8), 1082–1092. doi:10.1175/JTECH1902.1

Jiménez-Ruano, A., Rodrigues Mimbrero, M., & de la Riva Fernández, J. (2017). Understanding wildfires in mainland Spain. A comprehensive analysis of the fire conditions in a climate-human context. *Applied Geography (Sevenoaks, England)*, *89*, 100–111. doi:10.1016/j.apgeog.2017.10.007

Keane, R. E., Burgan, R. E., & van Wagtendonk, J. (2001). Mapping wildland fuel for fire management across multiple scales: Integrating remote sensing, GIS, and biophysical modeling. *International Journal of Wildland Fire*, *10*(4), 301–319. doi:10.1071/WF01028

Kunkel, S., Teumer, T., Dörnhofer, P., Schlachter, K., Weldeslasie, Y., Kühr, M., ... Repke, J.-U. (2018). Determination of heat transfer coefficients in direct contact latent heat storage systems. *Applied Thermal Engineering*, *145*, 71–79. doi:10.1016/j.applthermaleng.2018.09.015

Kurbatskiy, N. P. (1964). *Problem of forest fires. Emergence of forest fires*. Moscow, Russia: Nauka. (In Russian)

Larjavaara, M., Kuuluvainen, T., & Rita, H. (2005). Spatial distribution of lightning-ignited fires in Finland. *Forest Ecology and Management*, *208*(1-3), 177–188. doi:10.1016/j.foreco.2004.12.005

Lee, B. S., Alexander, M. E., Hawkes, B. C., Lynham, T. J., Stocks, B. J., & Englefield, P. (2002). Information systems in support of wildland fire management decision making in Canada. *Computers and Electronics in Agriculture*, *37*(1-3), 185–198. doi:10.1016/S0168-1699(02)00120-5

Li, F., & Bai, B. (2019). A model of heat transfer coefficient for supercritical water considering the effect of heat transfer deterioration. *International Journal of Heat and Mass Transfer*, *133*, 316–329. doi:10.1016/j.ijheatmasstransfer.2018.12.121

Martell, D. L. (2000). A Markov chain model of day to day changes in the Canadian Forest Fire Weather Index. *International Journal of Wildland Fire*, *9*(4), 265–273. doi:10.1071/WF00020

Matlab official web-site. Retrieved from https://matlab.ru/products/matlab

Melekhov, I. S. (1947). *Nature of the forest and forest fires.* Arkhangelsk, Russia. (In Russian)

Melluma, A. Z., Rungule, R. H., & Emsis, I. V. (1982). *Camping as a nature conservation problem.* Riga, Latvia: Zinatne. (In Russian)

Meng, Z., & Zhang, F. (2015). Data assimilation and predictability. Ensemble-based data assimilation. Encyclopedia of Atmospheric Sciences (2nd Ed.). 241-247.

Origin lab official web-site. Retrieved from https://www.originlab.com/

Pakshirajan, R. P. (2013). Probability theory / Texts and readings in mathematics book series (TRIM, volume 63). Springer.

Perminov, V. A., Loboda, E. L., & Reyno, V. V. (2014, November). Mathematical modeling of ignition of woodlands resulted from accident on the pipeline. In *Proceedings 20th International Symposium on Atmospheric and Ocean Optics: Atmospheric Physics (Vol. 9292, p. 929224).* International Society for Optics and Photonics.

Pew, K. L., & Larsen, C. P. S. (2001). GIS analysis of spatial and temporal patterns of human-caused wildfires in the temperate rain forest of Vancouver Island, Canada. *Forest Ecology and Management*, *140*(1), 1–18. doi:10.1016/S0378-1127(00)00271-1

Podolskaya, A. S., Ershov, D. V., & Shulyak, P. P. (2011). Application of the method for assessing the probability of occurrence of forest fires in ISDM-Rosleskhoz. *Modern problems of remote sensing of the Earth from space. 8,* 118 - 126. (In Russian)

Poulos, H. M., Villanueva Díaz, J., Cerano Paredes, J., Camp, A. E., & Gatewood, R. G. (2013). Human influences on fire regimes and forest structure in the Chihuahuan Desert Borderlands. *Forest Ecology and Management*, *298*, 1–11. doi:10.1016/j.foreco.2013.02.014

Preisler, H. K., Brillinger, D. R., Burgan, R. E., & Benoit, J. W. (2004). Probability based models for estimation of wildfire risk. *International Journal of Wildland Fire*, *13*(2), 133–142. doi:10.1071/WF02061

Pummakarnchana, O., Tripathi, N., & Dutta, J. (2005). Air pollution monitoring and GIS modeling: A new use of nanotechnology based solid state gas sensors. *Science and Technology of Advanced Materials*, *6*(3-4), 251–255. doi:10.1016/j.stam.2005.02.003

RAD Studio. Retrieved from https://www.embarcadero.com/ru/products/rad-studio

Rawlings, J. O., Pantula, S. G., & Dickey, D. A. (2015). *Applied regression analysis. A research tool / Springer texts in statistics book series (STS).* Springer.

Read, N., Duff, T. J., & Taylor, P. G. (2018). A lightning-induced wildfire ignition forecasting model for operational use. *Agricultural and Forest Meteorology*, *253-254*, 233–246. doi:10.1016/j.agrformet.2018.01.037

Ruffault, J., & Mouillot, F. (2017). Contribution of human and biophysical factors to the spatial distribution of forest fire ignitions and large wildfires in a French Mediterranean region. *International Journal of Wildland Fire*, *26*(6), 498–508. doi:10.1071/WF16181

Sakellariou, S., Tampekis, S., Samara, F., Sfougaris, A., & Christopoulou, O. (2017). Review of state-of-the-art decision support systems (DSSs) for prevention and suppression of forest fires. *Journal of Forestry Research*, *28*(6), 1107–1117. doi:10.100711676-017-0452-1

Samarskii, A. A., & Vabishchevich, P. N. (1995). Computational heat transfer: Vol. 1. *Mathematical modelling*. Chichester, UK: Wiley.

Samarskii, A. A., & Vabishchevich, P. N. (1995). Computational heat transfer: Vol. 2. *The finite difference method*. Chichester, UK: Wiley.

Srivas, T., de Callafon, R. A., Crawl, D., & Altintas, I. (2017). Data assimilation of wildfires with fuel adjustment factors in farsite using ensemble kalman filtering. *Procedia Computer Science*, *108C*, 1572–1581. doi:10.1016/j.procs.2017.05.197

State Standard GOST R 22.1.09-99. Monitoring and prediction of forest fires. General requirements. (Moscow, Gosstandart of the Russian Federation, 1999) (In Russian)

Taflove, A., & Hagness, S. (2005). *Computational electrodynamics: The finite-difference time domain method*. Boston, MA: Arthech House Publishers.

Telitsin, G. P. (1984). Study of the relationship of forest attendance and the emergence of fires. [In Russian]. *Forest Science*, *1*, 59–63.

Telitsin, G. P. (1987). *Method of determining the fire danger of forest territory. Forest fires and the fight against them*. Moscow, Russia: VNIILM. (In Russian)

van Wagner, C. E. (1987). *Development and structure of the Canadian forest fire weather index system. Rep. No. Forest Technology Report 35*. Ottawa, Canada: Canadian Forestry Service.

Viegas, D. X., Bovio, G., Ferreira, A., Nosenzo, A., & Sol, B. (2000). Comparative study of various methods of fire danger evaluation in southern Europe. *International Journal of Wildland Fire*, *9*(4), 235–246. doi:10.1071/WF00015

WFAS Wildland Fire Assessment System. Official site. Retrieved from http://www.wfas.net

Wotton, B. M. (2009). Interpreting and using outputs from the Canadian forest fire danger rating system in research applications. *Environmental and Ecological Statistics*, *16*(2), 107–131. doi:10.100710651-007-0084-2

Yashwanth, B. L., Shotorban, B., Mahalingam, S., & Weise, D. R. (2015). An investigation of the influence of the heating modes on ignition and pyrolysis of woody wildland fuel. *Combustion Science and Technology*, *187*(5), 780–796. doi:10.1080/00102202.2014.973948

Ye, J., Wu, M., Deng, Z., Xu, S., Zhou, R., & Clarke, K. C. (2017). Modeling the spatial patterns of human wildfire ignition in Yunnan province, China. *Applied Geography (Sevenoaks, England)*, *89*, 150–162. doi:10.1016/j.apgeog.2017.09.012

You, W., Lin, L., Wu, L., Ji, Z., Yu, J., Zhu, J., ... He, D. (2017). Geographical information system-based forest fire risk assessment its spatiotemporal variability. *Ecological Indicators*, *77*, 176–184. doi:10.1016/j.ecolind.2017.01.042

ADDITIONAL READING

Adámek, M., Jankovská, Z., Hadincová, V., Kula, E., & Wild, J. (2018). Drivers of forest fire occurrence in the cultural landscape of Central Europe. *Landscape Ecology*, *33*(11), 2031–2045. doi:10.100710980-018-0712-2

Baranovskiy, N. (2018). Mathematical model to simulate anthropogenic loading around point source in one-dimensional statement. *International Multidisciplinary Scientific GeoConference Surveying Geology and Mining Ecology Management, SGEM, 18(3.2),* 857-862

Bowman, D. M. J. S., Moreira-Muñoz, A., Kolden, C. A., Chávez, R. O., Muñoz, A. A., Salinas, F., ... Johnston, F. H. (2019). Human–environmental drivers and impacts of the globally extreme 2017 Chilean fires. *Ambio*, *48*(4), 350–362. doi:10.100713280-018-1084-1 PMID:30128860

Clarke, H., Gibson, R., Cirulis, B., Bradstock, R. A., & Penman, T. D. (2019). Developing and testing models of the drivers of anthropogenic and lightning-caused wildfire ignitions in south-eastern Australia. *Journal of Environmental Management*, *235*, 34–41. doi:10.1016/j.jenvman.2019.01.055 PMID:30669091

Elia, M., Giannico, V., Lafortezza, R., & Sanesi, G. (2019). Modeling fire ignition patterns in Mediterranean urban interfaces. *Stochastic Environmental Research and Risk Assessment*, *33*(1), 169–181. doi:10.100700477-018-1558-5

Grishin, A. M., & Filkov, A. I. (2011). A deterministic-probabilistic system for predicting forest fire hazard. *Fire Safety Journal*, *46*(1-2), 56–62. doi:10.1016/j.firesaf.2010.09.002

Scheller, R., Kretchun, A., Hawbaker, T. J., & Henne, P. D. (2019). A landscape model of variable social-ecological fire regimes. *Ecological Modelling*, *401*, 85–93. doi:10.1016/j.ecolmodel.2019.03.022

Viedma, O., Urbieta, I. R., & Moreno, J. M. (2018). Wildfires and the role of their drivers are changing over time in a large rural area of west-central Spain. *Scientific Reports*, *8*(1), 17797. doi:10.103841598-018-36134-4 PMID:30542114

KEY TERMS AND DEFINITIONS

Anthropogenic Load: Different human activities on forested territories lead to forest fire occurrence and characterized by presence of fire sources.

Forest Fire: Uncontrolled aerothermochemical phenomenon characterized by step-by-step mechanism which includes following stages: inert heating, moisture evaporation, high temperature terpens evaporation, dry organic matter pyrolysis, flammable combustion and smoldering.

Mathematical Simulation: The production of a computer model of forest fire conditions and pre-requisites, especially for the purpose of study.

Monitoring: Monitoring refers to the periodic calculation of the parameters of forest fire danger with a portion of information available in real time.

Point Source: Source of anthropogenic load that characterized in virtual number of forest fire evolution in polar system of coordinates.

Prediction: Under the prediction of forest fires is the calculation of the parameters of forest fire danger with a certain projection in advance in order to have enough time to anticipate an emergency. The calculation in this case is carried out in a mode ahead of the real time of the development of the catastrophe - the occurrence of a forest fire.

Virtual Number of Forest Fires: Possible (calculated) number of forest fire around source of anthropogenic load which equal to real number of forest fire in some selected sites.

Chapter 4
Predicting Forest Fire Numbers Using Deterministic–Probabilistic Approach

Nikolay Viktorovich Baranovskiy
https://orcid.org/0000-0001-9202-8171
Tomsk Polytechnic University, Russia

ABSTRACT

The annual task of forecasting forest fire danger is becoming increasingly relevant, especially in the context of global warming. The forecast of surface fires is most important, as more than 80% of all vegetation fires are surface fires. Practically all crown fires develop from surface fires. This chapter discusses the deterministic-probabilistic method for predicting the number of forest fires in a controlled forest area. This methodology is based on the assumption that the number of registered and projected forest fires is related to the probability of their occurrence. The influence of forest fire retrospective data on the predicted number of forest fires for some sites of the Timiryazevskiy forestry of the Tomsk region was studied. This chapter presents the results of a comparative analysis of forecast data and statistics.

INTRODUCTION

The task of forecasting forest fire danger with each passing year is becoming increasingly relevant, especially in the context of global warming (Budyko, 1980). Forecast of surface fires are of the greatest importance, since more than 80% of all vegetation fires are surface fires, almost all crown fires develop from surface fires (Valendik, 1979). A deterministic-probabilistic method for determining the probability of forest fires (Baranovskiy, 2004) was proposed earlier, which takes into account the influence of meteorological conditions, thunderstorm activity, and anthropogenic load and is based on simple mathematical models of drying a layer of forest fuels (Grishin and Baranovskiy, 2003; Baranovskiy and Grishin, 2002). As a result of calculations, the numerical value of the probability of occurrence of forest fires in a controlled area with spatial detail at the level of the minimum forest taxation unit (site) is

DOI: 10.4018/978-1-7998-1867-0.ch004

determined. Despite the fact that such an assessment is both qualitative and to some extent quantitative, it does not give an answer to the question - how many forest fires should be expected at a given level of forest fire danger at a certain time interval in a controlled forest area?

The purpose of this work is to develop a deterministic-probabilistic method for predicting the number of forest fires in a controlled forest area and to carry out a parametric study of the effect of weather conditions, anthropogenic load, thunderstorm activity and fire retrospective data on the predicted number of forest fires, as well as comparative analysis with statistical data on real forest fires forestry.

BACKGROUND

Note that information on methods for predicting forest fire danger existing in Russia and abroad can be found in (Nesterov, 1949; Reyfsnyder, 1978; Kurbatskiy and Kostyrina, 1977). All techniques are based, as a rule, on the analysis of weather data.

Attempts were made in (Kataeva, 2000) to relate the moisture content of the forest fuel layer and the number of forest fires, but this is incorrect, since not always a high degree of fire danger due to weather conditions or forest vegetation conditions will lead to a high fire danger in the area. For example, the Bogorodskoye local forestry of the Timiryazevskiy forestry of the Tomsk region, in contrast to other forestry, even during dry periods is characterized by a small number of forest fires (Matsenko et al., 1999). The main reason is a weak anthropogenic load on this territory. In turn, work (Garcia Diez et al., 1993) presents a method for predicting the number of forest fires, which is based on taking into account the effect of instability in the atmospheric column on a controlled forest area on the forest fire danger. The authors of (Garcia Diez et al., 1993) note that their approach can be used simultaneously with other indices and, moreover, that their methodology can be built on the basis of other meteorological parameters. However, it does not take into account the real processes that affect forest fire danger, namely, drying of the forest fuel layer under the influence of external conditions, thunderstorm activity and the level of anthropogenic load.

In this work, as well as in (Baranovskiy et al., 2003), when developing a quantitative forecast method for the occurrence of forest fires, it is proposed to focus on the prospects for interactive interaction with computer programs that implement global or regional atmospheric models (for example, the semi-lagranzh model (Tolstykh, 2001).

Thus, at the moment in Russia and abroad there is virtually no physically meaningful method for predicting the number of forest fires and the relevance of developing such a technique is not in doubt.

PROBABILISTIC CRITERION

The formula was obtained to predict forest fire probability for time interval j of a forest fire season using the basic principles of the probability theory (Baranovskiy, 2017; Baranovskiy & Zharikova, 2014):

$$P_j = [P(A)P(A_j / A)P(FF / A / A_j) + P(L)P(L_j / L)P(FF / L / L_j)]P_j(D) \quad (1)$$

It is necessary to use the definition of probabilities by means of the frequency of events and the statistical data of the specific forestry to determine all the parts in formula (5) (Baranovskiy, 2017; Baranovskiy & Zharikova, 2014):

$$P(A) \approx \frac{N_A}{N_{FS}}, P(A_j / A) \approx \frac{N_{FD}}{N_{FW}}, \quad (2)$$

$$P(FF / A / A_j) \approx \frac{N_{FA}}{N_{FT}}, \quad (3)$$

$$P(L) \approx \frac{N_L}{N_{FS}}, P(L_j / L) \approx \frac{N_{LH}}{N_{LD}}, \quad (4)$$

$$P(FF / L / L_j) \approx \frac{N_{FL}}{N_{FT}}, \quad (5)$$

Obviously, the more cases are considered for the specific forest, the more precise formulas (2) - (5) will be. All the parameters of a fire season have therefore to be registered every year.

P_j is the forest fire probability for interval j in the controlled forest area;
$P(A)$ is the fire probability caused by the human activity;
$P(A_j/A)$ is the forest fire probability on weekday j;
$P(FF/A,A_j)$ is the forest fire probability caused by the human activity in the forest area;
$P(L)$ is the probability of dry thunderstorms in the forest area;
$P(L_j/L)$ is the probability of cloud-to-ground lightning discharge;
$P(FF/L,L_j)$ is the forest fire probability caused by the lightning provided that there is a dry thunderstorm in the forest area;
$P_j(D)$ is the probability that forest fuel layer will be dry;
index j corresponds to the day of a fire season.
N_A is the number of days during a fire season when anthropogenic load is enough for forest fuel ignition;
N_{FA} is the number of fires due to anthropogenic load;
N_{FT} is the total number of fires; N_L is the number of days when there was lightning (when dry thunderstorms occurred);
N_{FS} is the total number of days in a fire season;
N_{FL} is the number of fires due to lightning (when dry thunderstorms occurred);
N_{FD} is the number of fires on the specific day of week; N_{FW} is the total number of fires during week;
N_{LH} is the number of cloud-to-ground lightning discharges from 00:00 am;
N_{LD} is the total number of cloud-to-ground lightning discharges per day.

To determine all the terms in formula (1), it is necessary to use the definition of probabilities in terms of the frequency of events [3] and use the statistical data for a specific forestry. The drying time of the forest fuel layer makes it possible to determine the probability of occurrence of forest fires according to the weather conditions Pij (C). To determine the value of Pij (C), we have the formula (Baranovskiy, 2004):

$$P_{ij}(C) = \begin{cases} 0 \\ k_h \dfrac{F_i}{F} \exp[-(\Delta \bar{t}_{ij})^2] \end{cases}, \quad F = \sum_{i=1}^{N} F_i, \Delta \bar{t}_{ij} = \frac{t_{ij} - t_{ij}^*}{t_{ij}^*}, \tag{6}$$

where kh is the precipitation correction factor (Zhdanko and Gritsenko, 1980), 0 on the right-hand side of (6) corresponds to the case when there is no forest fuel on the i-th area of the forest territory (the surface of roads, rivers, lakes and water-saturated wetlands). In the derivation of (6), it was assumed that Pij (C) follows the normal (Gaussian) probability distribution law.

FOREST FIRE NUMBER AND PROBABILITY

It should be noted that the number of fires recorded during a particular day depends on:

1. meteorological (Garcia Diez et al., 1993) and other conditions that characterize the current day,
2. the cumulative long-term effect of the meteorological conditions of the previous few days (Garcia Diez et al., 1993) in the area. For example, drought or, on the contrary, prolonged rains undoubtedly influence the probability of forest fires during the current day. It will be logical to take this period for a period that is equal to the period during which there was very little rainfall, or vice versa it rained.

Thus, the number of forest fires actually depends on the probabilities of occurrence of Pj0 and PjN, where Pj0 is the probability of occurrence of forest fires on a particular day on which the forecast is made, and PjN is the average value of the probability of occurrence of forest fires for N previous days and is determined by the formula:

$$P_j^N = \frac{1}{N} \sum_{1}^{N} P_j^{\alpha}, \tag{7}$$

It should be noted that in formula (7) the probability values at 13-00 o'clock in the afternoon are implied, since according to statistics the greatest number of forest fires falls on these hours.

Thus, the number of fires NF (d) on a particular day will be logical to put equal to:

$$NF(d) = K_d P_j^N P_j^0, \tag{8}$$

where Kd is a certain proportionality coefficient (to know the value of which is not necessary, since in consequence it simply decreases).

Then the predicted number of forest fires for days d and d-1 can be determined by the formulas:

$$PNF(d) = K_d P_j^{N-1} P_j^0,\ PNF(d-1) = K_d P_j^N P_j^1 \tag{9}$$

Given that RNF (d-1) is the number of fires registered on the day (d-1), and assuming that the prediction was correct, the following equals are true:

$$PNF(d-1) = RNF(d-1) \text{ и } K_d P_j^N = RNF(d-1) / P_j^1. \tag{10}$$

As already mentioned, the time interval for which the PjN and PjN-1 values are calculated is significantly longer than one day, it can be assumed that the meteorological conditions during these periods did not change significantly (i.e., they almost coincide, since the difference is only one day) then the ratio will be fair enough

$$K_d P_j^N \approx K_d P_j^{N-1}. \tag{11}$$

Then the number of predicted forest fires is determined by the formula

$$PNF(d) = \frac{RNF(d-1) P_j^0(d)}{P_j^0(d-1)}. \tag{12}$$

Thus, the number of predicted fires for a particular day depends on the number of registered fires on the previous day multiplied by the ratio of the probability of occurrence of forest fires in the current and previous days.

STUDY AREA

Timiryazevskiy forestry of the Tomsk region is located between two large rivers, the Ob river and the Tom river on the territory of three administrative districts of the Tomsk region - Tomsk, Shegarsky and Kozhevnikovsky. The length of the territory of the forestry enterprise from North to South is 64 km, from West to East - 50 km. Timiryazevskiy forestry was established in 1966 on the basis of the order of the Ministry of Forestry of the RSFSR of July 8, 1966. No. 261. The forests are mainly represented by a single forest, except for the isolated near-alpine cedar forests of the villages of Zorkaltsevo, Nizhne Sechenovo and Gubino.

According to the forest vegetation zoning of Western Siberia, the territory of the Timiryazevsky forestry enterprise belongs to the zone of the Southern taiga (Ob-Tomsk cedar-pine forest vegetation district). According to the agroclimatic zoning of the Tomsk Region adopted by the Tomsk Branch of Sibgiprozema, the territory of the forestry enterprise is assigned to a moderately wet area. The growing season is 120 days.

The most common soils on the territory of the leshoz are podzolic and sod-podzolic (58%). The mechanical composition is dominated by sand and sandy (99%). Sod-podzolic soils make up 37.3%. On the territory of the leshoz, almost 20% are swamps, of which 35.5% are wet. The main dominant species is pine - 39.6%; Aspen - 26.2% and birch - 21.2%; cedar, larch, spruce and fir make up 13%.

RESULTS AND DISCUSSION

In modeling, a hypothetical forest area is considered, in which pine-type coniferous forest grows. In the ground cover most of the leaves are pine needles with a very small number of thin pine twigs. The forest territory is in close proximity to the settlement, which causes the predominant influence of anthropogenic load. This qualitatively reflects the forest fire situation in the Timiryazevskiy local forestry of the Timiryazevskiy forestry of the Tomsk region. In (Matsenko et al., 1999), data on forest fire conditions and a characteristic of forests of the Timiryazevskiy forestry of the Tomsk region for 1994-1998 are presented. Weather data are taken from the directory (Applied reference book on climate…, 1993).

It is of interest to conduct a retrospective analysis of the proposed methodology. The work (Matsenko et al., 1999) presents the results of processing statistics on forest fires in the Timiryazevsky forestry enterprise of the Tomsk region. Figure 1 presents the results of statistical processing and the forecast for the number of forest fires during the week on the territory of the Timiryazevsky forestry of the Timiryazevsky forestry enterprise of the Tomsk region for one forest fire season.

In Figure 1, the numbers indicate: 1 - statistical data on the number of forest fires per week for Timiryazevsky forest area, 2 - the predicted number of forest fires per week for Timiryazevsky forest area. The greatest number of fires occurs on Friday, Saturday, Sunday, and as an echo continues on Monday. The average relative prediction error was 10.4%, which is a good result.

As the results of calculations show, with a decrease in the flux density of solar radiation in clear-air conditions every day, the number of predicted forest fires decreases. These results are true, because with a decrease in the flux density of solar radiation, the drying time of the forest fuel layer increases, the probability of forest fires decreases and the predicted number of fires decreases, respectively. In conditions of average cloudiness, the situation does not fundamentally change, only the curves of changes in the predicted number of forest fires fall more closely, which is explained by smaller differences in the flux density of solar radiation.

Changes in ambient temperature also adequately affect the change in the number of predicted forest fires. With a decrease in the ambient temperature, the drying time of the forest fuel layer increases, the probability of occurrence decreases, which leads to a decrease in the number of predicted forest fires in the controlled forest area every day.

In addition, parametric calculations were carried out on the effect of the initial moisture content of the forest fuel layer, precipitation, anthropogenic load, and ground lightning discharges on the predicted number of forest fires. As the analysis of the obtained results showed, the technique adequately responds to changes in the initial moisture content and other parameters of forest fire danger.

A comparative (retrospective) analysis of the results of a numerical forecast using the methodology of quantitative forecasting of forest fires proposed in this work and the results of processing statistics on forest fires in the Timiryazevskiy local forestry of the Timiryazevskiy forestry of the Tomsk Region (dynamics of forest fires during the week) was carried out. Statistical data for the period 1994-1998 were processed (Matsenko et al., 1993).

Figure 1. Comparative analysis of statistics and forecast for the number of forest fires in the territory of Timiryazevskiy local forestry of Timiryazevskiy forestry

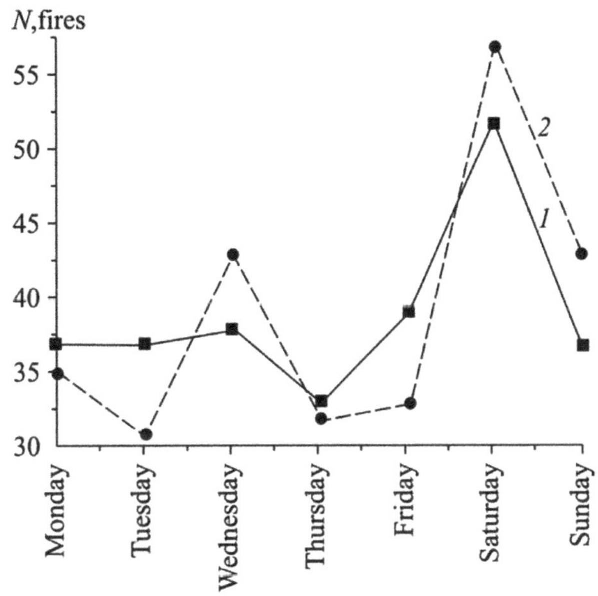

The average relative error was 10.4%, which is a fairly good result, given the accuracy of the input data. It should be noted that at the moment it is impossible to conduct a full-fledged retrospective analysis on the territory of the entire forestry, since the necessary ground data on ground cover is missing. However, in the future, trial operation of the technique on the territory of Timiryazevskiy forestry is expected.

CONCLUSION

Thus, according to the results of the study, the following conclusions can be made:

1. A new technique is presented for determining the number of forecasted forest fires, the scientific and practical importance of which is great, because for the first time, unlike the known methods developed abroad and Russia, a technique has been developed that takes into account meteorological data, thunderstorm activity, the level of anthropogenic load, type and the properties of the forest fuel, as well as statistical data of the forest fire retrospective.
2. The technique adequately responds to a parametric change in input data. A retrospective analysis of the number of forest fires in the Timiryazevskiy local forestry of the Timiryazevskiy forestry showed good agreement between the statistics and the numerical forecast.
3. Spatial detailing can be adopted on the basis of the necessary requirements for each specific case. In this paper, it is recommended to carry out an integrated assessment of forestry.
4. The application of the new methodology in the practice of protecting forests from fires will create objective conditions for a more accurate, physically sound and operational forecast of the number of forest fires and the timely adoption of preventive, preemptive measures to prevent the destruction of the state forest fund by fires.

5. A quantitative forecast of forest fires will allow you to optimally calculate the required number of units of combat personnel and auxiliary equipment, as well as justify the amount of funds requested for preventive measures.

Thus, in this work, an important scientific and practical problem was solved: a new deterministic-probabilistic method of quantitative forecasting of forest fires was developed and a parametric analysis of the influence of input data and a retrospective analysis of the number of forest fires in some areas of the Timiryazevskiy forestry of the Tomsk region was carried out (Data archive, 2010). Up-to-date computer technologies and methods can be used with suggested approach (Lee et al., 2002; Deeming et al. 1974, 1977; European Forest Fire Information System, 2018; Viegas et al., 1999; Information system…, 2019; Podolskaya et al., 2011; Bartalev et al., 2007; Larjavaara et al., 2005; Hernandez-Leal et al., 2008; Keane et al., 2001; GRASS GIS manual, 2018; Matlab, 2019; RAD Studio, 2019; Origin Lab, 2019; Grishin and Filkov, 2011). Suggested approach took into account some aspects of anthropogenic load and lightning activity (Ruffault and Mouillot, 2017; Adámek et al., 2018; Bowman et al., 2019; Clarke et al., 2019; Elia et al., 2019; Scheller et al., 2019; Viedma et al., 2018).

ACKNOWLEDGMENT

This work was implemented with the financial support of the Russian Foundation for Basic Research (Model to predict forest fires number) [Project N 17-29-05093]. Initial data was obtained during project that was supported by the Russian Foundation for Basic Researches and administration of Tomsk region [Project N 16-41-700831].

REFERENCES

Adámek, M., Jankovská, Z., Hadincová, V., Kula, E., & Wild, J. (2018). Drivers of forest fire occurrence in the cultural landscape of Central Europe. *Landscape Ecology*, *33*(11), 2031–2045. doi:10.100710980-018-0712-2

Applied reference book on climate of the USSR. Series 3. Long-term data. Parts 1-6. Issue 20. St. Petersburg, Russia: Gidrometeoizdat, 1993.

Baranovskiy, N., & Zharikova, M. (2014). A web-oriented geoinformation system application for forest fire danger prediction in typical forests of the Ukraine. *Lecture Notes in Geoinformation and Cartography, (199669),* pp. 13-22.

Baranovskiy, N. V. (2017). Forest fire danger assessment using SPMD-model of computation for massive parallel system. *International Review on Modelling and Simulations*, *10*(3), 193–201. doi:10.15866/iremos.v10i3.10570

Baranovskiy, N. V. (2004). The influence of anthropogenic load and lightning activity on the probability of forest fires. *Siberian Journal of Ecology, 6.* pp. 835-842.

Baranovskiy N. V., & Grishin A. M. (2002). Mathematical modeling of drying of a layer of forest combustible materials under the canopy of pine stand, *Computational Technologies, 7,* Part 1. (Joint issue Vestnik KazNU, N 4 (32)), pp. 320-324.

Baranovskiy, N. V., Grishin, A. M., & Loskutnikova, T. P. (2003). Information and prognostic system for determining the probability of occurrence of forest fires. *Computational Technologies, 2,* pp. 16-26.

Bartalev, S. A., Egorov, V. A., Loupian, E. A., & Uvarov, I. A. (2007). Multi-year circumpolar assessment the area burnt in Boreal ecosystems using SPOT-vegetation. *International Journal of Remote Sensing*, *28*(6), 1397–1404. doi:10.1080/01431160600840978

Bowman, D. M. J. S., Moreira-Muñoz, A., Kolden, C. A., Chávez, R. O., Muñoz, A. A., Salinas, F., ... Johnston, F. H. (2019). Human–environmental drivers and impacts of the globally extreme 2017 Chilean fires. *Ambio*, *48*(4), 350–362. doi:10.100713280-018-1084-1 PMID:30128860

Budyko, M. I. (1980). *The climate in the past and the future*. Leningrad, Russia: Gidrometeoizdat.

Clarke, H., Gibson, R., Cirulis, B., Bradstock, R. A., & Penman, T. D. (2019). Developing and testing models of the drivers of anthropogenic and lightning-caused wildfire ignitions in south-eastern Australia. *Journal of Environmental Management*, *235*, 34–41. doi:10.1016/j.jenvman.2019.01.055 PMID:30669091

Data archive on forest fires in the Timiryazevskiy forestry of the Tomsk region, CD. (2010). Department of Forestry of Tomsk Region. (in Russian)

Deeming, J. E., Burgan, R. E., & Cohen, J. D. (1977). The National Fire Danger Rating System - 1978. United States Department of Agriculture, Forest Service, General Technical Report INT-39, Intermountain Forest and Range Experiment Station, Odgen, UT.

Deeming, J. E., Lancaster, J. W., Fosberg, M. A., Furman, W. R., & Schroeder, M. J. (1974). The National Fire Danger Rating System. United States Department of Agriculture, Forest Service, Research Paper RM-84, Rocky Mountain Forest and Range Experiment Station, Fort Collins, CO.

Elia, M., Giannico, V., Lafortezza, R., & Sanesi, G. (2019). Modeling fire ignition patterns in Mediterranean urban interfaces. *Stochastic Environmental Research and Risk Assessment*, *33*(1), 169–181. doi:10.100700477-018-1558-5

European Forest Fire Information System. Official site. Retrieved from http://effis.jrc.ec.europa.eu

Garcia Diez, E. L., Labajo Salazar, J. L., & de Pablo, F. (1993). Some meteorological conditions associated with forest fires in Galicia (Spain). *International Journal of Biometeorology*, *37*(4), 194–199. doi:10.1007/BF01387522

GRASS GIS manual: i.atcorr: https://grass.osgeo.org/grass76/manuals/i.atcorr.html. Accessed: 2018-12-05.

Grishin, A. M., & Baranovskiy, N. V. (2003). Comparative analysis of simple models of drying of the forest fuel layer, including data from experiments and field observations. *Journal of Engineering Physics and Thermophysics*, *76*(5), 166–169. doi:10.1023/B:JOEP.0000003234.63878.7b

Grishin, A. M., & Filkov, A. I. (2011). A deterministic-probabilistic system for predicting forest fire hazard. *Fire Safety Journal*, *46*(1-2), 56–62. doi:10.1016/j.firesaf.2010.09.002

Hernandez-Leal, P. A., Gonzalez-Calvo, A., Arbelo, M., Barreto, A., & Alonso-Benito, A. (2008, December). Synergy of GIS and remote sensing data in forest fire danger modeling. *IEEE Journal of Selected Topics in Applied Earth Observations and Remote Sensing*, *1*(4), 240–247. doi:10.1109/JSTARS.2008.2009043

Information system for remote monitoring of forest fires ISDM-Rosleskhoz. Official site. Retrieved from https://nffc.aviales.ru/main_pages/index.shtml

Kataeva, L. Yu. (2000). Application of methods and concepts of fluid and gas mechanics for mathematical modeling of some urgent problems of ecology. (Dissertation), Cand. Phys.-Mat. Sciences, Tomsk, Russia: TSU.

Keane, R. E., Burgan, R. E., & van Wagtendonk, J. (2001). Mapping wildland fuel for fire management across multiple scales: Integrating remote sensing, GIS, and biophysical modeling. *International Journal of Wildland Fire*, *10*(4), 301–319. doi:10.1071/WF01028

Kurbatskiy, N. P., & Kostyrina, T. V. (1977). *US National Fire Danger Calculation System / Detection and analysis of forest fires* (pp. 38–90). Krasnoyarsk, Russia: ILID SO AN USSR.

Larjavaara, M., Kuuluvainen, T., & Rita, H. (2005). Spatial distribution of lightning-ignited fires in Finland. *Forest Ecology and Management*, *208*(1–3), 177–188. doi:10.1016/j.foreco.2004.12.005

Lee, B. S., Alexander, M. E., Hawkes, B. C., Stocks, B. J., & Englefield, P. (2002). Information systems in support of wildland fire management decision making in Canada. *Computers and Electronics in Agriculture*, *37*(1–2), 185–198. doi:10.1016/S0168-1699(02)00120-5

Matlab official web-site. (n.d.). Retrieved from https://matlab.ru/products/matlab

Matsenko, V. V., Sokolov, A. Ya., Kalinin, S. I., Andrianova, F. I., Andreeva, T. A., Ananyin, S. V., ... Kazantseva, L. V. (1999). *Forest fire protection master plan* (Vol. 1). Barnaul, Russia: Rosgiproles. Altay Branch.

Nesterov, V. G. (1949). Combustibility of forests and methods for its determination. Moscow, Russia: Goslesbumizdat. Retrieved from https://www.originlab.com/

Podolskaya, A. S., Ershov, D. V., & Shulyak, P. P. (2011). Application of the method for assessing the probability of occurrence of forest fires in ISDM-Rosleskhoz / Modern problems of remote sensing of the Earth from space, 8. pp. 118-126. (in Russian) RAD Studio. Retrieved from https://www.embarcadero.com/ru/products/rad-studio

Reifsnyder, W. E. (1978). *Systems for evaluating and prediction.*

Ruffault, J., & Mouillot, F. (2017). Contribution of human and biophysical factors to the spatial distribution of forest fire ignitions and large wildfires in a French Mediterranean region. *International Journal of Wildland Fire*, *26*(6), 498–508. doi:10.1071/WF16181

Scheller, R., Kretchun, A., Hawbaker, T. J., & Henne, P. D. (2019). A landscape model of variable social-ecological fire regimes. *Ecological Modelling*, *401*, 85–93. doi:10.1016/j.ecolmodel.2019.03.022

Tolstykh, M. A. (2001). High resolution semi-lagrangian atmospheric model for numerical weather prediction, *Meteorology and Hydrology, 4*. pp. 5-15.

Valendik, E. N., Matveev, P. M., & Sofronov, M. A. (1979). *Large forest fires*. Moscow, Russia: Science.

Viedma, O., Urbieta, I. R., & Moreno, J. M. (2018). Wildfires and the role of their drivers are changing over time in a large rural area of west-central Spain. *Scientific Reports*, *8*(1), 17797. doi:10.103841598-018-36134-4 PMID:30542114

Viegas, D. X., Bovio, G., Ferreira, A., Bovio, G., & Nosenzo, A. (1999). Comparative study of various methods of fire danger evaluation in Southern Europe. *International Journal of Wildland Fire*, *10*(4), 235–246. doi:10.1071/WF00015

Zhdanko, V. A., & Gritsenko, M. V. (1980). Analysis method for forest fire seasons, Practical recommendations. Leningrad.

ADDITIONAL READING

Andreev, Yu. A. (1986). *Laws of distribution of forest fires. Methods and means of fighting forest fires*. Moscow: VNIILM. (In Russian)

Baranovskiy, N. V. (2017). Forest fire danger assessment using SPMD-model of computation for massive parallel system. *International Review on Modelling and Simulations.*, *10*(3), 193–201. doi:10.15866/iremos.v10i3.10570

Baranovskiy, N. V. (2018). Mathematical simulation of anthropogenic load from a linear source using partial differential equations in the context of the forest fires occurrence. *Advances in Differential Equations and Control Processes.*, *19*(3), 237–262. doi:10.17654/DE019030237

Baranovskiy, N. V., & Kuznetsov, G. V. (2017). *Forest fire occurrences and ecological impact prediction: monograph*. Novosibirsk: Publishing House of the Siberian Branch of the Russian Academy of Science. doi:10.15372/FOREST2017BNV

Baranovskiy, N. V., & Yankovich, E. P. (2015). Geoinformation Monitoring of Forest Fire Danger on the Basis of Remote Sensing Data of Surface by the Artificial Earth Satellite. *Journal of Automation and Information Sciences*, *47*(8), 11–23. doi:10.1615/JAutomatInfScien.v47.i8.20

Canadian Wildland Fire Information System. Official site. Access: http://cwfis.cfs.nrcan.gc.ca/home (Available 13 May 2019)

Deeming, J. E., Burgan, R. E., & Cohen, J. D. (1977). *The National Fire Danger Rating System - 1978*. United States Department of Agriculture, Forest Service, General Technical Report INT-39, Intermountain Forest and Range Experiment Station, Odgen, Utah.

European Forest Fire Information System. Official site. Access: http://effis.jrc.ec.europa.eu (Available 13 May 2019)

Ganteaume, A., Camia, A., Jappiot, M., San-Miguel-Ayanz, J., Long-Fournel, M., & Lampin, C. (2013). A review of the main driving factors of forest fire ignition over Europe. *Environmental Management*, *9*(3), 651–662. doi:10.100700267-012-9961-z PMID:23086400

KEY TERMS AND DEFINITIONS

Anthropogenic Load: Different human activities on forested territories lead to forest fire occurrence and characterized by presence of fire sources.

Forest Fire: Uncontrolled aerothermochemical phenomenon characterized by step-by-step mechanism which includes following stages: inert heating, moisture evaporation, high temperature terpens evaporation, dry organic matter pyrolysis, flammable combustion and smoldering.

Mathematical Simulation: The production of a computer model of forest fire conditions and prerequisites, especially for the purpose of study.

Monitoring: Monitoring refers to the periodic calculation of the parameters of forest fire danger with a portion of information available in real time.

Number of Forest Fires: Calculated (predicted) number of forest fire over controlled forest covered territory due to anthropogenic load and lightning activity.

Prediction: Under the prediction of forest fires is the calculation of the parameters of forest fire danger with a certain projection in advance in order to have enough time to anticipate an emergency. The calculation in this case is carried out in a mode ahead of the real time of the development of the catastrophe - the occurrence of a forest fire.

Chapter 5
Dynamic Spatial-Distributed Fire Risk Analysis

Maryna Vitalievna Zharikova
https://orcid.org/0000-0001-6144-480X
Kherson National Technical University, Ukraine

ABSTRACT

The chapter develops a dynamic spatial-distributed model of forest fire risk. The structure of forest fire risk is represented by two interrelated groups of components describing a potential of forest fire and describing valuable objects being under fire influence. The concept of fire risk which contains the probability of forest fire occurrence, its intensity, and effect, is extended using the threat as the additional prognostic spatial-temporal component that has a predictive property and allows forecasting the possibility of losses at any time. The model of fire risk is based on three stages: potential risk, the source of which is described by fire danger; risk of threat of active forest fire which hasn't covered valuable object yet; risk of destruction when the active fire has covered valuable object. Representation of risk as a process based on three stages allows describing the dynamics of risk in real-time systems, getting insight into risk nature, as well as diagnosing the situation in real time.

BACKGROUND

The first decade of the XXI century is characterized by activation of forest fires that cause damages, injuries, and even deaths. Expanding the scope of forest fires occurs due to such factors as increasing the population, urbanization, environmental degradation, global warming, etc. Thus, at present, the problem of timely and grounded decision making in real-time forest fire response, which allows mitigating fire damage, is in the focus of attention of researchers (Sherstjuk, V., Zharikova, M. & Sokol, I., 2018).

Decision making in a forest fire response is directly related to forest fire risk assessment. So, there is a growing interest in developing risk-oriented decision support systems for real-time forest fire response. Fire risk is a potential for a valuable object to take losses as a result of a forest fire.

DOI: 10.4018/978-1-7998-1867-0.ch005

Figure 1. Fire management cycle

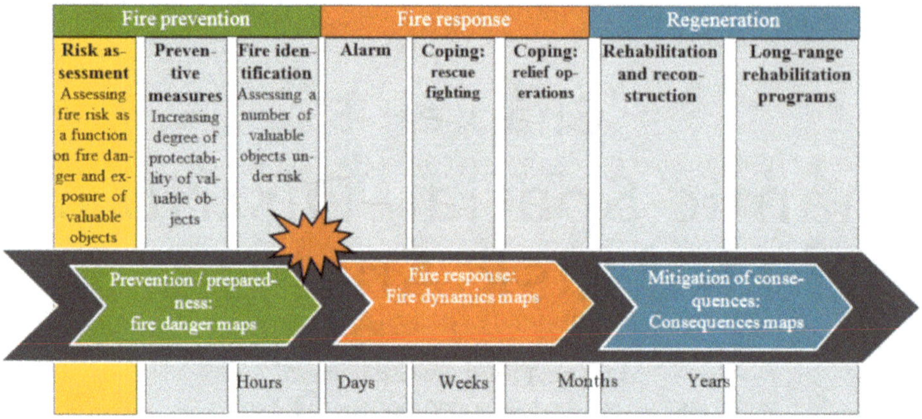

Taking into account the fact that risk is a basis for decision making for fire response, it's interesting to consider a place of risk assessment in a full fire management cycle (Fig. 1) (Kappes, 2011). This cycle contains three main phases:

1. Fire prevention;
2. Fire response;
3. Regeneration after a fire.

Currently, risk assessment is understood as a static value describing a potential effect from the impact of fire on the valuable objects. Accordingly, risk assessing is performed only in the frame of the first phase of the fire management cycle (Fig. 1). Whereas in real-time systems, end-to-end assessment of fire risk not only in the prevention phase but also in the respond and regeneration phases is of great interest.

There is a vast literature on forest fire risk assessment. The majority of the authors propose to assess risk quantitatively, using probabilistic methods of simulation modeling not suitable for the assessment of fire risk in real time, where probability is out of the question (Zharikova, M., 2018). Such methods of risk assessment lead to a high computational complexity and reduce a performance of a decision support system significantly, which is not allowed in the real-time systems. To increase the representativeness of statistical samples the big areas and the big time intervals are taken into account. On top of it, in existing works, no attention is paid to the assessment of fire risk dynamics in real time. Instead of this, a static assessment of risk components is considered.

Risk is widely viewed as a combination of probability of occurrence and the consequences of indefinite future events (Thompson, M., Haas, J., Gilbertson-Day, J., Scott, J., Langowski, P., Bowne, E. & Calkin, D., 2015; Thompson, M., Calkin, D., Finney, M., Ager, A. & Gilbertson-Day, J., 2011; Thompson, M., Zimmerman, T., Mindar, D. & Taber M., 2016).

Although the terminology related to fire risk has experienced a long term evolution and has its traditions, it is still controversial.

There is no unique definition of the term "risk" and the notions coming from it such as "danger", "threat", etc. Many researchers focus on the losses describing risk as to the interrelation between the physical probability of fire occurrence and potential losses for people and the environment (Mimbrero, M.,

2015). For example, in (Kaplan, S., Garrick, B.J., 1981) risk is defined as a combination of uncertainty related to fire occurrence and losses; in (Hohenemser, C., Kates R. & Slovic, P., 1985) risk is represented as a quantitative measure of the consequences of danger occurrence expressed using conditional probability of losses occurrence; in (Royal Society, 1992) risk is measured as a probability of occurrence of a definite adverse event over a definite period of time.

Society for risk analysis (Aven, T., Ben-Heim, Y., Andersen, H., Cox, T., Droguett, E., Greenberg, M., Guikema, S., Kroeger, W., Renn, O., Thompson, K. & Zio, E., 2015) gives the following definitions of the notion of risk:

1. Risk is a potential for realization of unwanted, negative consequences of an event;
2. Risk is an uncertainty about the unwanted event and severity of the consequences of its activity concerning something that humans value.

Based on the above stated, any risk is connected with the potential consequences of adverse event. In all above-mentioned works, the losses are in the focus of attention but the issues of their theoretical description are neglected. Such notions as losses and adverse consequences assume definite value attached to the objects at risk.

The objects having value (valuable objects) are buildings, the elements of infrastructure, the elements of ecosystem, and even human lives (Thompson, M., Calkin, D., Finney, M., Ager, A. & Gilbertson-Day, J., 2011; Thompson, M., Haas, J., Gilbertson-Day, J., Scott, J., Langowski, P., Bowne, E. & Calkin, 2015; Thompson, M., Zimmerman, T., Mindar, D. & Taber M., 2016). In (Rose, E., 1998) risk is represented as a situation or event where something that has value for people is put at threat with an uncertain result. This work has extended understanding of risk concept. The issue of assessing the object value as a component of the total risk assessment has been studied also in the works (Aven, T., Ben-Heim, Y., Andersen, H., Cox, T., Droguett, E., Greenberg, M., Guikema, S., Kroeger, W., Renn, O., Thompson, K. & Zio, E, 2015; Aven, T., 2011; Aven, T., 2013; Aven, T., 2015; Boholm, A. & Coevellec, H., 2011; Hansson, S., 2010; Hardy, C., 2007).

The key feature of the risk related to a forest fire is a consideration of fire characteristics. The analysis of literature (Gallina, V., 2016; Gregory, V. & Vidal, A, 1999) has allowed distinguishing two groups of components of the risk concept:

1. The components describing a potential of risk source, namely forest fire;
2. The components describing the valuable objects being under the influence of forest fire.

The vision of these two groups of components has been changing over time determining the evolution of fire risk concept. In the easiest understanding fire risk is defined based on the potential of fire occurrence and spreading without a focus on potential losses. For example, Food and Agriculture Organization of the United Nations defines forest fire risk as a probability of fire occurrence based on the existence of causal factors, such as lightning or anthropogenic load (Food and Agriculture Organization of the United Nations, 2017). This term doesn't allow to consider potential losses from a forest fire.

The next stage of developing the terminology connected with forest fire risk was two-component risk assessment scheme containing the potential of fire combustion and spreading (fire danger) and potential losses that were not taken into account within the previous approach. Two-component scheme of forest fire risk assessment took into account the valuable resources influenced by fire. Defense of these valu-

Figure 2. Fire risk structure

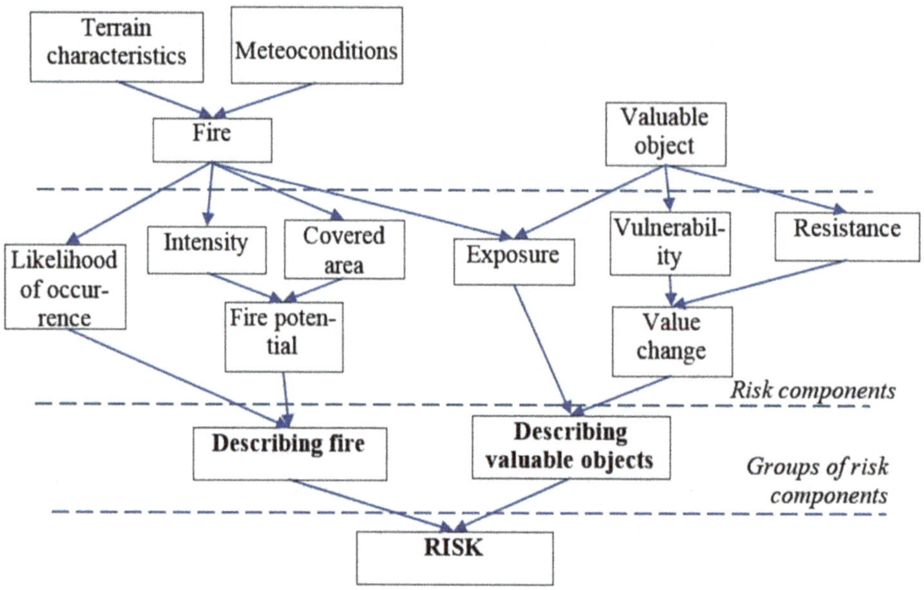

able resources is crucial in decision making for forest fire response, especially within urbanized areas. At that, a potential of fire beginning and spreading is assessed using probability and is called fire danger (Bachmann, A. & Allgower, B., 1998; Bachmann, A. & Allgower, B., 2001; Mimbrero, M., 2015).

This scheme was extended by an additional component that described fire intensity. According to the new scheme, the risk was also assessed in the context of valuable resources being influenced by fire but as a combination of three components such as fire intensity, a probability of burning with a certain intensity and a potential influence of fire with this intensity. A basis of three-component scheme was established by Finney (Finney, M., 2005; Hyde, K., Dickinson, M., Bohrer, G., Calkin, D., Evers, L., Gilbertson-Day, J., Nikolet, T., Ryan, K. & Tague, C., 2013). The special aspect of this approach was the fact that risk was assessed quantitatively using statistical analysis methods. The risk for a definite valuable object was assessed summing the products of probability of fire of a definite intensity and a potential influence of this fire on the valuable object expressed as a change of object's value that can be whether positive or negative. The three-component scheme is widely used currently in different modifications. For example, the authors of some modern works in addition to fire intensity take into account other fire characteristics such as flame height (Thompson M.P., 2015).

Currently, the above-mentioned groups of components are considered comprehensively (Fig. 2).

The first group of risk components describing fire contains:

- Likelihood (possibility or probability) of fire occurrence or spreading.
- Fire potential (intensity and terrain coverage).

The second group of risk components describing valuable objects contains:

- Valuable objects' exposure to fire that is the spatial juxtaposition of forest fire likelihood and intensity metrics with the location of valuable objects (Goldammer, J., Mitsopoul os, I., Mallinis, G. & Woolf, M., 2017).
- Valuable objects' vulnerability that expresses the potential damage from a forest fire (Goldammer, J., Mitsopoulos, I., Mallinis, G. & Woolf, M., 2017).
- Valuable objects' resistance that describes a potential of valuable object to save its value under the influence of fire.
- Dynamics of objects' value under the influence of fire.

The likelihood is typically interpreted as a probability based on the statistics of fire occurrence and spreading, as well as meteorological, anthropogenic and other conditions. A probability of spreading depends on a probability of occurrence, as well as on fire dynamics described typically using simulation modelling methods (Miller, C. & Ager, A., 2013; Hyde, K., Dickinson, M., Bohrer, G., Calkin, D., E vers, L., Gilbertson-Day, J., Nikolet, T., Ryan, K. & Tague, C., 2013).

The above-mentioned review of risk assessment approaches allows for the conclusion that existing approaches enable obtaining static risk assessment only at the stage being preliminary to the fire and can't be applied in real-time decision support systems. Most currently used methods of fire risk assessment are quantitative based on statistical data analysis. Since acquisition of a sufficient amount of statistical data describing real fires is impossible, data samples are taken within big areas and in big periods. Non-representativeness of data samples is compensated by simulation modeling methods connected with conducting a big number of computational experiments. For instance, the Monte-Carlo method is usually used for assessing the probability of fire occurrence. The methods of quantitative risk assessment have one main drawback, namely big computational complexity.

Risk assessment requires a big number of steps to process geographical information, at each of which we need to calculate fire characteristics and valuable objects' vulnerability. Big amount of calculations generates a potential for mistakes, and the time required for calculations complicates using these characteristics in real-time decision support systems. As a result, the risk assessments proposed in the current works are static values calculated in advance, before the fire, and can be used for fire prevention rather than for real-time decision making, where probability is out of the question, and decisions are made in time limit conditions.

In general, the models, in which fire model goes together with risk dynamics model, and the risk is represented as a spatially-distributed process, where its end-to-end assessment is performed, are absent for now.

Although some scientific works exist concerning fire risk assessment, we don't have a comprehensive investigation of this problem. The current fire risk concept is far from giving a complete picture of fire risk as a spatially-distributed process.

In this chapter, the approach to dynamic spatially-distributed risk analysis for real-time decision support systems is considered.

FIRE RISK CONCEPT

Fire risk analysis is considered within a certain territory. Let us consider the territory as an open connected subspace X of three-dimensional Euclidean space endowed with the topological properties (Zharikova, M., 2018). The subspace X is discretized by a grid C of isometric cubic cells $c \in C$. Assume that a cell $c \in C$ is a spatial homogeneous object of minimal size (Zharikova, M., 2018; Allam, A., Bakeir, M. & Abo-Tabl, E., 2008; Zharikova, M. & Sherstjuk, V., 2015; Zharikova, M., Sherstjuk, V. & Baranovskij, N., 2015).

A model of the territory with induced norm and metrics is called territorial system (TS) Ξ. A set of natural and artificial cartographic objects $O = \{o_1, o_2, \ldots, o_n\}$ is distinguished within the TS. Each object at each time moment is characterized by its state in the form of a set of parameters' values. Among cartographic objects we consider valuable objects being of value for people. A state of the object o_i at time moment t is denoted by $State^t_{o_i}$. From outside the TS is affected by environment E, a state of which at time moment t is denoted by $State^t_E$. A manifestation of the disruptive influence of the environment is a set of fire scenarios F. In fire conditions, the objects can take losses.

Fire scenarios F_j, $j = 1,.., J$ can be the sources of risk for valuable objects. Each scenario describes a spread of fire from a certain source u_j within TS at the certain parameters of the environment. Each scenario at a certain time moment can be potential or active. At the moment of activation (realization) of the scenario F_j a some valuable objects can be affected by it resulting in decreasing of their value. Such objects are called vulnerable and they are the receivers of risk from the source F_j.

Let us suppose that at time moment t a set F consists of the individual fire scenarios $F = \{F_j\}$ each of which are characterized by the state $State^t_{F_j}$. Fire state at time moment t is a set of values of its parameters describing its spatial location and potential. A model of TS with the models of fire scenarios within this system is called natural-territorial system (NTS).

NTS containing three active scenarios F_1, F_2 and F_3 with the sources u_1, u_2 and u_3 respectively, two potential fire sources u_4 and u_5, as well as three valuable objects o_1, o_2 and o_3, one of which is engulfed in fire, is represented in Fig. 3.

The objects are connected with the fire with the help of a set of relations $\Re$ forming the internal structure of NTS. Thus, in fire conditions NTS can be represented as a quadruple:

$$\Omega = \left\langle \Xi, O, \Re, E, F \right\rangle.$$

NTS state at time moment t is a set of the states of objects, fire scenarios, and environment:

$$State^t_\Omega = \left\{ \bigcup_{o_i \in O} State^t_{o_i}, \bigcup_{F_j \in F} State^t_{F_j}, State^t_E \right\}.$$

Let us introduce a notion of "value" ν_i for each cartographic object $o_i \in O$ that defines its utility. Let us assume that a certain level of interest c^* is set for a decision maker. At that, all objects o_i having value assessment $\nu_i < c^*$ at time moment t are excluded from the consideration.

Figure 3. Components of NTS

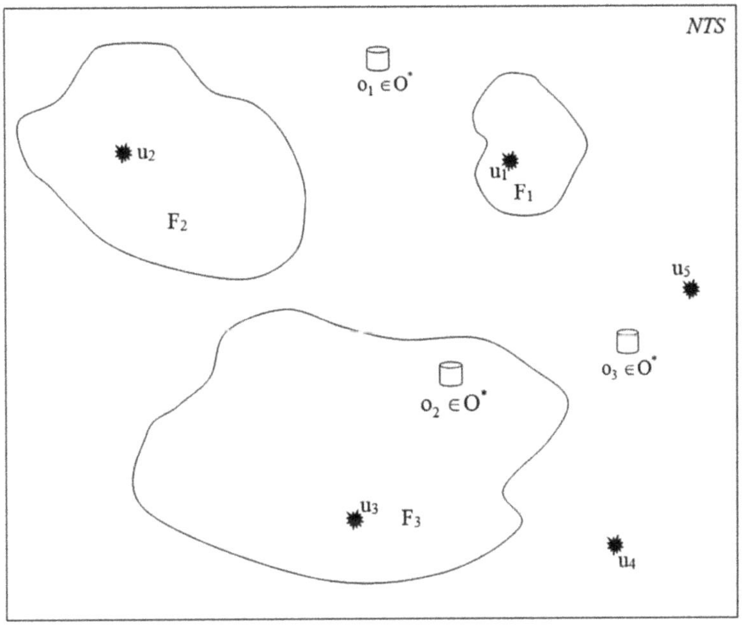

The target object is an object $o_i \in O$ having a value $\nu_i(t)$ such that $\nu_i \geq c^*$ at time moment t. Decision-making task at time moment t is solved concerning a set of target objects $O^*(t) = \{o_i\}_{i=1}^{m} \in O$ in NTS such that

$$(\forall o_i \in O)\left(\left(o_i \in O^*(t)\right) \Leftrightarrow \left(\nu_i(t) \geq c^*\right)\right).$$

Each potential fire source j at time moment t is characterized by danger $\mu_j(t)$. When danger is materialized in the form of fire F_j with a potential S_j a threat is created for each valuable object o_i with the value $\nu_i(t)$ (as well as for each area of the terrain) being at a certain distance from the fire contour. At that, a notion of danger is out of sense. A threat is characterized by fire potential, as well as by spatial and temporal distance between a valuable object and fire contour that characterizes a level of threat for this valuable object $\varsigma_{ji}(t)$.

Risk in the context of this work is an assessment of interrelation between fire scenarios F_j (risk sources), and vulnerable objects $o_i \in O^*$ (risk receivers), and its assessment is used for diagnosis of the situation in the NTS. The risk from one fire source is individual. Individual risk is assessed for a definite valuable object $o_i \in O^*$ and is characterized by a possibility for the object to take losses from the fire F_j with the source u_j.

A set of threat levels is described by spatial and temporal metrics ξ_X and ξ_T, and can create zones around fire contour describing the distribution of the threats according to their metrics.

Figure 4. Types of influences over valuable objects

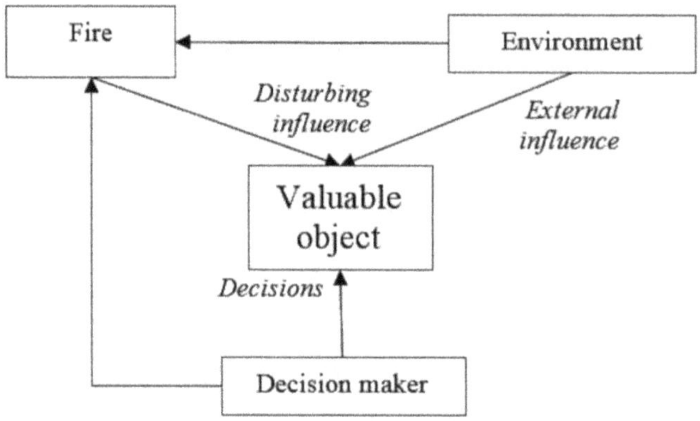

A vulnerable object concerning a certain scenario is a valuable object that is a subject for decreasing the value as a result of this scenario realization. This means that the valuable object is situated at such distance from the scenario that it can be affected by the scenario in a certain time.

A set of vulnerable objects with respect to the scenario F_j is denoted by O^{*j}, at that $(\forall j = 1,..,J) O \subseteq O^{*} \subseteq O^{*j}$. These objects $o_j \in O^{*j}$ require defense while the scenario F_j is realizing, consequently, risk assessment R_{ij} should be considered for them.

Under the conditions of realization of the scenario F_j the states of the objects $o_i \in O^{*j}$ change dynamically influenced by fire. Influence over the object is a certain value of a certain parameter describing the environment or fire scenario, to which the object responds changing the values of some parameters of its state. Influences over objects in NTS can be generated by the following sources (Fig. 4) (Zharikova, M., 2018):

1. Environment (external influence);
2. Fire (disturbing influence);
3. Decision making (decisions aimed at a compensation of disturbing influence).

Thus, for each vulnerable target object o_i being in fire conditions at each time moment t we can determine a level of threat $\varsigma_{ji}(t)$ and value $\nu_i(t)$. From knowing fire potential S_j, it's possible to assess risk $R_{ji}(t)$ for this valuable object in the form of function $R_{ji}(t) = f(\varsigma_{ji}(t), \nu_i(t), S_j)$.

If the object o_i isn't in fire conditions, the risk from potential source *j* is assessed taking into account danger $\mu_j(t)$. In general, the risk for valuable object o_i from the fire F_j at time moment t is a function:

$$R_{ji}(t) = f(\mu_j(t), \varsigma_{ji}(t), \nu_i(t), S_j),$$

where $R_{ji}(t)$ is a risk for the object *i* from the fire *j* at the time moment t, $\varsigma_{ji}(t)$ is a threat created by the fire *j* to the object *i* at the time moment t.

Figure 5. Components of individual risk for valuable object from potential scenario

Figure 6. Components of individual risk of threat for valuable object

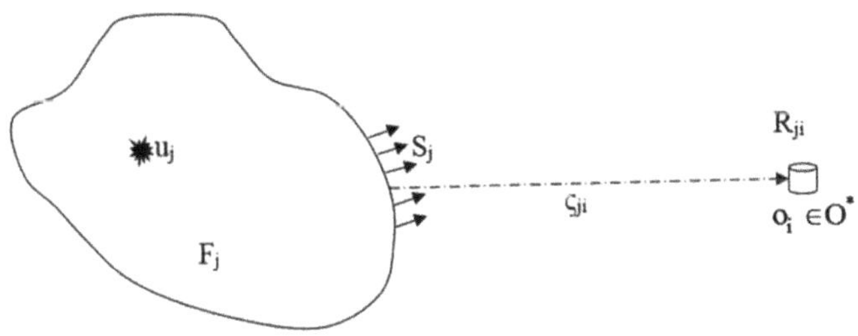

Fire source u_j and the scenario associated with it can be potential described by danger μ_j characterizing a possibility of fire occurrence in the point u_j. The risk from a potential source is called potential (Fig. 5).

In case of fire F_j situation a set of vulnerable objects $O^{*j} \in O^*$ are allocated. For these objects a possibility to be affected by fire exists. At that, a notion of danger is out of sense. For each vulnerable object $o_i \in O^{*j}$ a dynamic characteristic of threat ς_{ji} is determined. Losses for the object o_i as a result of fire F_j depend on a fire potential. Risk from active scenario for vulnerable object that has not engulfed in flame yet is called risk of threat (Fig. 6).

When a valuable object is engulfed in fire scenario, a notion of threat is out of sense, and the object is under the adverse influence of scenario resulting in decreasing of its value. Such risk is called risk of destruction (Fig. 7).

Thus, at any time moment a certain scenario F_j can be potential or active. A source of potential scenario is described by danger. At that, there is no threat for valuable objects, and a potential risk exists. The active scenario arises when a danger is materialized in the form of fire spreading from a source

Figure 7. Components of individual risk of destruction for valuable object

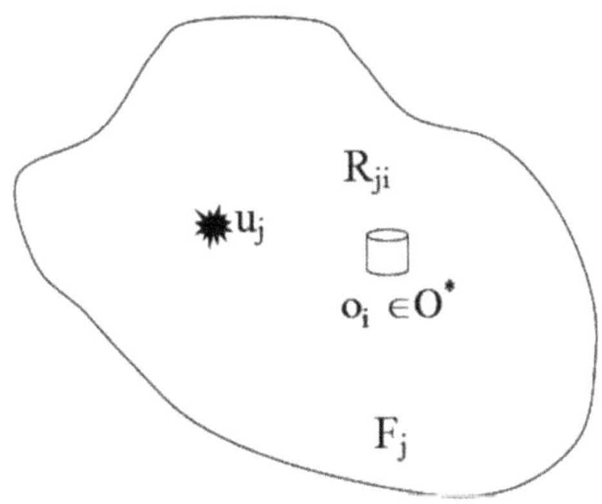

and can pose a risk of threat or destruction for a certain vulnerable object. A scenario that doesn't cover vulnerable object poses a threat for the object and, as a result, a risk of threat. Scenario covering vulnerable object leads to a change of value for the object and poses a risk of destruction.

Thus, the dynamics of individual risk for a certain valuable object in a general case goes through three stages:

1. Potential risk
2. Risk of threat
3. Risk of destruction

A task of decision making for fire response is to minimize the influence of the last stage of risk dynamics.

A potential risk is characterized by danger μ. The danger is a risk component describing a potentiality of occurrence of fire of a certain type and intensity, within a certain territory, as well as at a certain time moment. The danger becomes active (materializes) when fire arises within a certain area of terrain. After fire occurrence, a notion of danger loses its meaning for this area of terrain. Danger assessment allows determining the areas of terrain where the occurrence of fire of a certain type and intensity is most likely at a certain time moment. The danger is not connected with objects' value. It determines the temporal and spatial probability of fire occurrence.

Danger has the following characteristics:

- Danger describes a potential source of fire of a certain type and intensity within a certain area of terrain at certain time moment.
- The danger is expressed as a probability.
- The time of danger existing – before fire occurrence.

The materialized danger for a certain source means that a scenario connected with this source has become active and poses a threat for vulnerable objects or causes a process of changing the value (destruction of the object). A threat ς_{ji} for the object o_i from the fire F_j arises when there is a possibility for this object to be affected by this fire. The objects for which it's reliably known that they will not be affected by a certain fire are invulnerable. There is no threat to them.

Thus, a threat is a dynamic risk component describing spatiotemporal relations between fire contour and vulnerable object. A threat arises at the moment of danger materialization and loses its sense at the moment when the fire covers the valuable object.

Threat characteristics:

- The threat is an assessment of spatiotemporal relations between a certain fire scenario and a certain vulnerable object.
- The time of threat existing – from the moment of fire occurrence to the moment when fire covers the valuable object.

A threat is created by fire scenario and is represented in the form of a set of spatial zones differentiated concerning the threat levels and situated around fire contour. Each zone determines the boundaries of that part of NTS with a threat of a certain level. Let us denote a set of threats by $\{D_1, D_2, \ldots, D_k\}$, where k is a number of the threat zones ranked in the order of increasing the threat level. Thus, a threat is a risk component by which risk concept proposed in this chapter differs from existing concepts.

Risk R_{ji} is an integral assessment characterizing potentiality for the object o_i to take losses at the time of the fire F_j of a certain type and intensity. Threats and risks are assessed for vulnerable valuable objects that require defense in fire conditions. For the other objects, there is no need to consider these assessments especially in time limit conditions.

Risk characteristics:

- Risk is an assessment of the possible negative influence of a certain fire on a certain valuable object.
- Risk is a function of danger, threat, fire intensity (fire potential) and potential losses for a valuable object.
- In the course of its dynamics, the risk goes through three stages such as potential risk, risk of threat and risk of destruction.

A stage of potential risk is characterized by danger connected with a potential fire source and exists before the moment of fire occurrence. Risk of threat is characterized by threat describing interrelation between active fire and vulnerable valuable object and exists from the moment of fire occurrence until the moment when the object is covered by the fire. Risk of destruction exists from the moment when the object is covered by the fire until the moment of the total destruction of the object and is characterized by a decrease in object value. Dynamics of object value decrease depend on fire potential, as well as on vulnerability and resistance of the object.

In the case of several fire sources simultaneously influencing a vulnerable object, multi-threat and multi-risk are created for this object:

$$\varsigma_i = \mathop{\oplus}_{j=1}^{n_j} \varsigma_{ji}\left(t\right),\ R_i = \mathop{\oplus}_{j=1}^{n_j} R_{ji}\left(t\right),$$

where n_j is a number of the risk sources for the object i, $\oplus$ is an absorption operation.

Fig. 8 shows five fire scenario sources ($u_1, \ldots, u_5$), three of which are active (u_1, u_2, u_3) and two are potential (u_4, u_5), as well as two valuable objects (o_1 and o_2). At that, both valuable objects belong to the sets of vulnerable valuable objects concerning the notified five scenarios: $o_1, o_2 \in \bigcap_{j=1}^{5} O^{*j}$.

Three active scenarios F_1, F_2 and F_3 with the sources u_1, u_2 and u_3 respectively pose threats for the objects o_1 and o_2, where ς_{ji} is a threat posed by active scenario j for the object i. Potential sources u_4 and u_5 pose a danger of occurrence of active scenarios μ_4 and μ_5 respectively.

Figure 8. Multi-risk components

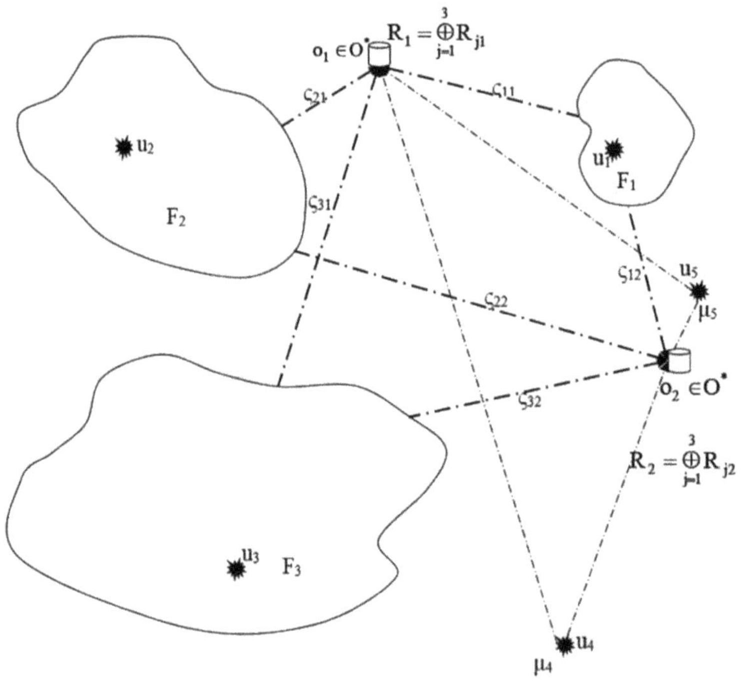

Let us consider a set of target objects $O^*(t) = \{o_i\}$ in fire conditions such that:

1. a set $O^*(t)$ in fire conditions is a changeable part of NTS (a structure of the set can change dynamically); a condition of $O^*(t)$ membership for the object o_i at time moment t is $\nu_i(t) \geq c^*$;
2. each o_i has cartographical referencing and the state changing dynamically;
3. each o_i has the value $\nu_i(t)$ that also changes dynamically;
4. in the conditions of the fire F_j, the object o_i at each time moment falls in some threat zone, in case of a set of fires threat zones can superimpose on one another and form multi-threat zones;
5. danger, value and multi-threat assessments for each o_i allow determining multi-risk R_i for it in conditions of a set of fires $F = \{F_j\}$, where each fire F_j has a certain intensity S_j.

For each valuable object at time moment t the risk can be represented as a point in k-measured space of parameters. A trajectory of risk $R_i(t)$ for a valuable object o_i is infinite sequence $\{R_i(t_1), R_i(t_2),\ldots, R_i(t_j),\ldots\}$ of multi-risk values at time moments $\{t_1, t_2,\ldots t_j,\ldots\}$.

Spatially-Distributed Dynamic Risk Analysis

Usually, the quantitative risk assessment is represented as a multiplication of event probability and losses. At that, the risk probability is calculated based on the frequency of event occurrences (Andreu, A. & Hermansen-Baez, L., 2008). In the context of this chapter, the subject of risk analysis is the estimation of the chances of losses resulting from the involvement of the valuable object in the fire that has been

Table 1. Ordinal scale for risk assessment

ς_{R0}	«absence of risk»
ς_{R1}	«low level of risk»
ς_{R2}	«high level of risk»
ς_{R3}	«critical level of risk»

occurred by the time of consideration and is spreading in time and space. In this case, the frequency of events occurrence is not a reliable base for the estimation of the chance. This leads to the conclusion that the conditional probability estimation based on the events frequency is not appropriate for the real-time systems. The possibility of assessments based on fuzzy or rough methods is much more reliable. In this case, the chance estimation can be represented as a conditional possibility that a target object is on fire.

Another problem is the estimation of losses for the valuable objects, which can not be estimated in monetary terms. They include the losses of historical, architectural, cultural values. In (Zharikova, M. & Sherstjuk, V., 2017; Zharikova, M, 2018) the authors propose the solution of this problem using quantification of quantitative estimations with the help of ordered scales and switch to the qualitative assessment. In the same way, the qualitative danger and threat assessments can be obtained.

To obtain a qualitative risk assessment for an object o_i from a fire F with a source u_F, we should first obtain a standardized quantitative risk assessment as a product of the normalized component values: $\breve{\mu}_F, \breve{S}_F, \breve{\varsigma}_{o_iF}, \Delta \breve{V}_{o_iF}$. It's difficult to present a threat assessment in the form of one normalized value. It can be presented as an interval value $\hat{\breve{\varsigma}}_{o_iF} = \left[\underline{\hat{\breve{\varsigma}}_{o_iF}}, \overline{\hat{\breve{\varsigma}}_{o_iF}}\right]$.

Taking into account the fact that the threat assessment is an interval value, the risk assessment will also be an interval value (Zharikova, M. & Sherstjuk, V., 2017; Zharikova, M, 2018):

$$\hat{\breve{R}}_{o_iF} = \breve{\mu}_F \times \breve{S}_F \times \hat{\breve{\varsigma}}_{o_iF} \times \Delta \breve{V}_{o_iF} = \left[\breve{\mu}_F \times \breve{S}_F \times \underline{\hat{\breve{\varsigma}}_{o_iF}} \times \Delta \breve{V}_{o_iF}, \breve{\mu}_F \times \breve{S}_F \times \overline{\hat{\breve{\varsigma}}_{o_iF}} \times \Delta \breve{V}_{o_iF}\right]$$

In the case of several fire sources that simultaneously affect the valuable object o_i, a multi-risk is created for this valuable object:

$$\hat{\breve{R}}_{o_i} = \bigoplus_{j=1}^{n_i} \hat{\breve{R}}_{o_iF_j},$$

where n_i is a number of risk sources for the valuable object o_i.

Figure 9. Surfaces of risk and value

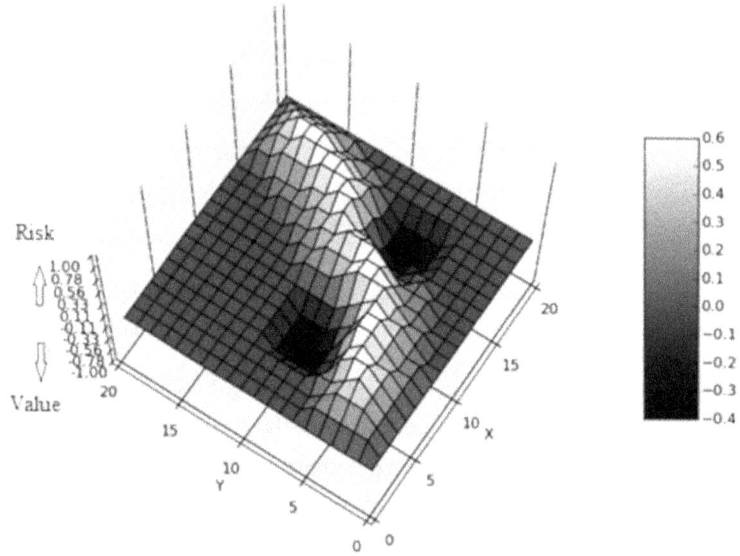

The normalized interval quantitative values of the multi-risk can be converted into qualitative values by defining the corresponding ordinal scale $\c{C}_R = \{\varsigma_{R0}, \varsigma_{R1}, \ldots \varsigma_{Rz}\}$ (Table 1) with the partial-order relation $\preccurlyeq_R$ induced on it, such that $\varsigma_{R0} \preccurlyeq_R \varsigma_{R1} \preccurlyeq_R \ldots \preccurlyeq_R \varsigma_{Rz}$ (Zharikova, M, 2018).

For the direct transformation of the normalized interval quantitative estimation of multi-risk into a qualitative assessment, we define a function $\Re$ that implements the mapping $\Re : \hat{\bar{R}} \rightarrow \c{C}_R$. Using this mapping, the quantitative interval value of the multi-risk $\hat{\bar{R}}_{o_i}$ for the valuable object o_i uniquely matches the qualitative assessment R'_{o_i} concerning the scale $\c{C}_R$.

A similar approach can be used to assess the multi-risk for any cell A_m of the universe ! including the cells that do not belong to any of the objects of a set of valuable objects $O*$ (in this case, the value of a cell can be determined by the elements of its state vector, for example, a cell with coniferous vegetation can have a value equal to the cost of wood, based on the product of the existing volume of wood in the cell at its price). Multi-risk can be expressed quantitatively by the interval $\hat{\bar{R}}_m$ or qualitatively as R'_m.

A multi-risk trajectory for a certain cell c_m during time interval $\left[t_j, \ldots t_k\right] \in T$ is a continuous sequence $\left[\hat{\bar{R}}_m(t_j), \ldots, \hat{\bar{R}}_m(t_k)\right]$.

Thus, the NTS at the time *t* is characterized by an integrated dynamic spatially-distributed estimation of the multi-risk in fire conditions:

$$R_\Omega^t = \left\{R_i(t) \middle| \forall o_i \in O^*(t)\right\}.$$

Figure 10. Cut surface for allocation of valuable objects

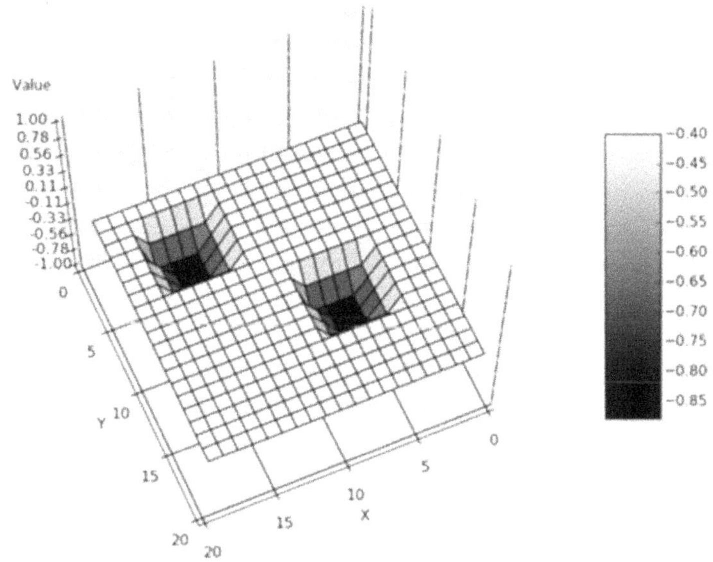

To visualize information about the integrated spatially-distributed estimation of the multi-risk in the NTS for a certain time t, a risk surface that reflects the normalized assessment of the level of risk for each cell can be represented. This surface visually reflects the convex areas of the terrain with a maximal risk. However, in some situations, the areas with the highest risk do not need the attention of the decision maker. The areas with the maximum value among them should be selected. Therefore, we have drawn also the value surface in the same coordinate system, but the increase in value will be displayed in reverse order: from 0 to -1. The areas with maximum value are displayed with the concave areas on the surface (Fig. 9).

To select areas requiring the attention of the decision maker, it is necessary to draw two cut surfaces that are perpendicular to the *Z* axis and establish the critical values of value and risk.

Figure 10 shows the cut surface for the surface of values. All areas with the value below this surface, have critical value. The same cut surface is drawn for the surface of risk. All areas with the risk assessment being above this surface are in critical risk conditions. The intersection of a set of areas having a critical value and ones being in critical risk requires the attention of the decision maker. By varying the arrangement of cutting surfaces, we can narrow or expand the set of valuable objects $O^*(t)$ that need protection.

Experiment Results

The developed model of spatial-distributed dynamic fire risk analysis was used in the spatial real-time decision support system (DSS) providing the geospatial risk analysis in real time forest fire situations. To examine the validity and the efficiency of the proposed hybrid diagnosis method, an experiment has been conducted. It has been based on the collected retrospective information on the series of large-scale forest fires, which had been taken place in Tsyurupinsk and Golopristan forestries, Ukraine, on July

Figure 11. Influence of the cell size on the diagnosis time

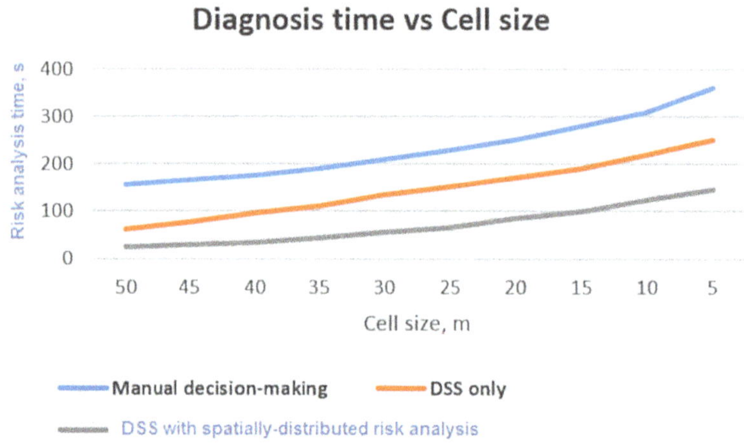

20-31, 2007. The proposed method has been implemented using Visual C++ and tested on a computer based on the Pentium i5-7400 3-3,5 GHz processor and 16 GB RAM. The obtained software has been integrated in the spatial real-time DSS Forest Project aimed to providing a geospatial analysis of risks in real time fire situations.

We have modeled the ongoing processes of the situation diagnosis during the forest fire propagation performed by the decision maker manually (a), as well as performed by own tools of the spatial DSS (b), and using the spatial-distributed risk analysis software integrated into the spatial DSS (c).

During the experiment, we have evaluated the total time needed for risk analysis. The experiment results depend on the cell size within the spatial model. Therefore, we have varied the cell size to investigate its impact on the risk analysis time and compared the results obtained using DSS with the results obtained using manual decision-making. The results of the experiment are depicted in Fig. 11.

Thus, it can be concluded that the proposed method provides sufficient performance for the cell size 4 m and above, so it is acceptable for solving the practical forest fire fighting problems in the real time GIS-based DSS.

FUTURE RESEARCH DIRECTIONS

The direction of further research is related to the use of modern methods of using a heterogeneous team of unmanned aerial vehicles and remote sensing techniques, which will allow to significantly improve the quality of fire recognition and the accuracy of its parameters estimation for real-time fire risk assessment. Multi-UAV-based approach will allows not only estimate the spatial position of the fire front but using UAV based remote sensing provides indirect estimates of valuable parameters necessary for the fire-spreading model.

CONCLUSION

In the chapter, a dynamic spatial-distributed model of forest fire risk is proposed. The structure of forest fire risk is represented by two interrelated groups of components such as the components describing a potential of forest fire and the components describing valuable objects being under forest fire influence.

The concept of fire risk which contains probability of forest fire occurrence, its intensity and effect, is extended using threat as additional prognostic spatial-temporal component that has predictive property and allows forecasting the possibility of losses at any time, as well as allows assessing the risk to certain valuable objects in forest fire conditions in real time.

A threat is represented in the form of a set of spatial zones differentiated concerning the threat levels and situated around fire contour. Each zone determines the boundaries of that part of NTS with a threat of a certain level. This allows diagnosing the situation in NTS in fire conditions for decision making.

The model of fire risk is linked to valuable objects and is based on three following stages: potential risk, the source of which is described by fire danger; risk of threat of active forest fire which hasn't covered valuable object yet; risk of destruction when active forest fire has covered valuable object. Representation of risk as a process based on three stages allows describing the dynamics of risk in real-time systems, getting insight into risk nature, as well as diagnosing situation in real-time decision support systems which stimulates making more grounded decisions in forest fire conditions.

The concept and the model proposed in the chapter can be taken as a basis for real-time risk-oriented geoinformation decision support systems for forest fire response.

REFERENCES

Allam, A., Bakeir, M., & Abo-Tabl, E. (2008). Some methods for generating topologics by relations. *Bulletin of the Malaysian Mathematical Sciences Society*, *2*(31), 35–45.

Andreu, A., & Hermansen-Baez, L. (2008) Fire in the South 2: the southern wildfire risk assessment. A report by Southern Group of State Forester.

Aven, T. (2011). On different types of uncertainties in the context of the precautionary principle. *Risk Analysis*, *31*(10), 1515–1525. doi:10.1111/j.1539-6924.2011.01612.x PMID:21477085

Aven, T. (2013). On how to deal with deep uncertainties in a risk assessment and management context. *Risk Analysis*, *33*(12), 2082–2091. doi:10.1111/risa.12067 PMID:23656628

Aven, T. (2015). Comments to the short communication by Jan Erik Vinnem and Stein Haugen titled "Perspectives on risk and the unforeseen". *Reliability Engineering & System Safety*, *137*, 69–75. doi:10.1016/j.ress.2015.01.001

Aven, T. (2015). Implications of black swans to the foundations and practice of risk assessment and management. *Reliability Engineering & System Safety*, *134*, 83–91. doi:10.1016/j.ress.2014.10.004

Aven, T., Ben-Heim, Y., Andersen, H., Cox, T., Droguett, E., Greenberg, M., . . . Zio, E. (2015). Society for Risk Analysis Glossary.

Aven, T., & Zio, E. (2013). Model output uncertainty in risk assessment. *International Journal of Performability Engineering*, *9*(5), 475–486.

Bachmann, A., & Allgower, B. (1998) A framework for wildfire risk analysis. In *Proceedings of the 14th Conference on Fire and Forest Meteorology*; Nov. 16-20, Coimbra, Portugal. Vol. 2. Coimbra, Portugal: ADAI. pp. 2177-2190.

Bachmann, A., & Allgower, B. (2001). A consistent wildland fire risk terminology is needed! *Fire Management Today*, *61*(4), 28–33.

Boholm, A., & Coevellec, H. (2011). Relational theory of risk. *Journal of Risk Research*, *14*(2), 175–190. doi:10.1080/13669877.2010.515313

Finney, M. (2005). The challenge of quantitative risk analysis for wildland fire. *Forest Ecology and Management*, *211*(1-2), 97–10. doi:10.1016/j.foreco.2005.02.010

Food and Agriculture Organization of the United Nations. (2017) Child labour in agriculture in Lebanon. A guide for practitioners. p. 88.

Gallina, V., Torresan, S., Critto, A., Sperotto, A., Glade, T., & Marcomini, A. (2016). A review of multi-risk methodologies for natural hazards: Consequences and challenges for a climate change impact assessment. *Journal of Environmental Management*, *168*, 123–132. doi:10.1016/j.jenvman.2015.11.011 PMID:26704454

Goldammer, J., Mitsopoulos, I., Mallinis, G., & Woolf, M. (2017). *Wildfire hazard and risk assessment* (p. 9). UNISDR.

Gregory, V., & Vidal, A. (1999). Fuzziness in Chang's fuzzy topological spaces. *Rend. Instit. Mat. Univ. Trieste*, *XXX*(Suppl.), 111–121.

Hansson, S. (2010). Risk: objective or subjective, facts or values. *Journal of Risk Research*, *13*(2), 231–238. doi:10.1080/13669870903126226

Hardy, C., & Hardy, C. E. (2007). Fire danger rating in the United States of America: An evolution since 1916. *International Journal of Wildland Fire*, *16*(2), 217–231. doi:10.1071/WF06076

Hohenemser, C., Kates, R., & Slovic, P. (1985). *A casual taxonomy. Perilous progress: managing the hazards of technology* (pp. 67–125). Boulder, CO: Westview Press.

Hyde, K., Dickinson, M., Bohrer, G., Calkin, D., Evers, L., Gilbertson-Day, J., ... Tague, C. (2013). Research and development supporting risk-based wildfire effects prediction for fuels and fire management: Status and needs. *International Journal of Wildland Fire*, *22*(1), 37–50. doi:10.1071/WF11143

Kaplan, S., & Garrick, B. J. (1981). On the quantitative definition of risk. *Risk Analysis*, *1*(1), 11–27. doi:10.1111/j.1539-6924.1981.tb01350.x PMID:11798118

Kappes, M. (2011). *Multi-hazard risk analyses: a concept and its implementation.* (Dissertation). Universitat Wien, Austria.

Miller, C., & Ager, A. (2013). A review of recent advances in risk analysis for wildfire management. *International Journal of Wildland Fire*, *22*(1), 1–14. doi:10.1071/WF11114

Mimbrero, M. (2015). *Review and new methodological approaches in human-caused wildfire modeling and ecological vulnerability: Risk modeling at mainland Spain.* (Doctoral thesis), University of Zaragoza, Spain.

Rose, E. (1998) Metatheoretical foundations for post-normal risk. *Journal of Risk Research, 1,* pp. 15-41.

Royal Society. (1992). *Risk: Analysis, perception and management.* London, UK: Royal Society. p. 201.

Sherstjuk, V., Zharikova, M., & Sokol, I. (2018). Forest fire-fighting monitoring system based on UAV team and remote sensing. In *Proc. of the 38th Int. Conf. "Electronics and Nanotechnology",* pp. 99-106. Kyiv, Ukraine. 10.1109/ELNANO.2018.8477527

Thompson, M., Calkin, D., Finney, M., Ager, A., & Gilbertson-Day, J. (2011). Integrated national-scale assessment of wildfire risk to human and ecological values. *Stochastic Environmental Research and Risk Assessment, 25*(6), 761–780. doi:10.100700477-011-0461-0

Thompson, M., Haas, J., Gilbertson-Day, J., Scott, J., Langowski, P., Bowne, E., & Calkin, D. (2015). Development and application of a geospatial wildfire exposure and risk calculation tool. *Environmental Modelling & Software, 63,* 61–72. doi:10.1016/j.envsoft.2014.09.018

Thompson, M., Zimmerman, T., Mindar, D., & Taber, M. (2016) Risk terminology primer: Basic principles and glossary for the wildland fire management community. *Gen. Tech. Rep. RMRS-GTR-349. Fort Collins, CO: U.S. Department of Agriculture, Forest Service, Rocky Mountain Research Station.*

Thompson, M. P., Haas, J. R., Gilbertson-Day, J. W., Scott, J. H., Langowski, P., Bowne, E., & Calkin, D. E. (2015). Development and application of a geospatial wildfire exposure and risk calculation tool. *Environmental Modelling & Software, 63,* 61–72. doi:10.1016/j.envsoft.2014.09.018

Zharikova, M. (2018) *Methodological basis of geoinformation technology of decision support in combined natural and man-made systems in destructive processes conditions.* (Doctoral Thesis), KNTU, Ukraine. p. 503.

Zharikova, M., & Sherstjuk, V. (2015). Development of the model of natural emergencies in decision support system. *Eastern European Journal of Enterprise Technologies, 4*(1), 62–69. doi:10.15587/1729-4061.2015.37801

Zharikova, M., & Sherstjuk, V. (2017). Threat assessment method for intelligent disaster decision support system. *Advances in Int. Systems and Computing, 512,* 81–99.

Zharikova, M., Sherstjuk, V., & Baranovskij, N. (2015). The plausible wildfire model in geoinformation decision support system for wildfire response. In Proc. of the 15th Int. Multidisciplinary Sc. Geoconference SGEM-2015, 2(3), pp. 575–583. Albena, Bulgaria.

ADDITIONAL READING

Merino, L., Caballero, F., Martinez-de-Dios, J., Maza, I., & Ollero, A. (2012). An Unmanned Aircraft System for Automatic Forest Fire Monitoring and Measurement. *Journal of Intelligent & Robotic Systems*, *65*(1-4), 533–548. doi:10.100710846-011-9560-x

Merino, L., Martínez de Dios, J., & Ollero, A. (2015). Cooperative Unmanned Aerial Systems for Fire Detection, Monitoring, and Extinguishing. In *Handbook of Unmanned Aerial Vehicles* (pp. 2693–2722). Springer Netherlands. doi:10.1007/978-90-481-9707-1_74

Olsson, H., Egberth, M., Engberg, J., Fransson, J., & Pahlén, T. (2005). Current and Emerging Operational Uses of Remote Sensing in Swedish Forestry. In *Proc. of the 5th Annual Forest Inventory and Analysis Symposium*, pp. 39–46. US Forest Service, Portland, USA.

Qin, C., Wu, X., Jiang, J., & Zhu, A. (2016). Case-based knowledge formalization and reasoning method for digital terrain analysis – application to extracting drainage networks. *Hydrology and Earth System Sciences*, *20*(8), 3379–3392. doi:10.5194/hess-20-3379-2016

Wardihani, E., Ramdhani, M., Suharjono, A., Setyawan, T., Hidayat, S., Widodo, S., ... Saifullah, F. (2018). Real-time forest fire monitoring system using unmanned aerial vehicle. *Journal of Engineering Science and Technology*, *13*(6), 1587–1594.

Yuan, C., Liu, Z., & Zhang, Y. (2017). Aerial Images-Based Forest Fire Detection for Firefighting Using Optical Remote Sensing Techniques and Unmanned Aerial Vehicles. *Journal of Intelligent & Robotic Systems*, *88*(2-4), 635–654. doi:10.100710846-016-0464-7

Yuan, C., Zhang, Y., & Liu, Z. (2015). A Survey on Technologies for Automatic Forest Fire Monitoring, Detection and Fighting Using UAVs and Remote Sensing Techniques. *Canadian Journal of Forest Research*, *45*(7), 783–792. doi:10.1139/cjfr-2014-0347

Zharikova, M., Sherstjuk, V., & Baranovskij, N. (2015). The Plausible Wildfire Model in Geoinformation Decision Support System for Wildfire Response. In Proc. of the 15th Int. Multidisciplinary Sc. Geoconference SGEM-2015, 2(3), pp. 575–583. Albena, Bulgaria.

KEY TERMS AND DEFINITIONS

Fire Danger: A potential of fire beginning within a certain area of terrain.

Fire Intensity: The rate of heat energy released by the fire.

Fire Potential: Fire intensity and the area of terrain covered by fire.

Fire Risk: A potential for a valuable object to take losses as a result of a forest fire.

Fire Threat: Assessment of spatial and temporal distance between a valuable object and fire contour.

Valuable Object: Object that is of value for people.

Vulnerable Object: A valuable object that is a subject for decreasing the value as a result of a certain fire scenario realization.

Section 3

Chapter 6
Assessing the Territory According to the Degree of Thunderstorm Danger

Svetlana Yurevna Karanina
Gorno-Altaisk State University, Russia

Nina Alekseevna Kocheeva
Gorno-Altaisk State University, Russia

Andrey Vladimirovich Karanin
Gorno-Altaisk State University, Russia

Marina Yurevna Belikova
Gorno-Altaisk State University, Russia

ABSTRACT

This chapter studies the dynamics of lightning activity in the mountains of southern Siberia. The information base for the studies was data on lightning discharges for 2011-2016, recorded by WWLLN. Daily and seasonal courses in the number of lightning discharges are analyzed. Maps show the average annual density and the coefficient of variation of the number of lightning discharges for the plots with dimensions of 10 × 10 km. A cluster analysis of lightning discharges during the passage of individual thunderstorms in the territories of Altai Republic and the Republic of Buryatia was performed. During the study, the daily course of the number of lightning discharges was detailed. Sites with a high average annual density of lightning discharges were identified as lightning-hazardous areas of the mountains of southern Siberia. An analysis of the spatial distribution of lightning-discharge clusters made it possible to trace the dynamics of thunderstorm focus.

DOI: 10.4018/978-1-7998-1867-0.ch006

INTRODUCTION

This chapter presents the assessing method of thunderstorm danger on a vast, low-density populated territory, where there is a lack of monitoring tools and systems, except global lightning detection networks. In some regions of the world, especially in sparsely populated areas, thunderstorms are the main cause of wildfires. For example, in the boreal zone of North America (Johnson, 1992), in the northern part of Canada (Wierzchowski, Heathcott, & Flannigan, 2002), in the sparsely populated northern regions of Siberia, the proportion of fires caused by lightning can reach 70–90% (Kozlov, Mullayarov, & Solovev, 2009; Ivanov & Ivanova, 2010; Tarabukina & Kononova, 2018). In areas where wildfires mainly occur due to careless handling of fire by the local population, sometimes thunderstorms significantly increase the complex fire hazard situation (Ganteaume, 2013; Conedera, Cesti, Pezzatti, Zumbrunnen, & Spinedi, 2006).

The most significant cause of the outbreak of wildfires is intense dry thunderstorms, which are formed and develop most often in drought conditions. However, the passage of absolutely dry thunderstorms is quite rare and most often thunderstorm processes are accompanied by precipitation. It should be noted that fires caused by thunderstorms most often occur during air mass thunderstorms (or heat thunderstorms), which are characterized by small precipitation. With the passage of frontal thunderstorms precipitation is uneven - from its complete absence to heavy rain. It is established that even heavy rain is not able to eliminate the fire caused by lightning (Ivanov & Ivanova, 2010). The fire may be unextinguished inside the deadwood or under the canopy of trees. In good weather such a weak hidden burning can turn into a strong fire. Therefore, there is a possibility of wildfires in the five-day and ten-day intervals after the passage of thunderstorms (Kozlov et al., 2009; Ivanov & Ivanova, 2010).

Strong ground lightning discharges predominantly have a burning effect (Larjavaara, Pennanen, & Tuomi, 2005; Solovyev, Kozlov, Karimov, & Vasiliev, 2010). The greatest current strength is characteristic of positive lightning discharges (Rakov & Uman, 2003; Mullayarov, Toropov, Kozlov, & Karimov, 2009). Such discharges occur on the outskirts of thunderstorm fronts and are accompanied by weak precipitation. In defining the possible places of wildfires, it is necessary to know the conditions of a territory's lightning hazard.

Thus, the study of the nature of the spatial-temporal distribution of lightning discharges is one of the important components in understanding the processes accompanying the occurrence of a wildfire caused by the activity of a thunderstorm discharge. The chapter will be focused on the description of ways and methods for assessing the territory according to the degree of thunderstorm danger.

On the example of the territory of the south of Siberia, the study of the spatial and temporal distribution of lightning activity is presented. It is worth noting that the territory of the south of Siberia is a mountainous country, which includes the territory of the Altai-Sayan mountainous country (ASMC) and the territory of the Republic of Buryatia. For mountainous countries poor development of road infrastructure is typical and it is in remote areas that thunderstorms are most often the main cause of wildfires. Due to the late detection and inability to use the equipment, fires cover large areas and cause significant damage. Therefore, for mountainous countries, it is important to define areas of the most susceptible to the occurrence of wildfires caused by lightning discharges (Wierzchowski et al. 2002; Conedera et al., 2006; Kozlov et al., 2009; Ivanov & Ivanova, 2010; Ganteaume, 2013; Müller & Vacik, 2017).

BACKGROUND

Studies on the character of thunderstorms are based on visual-auditory observation of thunderstorms by a meteorological network or on instrumental measurements of electromagnetic radiation arising from lightning discharges.

The meteorological network is a set of meteorological stations conducting observations under a single program and within strictly set time limits for studying the weather, climate and solving other applied and scientific problems. The observer records only those thunderstorms that can be seen or heard in the area of a meteorological station (World Meteorological Organization, 1956; Smith, 2008). Therefore, the results of observations of thunderstorms depend on the visual range of the station. In mountainous regions, the visibility range is much lower than in the plain areas. You can hear thunder at a distance of no more than 20 km from the place of observation (World Meteorological Organization, 1956). Meteorological station observers on the territory of the Russian Federation note the date, time of the beginning and end, the nature of the thunderstorm (strong, weak, medium, in the area, at the observation point, etc.). Based on the records of thunderstorms and thunder, the number of thunderstorm days and thunderstorm hours are calculated (World Meteorological Organization, 1956; Smith, 2008). The change of these two parameters from month to month characterizes the seasonal course of thunderstorm activity, from year to year - the long-term dynamics. Maps of the world, individual countries and regions by the number of days with a thunderstorm or by the duration of thunderstorms are of an overview character. They give an approximate estimate of thunderstorm activity in the areas between observation stations. The accuracy of this estimate depends on the number of weather stations and the distance between them.

Another way to evaluate thunderstorm activity is instrumental measurements of radio echo and electromagnetic signals that occur during lightning discharges. There are active, passive, active-passive methods of radiolocation of thunderstorms, a detailed review of which was made by V. Stepanenko and S. Halperin (1982), as well as L. Kachurin and L. Divinsky (1992). The current state and availability of instrumental observation systems for thunderstorm foci in Russia and abroad are summarized in the monograph by V. Gorbatenko and T. Ershova (2011), as well as A. Adzhieva and E. Bogachenko (2011). A review of methods for recording lightning discharges and a description of the physics of their occurrence was conducted by V. Rakov and M. Yuman (Rakov & Uman, 2003)

The active radiolocation of thunderstorm clouds is carried out using radar stations (RLS), which receive echo signals to the input of the receiving antenna, which are due to scattering of radio waves by hydrometeors, space charges and ionized channels when corona, brush, leader and main discharges occur (Stepanenko & Halperin, 1982; Kachurin & Divinsky, 1992). In addition, the input of the antenna receives the radio emission of these discharges. Meteorological network widely uses RLS, which operate in the centimeter range of radio waves. In Russia, such stations include meteorological radars (MRL-1, MRL-2, MLR-5), abroad Doppler meteorological radars (DRLR) are used. Recognition of thunderstorm foci among other forms of cloudiness using these devices is carried out with the help of a number of statistical characteristics and criteria (Stepanenko & Halperin, 1982; Kachurin & Divinsky, 1992; Gorbatenko & Ershova 2011). As a result of their use, the probability of the existence of a thunderstorm focus in a cumulonimbus cloud is calculated. MRL and DRLR do not detect lightning in a thundercloud. In this case, a thunderstorm focus means the region of space in which lightning is generated. To study the frequency of lightning discharges, their scale and duration of existence, the RLS of the decimeter and meter ranges were specially upgraded (Stepanenko & Halperin, 1982; Kachurin & Divinsky, 1992). With the help of such RLS in the meter range, the cellular structure of the thunderstorm focus, its intensity and evolution were reliably and clearly identified (Kachurin & Divinsky, 1992). Separate electrically active cells (zones of heterogeneity of thunderstorm activity) have a shorter duration and size compared with a thunderstorm focus, they appear, disappear, interact with each other, thereby determining the dynamics of lightning activity in the thunderstorm zone of the cloud.

The passive radio-technical means of thunderstorm detection includes systems based on the detection of electromagnetic radiation (EMR) of lightning flashes during its propagation in the Earth-ionosphere waveguide. The detection efficiency and location accuracy of the impulse components lightning flashes for these systems depends on the frequency range of the electromagnetic impulse and the method of its measurement. Overview of the majority of devices that carry out the passive location of lightning discharges are given in the monograph by A. Adzhiev and E. Bogachenko (2011). There are lightning warning and the thunderstorms location systems, as well as systems for the location of lightning discharges. The latter have more accurate characteristics for determining lightning discharges. Among them, there are direction finding (PSM / DFS), differential time of arrival (RDSM / DTOA) and hybrid systems (Rakov & Uman, 2003; Gorbatenko & Ershova, 2011). The PSM determines the values of the bearing to the radiator, at least from two points. An example of this type of system is the PSM of the company Lightning location and position (LLP) Inc., organized in the mid 1970s in the USA (Rakov & Uman, 2003). DTOAare mostly used to solve various research and applied business problems. Such systems are based on determining the time difference between the arrival of signals from three or more points. The accuracy of determining the coordinates of a lightning discharge in these systems depends on the configuration of the points where the registration points are located and on the number of such points. DTOA are demanding of the sampling of signals and the accuracy of their time synchronization. The accuracy of the positioning of lightning discharges is much higher. The Lightning position and tracking system (LPATS) in the USA and the Alves system in Russia are an example of differential-distance measurements systems (Rakov & Uman, 2003; Gorbatenko & Ershova, 2011; Mareev et al, 2016). Hybrid systems combine PSM and DTOA. Sensors of this type are IMPACT (Improved Accuracy from Combined technology) sensors, which are based on magnetic direction finder and time-of-arrival technologies.

Most methods of passive location of thunderstorms are focused on the emission of high-current discharges between the cloud and the ground. It is known that the proportion of cloud-to-ground outbreaks from the total number of lightning that occurs during a thunderstorm is not large and, according to some estimates, is 10–20% (Rakov & Uman, 2003). All types of flashes can be registered using ultrashort-wave (USW) positioning systems. There are two types of such systems. One of them is based on the technology of differential-distance measurements in spaced points (Proctor, Uytenbogaardt, & Meredith, 1988), the other on the interferometric measurement method (Kononov & Yusupov, 2004; Gorbatenko & Ershova, 2011; Adzhiev & Bogachenko, 2011). The Lightning Detection and Ranging System (LDAR) system functions based on the first technology. The interferometric SAFIR USW system is currently the most accurate and efficient system for the passive location of cloud-to-ground, cloud-to-cloud and intracloud lightning discharges. In Russia, an example of a system operating in the USW range is the System for registering lightning discharges — SRLD (Moskovenko, Znamenshchikov, & Zolotarev, 2012).

Currently, they are often combined into a single network of sensors with different lightning discharge location technology. Such networks are created not only in the regions, but also in individual countries and in several countries. North American Lightning Network - NALDN (Rakov & Uman, 2008) lightning positioning system operates in the US and in Canada. NORDic Lightning Information System (NORDLIS), includes sensors located in Finland, Norway, Sweden and Estonia (Tuomi & Mäkelä, 2008). The international lightning detection network LINET consists of 90 sensors in 17 European countries (Betz,

Schumann, & Laroche, 2008). The sensor network of this system also operates in Brazil, Australia and Central Africa. In addition, there is the European Cooperation for Lightning Detection - EUCLID network in Europe, which unites national lightning detection networks in 27 countries in Europe (EUCLID …, n.d.).

Along with regional, national and international lightning detection networks, global networks are developing which register lightning discharges across the Earth. Currently, there are three such networks World Wide Lightning Location Network - WWLLN (World Wide Lightning…, n.d.), the amateur network Blitzortung.org (Real-Time Lightning Map …, n.d.), Global Lightning Dataset of the company Vaisala - GLD360 (Unique Vaisala Global…, n.d.). WWLLN is a lightning direction-finding system based on international cooperation from more than 50 universities and research organizations. The main purpose of its creation is to study the global lightning activity, the contribution of lightning activity to climate change, as well as the global tracking of storms and cyclones. Blitzortung.org is an amateur network of over 1,200 stations, about 800 of which are located in Europe. The authors of this project focuse on private entertainment (non-scientific) goals. WWLLN and Blitzortung.org are significantly inferior in lightning detection efficiency and have a greater error in their positioning compared to some regional and national multipoint lightning location systems. The GLD360 network has a high detection efficiency for lightning discharges (more than 70%) and, on average, the accuracy of the location of their detection is 2–5 km. Vaisala provides GLD360 data on a commercial basis for meteorological services, airports, energy service companies, etc.

Along with multipoint lightning positioning systems, the development and improvement of single point lightning direction range finders - LDRF (Kononov, Krutoj, & Yusupov, 2013; Kozlov & Mullayarov, 2004) are topical. They are inferior in accuracy to the positioning of lightning discharges in comparison with multipoint difference-ranging systems, but they are simpler and more economical to operate, autonomous and can be mobile (installed aboard a ship, aircraft), relatively cheap.

In addition to global lightning-direction-finding systems, a high degree of territorial coverage is characterized by specialized imaging systems for observing thunderstorm activity located onboard spacecraft. Thunderstorm observations from an artificial satellite of the Earth (ASE) began in the mid-1960s with the registration of radio frequencies and optical radiation from lightning from Orbiting Solar observatory satellites - OSO-2, OSO-5, VELA, later from the satellites Defense Meteorological Satellite Program (DMSP) (Kirkland, Suszcynsky, Guillen, & Green, 2001). A detailed review of satellites and lightning detectors installed on them, which operated until the early 1990s, is presented in the monograph by R. Gurney et al. (1993).

The next step in the development of devices for monitoring lightning discharges from ASE was the optical imaging sensors (Optical Transient Detector and Lightning Imaging Sensor) of the Miclolab-1 satellite (Christian et al., 2003). With the help of OTD, lightning activity was monitored between 800 north and south latitude, every 100 minutes from 1995 to 2000. LIS was placed onboard the Tropical Rainfall Measuring Mission (TRMM) satellite from 1997 to April 2015. The shooting was carried out within the range of 35 ° south latitude - 35 ° north latitude. From February 2017, the LIS device was placed on the International Space Station (ISS), which allowed us to expand the range to 54 ° south and north latitude (Data User Guide…, n.d.). LIS detects the location of the lightning with a spatial resolution of 5–10 km and a time resolution of 2 ms. The efficiency of detection of lightning flashes is 70% at daytime and about 90% at night.

Another similar satellite device is the Geostationary Lightning Mapper (GLM) single-channel optical sensor operating in the near-infrared range and located in March 2017 on the GOES-16 (Geostationary Lightning Mapper …, n.d.). This tool is capable of real-time tracking of changes in the optical properties of recorded scenes, detecting lightning discharges. GLM captures lightning activity over America and neighboring oceanic regions, the spatial resolution of the sensor ranges from 8 km (when shooting in a nadir) to 14 km (edge of the scene). During the registration process, information on the frequency, location of thunderstorms and their distribution is collected, the intensity of thunderstorms and tropical cyclones is measured. Shooting is conducted around the clock, new scenes are recorded every 2 ms and the resulting product is formed every 20 seconds. The coverage area is limited to 52 degrees of latitude, in favorable conditions up to 90% of lightning discharges are recorded.

The FORTE low-orbit satellite, sponsored by the US Department of Energy, is also worth mentioning. The radio receivers aboard FORTE observe very high frequency (VHF) emissions from the air-breakdown process preceding (and sometimes accompanying) a lightning current (Kirkland et al., 2001; Jacobson et al., 2000; Suszcynsky et al., 2000). The goal is to develop an understanding of the correlation between the optical flash and the VHF (30-300 MHz) emissions from lightning (Jacobson et al., 2000; Suszcynsky et al., 2000). FORTE also conducts ionospheric physics experiments (Smith et al., 2004).

Area coverage of thunderstorms by a network of weather stations, including visual-aural methods and methods of active, passive radar, varies for countries around the world. Thunderstorm observing systems may differ for territories within countries. Active and passive thunder radar allows to quickly track the occurrence and movement of thunderstorm foci. To highlight the most hazardous areas of the territory, they mainly use the density maps of lightning discharges, as well as conduct a study of the seasonal and daily nature of the number of lightning discharges Works of this kind are based, as a rule, on an information array of data on lightning activity over a long period of time (more than 10 years). Density maps of lightning discharges provide a more detailed and accurate description of the spatial distribution of lightning activity in comparison with maps of the number of thunderstorm days and the duration of thunderstorms.

In the case of the absence of lightning direction finding systems, the estimating equations for the dependence of the number of lightning discharges on the parameters of thunderstorm activity are used. Such equations are obtained at sites where thunderstorms are recorded by a network of weather stations and instrumental measurements of lightning flashes. The values of the parameters of thunderstorm activity is calculated by the number of lightning discharges for areas where there are no observations of lightning discharges (Shindo & Yokoyama, 1998; MacGorman, Maier, & Rust, 1984; Konstantinova, Gorbatenko, & Polyakov, 2017; Tarabukina, Kozlov, Karimov, & Mullayarov, 2017). When developing such estimating equations, the orographic features of the area are necessarily taken into account. The duration of thunderstorms is better related to the actual production of lightning discharges than the number of days with thunderstorms.

Satellite measurements of lightning discharges are carried out with a certain frequency and, in addition to measurements of GLM over the United States, do not provide a complete picture of lightning activity in the region under study. Satellite lightning monitoring data is used primarily to estimate their distribution on a planetary scale (Christian et al., 2003; Light, Davis, Boeck, Jacobson, & Suszcynsky, 2003; Bailey, Blakeslee, Buechler, & Christian, 2007).

ANALYSIS OF LIGHTNING ACTIVITY IN THE SOUTH OF SIBERIA BASED ON WWLLN DATA

Modern Measurements of Thunderstorm Activity in Russia

Thunderstorm monitoring in Russia is carried out by the Federal Service for Hydrometeorology and Environmental Monitoring (Rosgidromet), as well as by individual scientific and scientific-educational organizations. Rosgidromet monitors thunderstorms with the help of the network of weather stations, including active radar using MRL, DMRL and passive radiolocation with a lightning direction different time of arrival system "Alves", the interferometric SAFIR system, global networks WWLLN and Blitzortung.org.Instrumental lightning measurements using active and passive radar are carried out in the European part of Russia, including the territory of the Kabardino-Balkarian Republic (Mareev et al, 2016; Adzhiev & Bogachenko, 2011; Adzhieva, Shapovalov, Boldyrev, & Ignatyev, 2018). In addition, on the territory of the Nizhny Novgorod Oblast, the Institute of Applied Physicists of the Russian Academy of Sciences has arranged a multi-point lightning direction finding network based on Boltek Stormtracker lightning direction finders (Mareev et al, 2016). The Center for Physical and Technical Problems of the Power Industry of the North of the KSC of the Russian Academy of Sciences has installed the Boltek Stormtracker lightning direction finder on the territory of the Kola Peninsula (Burcev, Nevretdinov, & Sytina, 2014).

The network of meteorological stations of Rosgidromet in the territory of Western and Eastern Siberia is poorly equipped with MRL and DMRL. Sources of data on lightning activity in Siberia are the data of the registration system for lightning discharges (RSLD) "Vereya-MR", which works for the needs of Forest Protection Aviation (Moskovenko et al., 2012). The control zone of this network includes a vast territory from St. Petersburg to Vladivostok, excluding part of Altai and Buryatia. The data of this network has the level of official use by the staff of Forest Protection Aviation; for other users they are available for a fee on request.

In addition, on the basis of Tomsk State University in the city of Tomsk (Konstantinova & Gorbatenko, 2011) and on the basis of the Yu.G. Shafer Institute of Cosmophysical Research and Aeronomy of the Siberian Branch of the Russian Academy of Sciences (SHICRA of the Siberian Branch of the Russian Academy of Sciences) in the cities of Yakutsk, Mirny and Neryungri (Kozlov & Mullayarov, 2004; Kozlov, Mullayarov, Grigorev, & Tarabukina, 2014) one-point lightning direction finders with a range of 480 km were installed. In Yakutsk, there is also a lihtning direction range finder, a range finder, which records lightning discharges in a neighborhood of 1,500 km radius.

In recent years, WWLLN data has been actively used to study the spatial and temporal distribution of thunderstorms in the Russian Federation. Currently, receiving stations of this network are organized in the city of Yakutsk (SHICRA SB RAS), in the city of Vladivostok and in the village of Paratunka (Institute of Cosmophysical Research and Radio Wave Propagation of the Far Eastern Branch of the Russian Academy of Sciences) (Tarabukina et al., 2017) and in the city of Gorno-Altaisk (Gorno-Altaisk State University) (Karanina, Kocheeva, & Karanin, 2017).

Figure 1. The scheme of the study area

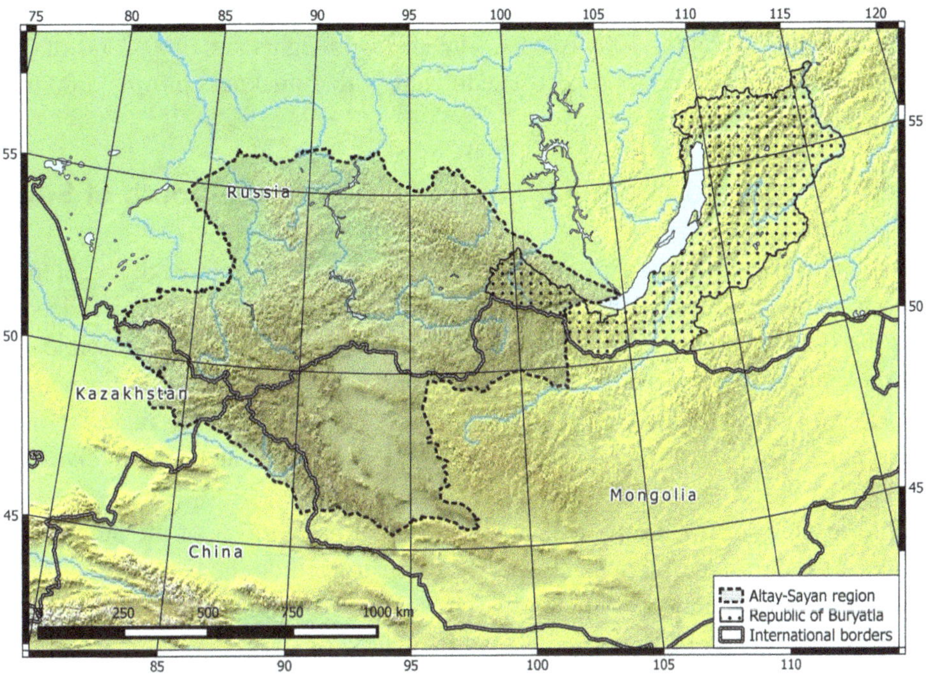

Study Area

The study area is located almost at equal distances from all oceans. From the west to the east it stretches for more than 2000 km - from the plains of Western Siberia to the eastern borders of the Republic of Buryatia (Figure 1). This territory is the western part of the watershed between the great Siberian rivers, which flow into the Arctic Ocean and the rivers that give their waters to the drainage area of Central Asia. Russian scientists define the southern border of Russia as the southern boundary of a mountainous country, the western border is the West Siberian Plain, the northern one is the Central Siberian Plateau.

The relief of the territory varies from plain in the boundaries of intermountain basin to alpine in the axial parts of the Altai ridges and includes: alpine highland, mid-mountain, low-mountain, ancient surfaces, basin-type relief.

The Altai-Sayan Mountain Country is the highest elevation in Northern Asia, the highest points of which are Belukha (4509) in the Katun mountain range and the mountain plexus of Tavan-Bogdo-Ula (with the largest peaks of 4374 and 4360 m). Even in 17 ridges and mountain ranges, the height of axial parts exceeds 3500 m. With distance from the Altai, the height of the ridges decreases. The absolute heights of the ranges in the Sayan Mountains are mainly 1800-2300 m, reaching a maximum of 3491 m (the city of Munku-Sardyk). Large intermountain basins Minusinsk, Kuznetsk, Cannes are located at altitudes from 150 to 300 m above sea level, Chui - at an altitude of 1750-1900 m above sea level.

The Republic of Buryatia - the eastern part of the study area, occupies a significant part of the south of Eastern Siberia. The area of the mountains is more than 4 times the area of the flat areas. The lowest level is the level of Lake Baikal - 456 m and the highest is the glacier-covered peak Munku-Sardyk in East Sayan - 3491 m above sea level.

In southern Buryatia, elevations of 1,000–1,800 meters prevail. In the northern Buryatia there are mountain ranges (South-Muya, North-Muya, Udokan, Kalar, etc.), which stretch over long distances and rise to maximum heights of 2500 - 3000 meters. The average heights are 1700-2200 m. For northern Buryatia, continuous distribution of permafrost sediments with a thickness of up to 600 meters occurring close to the surface is noted.

The Study of Thunderstorm and Lightning Activity in the South of Siberia

Thunderstorm activity in the 50s of the last century was estimated by the number of foci of thunderstorms in a thunderstorm season (Solovyev,1969). The number of thunderstorm foci characterized the qualitative picture of the space-time distribution of thunderstorms, but did not make it possible to proceed to the density of the number of thunderstorm discharges. The development of observations using lightning counters of various types (Lobodin, 1986) made it possible to obtain primary material on the number of lightning discharges for the territory of the former USSR and to conduct a qualitative assessment of thunderstorm activity based on objective data.

For the entire territory of the former USSR and modern Russia, maps of the number of days with thunderstorms and the duration of thunderstorms were constructed on the basis of long-term observations of thunderstorms by the network of meteorological stations (National atlas of Russia …, n.d.). The nature of thunderstorm activity was clarified for individual regions.

Based on WWLLN data for 2009-2010 for the first time, maps of the density of lightning discharges were constructed for the entire territory of Northern Asia (40 ° –80 ° N and 60 ° –180 ° E) (Kozlov, Mullayarov, & Karimov, 2011). Later, the data on discharges for 2009–2014 were corrected according to the spatial distribution of the network detection efficiency and an analytical expression was proposed for the spatial distribution of the density of thunderstorms for the specified area (Tarabukina et al., 2017). As a result, an integral characteristic of thunderstorm activity was obtained throughout North Asia for 2009–2014. and two powerful thunderstorm foci have been identified: in the area of the West Siberian Plain and between the Stanov Ridge and Little Khingan Mountains. Numerical equations for the density of lightning discharges were constructed and their dependence on the latitude and longitude of the terrain was determined.

In general, studies of thunderstorm and lightning activity show that the greatest gradient of thunderstorm activity is observed in the height interval of 20–200 m. The maximum thunderstorm activity is observed in mountainous areas at the height of 1,400–1,600 m. activity decreases, approaching at heights of 3-4 km to the values observed on the plain (Lobodin, 1986).

The study of thunderstorms and in particular lightning activity using instrumental measurements, in Yakutia and in the North Caucasus also covers mountainous areas (Kozlov & Mullayarov, 2004; Mullayarov et al., 2009; Kozlov et al., 2014; Adzhiev & Bogachenko, 2011; Adzhieva et al., 2018). For these territories, it has been established that the increase in thunderstorm activity, as a rule, occurs in the foothills, while confined to certain geomorphological elements - river valleys and basins (the absolute height varies widely, can be from 100 to 2000 m), as well as to territories with heights of about 500 m above sea level

The territory of the ASMC includes several administrative units, within which the analysis of the areal and temporal distribution of thunderstorms was carried out. Early in the 1970s, an increase in the number of thunderstorm event in the direction from north to south, in Western Siberia (Tyumen Oblast, Tomsk Oblast, Kemerovo Oblast, Altai Krai and Altai Republic) was revealed (Kashinskii, 1979; Gorbatenko

& Ershova, 2011; Dmitriev, Krechetova, & Kocheeva, 2011). At the same time, the latitudinal zonality of the distribution of thunderstorms is typical for the northern regions, above 63 ° N, for the rest of the territory, foci of thunderstorms are observed with an absolute maximum in the foothills of the ASMC. This spatial pattern was refined by constructing a map of the density of lightning discharges recorded by the Microlab-1 satellite in the period from1995 to 1999 (Gorbatenko & Ershova, 2011; Konstantinova et al., 2017). A number of studies have shown that the foci of thunderstorm activity are caused not only by the latitude of the place, but, to a large extent, by local terrain features, the direction of air flow, the state of the underlying surface (Gorbatenko & Ershova, 2011; Konstantinova et al., 2017), geological and geophysical factors and the activity of the Sun (Dmitriev et al., 2011). For the Gorno-Altaisk station, located on the border of the Altai Mountains and the plains of Western Siberia, the connection of thunderstorms and the number of days with the high trough (Kocheeva, Chankibaeva, Minaev, Sukhova, & Modorov, 2017) was revealed.

Research on thunderstorm activity was not conducted separately for Buryatia. According to generalized for the whole territory of Russia, data on the number of days with thunderstorms on the territory of Buryatia there are from 10 to 30 days with thunderstorms per year. The greatest activity of thunderstorms is observed in the west of the republic in the Eastern Sayan region, as well as along the southern administrative border.

Under the conditions of mountainous terrain, a rarefied network of weather stations in southern Siberia is not capable of providing satisfactory estimates of thunderstorm activity. The results of studies of the spatial-temporal distribution of the number of days with thunderstorms and the duration of thunderstorms are very approximate and satisfactorily describe thunderstorm activity only in the area of the weather station. At meteorological stations in southern Siberia there are no MRL and DMRL, this fact complicates the operational monitoring of the formation and development of thunderstorm foci.

The density maps of lightning discharges and the analysis of the seasonal variation in the number of lightning discharges made for the territory of North Asia also provide only a general picture of the distribution of lightning activity.

Thus, there is a problem of detailing the spatial distribution of thunderstorm foci that are stable over time, which are determined on the basis of long-term observations of thunderstorms. There is also the problem of operational monitoring of thunderstorm foci associated with the passage of individual thunderstorms.

ANALYSIS OF LIGHTNING ACTIVITY IN THE SOUTH OF SIBERIA BASED ON WWLLN DATA

In the south of Western Siberia, at present, the only source of continuous time information on lightning discharges is WWLLN data. This section presents the results of a study of lightning activity based on the data from this network for a six-year period from 2011 to 2016 in the south of Siberia. Also there is an example of using cluster analysis for distinction of thunderstorms during the passage of a separate thunderstorm.

Materials and Methods

Studies of lightning activity were carried out on the basis of WWLLN data for 2011-2016. The WWLLN network is based on the analysis of the difference in the time of arrival of the lightning signal - atmospheric (frequency range from 3 to 30 kHz) to the nearest network stations (Dowden, Brundell, & Rodger, 2002). A lightning discharge is detected by the system if its signal is detected by five or more stations. A detailed description of the registration of lightning discharges by this network is presented in a number of publications available on the WWLLN website (wwlln.net), as well as in (Kozlov et al., 2011). The estimated detection efficiency of the system for 2012 on the average is about 11-15% for all discharges, for discharges with a current above 40 kA - 25%, above 130 kA - more than 30% (Hutchins, Holzworth, Brundell, & Rodger, 2012; Virts, Wallace, Hutchins, & Holzworth, 2013). The error in determining the coordinates of a lightning discharge is the on average 5 km.

The WWLLN data on atmospherics contain following parameters: date, time, latitude, longitde, error rate and number of stations which detected the electromagnetic impulse. The net doesn't define the type of the lightning discharge. Since the maximum spectral density of atmospherics generated by ground-based lightning discharges falls in the low-frequency range of 3–30 kHz, that is, there is reason to assume that WWLLN detects mainly powerful impulses of ground-based lightning discharges (Tarabukina & Kozlov, 2018).

On the basis of the WWLLN data, the analysis was made of the daily and seasonal activity of lightning discharges in the ASMC and Buryatia. In this case, the average density of lightning discharges over the territory was calculated for each hour of the day and each month from May to August. To analyze the features of the areal distribution of thunderstorm events, lightning discharge density maps were constructed, which were calculated for areas of 10x10 km.

Seasonal and Daily Course of Lightning Activity

In the study area, the minimum activity of thunderstorms is recorded in 2011-2012. Within the ASMC range, the smallest number of lightning discharges, 86521, was registered in 2011, while in Buryatia the minimum activity of thunderstorms (23457 discharges) was shifted to 2012. The largest number of lightning discharges in the west and in the east of the study area was noted in 2014: ASMC - 208137, Buryatia - 38835. In the remaining years, the number of registered discharges changes in ASMC from 120,000 to 140,000, in Buryatia from 28,000 to 37,000.

The course of thunderstorm activity during the summer season corresponds to the trend identified earlier on the basis of data from weather stations: the number of lightning discharges begins to increase in May, the maximum falls in July, in August there is a decline in thunderstorm activity (Figure 2). In August, the number of lightning discharges exceeds their number, registered in May. During the years of minimum (2011-2012) and maximum (2014) of thunderstorm activity, the largest number of lightning discharges is observed in June.

The time of registration of the lightning discharge network WWLLN is made according to universal time. The study area is located in two time zones, therefore, to estimate the daily course of lightning discharges, the time for each discharge was recalculated from universal time to local solar time.

There is a ranging level of thunderstorm activity in the western and eastern parts of the study area. This determined the autonomous consideration of the daily course of the thunderstorm activity of these two parts.

The areas of the two study sites differ greatly. ASMC significantly exceeds the area of the Republic of Buryatia. The average density of lightning discharges in the territory in Buryatia is three times higher than in the ASMC (Figure 3). This may be due to the fact that the vast territories of Mongolia, over which thunderstorms are formed extremely rarely, make a decisive contribution to reducing the density of lightning within the ASMC.

Within the ASMC, the distribution of the number of lightning discharges during the day has a clear minimum in the range from 5 to 9 am. By the middle of the day, the number of discharges increases and, in the interval from 12 to 15 am, the greatest number of discharges is averagely recorded (Figure 3a). It is noteworthy that after a smooth decrease in thunderstorm activity in the range from 15 to 18 am, the second maximum appears at 19 am, after which the decline continues. This situation is noted every year of the study period. At the same time, in 2011, the second 19 am maximum number of lightning discharges is comparable to the midday maximum of 12-15 am.

In the daily course of lightning activity in the territory of Buryatia, two time periods are clearly distinguished: 24pm - 10 am and 11pm - 23am (Figure 3b). The first segment is characterized by the minimum activity and a relatively uniform distribution of the number of discharges by the time. The second segment (11pm -23am) is characterized by the sharp increase in the number of lightning discharges by noon local time, which slowly decreases to 16pm. Against the background of the further sharp decline, there is a surge in the number of lightning at 20 o'clock.

The analysis indicates that the years 2010-2012 are characterized by the presence of two daily maxima, of which the second ("evening") maximum exceeds the first ("lunch"). The second maximum occurs at 20 o'clock in all years of the study period, except for 2015, when it shifts to 21 o'clock. In 2010, 2015 and 2016, there were three maxima of the number of lightning discharges.

Thus, throughout the study, there are two maxima of lightning activity. It is important that everywhere the second "evening" maximum does not exceed the maximum shown in the first half of the day. In Buryatia, thunderstorm activity manifests itself more diversely. Its divers splashes draw attention, the temporal distribution of which is wider in Buryatia, more stable within the ASMC borders. An offset of 1-2 hours of maximum activity of lightning discharges in the eastern part of the study area was revealed

Lightning occurs within 24 hours. At night in Buryatia, their number is three times less than during the day maximum. Within the limits of the ASMC, this difference is two times. For the territory of the ASMC in the period 2014-2016. a higher intensity of night lightning discharges was established compared with previous years. In particular, in 2014, the number of lightning discharges at night practically coincides with that in the daytime hours for other years of observation.

Spatial Distribution of Lightning Discharges

The vast territory of the study and the high degree of diversity of its geographical conditions determine the spatial heterogeneity of the distribution of thunderstorm activity. The mountain structures located within the boundaries of the ASMC and Buryatia are characterized by differences in sketching, height and scale, which, according to previous studies, affects the character of the spatial distribution of thunderstorms. T. Lobodin (1986) showed that "the number of discharges ... depends in two ways on the height of the place of observation. Small elevations and even the edges of forests, enhancing dynamic turbulence,

Figure 2. The distribution of the average number of lightning discharges during a thunderstorm season: a) ASMC; b) Republic of Buryatia

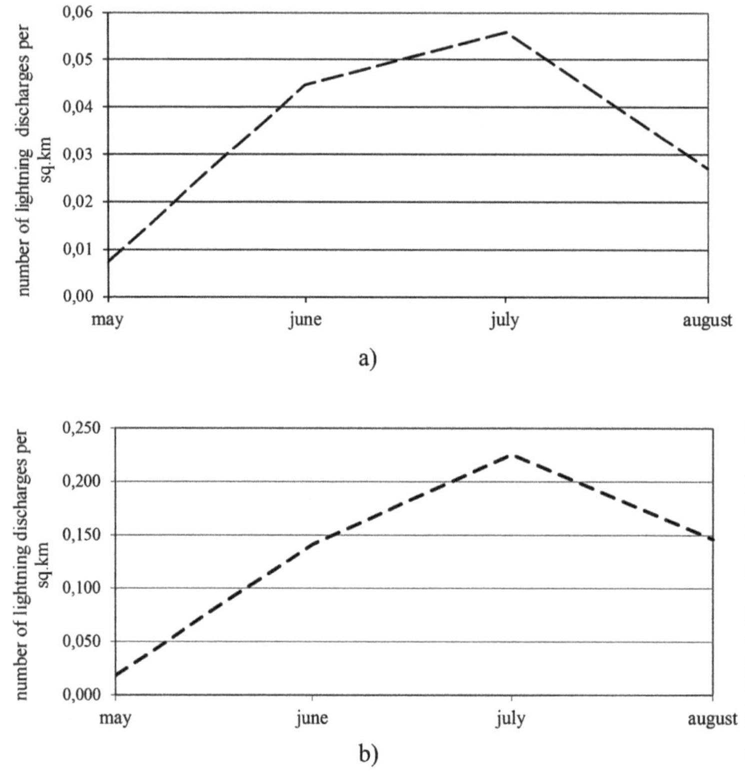

contribute to the strengthening of thunderstorm activity. The greatest gradient of thunderstorm activity is observed in the range of heights of 20–200m. The maximum magnitude of thunderstorm activity is observed in mountainous areas at the altitude of 1,400– 1,600 m "(p.73).

In the course of the study, the submeridional distribution of the density of thunderstorm discharges (Figure 4, 5) and the maximum density between the piedmont and low-mountain territories were established. When advancing deep into a mountainous country, the activity of lightning manifestation slightly decreases. In areas with altitudes from 800 m to 2500 m, the greatest number of lightning discharges is confined to the western and northwestern macroslopes of the ranges. Also, the windward slopes of the ridges, enclosing the intermontane basins in the North-Western, Northern, Central Altai, Tuva and Khakassia, are characterized by the highest density of discharges. The territory of the Eastern Sayan and the Mongolian part of the ASMC differs significantly in this indicator from the rest of the ASMC. The density of lightning discharges decreases, especially at altitudes above 2000 m, reaching the minimum in the deserted parts of Altai and Mongolia.

The density distribution of lightning throughout the study area is spatially uneven (Figures 4, 5). Most of the ASMC and Buryatia are characterized by low density of thunderstorms. Against this background, the south-western part of the Republic of Buryatia and the north-western part of the ASMC stand out, which coincides with the Eastern Sayan and Khamar-Daban ridges (Buryatia) and the low-mountain and mid-mountain part of the ASMC. In these regions, the maximum density of atmospheric atmospheres is

Figure 3. The daily course of lightning activity: a) ASMC; b) Republic of Buryatia

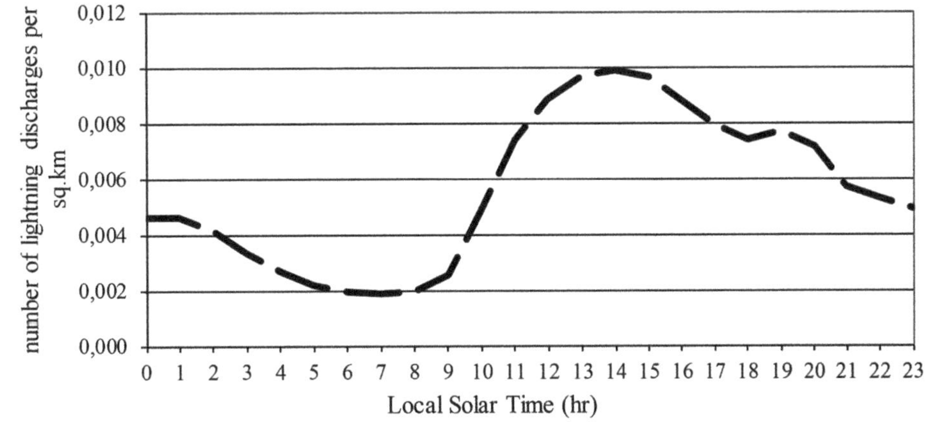

a)

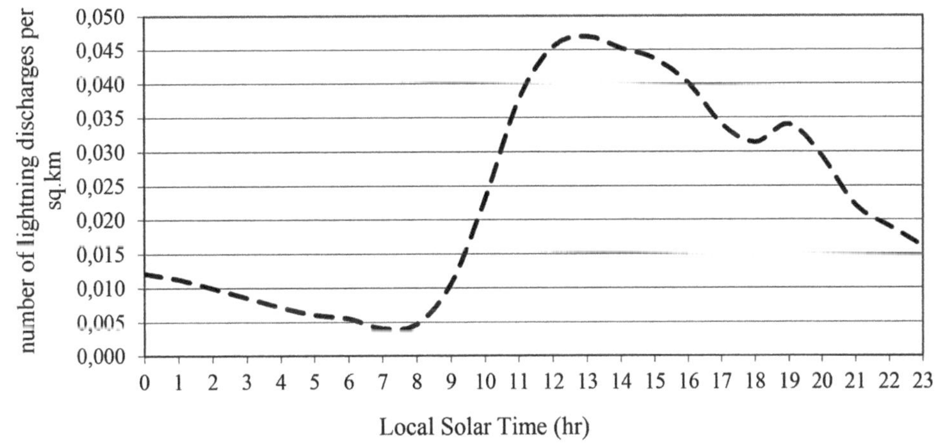

b)

noted in the basin of Uri river and in the head of Zun-Muren river (within the boundaries of Buryatia) and with the basins of the rivers Peschanaya-Anuy, Katun, Biya, Tom (within the boundaries of the ASMC). The average height of these territories is 1500 m.

To the east of Lake Baikal, the density of lightning discharges is very small. In this part of Buryatia, only in the northeast and southeast the density reaches 0.2 discharges per square km: in the northeastern part of the Vitim Plateau, as well as at the mouth of the Chikoi river (right tributary of Selenga river). The average height of this territory is about 1,300 m above sea level. In the basins of the northern part of Buryatia — Upper Angara basin, Muya basin, Barguzin basin and other smaller ones, the density of lightning discharges is minimal.

ASMC includes a part of the territory of Mongolia, which for the most part of the thunderstorm season is under the influence of the Mongolian anticyclone, does not contribute to the formation of thunderstorms. Therefore, in the Mongolian part, the density of lightning discharges is minimal compared to the rest of the study area (Figure 4, 5).

Figure 4. The density of lightning discharges on the territory of the ASMC - the average for 2011-2016

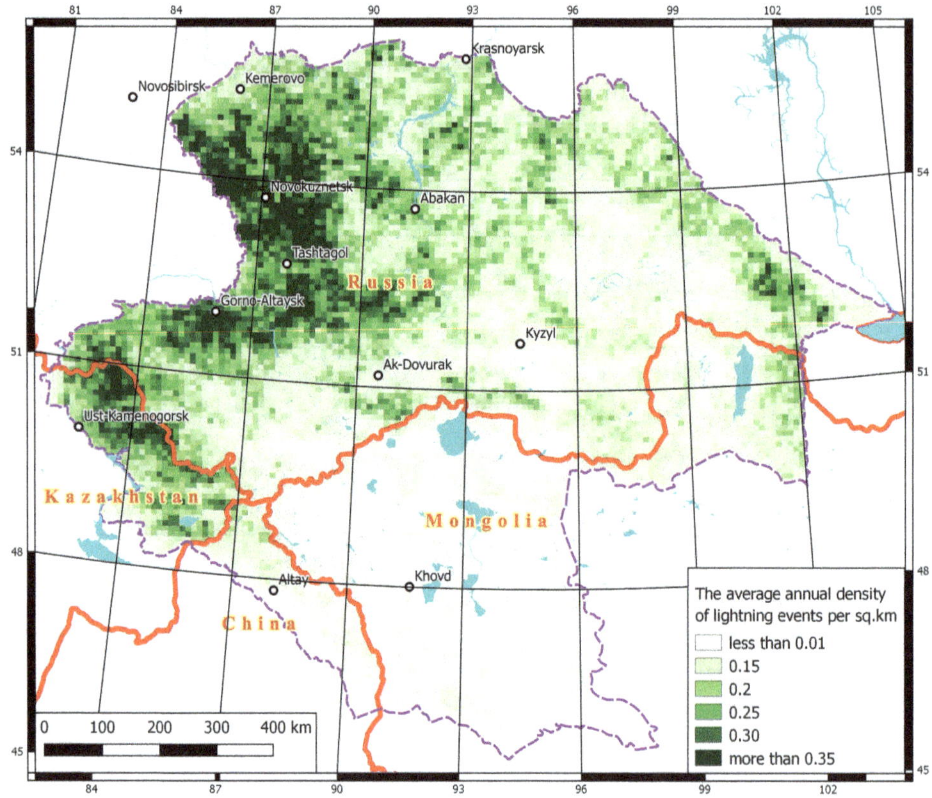

The number and density of lightning in the vast territory of the mountains of southern Siberia differs not only in space, but also in time. Discreteness of lightning causes temporary fluctuations in their number. However, there a reason to assume that there are years with an anomalous number of lightning in the study area.

The obtained numerical estimates of the atmospheric density are in good agreement with the results of the study of lightning activity on the WWLLN database for 2009–2014, presented for the entire territory of Northern Asia (40° – 80° N and 60° – 180° E) (Kozlov et al., 2011). Analysis of lightning activity using WWLLN data was made on the six-year interval. The results obtained during this study coincide with the patterns of thunderstorms, established on the fifty-year data array of the meteorological network (Dmitriev et al., 2011). In this case, the refinement of the spatial distribution of the most dangerous areas of the territory was carried out.

Selection of Individual Thunderstorm Foci

These lightning direction-finding networks allow additional monitoring of the occurrence, development and movement of active thunderstorm areas during the passage of a separate thunderstorm. To solve such problems, methods of grouping lightning discharges, which are characterized by temporal and spatial proximity, are used. Such groups of lightning discharges are associated with a convective cell in an active thunderstorm state (thunderstorm cell) (Shabaganova, Karimov, Kozlov, & Mullayarov, 2012).

Figure 5. The density of lightning discharges in the territory of Buryatia - average for 2011-2016.

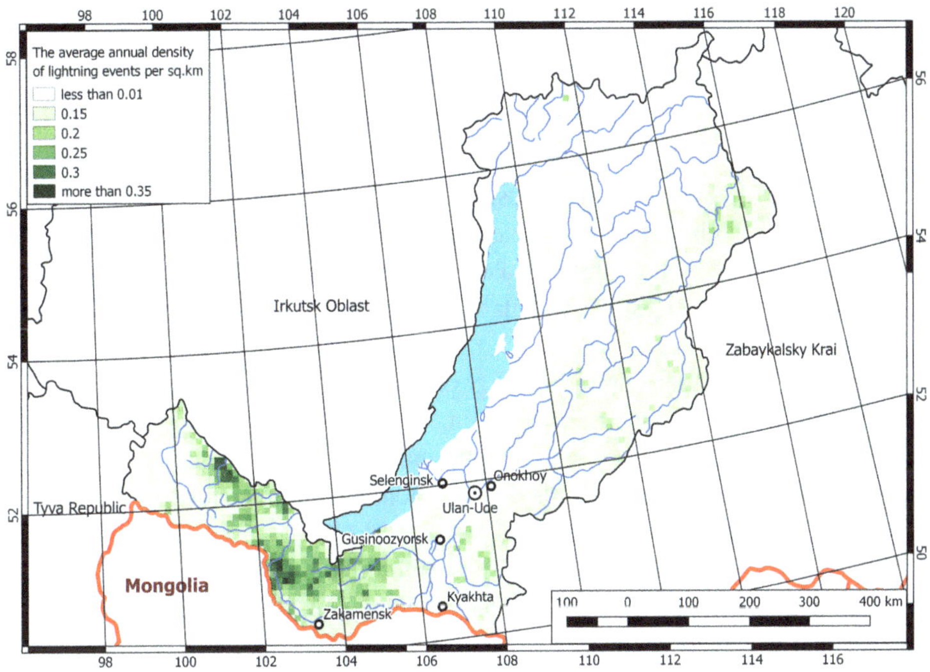

A thunderstorm cell is defined as an indivisible element of a thunder cloud. Thunderstorms consisting of a single convective cell are called a single-cell thunderstorm. If a thunderstorm consists of several cells moving as a whole, then it is called a multi-cell thunderstorm. Multi-cell thunderstorms are also called thunderstorms. The dimensions of a single-cell thunder cloud vary from 5 to 20 km. The duration of the existence of a convective thunderstorm cell at the stage of maturity in terms of the development of the electrical characteristics of the cloud takes from 20 minutes to 1 hour. The linear dimensions of a multi-cell thunder cloud vary from 10 to 1000 km. At the same time, thunderstorm cells in such a cloud exist independently and can be in different stages of development.

Methods of cluster analysis are used for grouping lightning discharges and obtaining models of convective thunderstorm cells (Shabaganova et al., 2012; Kononov & Yusupov, 2004; Hutchins, Holzworth, & Brundell, 2014). In our research clustering of lightning discharges was made with the help of FRiS-Tax algorythm (Zagoruiko, Borisova, Dyubanov, & Kutnenko, 2008), where grouping of lightning discharges is made on the following idea. If a discharge arises "close" from an already existing cluster, then it belongs to this cluster. Otherwise, it is the first discharge of the new cluster. The "proximity" of lightning discharges is usually determined by means of threshold values.

Let the results of the registration of lightning discharges be represented as the vector O = (x, y, t), where x is the north latitude, y is the east longitude, and t is the time of registration of the discharge. It is believed that lightning discharges $O_1 = (x_1, y_1, t_1)$ and $O_2 = (x_2, y_2, t_2)$ are "close" if the distance between them

$$\rho(O_1, O_2) = \sqrt{(x_1 - x_2)^2 + (y_1 - y_2)^2}$$

is less than some given value ρ * and the difference in recording time $\tau(O_1, O_2) = (t_1 - t_2($ does not exceed some value τ *. The values of ρ * and τ * determine the clustering parameters and set the thunderstorm-neighborhood $\varepsilon_{\rho\tau}$. A thunderstorm is "close" to cluster C, if in its neighborhood $\varepsilon_{\rho\tau}$ there are discharges of this cluster and the discharge is "close" to the center of the cluster, called the stolp (the concentration center of the cluster of close discharges). Also, if $\varepsilon_{\rho\tau}$ -neighborhood of lightning-discharge O contains objects of two already «existing» clusters C_1 и C_2, then clusters can belong to one class.

A separate cluster of lightning discharges selected as a result of the algorithm execution is a thunderstorm cell model. Combined set of clusters - a class may be associated with a thunderstorm focus or a multi-cell thunderstorm (thunderstorm). The class may consist of one cluster, in this case we can talk about a single-cell thunderstorm.

In the computational experiment, we adopted two conditions for the parameterization of a cluster of lightning discharges: ρ_1 * = 50 km, τ_1 * = 30 min. We selected the threshold value of the distance between the discharges equal to 50 km taking into account the WWLLN's registration error for the location of the lightning discharge and the possible displacement of a thunderstorm cell during its existence (30-40 minutes) by 20-30 km (Shabaganova et al., 2012).

For each cluster, the start time (the time of the first lightning discharge assigned to this cluster) and the end time (the registration time of the last discharge included in the cluster), and the duration of the cluster's existence (the difference between the start and end times) were calculated. In addition, for each cluster, the number of discharges and the area were determined (based on the method of convex hulls, built on the thunderstorm discharges of the cluster). For each combination of two or more clusters united in a class, the time of existence and area were also calculated. The duration of the class was the difference between the time of the last and the first lightning discharge. The class area was determined similarly to the cluster area, taking into account all discharges belonging to the class.

The Example of Lightning Discharge Clustering in the South of Western Siberia

For testing cluster analysis, the days of August 7–8, 2017 were selected, when the network of meteorological stations in the south of Western Siberia predicted and observed baric formations and associated thunderstorms. According to the Federal Service for Hydrometeorology and Environmental Monitoring, those days the thunderstorm caused serious social and material damage in the Tatar district of the Novosibirsk Oblast, in the village of Soloneshnoye, Altai Krai and in the village of Turochak, Altai Republic (Svedeniya o neblagopriyatnyh …, n.d.).

On the indicated days from 7 am (local time) on August 7 and until 7 am on August 8 the WWLLN network registered more than 3,934 lightning discharges, more than 80% of which originated in the territory of Altai Republic and in the northeast Kazakhstan. At the same time, the difference in time between two successive lightning discharges of the entire sample did not exceed 12 minutes.

As a result of cluster analysis, out of 3984 lightning discharges, 249 clusters with 2 or more discharges were obtained. As a "noise" (a cluster consisting of one discharge), 65 discharges were defined, which accounted for 2% of the total experimental sample. The average duration of one thunderstorm cluster was 1 hour, the median value was 45 minutes, the minimum duration was 5 minutes, the maximum duration was 4 hours and 17 minutes. The average area was 2871 km^2, the median is 1224 km^2., The maximum area is 34559 km^2, the minimum area is 2 km^2.

Figure 6. An example of thunderstorm foci (classes) of lightning discharges registered by WWLN on August 7–8, 2017 in the territory of the Republic of Altai
(numbers in the figure indicate clusters that make up the classes indicated in map legend)

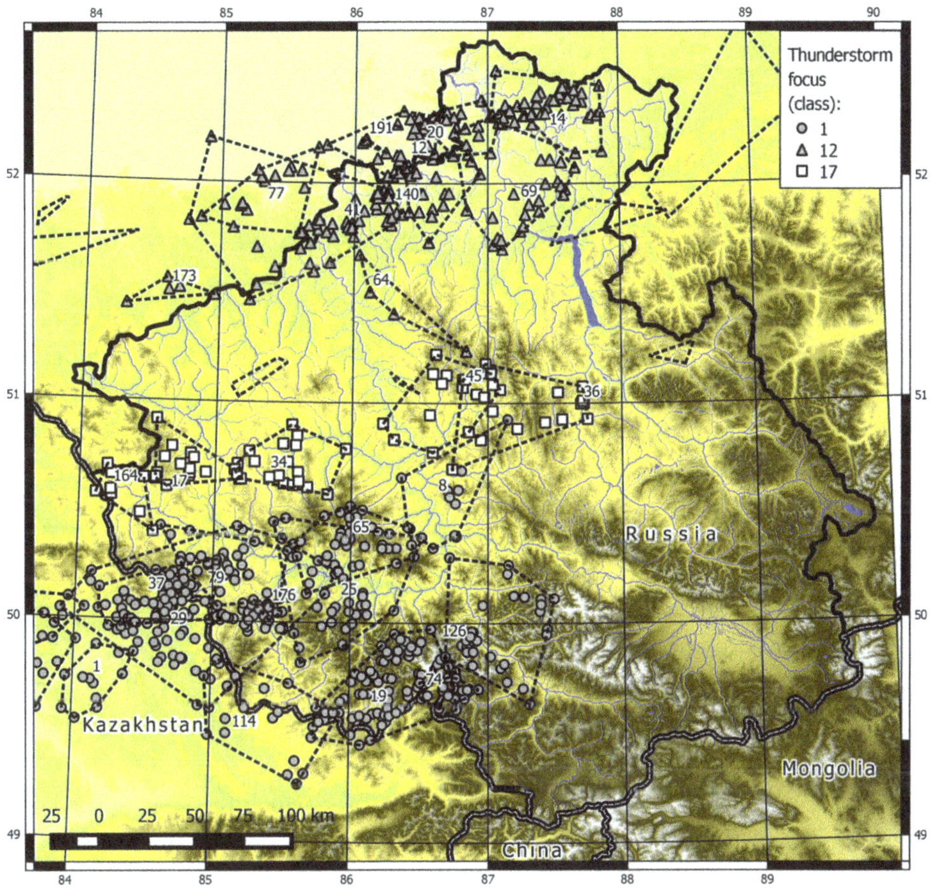

Using the clustering algorithm, the selected clusters were automatically combined into classes. 119 classes were obtained, 80% of which consisted of a single cluster. The reason for the weak integration of clusters of lightning discharges into classes is the low percentage of registration of lightning discharges by the WWLLN network. Those discharges that could be "connecting" for clusters, the network does not detect.

Thunderstorms, belonging to one class, mainly form a band 100–150 km wide, stretching from south-west to north-east to 300 km Within the boundaries of this band, the sequence of cluster formation (as a whole) has an orthogonal direction to the route of movement of air masses. The shift of active thunderstorm areas is traced as the thunderstorm develops from the southwest to the northeast.

As an example of the spatial location and operation of several thunderstorm foci, Figure 6 shows three classes of thunderstorm discharges identified using cluster analysis. Discharges of these classes form wide bands in space. The dotted lines indicate the clusters belonging to each class. Over time, clusters of lightning discharges are shifting inland of mountainous country. Of the thunderstorm foci presented in Figure 6, two operate almost simultaneously - these are the 12th and 17th classes (Table 1). An hour before the end of their existence, a new thunderstorm focus appears - class 1.

Table 1. Characteristics of the classes (thunderstorm foci), presented in Figure 6

Number of the Class (Lightning Focus)	Quantity of Clasters (Lightning Cells)	Cluster Start Time (UTC)	Cluster End Time (UTC)	Number of Discharges	Duration (minutes)
17	5	12:14:09	16:34:35	74	260
12	10	12:22:13	16:00:01	220	217
1	15	15:09:06	21:12:15	459	363

Example of Lightning Discharge Clustering on the Territory of Buryatia

For cluster analysis of lightning discharges for the territory of Buryatia, data on lightning discharges recorded on July 30-31, 2015 were used. These dates were chosen in connection with the observed aggravation of the fire hazard situation in the study area in late July and early August. During this period, about 280 wildland fires, caused by lightning discharges, were recorded in hard-to-reach mountainous areas (Tulokhonov & Puntsukova, 2016).

On July 30-31, 2015, around 3900 lightning discharges were recorded on the territory of Buryatia by the WWLLN network. During the computational experiment, 79 clusters were identified. The average duration of a thunderstorm cell was 2 hours and 10 minutes, the median value was 1 hour and 38 minutes, the minimum duration was 8 minutes, and the maximum duration was 8 hours and 10 minutes. The average area of thunderstorm cells was 3,994 square kilometers, the median is 2,186 square kilometers, the maximum area is 33,454 square kilometers, and the minimum area is 1 square kilometer. At the same time, the largest area corresponds to the longest thunderstorm cells with the longest existence (from 5 to 8 hours).

Using the clustering algorithm, 4 classes were identified for which the space-time conjugation of clusters was established. These classes consisted of a different number of thunderstorm clusters: one focus consisted of 8 thunderstorm clusters, the rest consisted of 2 to 5 clusters. Figure 7 shows the clusters and the two most long-lasting multi-cluster classes that operated from 0:46 to 13:20 (UTC) on July 31, 2015. Table 2 shows the characteristics of the highlighted areas for the activation of lightning discharges, shown in Figure 6. In the table, isolated clusters are shown in italics, which were not combined into a class. They are separate thunderstorm cells. Clusters with a duration of more than 3 hours are highlighted in bold.

With this example, we illustrate a more detailed order of time and space when clusters of thunderstorm discharges occur in the west and east of Buryatia, where almost all of the discharges were recorded during the considered period of time.

Lightning activity in the west of the region in the Eastern Sayan area is higher than in the east of the region. It is noted that there are areas where a long and intense thunderstorm cluster or a multi-cluster class is formed, on the periphery of which thunderstorm cells arise with less lightning activity.

A vivid example of the presence of a "central" cluster within a multi-cluster class is the 37th class (Table 2, Figure 7). In this class, the "central" place is occupied by the 41st cluster, which lasted more than 8 hours. The activation of the lightning-discharge process begins in the north from the 37th cluster, an hour before the appearance of the 41st cluster. Clusters, on the northern periphery of the 41st storm center, have a duration of more than 3 hours. The appearance of lightning here ceases an hour and a half before the fade of the 41st cluster.

Figure 7. Storm cells and thunderstorm focus operating on the territory of the Republic of Buryatia on July 31 from 0:46 am to 1:20 pm (UTC)
(the numbers in the figure indicate the clusters that make up the classes indicated in the map legend)

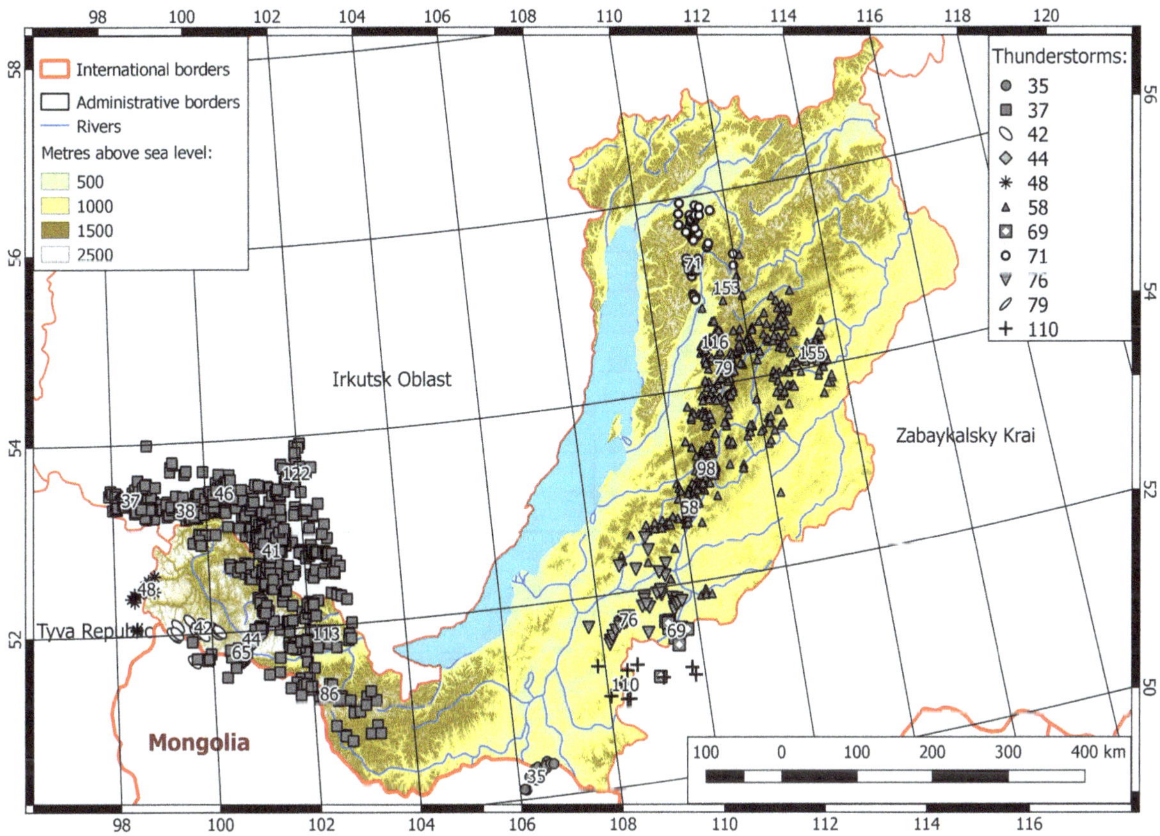

Half an hour after the start of the intensive 41st thunderstorm focus, its western and southwestern periphery is activated — the 42nd, 44th and 48th clusters. These clusters last about an hour, they are isolated from each other and located at the distance of more than 50 km from the 41st cluster operating simultaneously with them.

Next comes the activation of the southern periphery of the 41st cluster - the 65th, 86th and 113th clusters. At the same time, lightning discharges occur here simultaneously with discharges on the northern periphery. So, the 65th cluster in the south operates simultaneously with the 46th in the north, the 113th cluster with the 122nd cluster. Termination of the lightning discharge activity of the 41st thunderstorm focus and its periphery occurs with the fade of the 86th cluster.

In the eastern part of Buryatia, lightning discharges are confined mainly to the eastern spurs of the Ulan-Burgasy and Ikatsky ranges. The most active and long-lasting clusters of lightning discharges of the 58th, 98th, 116th and 115th, combined into a single multi-cell 58th class, occur at intervals of 1–2 hours from south to north of this part of the territory of Buryatia (table 2, figure 7).

On the southern periphery of the 58th class, lightning-discharge processes occur, outlined by the 69th, 76th and 110th individual clusters. Among these clusters, the 76th cluster occupies a "central" place, since the 69th and 110th clusters arise on its southeastern periphery. The 71st and 79th individual

Table 2. Characteristics of thunderstorm clusters (cells) and classes (thunderstorm foci), presented in Figure 7

Number of the Class (Lightning Focus)	Number of the Cluster (Lightning Cell)	Cluster Start Time (UTC)	Cluster End Time (UTC)	Number of Discharges	Duration (minutes)
South of Buryatia					
35	*35*	*0:46:47*	*2:46:27*	*19*	*119*
West of Buryatia					
37	**37**	**1:10:13**	**5:34:06**	**76**	**263**
37	**38**	**1:49:49**	**5:30:03**	**62**	**220**
37	**46**	**3:29:31**	**7:53:36**	**126**	**264**
37	122	6:54:57	8:53:00	36	118
37	**41**	**2:13:59**	**10:24:22**	**214**	**490**
42	*42*	*2:44:07*	*4:01:25*	*17*	*77*
44	*44*	*3:09:57*	*3:49:41*	*6*	*39*
48	*48*	*3:32:06*	*4:36:32*	*21*	*64*
37	65	4:38:49	6:54:13	45	135
37	**86**	**5:28:56**	**10:31:02**	**41**	**302**
37	113	6:36:57	8:35:35	27	118
East of Buryatia					
69	*69*	*4:41:59*	*5:36:00*	*13*	*54*
76	***76***	***5:10:07***	***9:06:30***	***48***	***236***
110	*110*	*6:29:19*	*8:07:17*	*11*	*97*
58	**58**	**4:04:04**	**7:34:54**	**56**	**210**
71	*71*	*4:43:40*	*7:29:37*	*68*	*165*
79	*79*	*5:16:05*	*6:10:48*	*9*	*54*
58	**98**	**5:47:17**	**10:15:50**	**137**	**268**
58	**116**	**6:44:17**	**10:59:42**	**134**	**255**
58	153	8:50:13	9:37:33	8	47
58	**155**	**8:53:48**	**13:20:14**	**109**	**266**

clusters can be attributed to the northern periphery of the 58th class, which end almost simultaneously with the 58th cluster - the first cluster of this class.

Thus, using cluster analysis, we managed to emphasize the areas of active formation of lightning discharges, determine the nature of their formation in time and space. Such information may be useful for the detailed determination of possible hazardous areas during the passage of a separate thunderstorm.

FUTURE RESEARCH DIRECTIONS

The results of the presented research can be used to develop methods and systems for predicting the occurrence of forest fires due to thunderstorms. The development of such methods requires the zoning of the territory according to the degree of its fire danger, as well as the comparison of the designated fire hazardous areas with the local level of thunderstorm activity. An important role is played by the identification of the time interval during which fires were detected after passing thunderstorms.

For this kind of research, it is important to synthesize several directions: obtaining and comparing the characteristics of thunderstorm activity and forest burning averaged over a long-term period; consideration and study of individual thunderstorms and their characteristics (lightning, amperage, time of appearance, etc.). For example, in periods of massive occurrence of fires, it is important to identify the trigger mechanism of its occurrence. In the latter case, cluster analysis of lightning discharges can be used to define thunderactive foci. Often the spatial dissociation of areas of maximum activity of thunderstorms and ground lightning discharges with areas with the highest fire rate significantly complicates not only the forecast, but also the detection of fires and consequently firefighting.

It should be reminded that on the territory of the Russian Federation for the operational monitoring of fires in Forest Protection Aviation from thunderstorms, a system for registering lightning discharges operates. This system covers the areas most exposed to thunderstorm fire hazards (40-80% of fires from thunderstorms), which include Amur river (Khabarovsk Krai), Tomsk Oblast, sparsely populated territories of Yakutia, Irtyish Land, southern and middle taiga of the Krasnoyarsk Krai (Ivanov & Ivanova, 2010). In addition, on the territory of Yakutia, a network of one-point detectors additionally operates and WWLLN data is used to analyze the fire-danger of forests, due to lightning discharges. When using these regional lightning direction systems, Forest Protection Aviation its time and routes of patrols in the potentially fire hazardous areas after thunderstorms.

In the mountains of southern Siberia, the percentage of forest fires caused by thunderstorms averages 15–20%. However, there have been cases of massive occurrence of fires in some years (Dmitriev et al., 2011). Thunderstorm fires were sometimes noted in inaccessible areas and due to the late detection of their area reached large sizes, which led to great economic damage. Such fire-hazardous situations are characteristic during periods of extreme weather conditions: a drought with the subsequent development of intense dry thunderstorms, or vice versa a dry period after the passage of intense thunderstorms. At the same time, the factors of the geological and geophysical environment and related processes are not excluded. For example, the increase in the frequency of ground lightning discharges in zones of tectonic disturbances due to the occurrence of induced magnetotelluric currents, deep degassing and changes in the structure of the pressure field (Dmitriev et al., 2011; Kutinov & Chistova, 2012). Against the background of the modern increase in the number of extreme weather conditions, including increased thunderstorm intensity and the intensification of geodynamic processes in the Altai-Sayan folded region, a comprehensive study of thunderstorm fire conditions, including vegetation, geographical and geological and physical factors is relevant.

The source of data on lightning activity, as was shown above, in the south of Siberia is WWLLN data. The percentage of discharges recorded by this network is small, but the network mainly records the strongest lightning discharges. It should be noted that it is precisely the strong lightning discharges of the cloud-to-ground type that lead to fires. Thus, even such a percentage of the registered discharges from all of their possible mabyfold arising during a thunderstorm, WWLLN data have practical significance in determining, under other necessary conditions, potentially fire hazardous areas. In this case, clusters of lightning discharges allocated on the basis of WWLLN, in our opinion, can serve as indicators of areas with active generation of ground lightning discharges. This makes further detailed and comprehensive studies of the conditions of occurrence of natural fires from thunderstorms using WWLLN data in the mountains of southern Siberia prospective and relevant.

CONCLUSION

The dependence of forest fires on thunderstorm activity is complex and requires taking into account many factors, from the state of the underlying surface to the time and nature of thunderstorms. However, the general connection of lightning activity and the number of forest fires cannot be denied. Therefore, the study of the spatial-temporal distribution of thunderstorms and lightning is of great importance for the successful prediction and timely localization of fire sites.

The instrument base for measuring lightning activity has significantly improved over the past 30 years, which makes it possible to clarify and specify the patterns of distribution of thunderstorms on a global and regional scale. The results of the analysis of the spatial-temporal distribution of the number of days with thunderstorms and the duration of thunderstorms, obtained on the basis of multi-year data from the network of meteorological stations, are confirmed by studies of lightning activity based on modern instrumental measurements. At the same time, along with the details of the areal distribution of lightning activity, the features of seasonal dynamics are additionally revealed. For the mountains of the south of Siberia, on the basis of WWLLN data, the dependence of the density of lightning discharges on the terrain elevation was confirmed. On average, the maximum thunderstorm and lightning activity occurs in the afternoon hours. On the basis of the WWLLN data, the second evening maximum of lightning discharges at 19-20 was detected. In addition, in some years, lightning activity in the evening and night hours was equal to the afternoon first maximum. In the course of the study, the most hazardous parts of the mountains of southern Siberia were detailed, which included areas with a high density of lightning discharges.

The cluster analysis of lightning discharges made it possible to trace the spatial dynamics of regions with the active generation of discharges during the passage of thunderstorms. At the same time, separate clusters of lightning discharges and their groups which have an intersection in time and space were associated with the thunderstorm focus. A cluster of lightning discharges is a model of a thunderstorm cell, operating for one to two hours. A thunderstorm focus, formed from a group of clusters, is a model of a multi-cell long-lasting thunderstorm. Taking into account the fact that WWLLN registers the strongest lightning discharges (mostly cloud-to-ground discharges), their clusters are a model of the "core" of a thunderstorm cell at the stage of its maturity (in the classical understanding of the physics of thunderstorm formation). Since ground lightning discharges cause fires of ground objects, WWLLN data are acceptable for identifying potentially fire hazardous areas. These data are particularly relevant in the absence of more efficient and accurate lightning direction finding systems.

It must be emphasized that lightning is a part of the global electrical circuit. Specific conditions for the localization of electrical discharges are created on the surface of the Earth, which indicates the interaction between several spheres of the Earth. Integral consideration of these characteristics is relevant at the present time from the standpoint of fundamental science and applied science - the prediction of the places of occurrence of forest fires caused by lightning strikes. For the further development of methods for fire prediction by thunderstorms, an integrated approach is needed, taking into account the combination of forest-growing, geographical and geological-geophysical conditions of the study region.

ACKNOWLEDGMENT

This work supported by Russian Foundation for Basic Researches. Scientific project N 17-29-05093.

The authors wish to thank the World Wide Lightning Location Network, a collaboration among over 50 universities and institutions, for providing the lightning location data used in this paper.

REFERENCES

Adzhiev, A. H., & Bogachenko, E. M. (2011). *Grozy Severnogo Kavkaza (Thunderstorms of the North Caucasus)*. Nalchik, Russian Federation: Poligrafservis i T.

Adzhieva, A. A., Shapovalov, V. A., Boldyrev, A. S., & Ignatyev, V. V. (2018, November). Characteristics of lightning activity in the North Caucasus according to the LS 8000 from 2009 to 2016. In *Remote sensing of clouds and the atmosphere XXIII* (Vol. 10786, p. 107861A). International Society for Optics and Photonics.

Bailey, J. C., Blakeslee, R. J., Buechler, D. E., & Christian, H. J. (2007). *Diurnal lightning distributions as observed by the Optical Transient Detector (OTD) and the Lightning Imaging Sensor (LIS)*. Retrieved from https://ntrs.nasa.gov/archive/nasa/casi.ntrs.nasa.gov/20070038367.pdf

Betz, H. D., Schumann, U., & Laroche, P. (Eds.). (2008). *Lightning: principles, instruments and applications: review of modern lightning research*. Springer Science & Business Media. doi:10.1007/978-1-4020-9079-0

Burcev, A. V., Nevretdinov, Yu. M., & Sytina, A. N. (2014). Opyt registracii grozovoj aktivnosti na territorii Kolskogo poluostrova (Experience of registration of storm activity in the territory of the Kola Peninsula). *Vestnik Kol'skogo nauchnogo centra RAN, (2), 130-137.*

Christian, H. J., Blakeslee, R. J., Boccippio, D. J., Boeck, W. L., Buechler, D. E., Driscoll, K. T., ... Stewart, M. F. (2003). Global frequency and distribution of lightning as observed from space by the Optical Transient Detector. *Journal of Geophysical Research, D, Atmospheres*, *108*(D1), ACL-4. doi:10.1029/2002JD002347

Conedera, M., Cesti, G., Pezzatti, G. B., Zumbrunnen, T., & Spinedi, F. (2006). Lightning-induced fires in the Alpine region: An increasing problem. *Forest Ecology and Management*, *234*(1), S68. doi:10.1016/j.foreco.2006.08.096

Data User Guide. International Space Station (ISS) Lightning Imaging Sensor (LIS) Provisional datasets. Retrieved from https://ghrc.nsstc.nasa.gov/pub/lis/iss/doc/isslis_dataset.pdf

Dmitriev, A. N., Krechetova, S. Yu., & Kocheeva, N. A. (2011). *Grozy i lesnye pozhary ot groz na territorii Respubliki Altaj (Thunderstorms and wildfires from thunderstorms in the territory of Altai Republic)*. Gorno-Altaysk, Russian Federation: Publishing House of the Gorno-Altaysk State University.

Dowden, R. L., Brundell, J. B., & Rodger, C. J. (2002). VLF lightning location by time of group arrival (TOGA) at multiple sites. *Journal of Atmospheric and Solar-Terrestrial Physics*, *64*(7), 817–830. doi:10.1016/S1364-6826(02)00085-8

EUCLID (EUropean Cooperation for LIghtning Detection). Retrieved from http://www.euclid.org

Ganteaume, A., Camia, A., Jappiot, M., San-Miguel-Ayanz, J., Long-Fournel, M., & Lampin, C. (2013). A review of the main driving factors of forest fire ignition over Europe. *Environmental Management*, *51*(3), 651–662. doi:10.100700267-012-9961-z PMID:23086400

Geostationary Lightning Mapper (GLM). Retrieved from https://ghrc.nsstc.nasa.gov/lightning/overview_glm.html

Gorbatenko, V. P., & Ershova, T. V. (2011). *Molnija kak zveno global'noj jelektricheskoj cepi (Lightning as link of a Global Electrical Circuit)*. Tomsk, Russian Federation: Izdatel'stvo TGPU.

Gurney, R. J., Foster, J. L., Gurney, R. J., & Parkinson, C. L. (1993). *Atlas of satellite observations related to global change*. Cambridge, UK: Cambridge University Press.

Hutchins, M. L., Holzworth, R. H., & Brundell, J. B. (2014). Diurnal variation of the global electric circuit from clustered thunderstorms. *Journal of Geophysical Research: Space Physics*, *119*(1), 620–629.

Hutchins, M. L., Holzworth, R. H., Brundell, J. B., & Rodger, C. J. (2012). Relative detection efficiency of the world wide lightning location network. *Radio Science*, *47*(6), n/a. doi:10.1029/2012RS005049

Ivanov, V. A., & Ivanova, G. A. (2010). *Pozhary ot groz v lesah Sibiri (Wildfires from thunderstorms in Siberia)*. Novosibirsk, Russian Federation: Nauka.

Jacobson, A. R., Cummins, K. L., Carter, M., Klingner, P., Roussel-Dupré, D., & Knox, S. O. (2000). FORTE observations of lightning radio-frequency signatures: Prompt coincidence with strikes detected by the National Lightning Detection Network. *Journal of Geophysical Research*, *105*, 15653–15662. doi:10.1029/2000JD900103

Johnson, E. A. (1996). *Fire and the vegetation dynamics: studies from the North American boreal forest*. Cambridge, UK: Cambridge University Press.

Kachurin, L. G., & Divinskij, L. D. (Eds.). (1992). Aktivno-passivnaya radiolokaciya grozovyh i grozoopasnyh ochagov v oblakah (Active-passive radar observation of thunderstorm and thunderstorm foci in the clouds). Izvestiya Akademii nauk SSSR. Fizika atmosfery i okeana, 24(6), 601-612.

Karanina, S. Yu., Kocheeva, N. A., & Karanin, A. V. (2017). Prostranstvennoe i vremennoe raspredelenie molnievyh razryadov po territorii Altae-Sayanskogo regiona (Spatial and temporal distribution of lightning discharges on the territory of the Altai-Sayan region). *Izvestiya vysshih uchebnyh zavedenij. Severo-Kavkazskij region. Estestvennye nauki,* (4-1 (196)).

Kashinskii, S. D. (Ed.). (1979). *Opasnye javlenija pogody na territorii Sibiri i Urala. Chast 1. Altajskij kraj, Kemerovskaja, Novosibirskaja i Tomskaja oblasti (The dangerous phenomena of weather in the territory of Siberia and the Urals. Part 1. Altai Krai, Kemerovo, Novosibirsk, and Tomsk regions)*. Leningrad, USSR: Gidrometeoizdat.

Kirkland, M. W., Suszcynsky, D. M., Guillen, J. L. L., & Green, J. L. (2001). Optical observations of terrestrial lightning by the FORTE satellite photodiode detector. *Journal of Geophysical Research, D, Atmospheres*, *106*(D24), 33499–33509. doi:10.1029/2000JD000190

Kocheeva, N. A., Chankibaeva, M. H., Minaev, A. I., Sukhova, M. G., & Modorov, A. A. (2017). Peculiarities of thunderstorms' occurrences on the border of the Western Siberia Plains and Altai Mountains. *Journal of Environmental Management & Tourism*, *8*(5), 21, 1061–1068.

Kononov, I. I., Krutoj, D. M., & Yusupov, I. E. (2013). Alternativnoe obespechenie grozobezopasnosti na ob'ektah vodnogo transporta (Alternative providing safety thunderstorm on objects of the water transport). *Vestnik gosudarstvennogo universiteta morskogo i rechnogo flota im. admirala S.O. Makarova, 3*(19), 116-121.

Kononov, I. I., & Yusupov, I. E. (2004). Cluster analysis of storm activity. [Radiotekhnika i elektronika]. *Journal of Communications Technology and Electronics*, *49*(3), 260–266.

Konstantinova, D. A., & Gorbatenko, V. P. (2011). Rezultaty registracii molnij nad yugo-vostochnoj territoriej Zapadnoj Sibiri (Results lightnings registration over the southeast territory of Western Siberia). *Izvestiya vysshih uchebnyh zavedenij Fizika*, *54*(11), 156–162.

Konstantinova, D. A., Gorbatenko, V. P., & Polyakov, D. V. (2017, November). Spatial distribution of the thunderstorm activity characteristics for the territory of Western Siberia. In *23rd International Symposium on Atmospheric and Ocean Optics: Atmospheric Physics* (Vol. 10466, p. 1046658). International Society for Optics and Photonics.

Kozlov, V. I., & Mullayarov, V. A. (2004). *Grozovaya aktivnost v Yakutii (Thunderstorm activity in Yakutia)*. Yakutsk, Russian Federation: YaF Izd-va SO RAN.

Kozlov, V. I., Mullayarov, V. A., Grigorev, Y. M., & Tarabukina, L. D. (2014). Parameters of thunderstorm activity and lightning discharges in Central Yakutia from 2009 to 2012. *Izvestiya. Atmospheric and Oceanic Physics*, *50*(3), 323–329. doi:10.1134/S0001433814030086

Kozlov, V. I., Mullayarov, V. A., & Karimov, R. R. (2011). Prostranstvennoe raspredelenie plotnosti grozovyh razryadov na Vostoke Rossii po dannym distancionnyh nablyudenij (The spatial distribution of the lightning in the east of Russia on remote sensing data). *Sovremennye problemy distancionnogo zondirovaniya Zemli iz kosmosa, 8*(3), 257.

Kozlov, V. I., Mullayarov, V. A., & Solovev, V. S. (2009). Lesnye pozhary v Yakutii ot groz (Wildfires from thunderstorms in Yakutia). *Sovremennye problemy distancionnogo zondirovaniya Zemli iz kosmosa, 2*(6), 388.

Kutinov, Yu. G., & Chistova, Z. B. (2012). Kompleksnaya model' protsessov mezhgeosfernogo vzaimodeystviya v tektonicheskikh uzlakh Severa Russkoy plity (Complex model of geospheres interaction processes in the tectonic nodes at the North of Russian Plate). *Prostranstvo i Vremya,* l(1).

Larjavaara, M., Pennanen, J., & Tuomi, T. J. (2005). Lightning that ignites forest fires in Finland. *Agricultural and Forest Meteorology*, *132*(3-4), 171–180. doi:10.1016/j.agrformet.2005.07.005

Light, T., Davis, S. M., Boeck, W. L., Jacobson, A. R., & Suszcynsky, D. M. (2003). *Global optical lightning flash rates determined with the Forte satellite (No. LA-UR-03-2614)*. Los Alamos, NM: National Laboratory.

Lobodin, T. V. (1986). Karta summarnogo chisla grozovyh razryadov dlya territorii SSSR. (The map of total number of lightning discharges for the territory of the USSR). *Tr. GGO,* (498), 72-76.

MacGorman, D. R., Maier, M. W., & Rust, W. D. (1984). *Lightning strike density for the contiguous United States from thunderstorm duration records (No. NUREG/CR--3759)*. National Oceanic and Atmospheric Administration.

Mareev, E. A., Stasenko, V. N., Bulatov, A. A., Dement'eva, S. O., Evtushenko, A. A., Il'in, N. V., ... Shatalina, M. V. (2016). Russian studies of atmospheric electricity in 2011–2014. *Izvestiya. Atmospheric and Oceanic Physics*, *52*(2), 154–164. doi:10.1134/S0001433816020080

Moskovenko, V. M., Znamenshchikov, B. P., & Zolotarev, S. V. (2012). Primenenie sistemy grozopelengacii Vereya-MR" v interesah elektroenergetiki Rossii (Application of a system of a fixing of lightnings "Vereya-MR" for the benefit of power industry of Russia). *Novoe v rossijskoj elektroenergetike,* (2), 15-23.

Mullayarov, V. A., Toropov, A. A., Kozlov, V. I., & Karimov, R. R. (2009). Patterns of spatial distribution of positive thunderstorm discharges in Eastern Siberia. *Russian Meteorology and Hydrology*, *34*(6), 364–370. doi:10.3103/S1068373909060041

Müller, M. M., & Vacik, H. (2017). Characteristics of lightnings igniting forest fires in Austria. *Agricultural and Forest Meteorology*, *240*, 26–34. doi:10.1016/j.agrformet.2017.03.020

National atlas of Russia. Thunderstorms. Retrieved from http://xn--80aaaa1bhnclcci1cl5c4ep.xn--p1ai/cd2/181/181.html

Proctor, D. E., Uytenbogaardt, R., & Meredith, B. M. (1988). VHF radio pictures of lightning flashes to ground. *Journal of Geophysical Research, D, Atmospheres*, *93*(D10), 12683–12727. doi:10.1029/JD093iD10p12683

Rakov, V. A., & Uman, M. A. (2003). *Lightning: physics and effects*. Cambridge, UK: Cambridge University Press. doi:10.1017/CBO9781107340886

Real-Time Lightning Map. Retrieved from http://www.lightningmaps.org]

Shabaganova, S. N., Karimov, R. R., Kozlov, V. I., & Mullayarov, V. A. (2012). Characteristics of storm cells from observations in Yakutia. *Russian Meteorology and Hydrology*, *37*(11-12), 746–751. doi:10.3103/S1068373912110088

Shindo, T., & Yokoyama, S. (1998). Lightning occurrence data observed with lightning location systems in Japan: 1992-1995. *IEEE Transactions on Power Delivery*, *13*(4), 1468–1474. doi:10.1109/61.714797

Smith, C. B. (2008). *Lightning: Fire from the Sky*. Dockside Consultants Inc.

Smith, D. A., Heavner, M. J., Jacobson, A. R., Shao, X. M., Massey, R. S., Sheldon, R. J., & Wiens, K. C. (2004). A method for determining intracloud lightning and ionospheric heights from VLF/LF electric field records. *Radio Science*, *39*(1), n/a. doi:10.1029/2002RS002790

Solovyev V. A. (1969). Prostranstvenno-vremennye harakteristiki grozovoj aktivnosti po dannym pelengovaniya atmosferikov (Space-time characteristics of storm activity by results of a bearing of atmospherics). *Tr. GGO* (242), 52-61.

Solovyev, V. S., Kozlov, V. I., Karimov, R. R., & Vasiliev, M. S. (2010). Kompleksnyj monitoring grozovoj aktivnosti i lesnyh pozharov po dannym nazemnyh i sputnikovyh nablyudenij (Simultaneous monitoring of thunderstorm activity and forest fires on data of ground-base and remote sensing observations). *Sovremennye problemy distantsionnogo zondirovaniya Zemli iz kosmosa. 7*(4), 218.

Stepanenko, V. D., & Galperin, S. M. (1983). *Radiotekhnicheskie metody issledovaniya groz (Radio Means for Thunderstorm Detection)*. Leningrad, USSR: Gidrometeoizdat.

Suszcynsky, D. M., Kirkland, M. W., Jacobson, A. R., Franz, R. C., Knox, S. O., Guillen, J. L. L., & Green, J. L. (2000). FORTE observations of simultaneous VHF and optical emissions from lightning: Basic phenomenology. *Journal of Geophysical Research, D, Atmospheres*, *105*(D2), 2191–2201. doi:10.1029/1999JD900993

Svedeniya o neblagopriyatnyh usloviyah pogody i opasnyh gidrometeorologicheskih yavleniyah, nanesshih social'nye i ekonomicheskie poteri na territorii Rossii (Information on adverse weather conditions and hydrometeorological hazards that caused social and economic losses in Russia). Retrieved from http://meteo.ru/data/310-neblagopriyatnye-usloviya-pogody-nanjosshie-ekonomicheskie-poteri

Tarabukina, L. D., & Kononova, N. K. (2018). Cirkulyaciya atmosfery, grozy i lesnye pozhary v Severnoj Azii v 2009-2016 gg (Circulation of the atmosphere, thunderstorm, and wildfires in Northern Asia in 2009-2016). *Zhizn Zemli*, *40*(1), 22–30.

Tarabukina, L. D., & Kozlov, V. I. (2018). Sravnenie izmerenij neskolkih sistem registracii molnievyh radioimpulsov (Comparison of measurements of several systems of registration lightning of radio impulses). *Vestnik Severo-Vostochnogo federal'nogo universiteta im. M.K. Ammosova,* 2(64), 77-86.

Tarabukina, L. D., Kozlov, V. I., Karimov, R. R., & Mullayarov, V. A. (2017). Spatial pattern of lightning strikes in North Asia. *Russian Meteorology and Hydrology*, *42*(2), 88–94. doi:10.3103/S1068373917020029

Tulokhonov, A. K., & Puntsukova, S. D. (2016) Forest fires in the Republic of Buryatia in a changing climate. *Society: Politics, economics, law,* (3), 72-78.

Tuomi, T. J., & Mäkelä, A. (2008). Thunderstorm climate of Finland 1998–2007. *Geophysica*, *44*(1-2), 67–80.

Unique Vaisala Global Lightning Dataset GLD360. Retrieved from https://www.vaisala.com/sites/default/files/documents/WEA-MET-GLD360%20Brochure-B211271EN.pdf

Virts, K. S., Wallace, J. M., Hutchins, M. L., & Holzworth, R. H. (2013). Highlights of a new ground-based, hourly global lightning climatology. *Bulletin of the American Meteorological Society*, *94*(9), 1381–1391. doi:10.1175/BAMS-D-12-00082.1

Wierzchowski, J., Heathcott, M., & Flannigan, M. D. (2002). Lightning and lightning fire, central cordillera, Canada. *International Journal of Wildland Fire*, *11*(1), 41–51. doi:10.1071/WF01048

World Meteorological Organization (WMO). (1956). World distribution of thunderstorms days [Geneva, Switzerland.]. *Publ.*, *21*, TP96.

World Wide Lightning Location Network (WWLLN). Retrieved from wwlln.net.

Zagoruiko, N. G., Borisova, I. A., Dyubanov, V. V., & Kutnenko, O. A. (2008). Methods of recognition based on the function of rival similarity. *Pattern Recognition and Image Analysis*, *18*(1), 1–6. doi:10.1134/S105466180801001X

ADDITIONAL READING

Betz, H. D., Schumann, U., & Laroche, P. (Eds.). (2008). *Lightning: principles, instruments and applications: review of modern lightning research.* Springer Science & Business Media.

Borisova, I. A. (2007). Clustering algorithm FRiS-Tax. Scientific bulletin of NGTU, 3, 3-12.

Ganteaume, A., Camia, A., Jappiot, M., San-Miguel-Ayanz, J., Long-Fournel, M., & Lampin, C. (2013). A review of the main driving factors of forest fire ignition over Europe. *Environmental Management*, *51*(3), 651–662.

Hutchins, M. L., Holzworth, R. H., & Brundell, J. B. (2014). Diurnal variation of the global electric circuit from clustered thunderstorms. *Journal of Geophysical Research: Space Physics*, *119*(1), 620–629.

Hutchins, M. L., Holzworth, R. H., Brundell, J. B., & Rodger, C. J. (2012). Relative detection efficiency of the world wide lightning location network. *Radio Science*, *47*(6).

Rakov, V. A., & Uman, M. A. (2003). *Lightning: physics and effects*. Cambridge University Press.

Smith, C. B. (2008). *Lightning: Fire from the Sky*. Dockside Consultants Inc.

KEY TERMS AND DEFINITIONS

Altai-Sayan Mountain Country: The highest elevation in Northern Asia, between the temperate regions of Western Siberia and the desert territories of Central Asia.

Cluster analysis: A method of uniting groups (clusters) of objects of research according to the principle of their proximity. The object is a point of multidimensional space, where its coordinates are given by the values of several marks. The measure of proximity of objects is given in different ways, for example, the Euclidean distance.

Density of Lightning Discharges: The ratio of the number of lightning discharges to the area of the site in which they are registered.

Lightning: An electrical spark in the atmosphere that usually occurs during a thunderstorm.

The Republic of Buryatia: An administrative unit within the Russian Federation.

Thunderstorm: An atmospheric phenomenon, in which electrical discharges occur within the clouds or between the clouds and the earth's surface - lightning accompanied by thunder.

Thunderstorm focus: A region of space with an active generation of lightning.

World-Wide Lightning Location Network (WWLLN): An international network for registering lightning discharges on the globe. WWLLN was based at the University of Washington in 2002 and currently includes more than 70 receiving stations.

Chapter 7
Cluster Analysis Algorithms for RS and WWLLN Data Processing

Marina Yurevna Belikova
Gorno-Altaisk State University, Russia

Alyona Viktorovna Glebova
Gorno-Altaysk State University, Russia

ABSTRACT

Thunderstorm activity is indirectly taken into account when using statistics on forest fires. Another way is to use the data of the lightning discharge finding sensors. This chapter proposes using the data of the World Wide Lightning Location Network (WWLLN). To assess the probability of forest fire occurrence, it is necessary to know the energy and the spatial distribution of lightning discharges. For the analysis of the data of the storm-direction-finding WWLLN, it is proposed to use clustering algorithms. For the computational experiment, the region covering the Timiryazevskiy forestry of the Tomsk region (55.93 - 56.86) on the north, (83.94 - 85.07), and the data on lightning discharges registered by the WWLLN network in this region in July 2014. The sample data were 273 lightning discharges. The results of clustering are presented, as well as the image of the silhouette index for each object, the average value of the ASW index for grouping solutions obtained using the complete, kmeans, and dbscan algorithms.

INTRODUCTION

Forest fires occur due to anthropogenic and natural causes [Baranovskiy & Kuznetsov, 2017]. The main reason among natural is thunderstorm activity [Wang, Zheng, Zhang & Liu, 2017]. It is necessary to take into account the thunderstorm activity [Deeming, Lancaster, Fosberg, Furman & Schroeder, 1972; Van Wagner, 1987; Grishin & Filkov, 2011] to develop forest fire forecasting systems. This can be implemented in various ways. For example, thunderstorm activity is indirectly taken into account when using statistics on forest fires [Nieto, Aguado, García & Chuvieco, 2012]. Typically, the causes of forest fires are indicated in books / logs [Data archive on forest fires, 2010]. Another way is to use archival monthly averages of thunderstorm activity in a controlled area [Scientific-applied reference book, 1993]. However,

DOI: 10.4018/978-1-7998-1867-0.ch007

the use of long-term averaged data can lead to errors in the forecast under severe weather changes. Another way is to use the data of the special sensors [Müller & Vacik, 2017]. In this paper, it is proposed to use data from the World Wide Lightning Location Network (WWLLN) [Rodger, Werner, Brundell, Lay, Thomson, Holzworth & Dowden, 2006]. It is necessary to know the energy and the spatial distribution of lightning discharges to assess the probability of forest fire occurrence. It is proposed to use clustering algorithms [Everitt, Landau, Leese & Stahl, 2011] to analyze the data of the storm-direction-finding WWLL Network. Cluster analysis has proved itself well in solving various problems, including the data processing of lightning-seeking systems. The purpose of the cluster analysis of lightning discharge data in most cases is to obtain information on the number of thunderstorms and their average duration. The result of the data clustering on lightning discharges in the "ideal case" should be a group of discharges that corresponds to one and only one thunderstorm. The purpose of this chapter is to compare the results of WWLLN data clustering with different algorithms and select an algorithm whose application to the WWLLN data gives results close to thunderstorm.

BACKGROUND

One of the causes of forest fires are lightning discharges. Lightning is an electrical discharge caused by separation into positive and negative charges in the clouds, which leads to a potential difference of the order of 10-100 MV [Latham & Williams, 2001]. In order to discharge is occur, the presence of water in all three phases is necessary (solid, liquid and gaseous) [Williams, 1989].

Thunderstorms are divided into intramass and frontal ones. Intra-mass thunderstorms over the continent arise as a result of local warming of air from the Earth's surface, which leads to the development in it of the ascending streams of local convection and to the formation of powerful cumulonimbus clouds. Frontal thunderstorms occur at the boundaries between warm and cold air masses [Kozlov & Mulyayarov, 2004]. Discharges may be classes of cloud-to-cloud and cloud-to-ground. Approximately 90% of the cloud-to-ground discharges are negative, and the nature of the remaining 10% of the positive discharges is not fully understood [Latham, 1991]. Discharges of the cloud-to-ground class [Ivanov, 1996] can lead to the forest fire. Energetic characteristics of the positive and negative cloud-to-ground lightning discharges are different. These differences are significant in terms of forest fuel ignition. Energy reaches the surface in one stroke as a result of the majority of positive discharges. In the same time, for the negative discharges a multi-stroke is characteristic [Uman, 1969].

Statistics on cloud-to-ground lightning discharges are collected within the framework of the National Lightning Detection Network in the United States [Cummins, Murphy, Bardo, Hiscox, Pyle, & Pifer, 1998]. This system can identify most of the cloud-to-ground discharges in the US and Canada with a spatial resolution of several kilometers and a time resolution accuracy of 1 ms. Data on the polarity of the discharge, the peak current, the impact complexity (single or multi-stroke) are archived as a result of the operation of the system [Latham, 1991].

Cloud-to-ground lightning discharges are recorded in other countries. Between 1992 and 2001, a series of observations on cloud-to-ground lightning discharges was conducted in one of the regions of Spain [Soriano, De Pablo & Tomas, 2005]. Characteristics of 4.3 × 106 discharges are recorded for the Pyrenean Peninsula region during the first ten years of operation of the thunderstorm monitoring network: annual and diurnal cycle, polarity, complexity, peak current of the first discharge. It is shown that the maximum thunderstorm activity is typical for the period from May to September. The maximum discharge density was 2.1 discharge / (km2year). The maximum thunderstorm activity is associated with the mountainous terrain. The share of single discharges was 53.6% and 89%, respectively, for negative and positive cloud-to-ground discharges. The average current for the first strike of the negative and positive discharges was 23.5 kA and 35.3 kA, respectively. For discharges of both polarities, the peak current is higher in summer than in winter.

Observations of lightning discharges are conducted, for example, in Japan [Mardiana, Kawasaki & Morimoto, 2002; Tantisattayakul, Masugata, Kitamura & Kontani, 2005], France [Seity, Soula & Sauvageot, 2000], Brazil [Pinto, Pinto & Naccarato, 2007; Campos, Saba, Pinto & Ballarotti, 2007; Pinto & Pinto, 2003; Pinto, Pinto, Diniz J.H. et al., 2003], Italy [Bernardi & Ferrari, 2004], the Russian Federation [Kozlov & Mulyayarov, 2004; Baranovskiy, Krechetova, Belikova & Perelygin, 2016; Gorbatenko & Ershova, 2011]. Between 1992 and 2000, in the Russian Federation, forest fires from thunderstorms with a relative amount of 8.8-17.5% accounted for 37 to 53% of the area covered by fire [Azmetov, Belyaev & Moskovenko, 2000]. Often, dry thunderstorms lead to very tense situation [Ivanov, 1989].

Improvement of lightning-seeking systems requires improving the methods of data processing of information recorded by these systems. One of the methods for obtaining information on thunderstorms from the data of systems recording lightning discharges is cluster analysis. The data on lightning discharges clustered almost all methods of cluster analysis: partitioning methods [Kononov & Yusupov, 2004; Adzhieva & Shapovalov, 2016], grid methods [Mezuman, Price & Galanti, 2014], graph methods [Bogushov & Panyukov, 2009; Kudryavtsev, Belikova, Krechetova, Geyman, Kocheeva & Kudin, 2014], density methods [Hutchins, Holzworth & Brundell, 2014] and combined method (nearest neighbor method and modal analysis) [Shabaganova, Karimov, Kozlov & Mullayarov, 2013]. The choice of the clustering algorithm is influenced by such properties of the initial data as the dimensionality of the feature space and the presence of a complex structure of the resulting clusters. As a rule, three variables (geographical coordinates and time) are essential for the clustering of data on lightning discharges and the Euclidean distance is used as a measure of proximity. It is necessary to normalize the variables, since they have different units of measure or determine the space-time function of the distance [Tork, 2012]. As a normalization parameter, a priori information is used about the mean lifetime of a thunderstorm cell and the average area of a thunderstorm cell [Shabaganova, Karimov, Kozlov & Mullayarov, 2013]. Another approach is to split the sample into time intervals (for lightning - 10-15 minutes [Kononov & Yusupov, 2004]) and data clustering only by geographic coordinates for each time interval [Tork, 2012].

The task of cluster analysis of lightning discharges does not initially look complicated and it is natural to solve it by choosing a “typical” algorithm based on an approach, and depends mainly on the researcher’s preferences. However, the results of clustering, and consequently, the conclusions based on these results, can significantly depend on the set of input parameters for clustering algorithms and the specified values of the normalization parameters (in the case when they are used).

One of the approaches to determinate the quality of the resulting lightning grouping can be used such informal methods as comparison with satellite images of clouds in the visible and infrared range [Mezuman, Price & Galanti, 2014], data from TRMM [Hutchins, Holzworth & Brundell, 2014], reference climatic data (mean thunderstorm duration, average annual thunderstorm), data from weather stations and meteosites (coincidence in the number of days with a thunderstorm, time and place of registration of a separate thunderstorm).

There are also formal ways to check the quality of clustering, which include, in particular, the calculation of various quality indicators (the index of "silhouette", the Dana index, etc.) [Kovács, Legány & Babos, 2005]. Formal evaluation of clustering results was not given in Refs [Kononov & Yusupov, 2004; Adzhieva & Shapovalov, 2016; Mezuman, Price & Galanti, 2014; Bogushov & Panyukov, 2009; Kudryavtsev, Belikova, Krechetova, Geyman, Kocheeva & Kudin, 2014; Hutchins, Holzworth & Brundell, 2014; Shabaganova, Karimov, Kozlov & Mullayarov, 2013], therefore it is actual to formally compare the clustering results with different lightning data algorithms and estimate the feasibility of using satellite cloud images to validate the results of clustering. Comparison of algorithms and comparison with cloud images will be performed using the language R [Gardener, 2012].

METHODOLOGY

The region covering the Timiryazevskiy forestry of the Tomsk region (55.93 - 56.86), (83.94 - 85.07), and the data on lightning discharges registered by the WWLLN network in this region July 2014 was taken for the computational experiment. The sample of data was 273 lightning discharges. A relatively small number of discharges allowed a visual analysis of clustering solutions obtained during clustering.

Three variables are essential in the task of grouping the data on lightning discharges: geographic coordinates (latitude, longitude) and time. Thus, lightning discharges can be considered as spatial-temporal events [Tork, 2012]. Grouping such events involves finding a group of discharges that are near and in time and in space. Thus, several discharges close in location and time can identify a convective cell producing lightning.

The paper [Tork, 2012] gives an overview of methods for clustering space-time events. The most commonly used strategies are: the use of a time characteristic as a threshold; description of the space-time distance function; splitting the sample of the objects under consideration into time intervals and clustering the data on each interval separately, usually using a sliding window, etc. For the clustering of lightning discharges, the timing was taken into account by importing time to spatial data, for example [Shabaganova, Karimov, Kozlov & Mullayarov, 2013], and performing the clustering task for each time interval of 10-15 minutes, for example [Kononov & Yusupov, 2004]. In this paper, an analysis will be made of the possibility to use these methods for clustering WWLLN network data.

Nonparametric clustering methods as hierarchical methods, partition methods, density methods, and grid methods are used as grouping methods (classification of clustering methods is given according to [Kovács, Legány & Babos, 2005]). For the analysis of data on lightning discharges, the basic algorithms of hierarchical agglomeration methods were chosen, and also based on the partitioning method and the density estimation method was used.

Hierarchical algorithms allow building a system of nested clusters, represented as a dendrogram [Jain, 2010; Xu & Wunsch, 2005]. Agglomerate algorithms in the first stage of work consider each object of the considered set to be a one-element cluster. In subsequent stages, close clusters are combined based on the selected intercluster distance, which most often uses the distance of the neighbor, the distance of the far neighbor, the group average distance, the distance of the ward, etc. The use of these algorithms for clustering lightning discharges is interesting from the position of determining the hierarchical structure of thunderstorms (multicell storms).

Partition methods are the most commonly used for solving various applied problems. The basis is an iterative procedure for constructing a partition of the original sample until the minimum value of some objective function is reached [Xu & Wunsch, 2005]. The k-means algorithm considers that the object belongs to the cluster if the distance between this element and the cluster center is the minimum of all distances between and the centers of the constructed clusters. The algorithm iteratively builds the partitioning of the initial set into clusters, which are described by the mean value vectors (centroids). In the process of constructing a partition, the sum of the distances from objects to the cluster centers to which they are assigned is minimized. Despite the known shortcomings of this algorithm [Xu & Wunsch, 2005], it nevertheless is often used to solve various applied problems, including that used for lightning discharge data [Adzhieva & Shapovalov, 2016].

The basic algorithm of clustering, based on density estimation, is the dbscan algorithm, which, based on the given density, selects "core objects", "boundary objects" and "noise" among the objects of the sample in question [Ester, Kriegel, Sander, & Xu, 1996]. Density is the number of objects within a sphere of a given radius Eps. A core object is an object x if the density around it is greater than MinPts (the minimum number of points in the cluster). A border object is an object x if the density around it is smaller than MinPts, but it is next to the core object. Noise is an object of x, if it is neither a core object nor a boundary object. Clusters consist of core and boundary objects. Note that there may be "noise" in the lightning discharge data, for example, for lightning discharges recorded by the WWLLN network, a single discharge (atmosphere) is considered noise and "a light thundercloud" is 2-3 close discharges [Hutchins, Holzworth & Brundell, 2014].

One of the methods for assessing the quality of clustering is quality indicators that assess the quality of the resulting clustering by comparing several cluster structures to each other without having a priori information about the structure of clusters and taking into account only information about the cluster structure and the clustered set. There are about 30 such indicators [Charrad, Ghazzali, Boiteau, & Niknafs, 2014]. One of the most commonly used indexes of clustering quality is the "silhouette" index. The silhouette of each cluster is determined through the measure of the similarity of the object belonging to the cluster and the measure of similarity of the same object with the nearest "alien" cluster [Rousseeuw, 1987]. As the first measure of similarity, the average distance from the object to other objects in the same cluster is calculated. As the second measure of similarity, the minimum distance to the objects of the "alien" cluster is calculated. The normalized difference of these proximity measures is the index of the silhouette of the cluster. The index of the silhouette of the object takes a value from -1 to 1. Values close to -1 indicate that the object is assigned to the wrong cluster. Positive values close to zero show that the object is close to the cluster boundary (and, correspondingly, close to the boundary of the nearest cluster). Values close to 1 indicate that the cluster is well defined. The average width of the silhouette ASW (average silhouette width) is the average value of the silhouettes of all clusters. In this case, the closer ASW to 1, the better the quality of the clusters obtained.

The construction of clustering solutions using selected clustering algorithms with quality estimation was performed using cluster analysis packages Nbclust [Charrad, Ghazzali, Boiteau & Niknafs, 2014] and dbscan [The Comprehensive R..., n. d.]. Nbclust contains "complex" functions that implement clustering algorithms, and, at the same time, allow to evaluate the quality of clustering using some indexes. The following functions were used, which return the optimal number of clusters according to the quality criterion ASW: Nbclust () of the Nbclust package for hierarchical algorithms (ward.D2 is the Ward method, single is the nearest neighbor method, the complete method is the average distance method) and the k-means algorithm. From the dbscan package, the function kNNdist () was used to determine the Eps parameter, to use the dbscan() function to cluster the data and to use the cluster function as a function of cluster.stats ().

The first computational experiment (Experiment 1) consisted in constructing clustering solutions by clustering algorithms for the normalized values of the lightning discharge sample being considered. At the first stage, the data is normalized. The following parameters were used as the normalizing parameters: for the discharge coordinates, the maximum distance between two discharges in the cluster is S=50 km; for the registration time - the maximum time interval between the registration of two discharges t=30 minutes [EUMETView…, n. d.]. In this case, the data were normalized by the indicated values and their squares. At the next stage, for the normalized data, a grouping solution was constructed using the selected clustering algorithms. For the function Nbclust() the following input parameters of the algorithms were set: normalized initial data; parameter distance = “euclidean”; minimum and maximum number of clusters min.nc = 2, max.nc = 20; The clustering method used is “single”, “complete”, “average”, “ward. D2”, “kmeans”; index of clustering quality index = “silhouette” (ASW). For the function dbscan(), the following input parameters of the algorithms were set: the matrix of pairwise distances of the normalized initial data; the eps parameter is the specified kNNdist (); minPts = 2.

The second computational experiment (Experiment 2) consisted in splitting the sample into 15-minute time intervals and grouping the data for each time interval with the considered clustering algorithms. At the same time, for each data set, matrices of pairwise distances were calculated only by geographic characteristics of the discharges. Input parameters used for clustering functions were used the same as in Experiment 1.

The results of clustering lightning discharges obtained as a result of computational experiments were also compared with satellite images of clouds obtained with the help of geostationary satellites, which make it possible to obtain data with a high temporal resolution. To this end, we used the EUMETView web service [EUMETView…, n. d.] of the website of the European Organization for the Exploitation of Meteorological Satellites (EUMETSAT). EUMETView provides free-mode interactive access for all users to information received from Meteosat, Metop and Copernicus Sentinel-3 marine satellites. For the area of interest, images obtained by Meteosat 41.5 Degrees (Indian Ocean Data Coverage (IODC) service) can be used. The EUMETView web service provides access to both high-rate level 1.5 image data in IR 03.9, IR 10.8, VIS 0.6, which are geolocated and radiometrically pre-processed image data, and visualized products, including products such as Cloud Analysis Image (CLAI), Cloud Top Height (CTH), Multi-Sensor Precipitation Estimate (MPE). And also rgb composites including the following products: Airmass product, Convection RGB product, Day Microphysics RGB product, Natural Color RGB product [Kerkmann, Lutz, König, Prieto, Pylkko, Roesli, Rosenfeld, Schmetz & Zwatz-Meise, n. d.]. To visualize the results of clustering, scatterplot3d, factoextra, raster and rgdal packages were used [The Comprehensive R…, n. d.].

RESULTS AND DISCUSSION

Tables 1 and Figures 1-12 show the results of the computational experiment 1. Table 1 shows the number of clusters and the value of the ASW index for the above a priori parameters and clustering algorithms. For the dbscan algorithm, the value of the parameter Eps = 2, obtained with the function kNNdist () for the corresponding normalizing parameter in time, and the number of noise points defined as “noise” are additionally indicated.

Table 1. Results of clustering

Algorithms	Quantity of Clasters (Lightning Cells)			
	d_A, t_A		d_A^2, t_A^2	
	k	ASW	k	ASW
single	2	0.62	6	0.3
average	2	0.62	19	0.5
complete	3	0.59	20	0.46
ward.D2	4	0.58	6	0.50
k-means	2	0.59	6	0.51
dbscan	$k = 45$ $noise = 50$	0.52	$k = 30$ $noise = 21$	0.34

The clustering results presented in Table 1 show that when using the normalization parameters, cluster structures as a whole have more clusters than when squares of the normalization parameters are used. In this case, a larger number of clusters in the sample are determined for the squares of the normalization parameters using the kmeans algorithm and hierarchical algorithms, for dbscan - on the contrary.

According to [Rousseeuw, 1987], the values of the ASW index in the range from 0.25 to 0.50 indicate the presence of a weak structure in the data, and the values of the index ASW in the range from 0.51 to 0.70 - about the existence of a reasonable structure of clusters. A reasonable structure of clusters is obtained for normalizing parameters and all considered clustering algorithms. For squares of normalization parameters, the lowest quality of clustering has a hierarchical algorithm that calculates inter-cluster distance by the nearest-neighbor method (single) and the dbscan algorithm. However, ASW values greater than 0.5 do not always indicate that we obtained the clustering results that we expected.

Figures 1-12 shows the clustering results, as well as the silhouette index image for each object, the average value of the ASW index for the grouping solutions obtained using the complete, kmeans, and dbscan algorithms. The choice of only the complete algorithm from hierarchical algorithms is due to the preliminary analysis, which showed the best results obtained by this algorithm in comparison with others on the sample of data in question [Kerkmann, Lutz, König, Prieto, Pylkko, Roesli, Rosenfeld, Schmetz & Zwatz-Meise, n. d.; Baranovskiy, Krechetova, Belikova, Kocheeva & Yankovich, 2016]. In general, when using the normalization parameters, all algorithms except dbscan share the considered discharges mainly on the time characteristic. In different clusters, discharges, mainly belonging to different time intervals, are combined. In this case, separate clusters for spatial characteristics are not singled out. Clustering using squares of normalization parameters, allocates the "vertical" structure of clusters, that is, in one cluster, as a rule, the selected algorithms combine discharges, which are predominantly spatially close. For hierarchical algorithms, this information can be obtained from the dendrogram of classes, and for the k-means algorithm, one more stage of clustering will be needed.

Note that the grouping solutions obtained by the complete and dbscan algorithms for both variants of the normalization parameters contain objects incorrectly assigned to clusters (negative values of the silhouette index). Despite the fact that only the cluster structure obtained by the k-means algorithm (Figure 2.a) does not have "incorrectly clustered" objects for the clusters obtained, it is hardly possible to regard this cluster structure as corresponding to two different thunderstorm events. So in a single cluster (cluster number 2) are combined discharges, the difference in time between which is more than an hour, it would be more logical to consider them as belonging to different storm events. The dbscan algorithm obtained a more detailed cluster structure, but even in the definition of "noise objects", not all the clusters obtained satisfy the compactness condition.

In general, the analysis of the clustering results obtained with the help of basic and most frequently used algorithms and the space-time distance function (normalized characteristics of space-time characteristics) for lightning discharges shows that in all cases we cannot obtain well-interpreted results specifying information on the spatial position of convective cells producing a thunderstorm.

Figures 13-24 and Figures 25-36 show the results of Experiment 2 for the objects of the sample in question, falling in time intervals from 13:30 to 13:45 (.. lightning discharges) and from 13:45 to 14:00 (.. lightning discharges) and clustered by hierarchical algorithms and algorithms kmeans and dbscan. The resulting grouping solutions for lightning discharge data corresponding to the selected time intervals are compared with images of the Natural Color RGB product obtained at 13:45 (Figures 13-24) and 14:00 (Figures 25-36).

For the first time interval (Figures 13-24), the same cluster structure with the largest number of clusters is determined by algorithms with the hierarchical algorithms average, complete and ward.D2. In the grouping solution, there are objects that lie on the cluster boundary (the value of the silhouette index is close to or equal to zero). The same objects exist for the grouping solution obtained with the single and kmeans algorithms.

Figure 1. Results of the computational experiment using complete algorithm for data normalization d_A, t_A

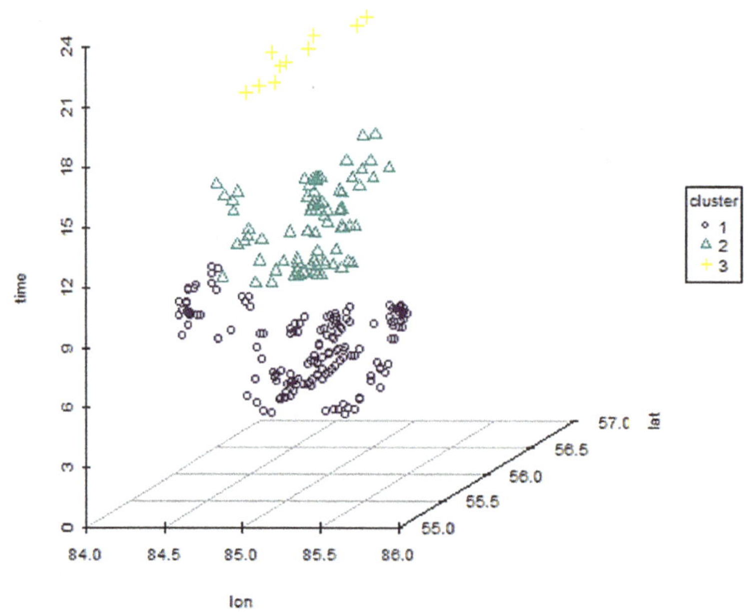

Figure 2. Results of the computational experiment using complete algorithm for data normalization $d_A^{\ 2}$, $t_A^{\ 2}$

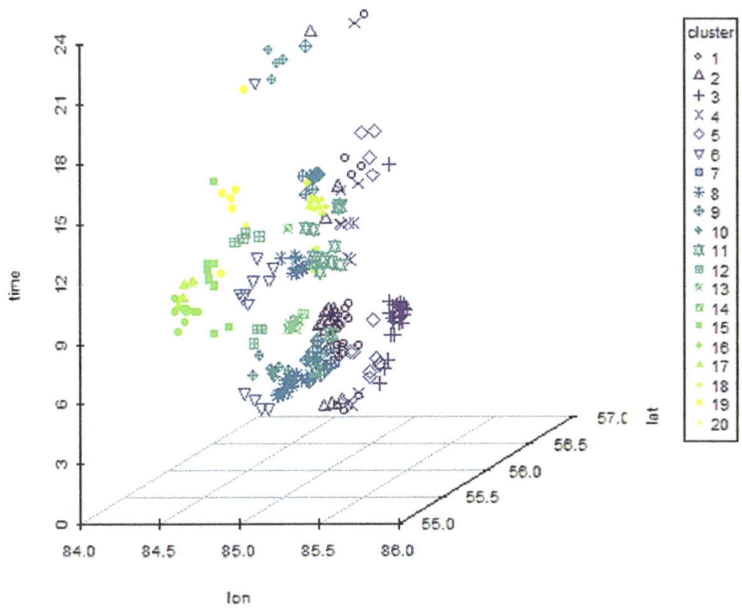

Figure 3. Silhouette Index of the computational experiment using complete algorithm for data normalization d_A, t_A

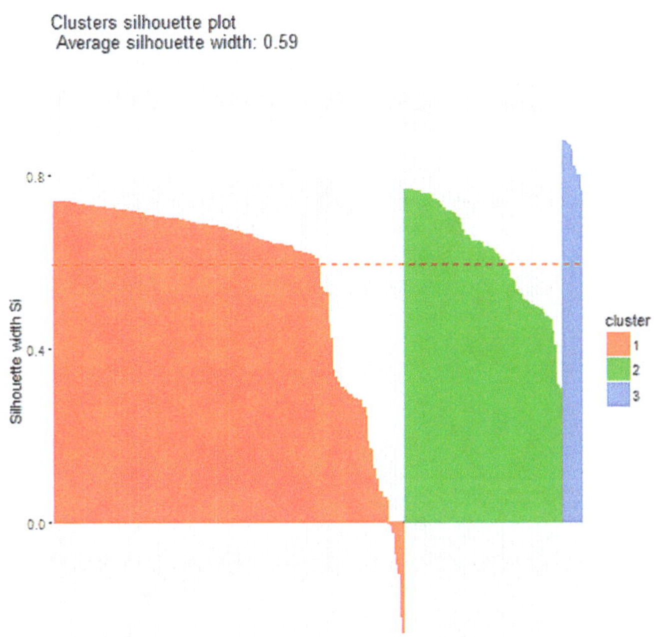

Figure 4. Silhouette Index of the computational experiment using complete algorithm for data normalization d_A^2, t_A^2

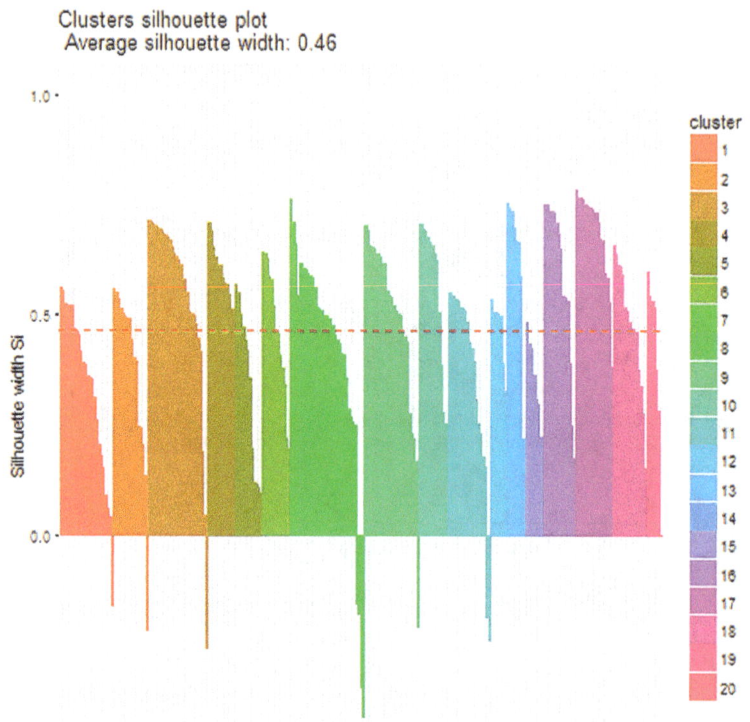

Figure 5. Results of the computational experiment using k-means algorithm for data normalization d_A, t_A

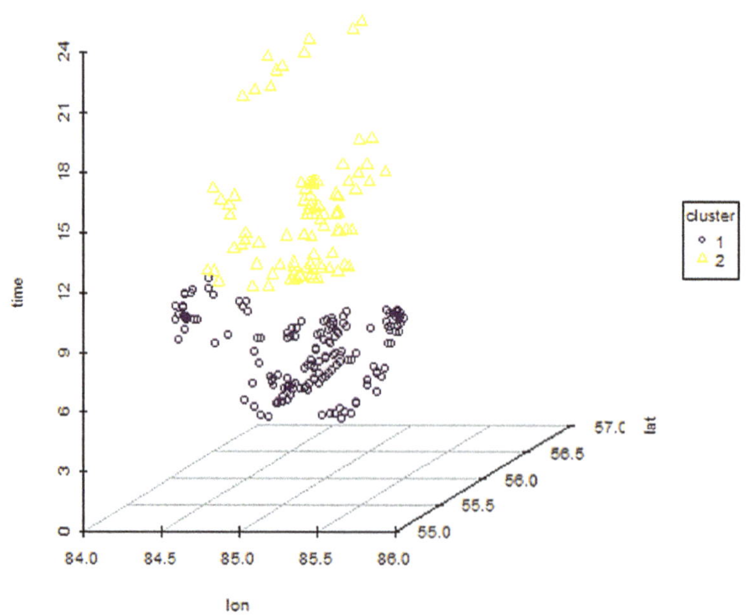

Figure 6. Results of the computational experiment using k-means algorithm for data normalization d_A^2, t_A^2

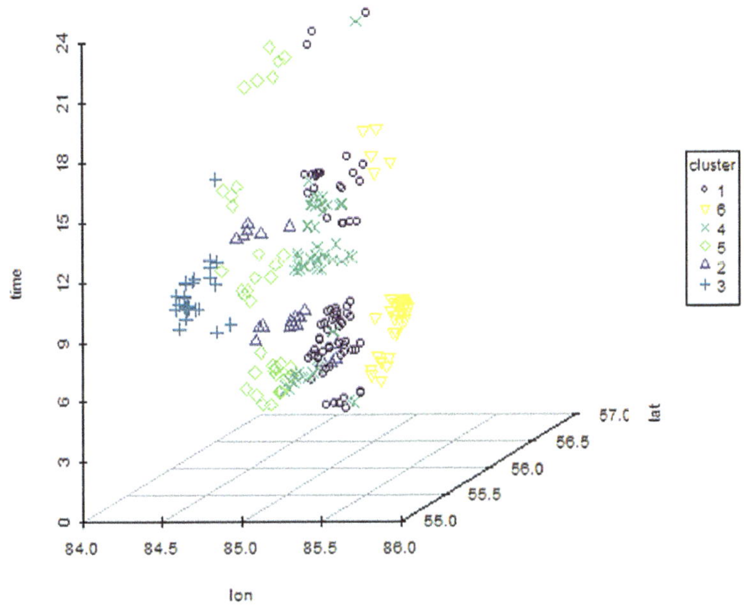

Figure 7. Silhouette Index of the computational experiment using k-means algorithm for data normalization d_A, t_A

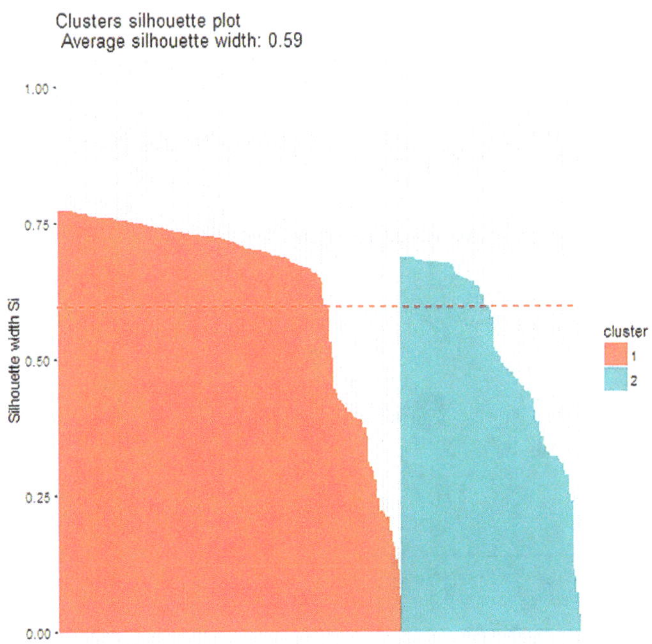

Figure 8. Silhouette Index of the computational experiment using k-means algorithm for data normalization d_A^2, t_A^2

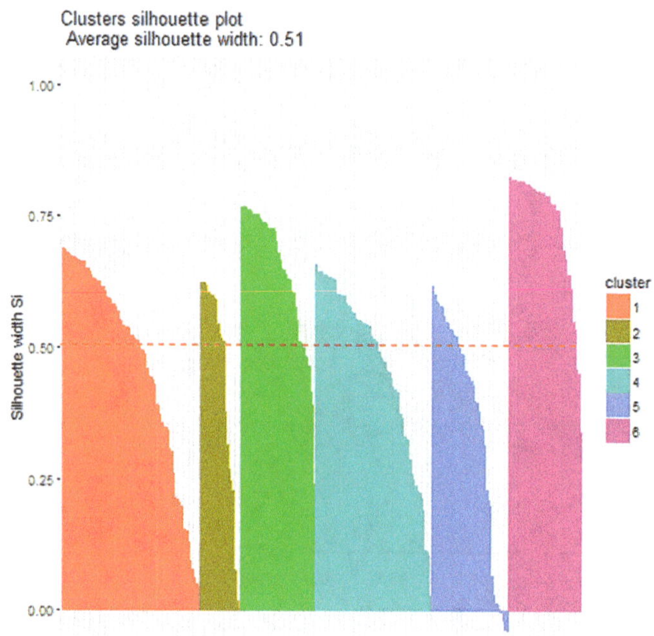

Figure 9. Results of the computational experiment using dbscan algorithm for data normalization d_A, t_A

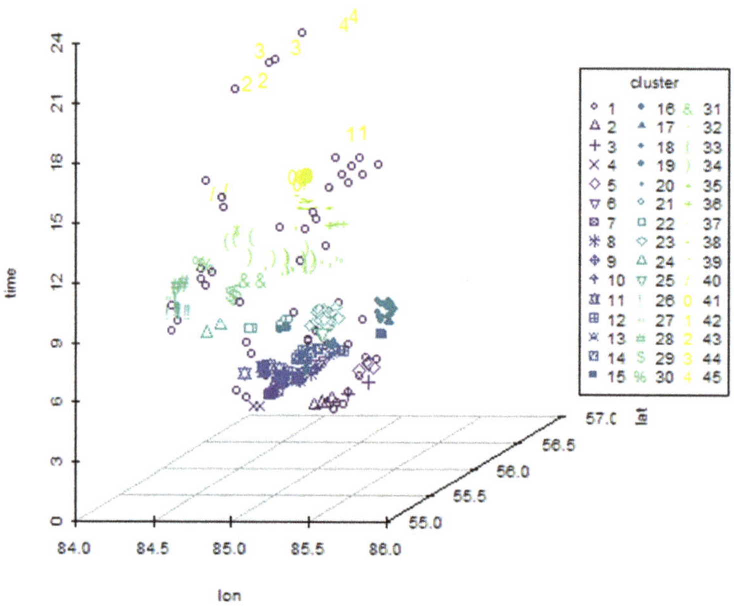

Figure 10. Results of the computational experiment using dbscan algorithm for data normalization d_A^2, t_A^2

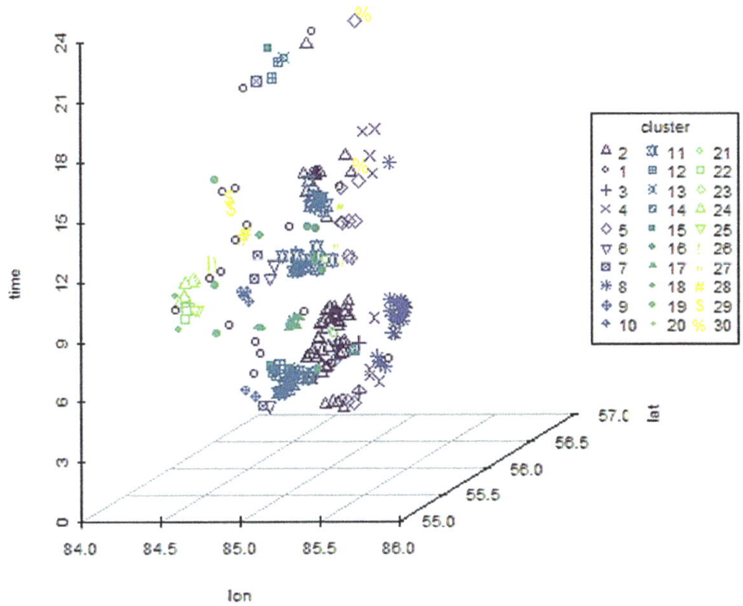

Figure 11. Silhouette Index of the computational experiment using dbscan algorithm for data normalization d_A, t_A

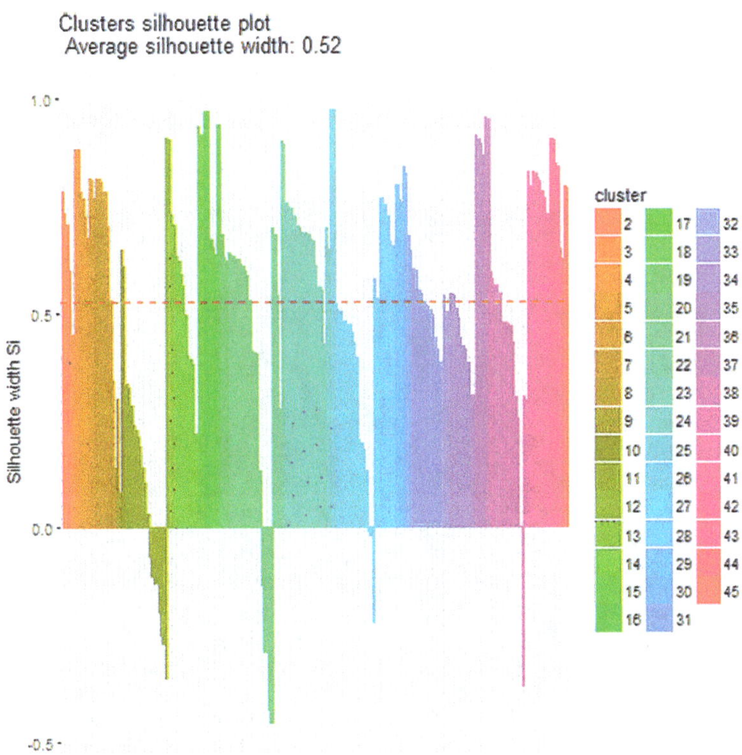

Figure 12. Silhouette Index of the computational experiment using dbscan algorithm for data normalization d_A^2, t_A^2

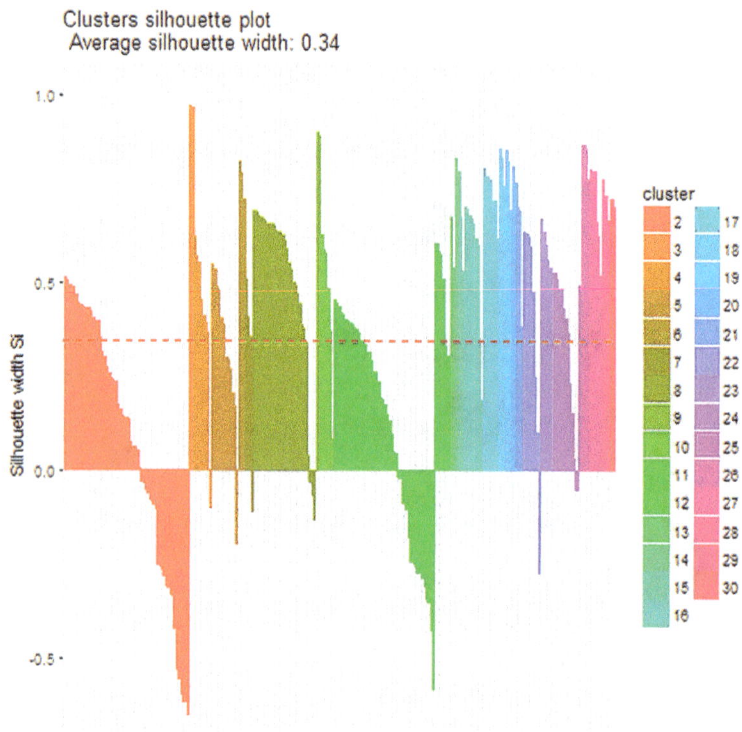

Figure 13. Comparison single algorithm results with EUMETSAT images

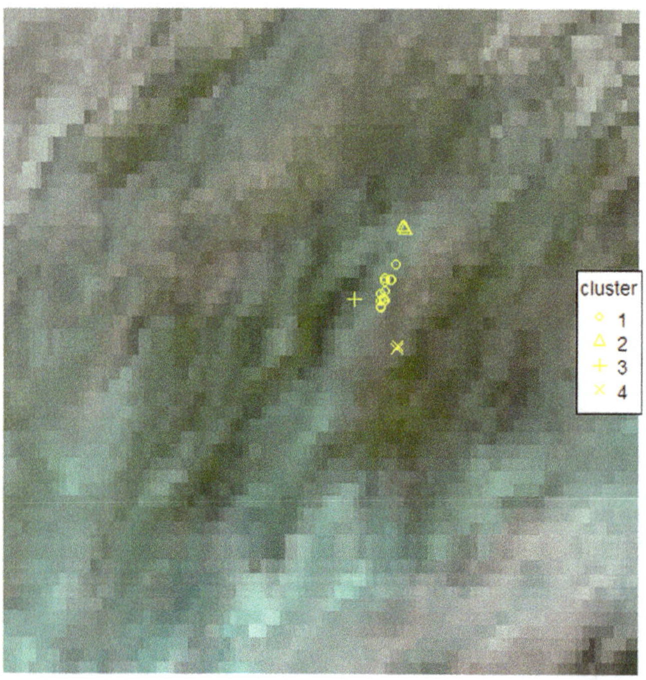

Figure 14. Silhouette Index of the single algorithm results

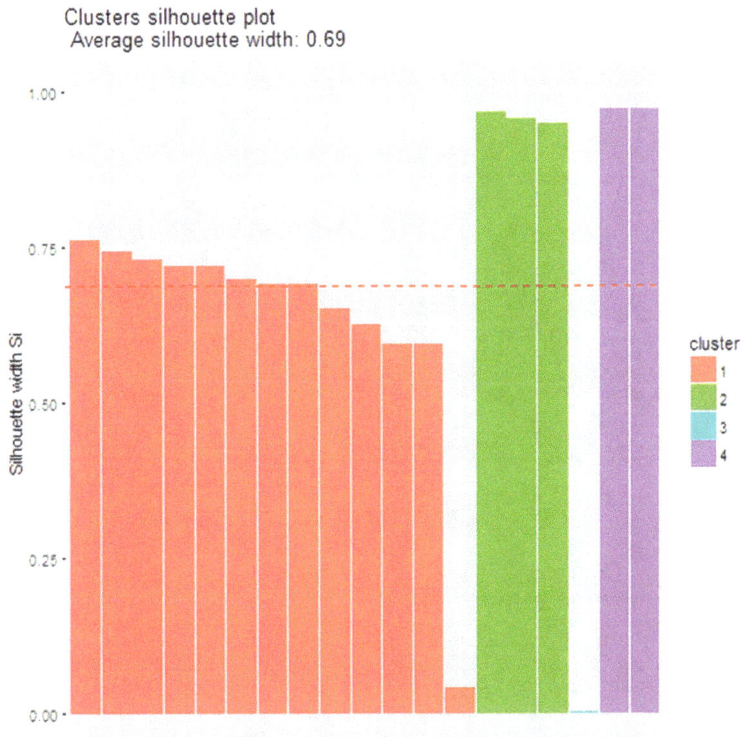

Figure 15. Comparison average algorithm results with EUMETSAT images

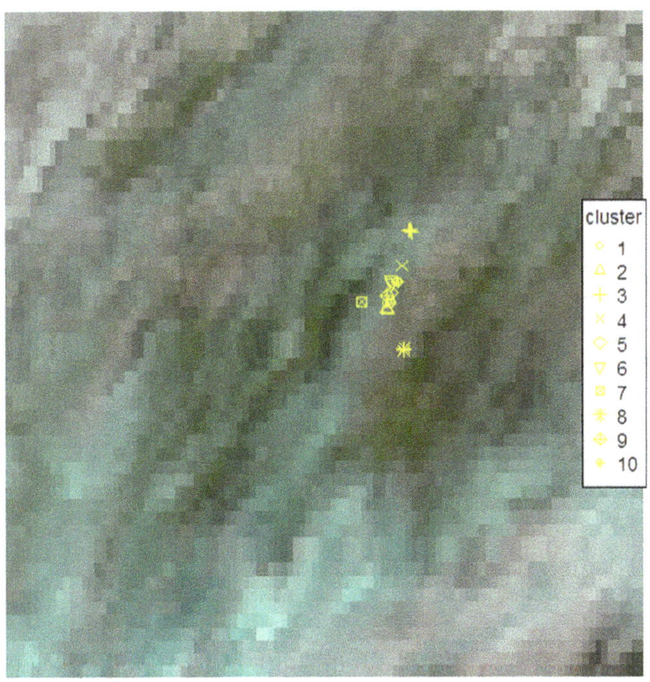

Figure 16. Silhouette Index of the average algorithm results

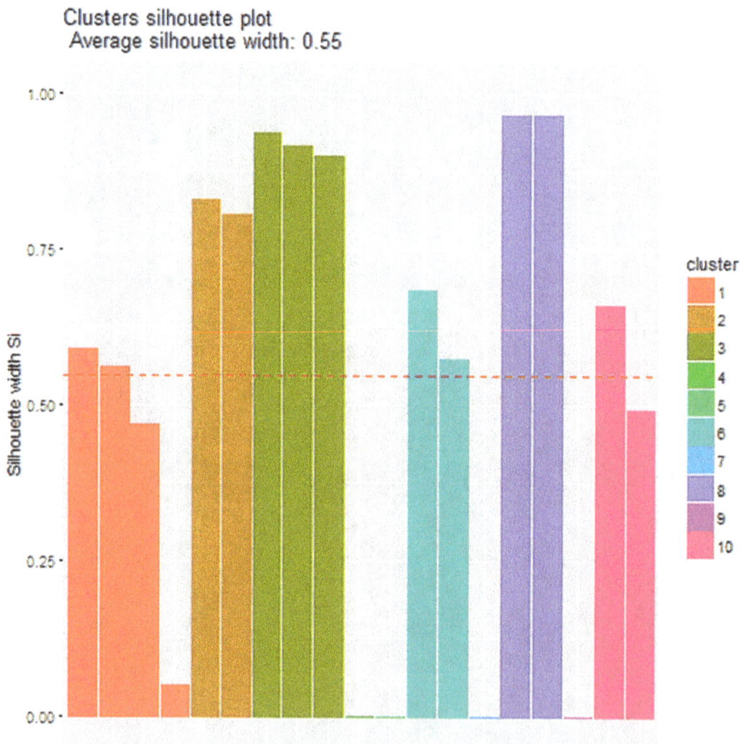

Figure 17. Comparison complete algorithm results with EUMETSAT images

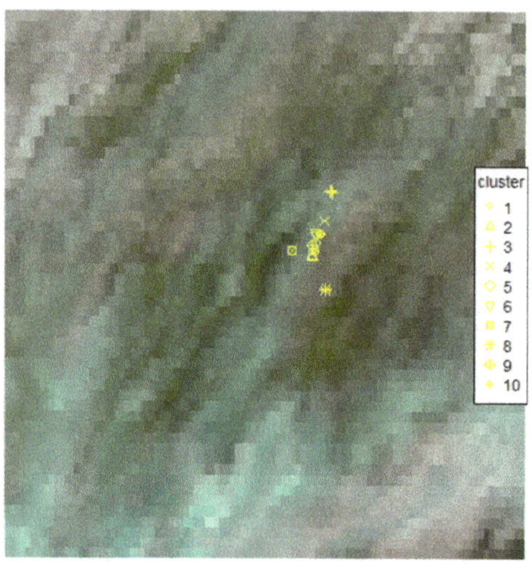

Figure 18. Silhouette Index of the complete algorithm results

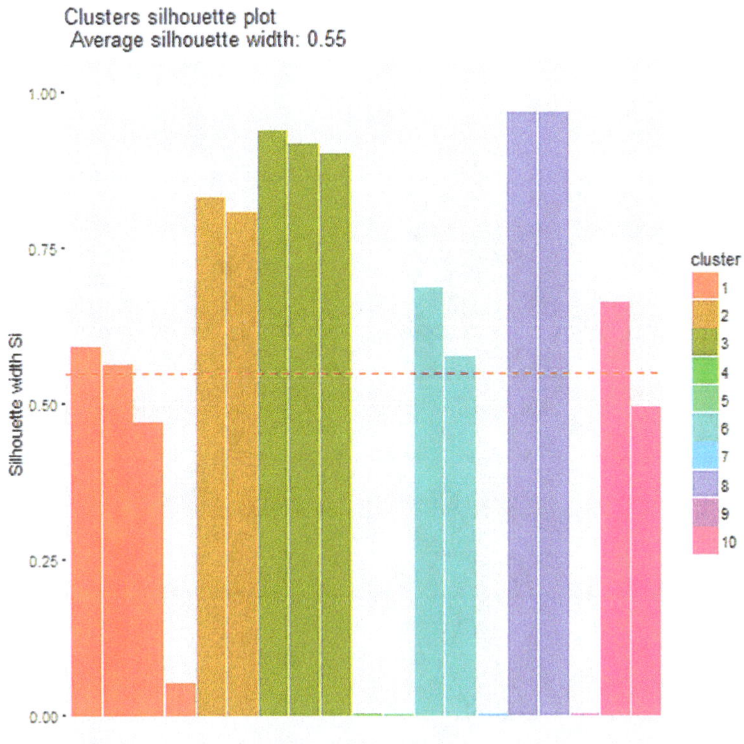

Figure 19. Comparison ward.D2 algorithm results with EUMETSAT images

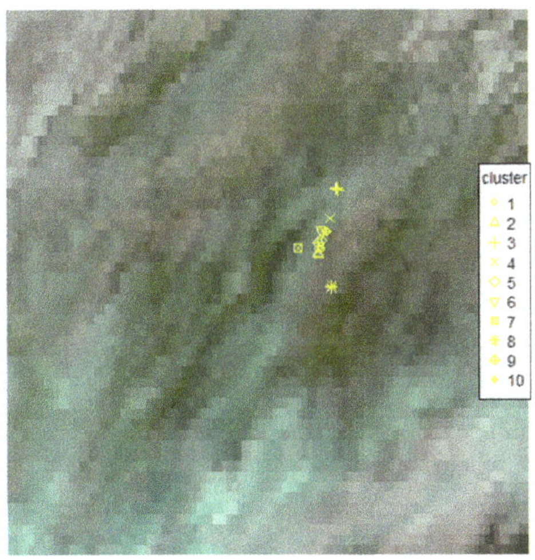

Figure 20. Silhouette Index of the ward.D2 algorithm results

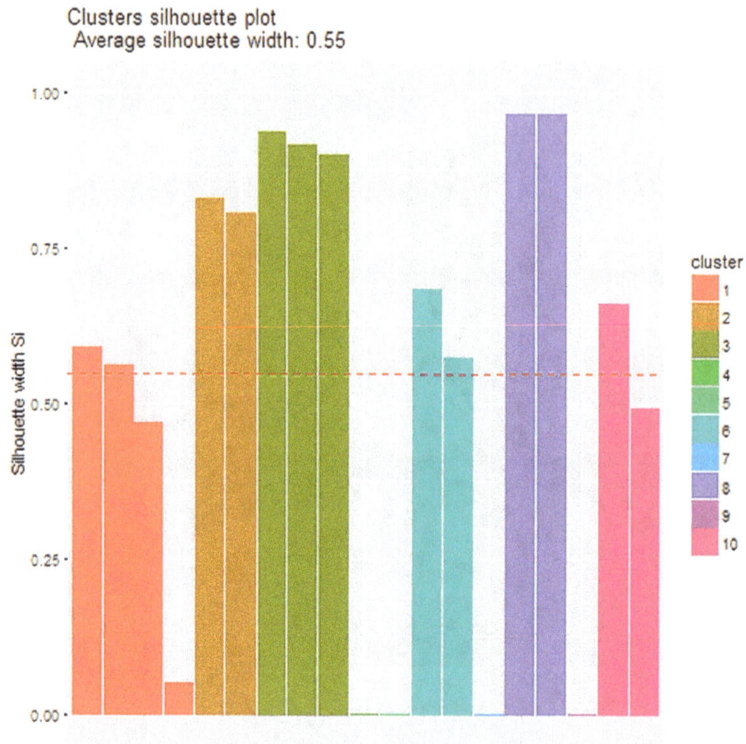

Figure 21. Comparison kmeans algorithm results with EUMETSAT images

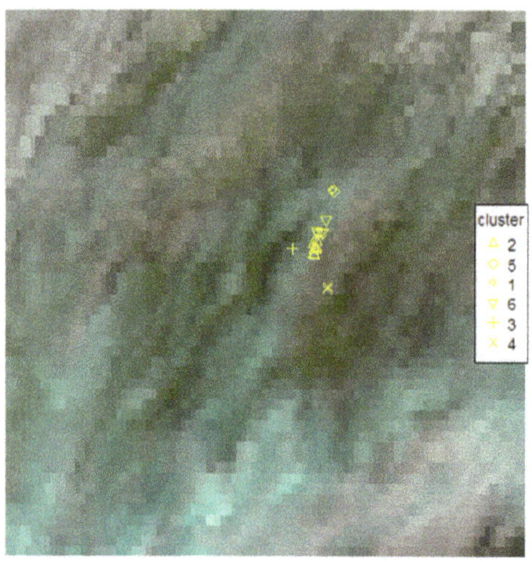

Figure 22. Silhouette Index of the kmeans algorithm results

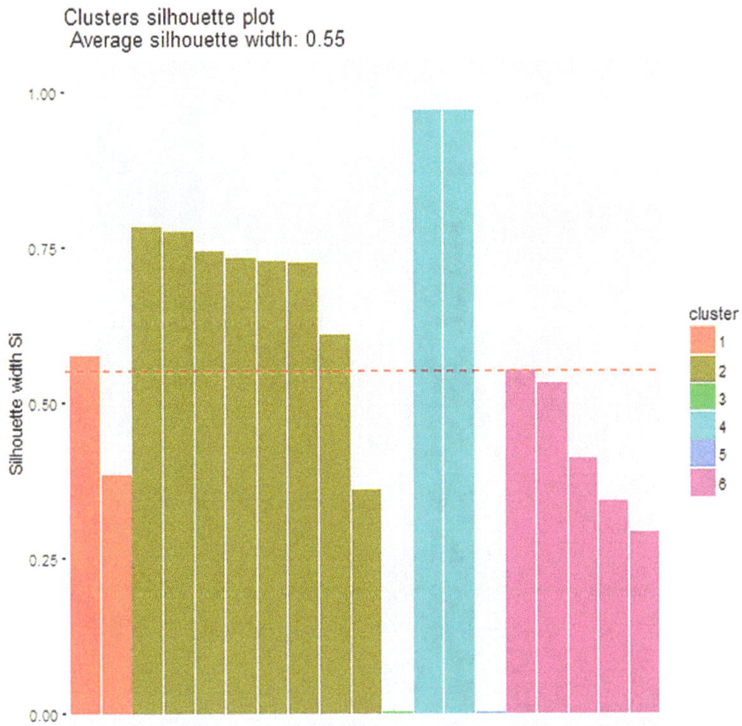

Figure 23. Comparison dbscan algorithm results with EUMETSAT images

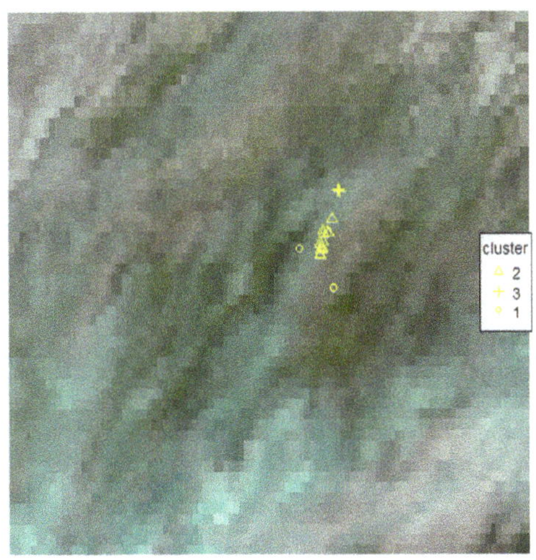

Figure 24. Silhouette Index of the dbscan algorithm results

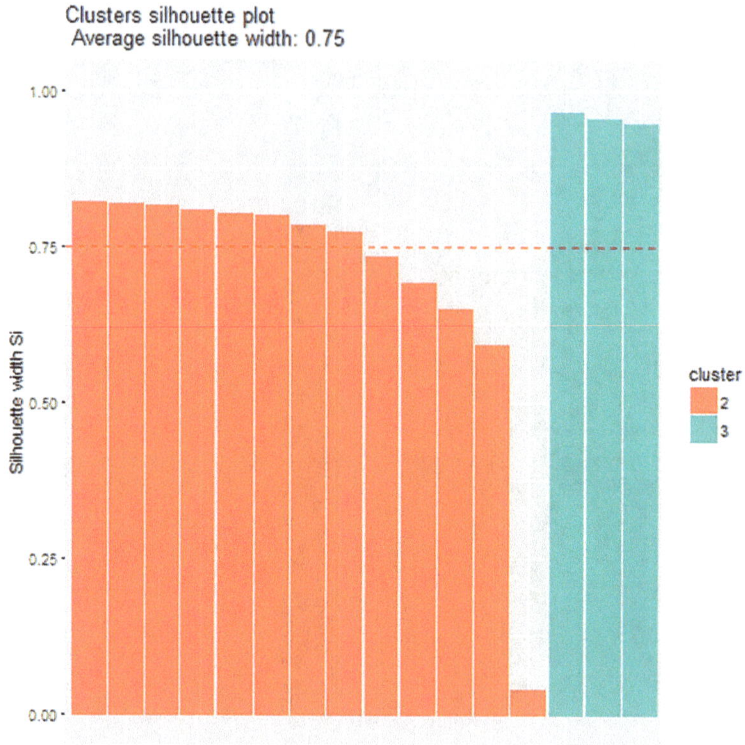

For the second time interval (Figures 25-36), the same cluster structure with the largest number of clusters was obtained by single and complete algorithms. Grouping solutions obtained by the algorithms single, complete and kmeans contain objects lying or close to the boundaries of clusters. The same number of clusters is obtained by the complete, ward.D2, and dbscan algorithms. In this case, a larger value of the index ASW is obtained for the grouping solution of the dbscan algorithm.

Figure 25. Comparison single algorithm results with EUMETSAT images

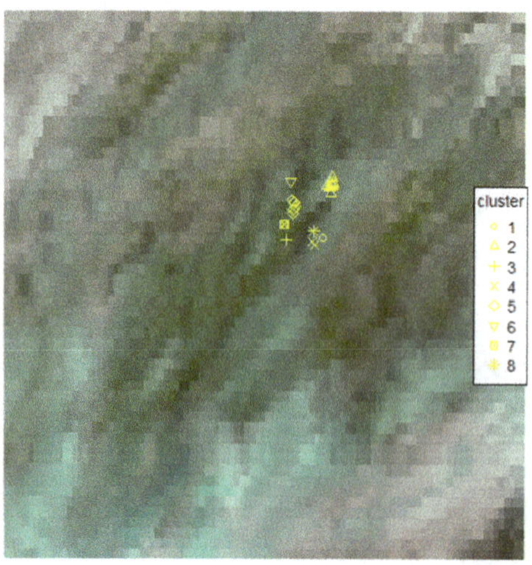

Figure 26. Silhouette Index of the single algorithm results

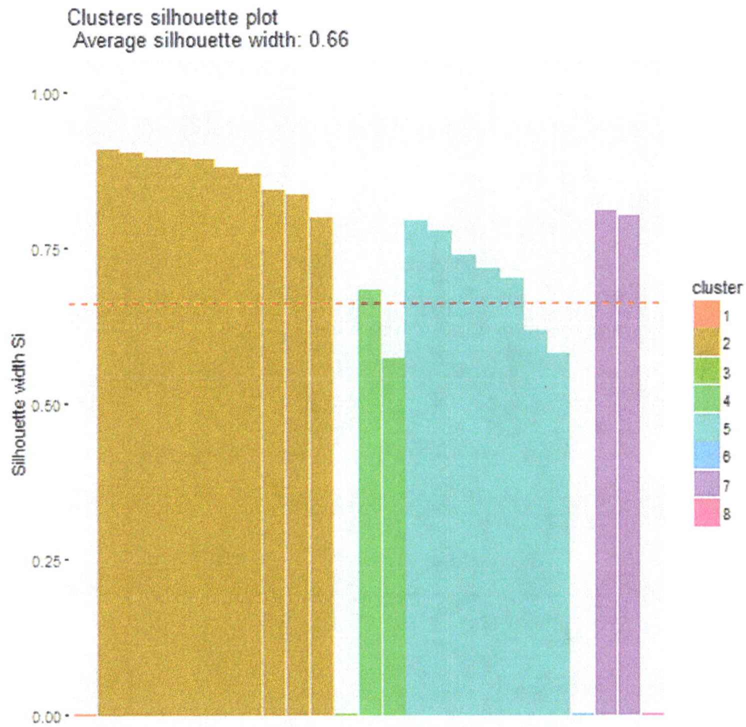

Figure 27. Comparison average algorithm results with EUMETSAT images

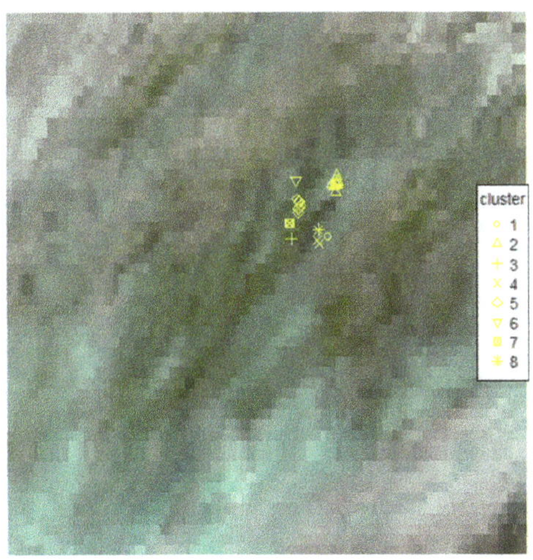

Figure 28. Silhouette Index of the average algorithm results

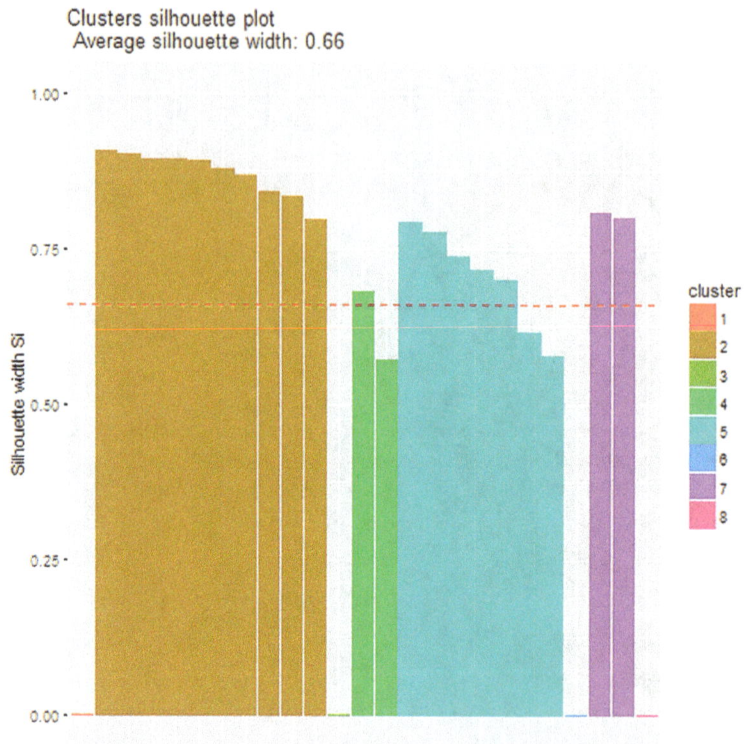

Figure 29. Comparison complete algorithm results with EUMETSAT images

Figure 30. Silhouette Index of the complete algorithm results

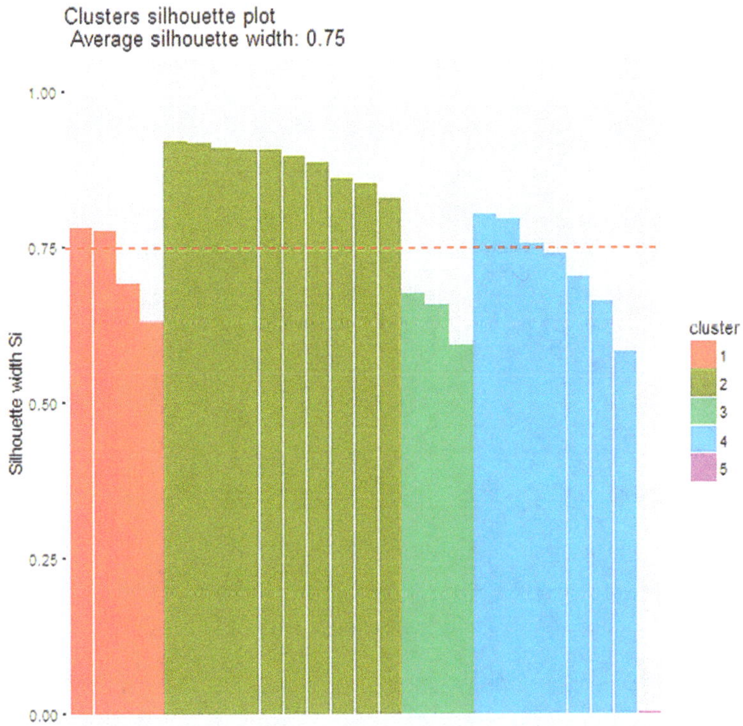

Figure 31. Comparison ward.D2 algorithm results with EUMETSAT images

Figure 32. Silhouette Index of the ward.D2 algorithm results

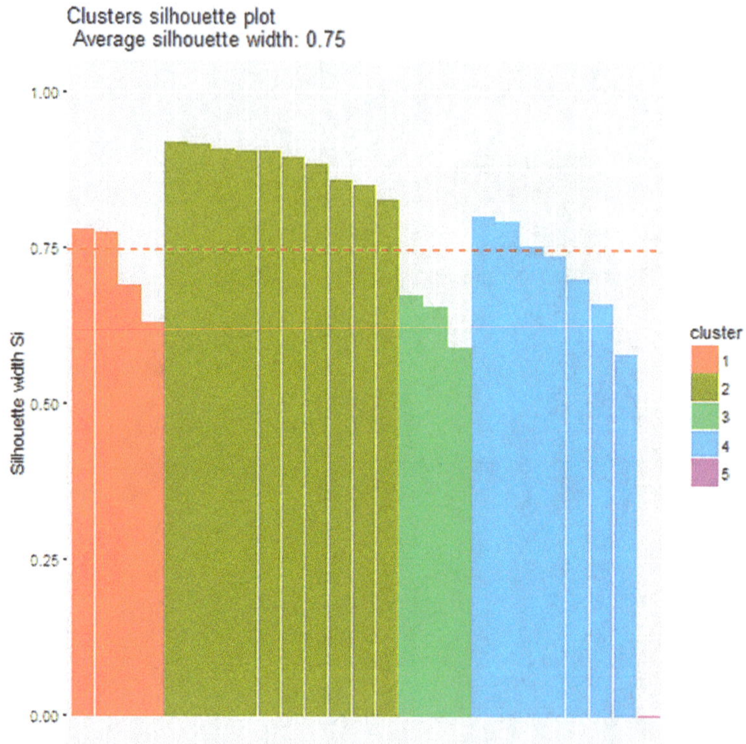

Figure 33. Comparison kmeans algorithm results with EUMETSAT images

Figure 34. Silhouette Index of the kmeans algorithm results

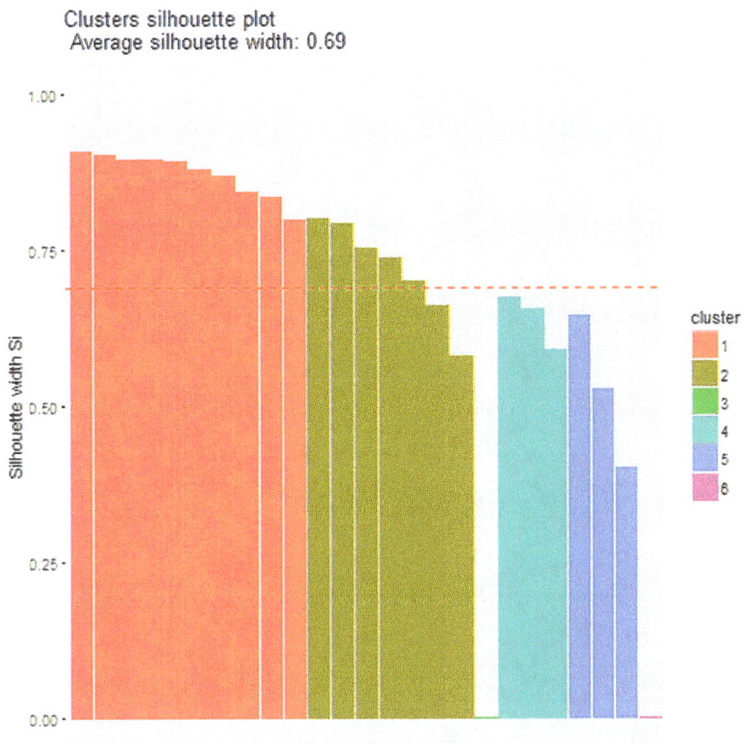

Figure 35. Comparison dbscan algorithm results with EUMETSAT images

Figure 36. Silhouette Index of the dbscan algorithm results

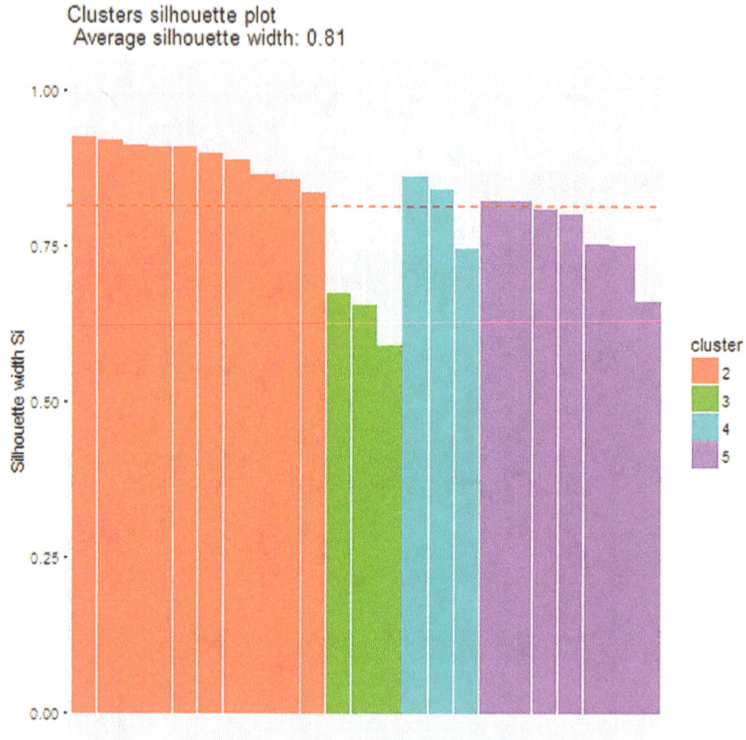

The best cluster solution for both time slots is obtained by the dbscan algorithm. However, for this solution, the definition of "noise objects" within the clouds, as well as within the clouds containing ice particles, raises doubts that these are "noise discharges" rather than individual lightning discharges, information about which can also be important. It is known that the WWLLN network does not register all lightning discharges [Hutchins, Holzworth, & Brundell, 2014] and for territories on which the network is the only source of instrumental observations of thunderstorms, the loss of information on single lightning discharges may be undesirable. Joint images of clusters of lightning discharges and clouds for selected consecutive time intervals is an example of images on which it is possible to trace the dynamics of the development of thunderclouds.

CONCLUSION

Comparison of the results of lightning data clustering with k-means, dbscan algorithms and hierarchical agglomerate algorithms using the space-time distance function (normalized space-time characteristics) for lightning discharges with recommended parameters for coupling spatial and temporal characteristics, and despite the positive experience of using These normalization parameters and the approach as a whole for lightning cluster clustering [Shabaganova, Karimov, Kozlov, & Mullayarov, 2013], on the selected data set were not allowed get a well-interpreted cluster structure corresponding to thunderstorms. The resulting cluster structures do not reflect the spatiotemporal dynamics of convective cells producing thunderstorms.

A comparison of clustering results with hierarchical agglomerate algorithms and k-means and dbscan algorithms for lightning data belonging to 15-minute time intervals showed that the best cluster structure for selected time intervals was obtained by the dbscan algorithm.

Using, along with informal methods of evaluating the results of clustering (RGB images of clouds), as well as formal methods of estimation, in particular, the ASW index, makes it possible to more reliably estimate the results of lightning cluster clustering by the chosen algorithm. It is better to use cloud images of geostationary satellites to determine the consistency of the results of lightning cluster clustering with single-cell and multi-cell storm clouds, since they have acceptable spatial resolution and good temporal periodicity, reflecting the dynamics of thunderstorm clouds. The EUMETView WMS Interface Standard provided by the web service will be used to download and display EUMETSAT products in the developed web-based system.

ACKNOWLEDGMENT

This work is implemented at financial support of Russian Foundation for Basic Research and administration of Tomsk region. The grant N 16-41-700831.

The authors wish to thank the World Wide Lightning Location Network, a collaboration among over 50 universities and institutions, for providing the lightning location data used in this paper.

REFERENCES

Adzhieva, A. A. & Shapovalov, V. A. (2016). Cluster analysis in the automatic detection and support of thunderstorm sites according to the thunderstorming network data. *Engineering Bulletin of the Don. 2*, 8.

Azmetov, R. R., Belyaev, A. I., & Moskovenko, V. M. (2000). *Prospects for the creation of the Russian system of electromagnetic monitoring of thunderstorms for the needs of forest protection from fires, energy, aviation, meteorology and disasters forecasting.* Paper presented at Proceedings of International Conference Conjugated Problems of Mechanics and Ecology: (pp. 9-11). Tomsk, Russia: Publishing house of Tomsk University.

Baranovskiy, N., Krechetova, S., Belikova, M., & Perelygin, A. (2016, December). WWLLN Data cluster analysis methods for lightning-caused forest fires monitoring. [IJECE]. *Iranian Journal of Electrical and Computer Engineering*, *6*(6), 3112–3120.

Baranovskiy, N. V., Krechetova, S. Y., Belikova, M. Y., Kocheeva, N. A., & Yankovich, E. P. (2016). Joint processing of RS and WWLLN data for forest fire danger estimation: New concept. *Proceedings of SPIE - The International Society for Optical Engineering, 10001, Article 1000113*, 6.

Baranovskiy, N. V., & Kuznetsov, G. V. (2017). *Forest fire occurrences and ecological impact prediction: monograph.* Novosibirsk, Russia: Publishing House of the SB RAS. doi:10.15372/FOREST2017BNV

Bernardi, M., & Ferrari, D. L. (2004). Evaluation of the LLS efficiency, on the ground flash density, using the Italian lightning detection system SIRF. *Journal of Electrostatics*, *60*(2-4), 131–140. doi:10.1016/j.elstat.2004.01.013

Bogushov, A. K., & Panyukov, A. V. (2009) Location of interconnected objects under uncertainty. *IV All-Russian Conference Optimization Problems and Economic Applications: Conference Materials, Omsk, Russia, June 29-July 4, 2009, Omsk Branch of the Institute of Mathematics of the SB RAS* (p. 113). Omsk, Russia: Polygraph. Center for CAS.

Campos, L. Z. S., Saba, M. M. F., Pinto, O. Jr, & Ballarotti, M. G. (2007). Waveshapes of continuing currents and properties of M-components in natural negative cloud-to-ground lightning from high-speed video observations. *Atmospheric Research*, *84*(4), 302–310. doi:10.1016/j.atmosres.2006.09.002

Charrad, M., Ghazzali, N., Boiteau, V., & Niknafs, A. (2014). NbClust: An R Package for Determining the Relevant Number of Clusters in a Data Set. *Journal of Statistical Software*, *61*(6), 1–36. doi:10.18637/jss.v061.i06

Cummins, K. L., Murphy, M. J., Bardo, E. A., Hiscox, W. L., Pyle, R. B., & Pifer, A. E. (1998). A combined TOA / MDF technology upgrade of the U.S. national lightning detection network. *Journal of Geophysical Research*, *103*(D8), 9035–9044. doi:10.1029/98JD00153

Deeming, J. E., Lancaster, J. W., Fosberg, M. A., Furman, R. W., & Schroeder, M. J. (1972). The national fire-danger rating system. Ft. Collins, CO: USDA Forest Serv. Res. Pap. RM-84 doi:10.5962/bhl.title.98933

Department of Forestry of Tomsk region. (2010). Data archive on forest fires in the Timiryazevsk forestry of the Tomsk region. CD.

Ester, M., Kriegel, H.-P., Sander, J., & Xu, X. (1996). A density-based algorithm for discovering clusters in large a spatial database. *Proc. 1996 Intern. Conf. on Knowledge Discovery and Data Mining*, 226-231.

EUMETView — EUMETSAT. Retrieved from http://eumetview.eumetsat.int/mapviewer/

Everitt, B. S., Landau, S., Leese, M., & Stahl, D. (2011). *Cluster analysis* (5th ed.). Hoboken, NJ: John Wiley & Sons. doi:10.1002/9780470977811

Gardener, M. (2012). *Beginning R: The statistical programming language*. Hoboken, NJ: John Wiley & Sons.

Gorbatenko, V. P., & Ershova, T. V. (2011). *Lightning as a link of the global electrical circuit: monograph.* Tomsk, Russia: Publisher of TSPU.

Grishin, A. M., & Filkov, A. I. (2011). A deterministic-probabilistic system for predicting forest fire hazard. *Fire Safety Journal*, *46*(1-2), 56–62. doi:10.1016/j.firesaf.2010.09.002

Hutchins, M. L., Holzworth, R. H., & Brundell, J. B. (2014). Diurnal variation of the global electric circuit from clustered thunderstorms. *Journal of Geophysical Research*, *199*, 620–629.

Ivanov, V. A. (1987). *Lightning activity and forest fires. Forest fires and fighting against them* (pp. 208–217). Moscow, Russia: VNIILM.

Ivanov, V. A. (1996). *Forest fires from thunderstorms on the Yenisei plain.* (Abstract of doctoral dissertation). Sukachev Institute of Forest SB RAS, Krasnoyarsk.

Jain, A. K. (2010). Data clustering: 50 years beyond K-means. *Pattern Recognition Letters, 31*(8), 651–666. doi:10.1016/j.patrec.2009.09.011

Kerkmann, J., Lutz, H. J., König, M., Prieto, J., Pylkko, P., Roesli, H. P., . . . Zwatz-Meise, V. *MSG channels Interpretation Guide Weather, surface conditions and atmospheric constituents.* Retrieved from http://www.eumetrain.org/IntGuide

Kononov, I. I., & Yusupov, I. E. (2004). Cluster analysis of thunderstorm activity. *Radio Engineering and Electronics, 49*(3), 283–291.

Kovács, F., Legány, C., & Babos, A. (2005, November). Cluster validity measurement techniques. In *Proceedings 6th International Symposium of Hungarian Researchers on Computational Intelligence* (p. 35).

Kozlov, V. I., & Mulyayarov, V. A. (2004). *Thunderstorm activity in Yakutia*. Yakutsk, Russia: YaP Publishing House of SB RAS.

Kudryavtsev, N. G., Belikova, M. Yu., Krechetova, S. Yu., Geyman, T. N., Kocheeva, N. A. & Kudin, D. V. (2014). Construction of clusters of lightning discharges on the basis of data of the global network of thunderstorm signaling WWLLN and determination of their parameters. *New information technologies in the study of complex structures in the village.* Katun, Russia. June 9-11, 21.

Latham, D., & Williams, E. (2001). Lightning and forest fires. In E. A. Johnson & K. Miyanishi (Eds.), *Forest fires: Behavior and ecological effects* (pp. 375–418). Academic Press. doi:10.1016/B978-012386660-8/50013-1

Latham, D. J. (1991). Lightning flashes from a prescribed fire-induced cloud. *Journal of Geophysical Research, 96*(D9), 17151–17157. doi:10.1029/91JD01808

Mardiana, R., Kawasaki, Z.-I., & Morimoto, T. (2002). Three-dimensional lightning observations of cloud-to-ground flashes using broadband interferometers. *Journal of Atmospheric and Solar-Terrestrial Physics, 64*(1), 91–103. doi:10.1016/S1364-6826(01)00099-2

Mezuman, K., Price, C., & Galanti, E. (2014). On the spatial and temporal distribution of global thunderstorm cells. *Environmental Research Letter, Vol. 9. Paper, 124023*, 9.

Müller, M. M., & Vacik, H. (2017). Characteristics of lightnings, igniting forest fires in Austria. *Agricultural and Forest Meteorology, 240-241*, 26–34. doi:10.1016/j.agrformet.2017.03.020

Nieto, H., Aguado, I., García, M., & Chuvieco, E. (2012). Lightning-caused fires in Central Spain. *Agricultural and Forest Meteorology, 162-163*, 35–43. doi:10.1016/j.agrformet.2012.04.002

Pinto, I. R. C. A., & Pinto, O. Jr. (2003). Cloud-to-ground lightning distribution in Brazil. *Journal of Atmospheric and Solar-Terrestrial Physics, 65*(6), 733–737. doi:10.1016/S1364-6826(03)00076-2

Pinto, O. Jr, Pinto, I. R. C. A., Diniz, J. H., Cherchiglia, L. C. L., Filho, A. C., & Carvalho, A. M. (2003). A seven-year study of the negative cloud-to-ground flash characteristics in Southeastern Brazil. *Journal of Atmospheric and Solar-Terrestrial Physics, 65*, 739–748. doi:10.1016/S1364-6826(03)00077-4

Pinto, O. Jr, Pinto, I. R. C. A., & Naccarato, K. P. (2007). Maximum cloud-to-ground lightning flash densities observed by lightning location systems in the tropical region: A review. *Atmospheric Research*, *84*(3), 189–200. doi:10.1016/j.atmosres.2006.11.007

Rodger, C. J., Werner, S., Brundell, J. B., Lay, E. H., Thomson, N. R., Holzworth, R. H., & Dowden, R. L. (2006). Detection efficiency of the VLF World-Wide Lightning Network Location Network (WWLLN): Initial case study. *Annales Geophysicae*, *24*(12), 3197–3214. doi:10.5194/angeo-24-3197-2006

Rousseeuw, P. J. (1987). Silhouettes: A graphical aid to the interpretation and validation of cluster analysis. *Journal of Computational and Applied Mathematics*, *20*, 53–65. doi:10.1016/0377-0427(87)90125-7

Scientific-applied reference book on the climate of the USSR. Parts 1-6. Issue. 20. (Tomsk, Novosibirsk, Kemerovo regions and Altai Territory). (1993). St. Petersburg, Russia: Gidrometeoizdat.

Seity, Y., Soula, S., & Sauvageot, H. (2000). Radar observation and lightning detection in coastal thunderstorms. *Physics and Chemistry of the Earth. Part B: Hydrology, Oceans and Atmosphere*, *25*(10-12), 1107–1110. doi:10.1016/S1464-1909(00)00161-1

Shabaganova, S. N., Karimov, R. R., Kozlov, V. I., & Mullayarov, V. A. (2013). Characteristics of the solar cells from observations in Yakutia. *Russian Meteorology and Hydrology*, *37*(11-12), 746–751. doi:10.3103/S1068373912110088

Soriano, L. R., De Pablo, F., & Tomas, C. (2005). Ten-year study of cloud-to-ground lightning activity in the Iberian Peninsula. *Journal of Atmospheric and Solar-Terrestrial Physics*, *67*(16), 1632–1639. doi:10.1016/j.jastp.2005.08.019

Tantisattayakul, T., Masugata, K., Kitamura, I., & Kontani, K. (2005). Broadband VHF sources locating system using arrival-time differences for mapping of lighting discharge process. *Journal of Atmospheric and Solar-Terrestrial Physics*, *67*(11), 1039–1039. doi:10.1016/j.jastp.2005.04.002

The Comprehensive R Archive Network | Submitting to CRAN Retrieved from https://cran.r-project.org

Tork, H. F. (2012). Spatio-temporal clustering methods classification. *Doctoral Symposium on Informatics Engineering*. 199-209.

Uman, M. N. (1984). *Lightning*. New York, NY: Dover Publications.

Van Wagner, C. E. (1987). *Development and structure of the Canadian Forest Fire Weather Index System*. Can. For. Serv., Ottawa, Canada: For. Tech. Rep. 35.

Wang, C., Zheng, D., Zhang, Y., & Liu, L. (2017). Relationship between lightning activity and vertical airflow characteristics in thunderstorms. *Atmospheric Research*, *191*, 12–19. doi:10.1016/j.atmosres.2017.03.003

Williams, E. R. (1989). The triple structure of thunderstorms. *Journal of Geophysical Research*, *94*(D11), 13151–13167. doi:10.1029/JD094iD11p13151

Xu, R., & Wunsch, D. II. (2005). Survey of clustering algorithms. *IEEE Transactions on Neural Networks*, *16*(3), 645–678. doi:10.1109/TNN.2005.845141 PMID:15940994

ADDITIONAL READING

Betz, H. D., Schumann, U., & Laroche, P. (Eds.). (2008). *Lightning: principles, instruments and applications: review of modern lightning research*. Springer Science & Business Media. doi:10.1007/978-1-4020-9079-0

Borisova, I. A. (2007). Clustering algorithm FRiS-Tax. *Scientific bulletin of NGTU, 3*, 3-12.

Ganteaume, A., Camia, A., Jappiot, M., San-Miguel-Ayanz, J., Long-Fournel, M., & Lampin, C. (2013). A review of the main driving factors of forest fire ignition over Europe. *Environmental Management, 51*(3), 651–662. doi:10.100700267-012-9961-z PMID:23086400

Hutchins, M. L., Holzworth, R. H., & Brundell, J. B. (2014). Diurnal variation of the global electric circuit from clustered thunderstorms. *Journal of Geophysical Research: Space Physics, 119*(1), 620–629.

Hutchins, M. L., Holzworth, R. H., Brundell, J. B., & Rodger, C. J. (2012). Relative detection efficiency of the world wide lightning location network. *Radio Science, 47*(6), n/a. doi:10.1029/2012RS005049

Rakov, V. A., & Uman, M. A. (2003). *Lightning: physics and effects*. Cambridge University Press. doi:10.1017/CBO9781107340886

Smith, C. B. (2008). *Lightning: Fire from the Sky*. Dockside Consultants Inc.

KEY TERMS AND DEFINITIONS

Cluster Analysis: A method of uniting groups (clusters) of objects of research according to the principle of their proximity. The object is a point of multidimensional space, where its coordinates are given by the values of several marks. The measure of proximity of objects is given in different ways, for example, the Euclidean distance.

Lightning: An electrical spark in the atmosphere that usually occurs during a thunderstorm.

The density of Lightning Discharges: The ratio of the number of lightning discharges to the area of the site in which they are registered.

The Republic of Buryatia: An administrative unit within the Russian Federation.

Thunderstorm: An atmospheric phenomenon, in which electrical discharges occur within the clouds or between the clouds and the earth's surface - lightning accompanied by thunder.

Thunderstorm Focus: A region of space with an active generation of lightning.

World Wide Lightning Location Network (WWLLN): An international network for registering lightning discharges on the globe. WWLLN was based at the University of Washington in 2002 and currently includes more than 70 receiving stations.

Section 4

Chapter 8
Forest Fire Danger Assessment Using Meteorological Trends:
Case Study

Nimazhap Bayarzhapovich Badmaev
Institute of General and Experimental Biology of the Siberian Branch, Russian Academy of Sciences, Russia

Aleksandr Vladimirovich Bazarov
Institute of Physical Material Science of the Siberian Branch, Russian Academy of Sciences, Russia

Roman Sergeevich Sychev
Institute of Physical Material Science of the Siberian Branch, Russian Academy of Sciences, Russia

ABSTRACT

The chapter presents the results of research in the Republic of Buryatia, where the number and area of fires have increased over the past 20 years due to the rise in temperature and aridity. Most of the fires are registered in the large river valleys where pine forests are formed, which have low soil moisture capacity. Fewer fires occurred on the Eastern Sayans, Khamar-Daban ridges, and the Stanovoye Highlands, where the precipitation maximum falls. A correlation analysis was carried out between meteorological parameters and fires in climate-contrasting forests. The lack of precipitation at the end of the previous summer, combined with the hot and dry spring weather of the current year, have a significant impact on fires in the arid ecosystems of the Transbaikal middle mountains. In the humid coastal climate of the Eastern Baikal region, the high temperature of the air determines the fires, but there is no precipitation.

INTRODUCTION

A forest fire is a spontaneous, uncontrolled spread of fire across forest areas. Fires have serious negative consequences not only for the biosphere and the ecology of the Earth, but also for human health (Teakles et al., 2017). They destroy forest (Wastl et al., 2013, Barbero, Abatzoglou, Larkin, Kolden, & Stocks,

DOI: 10.4018/978-1-7998-1867-0.ch008

2015, De Vicente & Crespo, 2012), burning a huge mass of organic matter, emit a significant amount of smoke (Yao et al, 2018), combustion gases (Adame et al., 2018) and aerosols (Zhao, Kooperman, Pritchard, Russell, & Somerville, 2014), thereby increasing the "greenhouse effect." Besides the fact that fires adversely affect the biota covered, they can change the physicochemical characteristics of the soil (Kelly, Montgomery, & Reid, 2008) and cause significant economic damage (San-Miguel-Ayanz, Moreno, Camia, 2013).

The problem of forest fires has a global scale. The study of this problem has attracted the attention of scientists from Spain (Bedia, Herrera, & Gutierrez, 2014, De la Cueva, Quintana, & Canellas, 2012), Greece (Papadopoulos, Paschalidou, Kassomenos, & McGregor, 2013, Koutsias et al., 2013), Switzerland (Wastl, Schunk, Leuchner, Pezzatti, & Menzel, 2012), Portugal (Parente, Pereira, Amraoui, & Fischer, 2018), India (Kale et al., 2017), Bolivia (Heyer, Power, Field, & van Marle, 2018), South Africa (Strydom & Savage, 2017), USA and Canada (Feltman, Straka, Post, & Sperry, 2012, Shabbar, Skinner, & Flannigan, 2011, Girardin, 2010), Germany (Holsten, Dominic, Costa, & Kropp, 2013), South Korea (Won, S. Lee, M. Lee, & Ohga, 2010), China (Liu et al., 2016), of course, Russia (Zdereva, & Vinogradova, 2009) and others.

Recently, there has been a surge of extremely devastating fires with corresponding social upheavals and significant economic costs. Although most fires are anthropogenic in nature (Kale et al., 2017, Heyer et al., 2018), favorable conditions are necessary for ignition. The occurrence of fires most often falls on the warm period of the year. For a relevant fire danger assessment, data on temperature and relative air humidity (Matsoukis, Kamoutsis, & Chronopoulos, 2018), precipitation, dry days (Kale et al., 2017), heat waves (Parente et al, 2018), wind (Rolinski et al., 2016), and, in addition, soil drought (Zink et al., 2016) is necessary. A favorable combination of these parameters can provide forest dry fuel and necessary conditions for fire, often caused by lightning of dry thunderstorms (Liu et al., 2016) and anthropogenic impact (Guo et al., 2015).

Thus, the weather is the most volatile and the biggest driving force before the occurrence and spread of fires (Abatzoglou, & Kolden, 2013). The optimal conditions for fires are hot weather, dry fuel and sources of ignition (Moritz, Morais, Summerell, Carlson, & Doyle, 2005).

BACKGROUND

In recent decades, global temperatures have risen at a rate of ~ 0.2 °C per decade (Hansen, Ruedy, Sato, & Lo, 2010). Climate change affects global fire fluctuations (Flannigan, Krawchuk, de Groot, Wotton, & Gowman, 2009) and is expected to increase the severity of fires in the coming decades (Flannigan et al., 2013). Three intervals are distinguished: the warming of 1910–1945, the weak cooling of 1946–1975, and the most intense warming since the 1970s. In Russia, changes towards higher air temperatures are more pronounced than on the planet as a whole (Figure 1), and on the territory of Buryatia even more, by 2.5 ° C (A. Kulikov, M. Kulikov, & Smirnova, 2008).

Forced warming in the Prebaikalia and Transbaikalia is also emphasized by other researchers (Federal Hydrometeorology and Environmental Monitoring Service (2008a), Federal Hydrometeorology and Environmental Monitoring Service (2008b), Gruza & Rankova, 2004). It is noteworthy that in the last decade, the average annual temperature in Transbaikalia crosses the zero boundary, becoming positive, which was previously noted sporadically (Figure 2).

Figure 1. Inhomogeneity of modern climate warming: the thin line - the results of summarizing observations at stations in Russia; bold red and black lines - 11-year moving averages of air temperature (Federal Hydrometeorology and Environmental Monitoring Service (2008a), Federal Hydrometeorology and Environmental Monitoring Service (2008b))

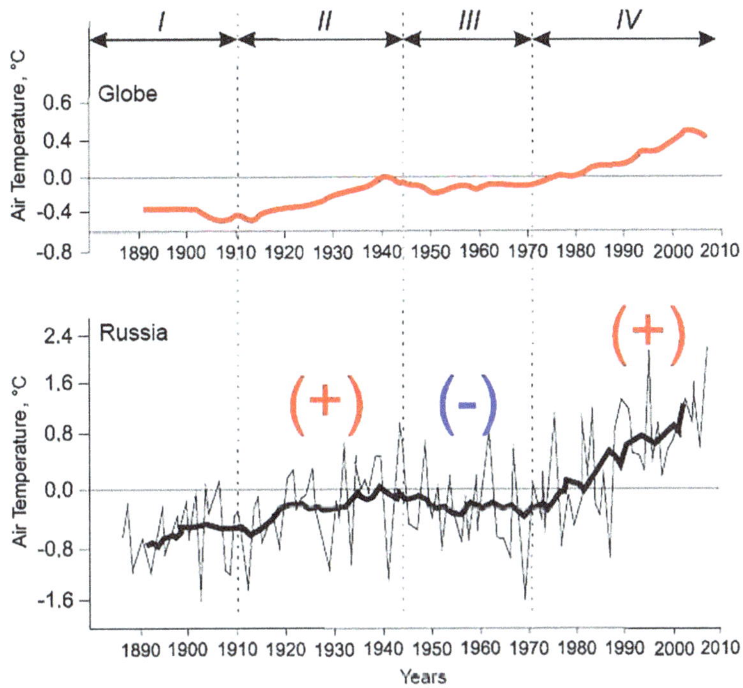

Figure 2. Long-term dynamics and trends of average annual air temperature in Transbaikalia Row 1 - 1900-2015; Row 2 (Ulan-Ude) - 1970-2015 (Garmaev et al., 2017)

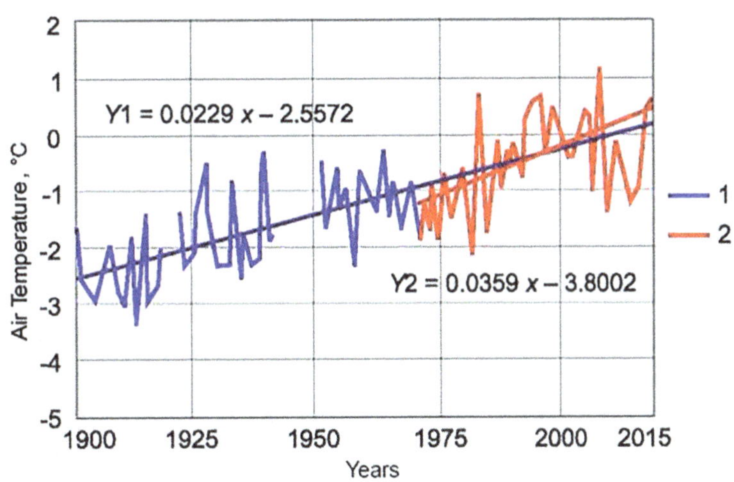

Figure 3. Dynamics of atmospheric precipitation sums over climatic cycles and an increase in the instability of atmospheric moistening over time: 1 — the average perennial precipitation total over climatic cycles; 2 - the average long-term value of precipitation for the entire period; 3 - the range of changes in the mean deviation and the mean deviations of the climatic cycles
(Garmaev et al., 2017)

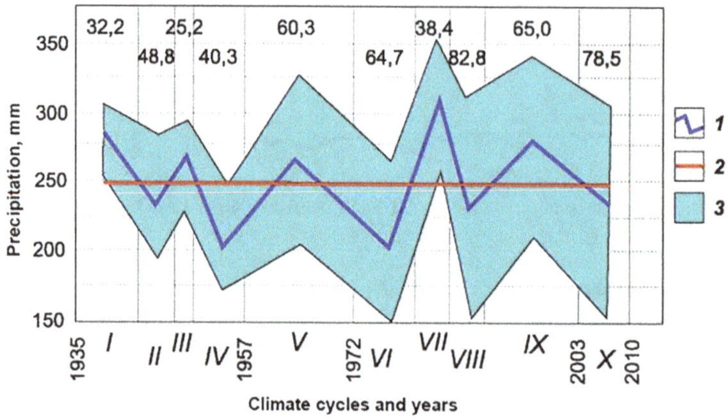

The authors (Kulikov et al., 2008) obtained the linear trend equations for average annual air temperature (Y). According to the weather station of Ulan-Ude, we have

Y = 0.0229x - 2.5572 (1900-2011);

Y = 0.0359x - 3.8002 (1970-2011).

The regression coefficient indicates an air temperature rise rate of 0.23 °C/10 years. Since the 1970s, the rate of warming has risen to 0.36 °C/10 years. In Novoselenginsk, warming is at a rate of 0.18 °C/10 years, and in Kyakhta - 0.16 °C/10 years.

Variations in the amount of precipitation over the years are very significant, the difference reaches about 100-150 mm. In the city of Ulan-Ude with extremes of 413.3 mm (1959) and 109.6 mm (1989), the extreme amplitude is 303.7 mm. The variability of precipitation increases with time, i.e. the extreme conditions for this indicator increase. So, if the segment 1935-1942, the standard deviation of the annual precipitation amount was ± 32.2 mm, now the deviations from the climatic norm have become sharper and the mean deviations are ± 65-83 mm, i.e. deviations from the mean multiyear sum (256 mm) increased from 13 to 25-32% (Figure 3).

Thus, on the territory of the Republic of Buryatia, statistically significant trends of increase in temperature and decrease in precipitation with identification of wet and dry periods are established. Due to global climate change (temperature increase, decrease in precipitation), the number of fires has increased. Forest fires on the territory of the Republic of Buryatia every year in the spring-summer period are of a massive character, which is primarily due to the high aridity of the territory and the human factor (Sidorov & Sanzhieva, 2018). Due to global changes in the territory of the Republic of Buryatia, significant transformations of the environment are observed. Over the past 20 years, due to the abnormally high air temperatures and the increasing aridity of the territory, the frequency and area of fires have increased.

Figure 4. The number of forest fires (a) in Buryatia and their area, ha, (b) in the period from 1966 to 2016 (Sidorov & Sanzhieva, 2018)

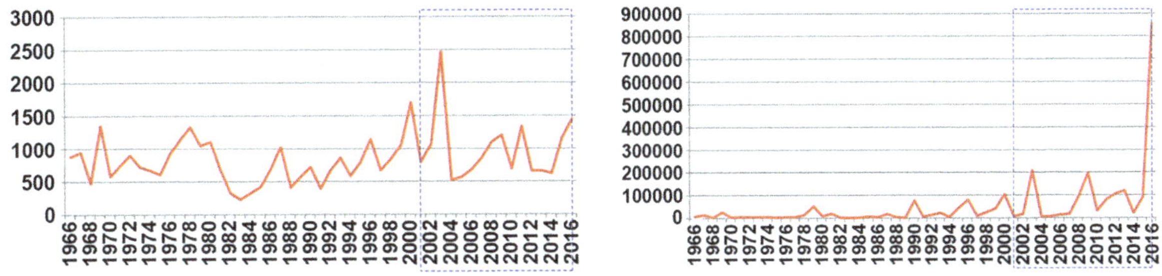

(a) (b)

The annual trend towards an increase in the number of fires and the area of forest areas exposed to fires is a serious regional environmental problem for Buryatia. Despite the local forest fires and the regional scale of this problem, its consequences are more significant and global. This is due to the destruction of a huge array of forest ecosystems in the unique Baikal natural territory. In addition, forest fires are accompanied by air pollution with harmful and toxic gases, vapors and aerosols spreading outside the republic. The forests of Buryatia are distinguished by extremely high burning capacity, determined by the climatic features of the region, as well as by the species composition of the prevailing coniferous tree species (Antropov, Seredkin, & Schepin, 2013, Sidorov, & Khankhunov, 2015, Sidorov, & Khankhunov, 2016).

A review of sources (Sidorov & Sanzhieva, 2018, Antropov, Seredkin, & Schepin, 2013, Ministry of Natural Resources of the Republic of Buryatia, (2008-2018)) showed quantitative data on the frequency and area of fires over the past 50 years, as well as the existing problems on the inconsistency of statistical data on the situation on forest fires between the executive authorities (the Center for Forest Protection, the Main Directorate of the EMERCOM of Russia for the Republic of Buryatia, the Ministry of Natural Resources of the Republic of Buryatia). It is also necessary to note the weak coordination of work between these ministries and departments with scientific organizations of the republic and the weak use of modern methods for analyzing cartographic information and remote sensing data, as well as the imperfection of the existing protection system (Tulokhonov & Puntsukova, 2016).

Having studied the chronology of forest fires in the Republic of Buryatia since the 60s of the twentieth century, we will see a significant increase in their number and area in certain years. The sharp increase in the number of fires and, consequently, the areas exposed to fire, has been observed from 1999 to the present. The increase in both the number and total area of fires leads to a significant loss of forests in vast areas in the Republic of Buryatia. According to (Sidorov & Sanzhieva, 2018), the increase in the number of forest fires from 1966 to 1996 occurred with a certain cyclical nature every 6–8 years (Figure 4 a, b).

In addition, there is an increase in the frequency of fires and their area on the territory of Buryatia over the past 20 years (Figure 4 a, b). In the period from 1966 to 1995, fires covered 359,537 hectares of forests in Buryatia, which on average is 11,984.6 hectares for one year. The increase in the area of fires in the period from 1996 to 2016 compared to the previous period occurred almost 6 times. A record number of fires occurred in 2003, the area of forests engulfed in fire - in 2015.

The problem of protecting forests from fires is particularly acute. In Buryatia, as in the Baikal region as a whole, this problem is given special attention, (Belyakin & Volokitina, 2010, Evdokimenko, 2013, Ukraintsev, & Plyusnin, 2015, Makarenko, 2016) but far from being sufficient (Dorzhiev et al., 2017). It should be noted that Buryatia is an extremely rugged territory, where literally in the neighborhood there can be and are quite climatically contrasting landscapes. This chapter discusses the conditions of occurrence and spread of forest fires, depending on previous meteorological parameters at the level of forest areas of the Republic of Buryatia. To effectively deal with them, it is necessary to know the landscape features of the spatial and temporal dynamics of forest fires, the specific causes of their occurrence in various conditions, etc.

NATURAL CONDITIONS AND GEOGRAPHICAL POSITION OF THE REPUBLIC OF BURYATIA

The Republic of Buryatia is located in the center of the Asian continent, between the taiga spaces of Eastern Siberia and the vast steppes of Mongolia. The distance by rail from Ulan-Ude to Moscow is 5,519 km, to the Pacific Ocean - 3,500 km. The territory of the Republic of Buryatia is washed by the waters of Lake Baikal from the north and west, bordered by Trans-Baikal Territory in the east, by the Irkutsk Region in the west and north, by the Republic of Tyva in the south-west, and by the Mongolia in the south. The Republic of Buryatia is divided into 21 municipal districts, it has 6 cities, 29 urban-type settlements, and 226 rural administrations. As at the end of 2017, the resident population amounted to 985 thousand people (5% of the population of the Siberian Federal District; 0.7% of the population of the Russian Federation). The average population density is 2.8 people. on 1 square. km The administrative center - the city of Ulan-Ude - was founded in 1666. The Republic of Buryatia has a favorable geographical position in the system of relations between the Russian Federation and the countries of the Asia-Pacific region.

The territory of the Republic of Buryatia enters the mountain system, which occupies a significant part of the south of Eastern Siberia, and is characterized by powerful mountain ranges and extensive, deep and sometimes almost closed intermountain hollows. Virtually the entire territory is dominated by heavily dissected mountains, flat surfaces are found only in tectonic depressions and valleys of large rivers. The area of the mountains is more than 4 times the area occupied by the lowlands. For the Republic of Buryatia characterized by significant elevation above sea level. The lowest indicator is the level of Lake Baikal - 456 meters at the Pacific mark, and the highest is the Munku-Sardyk peak covered with glaciers in the Eastern Sayan Mountains - 3,491 meters above sea level.

The southern part of the Republic of Buryatia, represented by the Selenga middle mountain, covers a significant part of the Selenga river basin, the main waterway of Lake Baikal, including all its major tributaries, and is characterized by a predominance of mountains of medium height (1000-1500 meters above sea level).

High Baikal ridges of the Baikal with wide intermountain hollows separating them adjoin Lake Baikal. Their belt includes the highlands of the Eastern Sayan - a kind of uplift, stretching from the north-west to the south-east for a distance of about 1000 km with a width within 200-300 km and rising in the central part of the ridges for more than 2500-3000 meters.

The mountain ranges on the territory of Buryatia have the form of elongated ridges, mainly from the south-west to the northeast, in the direction of the main tectonic faults of the earth's crust. The belt of the Baikal mountains is continued by the Khamar-Daban, Ulan-Burgasy, Barguzinsky, Ikatsky and Baikal ranges. The watersheds of the Barguzinsky Range are the classic alpine landforms. Further north of the Baikal region, they are continued by the ranges of the Stanovoye highland: North and South Muysk, Verkhneangarsk, Delun-Uran, Kodar. The Vitim plateau adjoins the northeast of the Baikal region. For the entire northern Baikal region, continuous distribution of permafrost, occurring at a depth of 0.5 meters, with a capacity of up to 500-600 meters, is characteristic.

The diversity of soils of Buryatia is due to the difference in the climatic conditions of its territory. On this basis, the soil covers of steppe, forest-steppe, taiga territories and loaches is distinguished. At the same time, soils of mountain areas and intermontane depressions are distinguished.

In the mountainous parts of the soil are often formed on permafrost. The main soil types are Holtz-desert, Holtz-tundra, mountain frozen, as well as taiga and frozen-frost, frozen-meadow-forest, deep-frozen alpine podzol, mountain-gray forest soils, chestnut and chernozem soil.

In intermountain depressions, soil types are much more diverse, often more deeply frozen. Here, the main soil types are permafrost-taiga, forest gray, chestnut and chernozem, meadow-frost, meadow chestnut, meadow-saline, permafrost-marsh, alluvial, salt marsh and salt marshes.

The rivers of Buryatia form a rather dense network. In the Republic of Buryatia there are, according to forest management expeditions, up to 30 thousand rivers, streams and streams that belong to the basins of the Angara and Lena rivers and the basin of Lake Baikal. The rivers are mainly mountainous, shallow, with a rapid flow, sandy and rocky bottom, in the upper reaches there are many waterfalls and rapids. The Selenga River is the main waterway of the Republic of Buryatia. Its tributaries Chikoi and Khilok were navigable, but the passage of riverboats from the Selenga to Lake Baikal is currently difficult due to the fact that in the past, along the banks of rivers and streams flowing into the main tributaries of the Selenga river, logging was conducted. In the 60s, the large logging enterprise "Elovsky" was organized with an annual logging volume of up to 80 thousand m3, and since the 80s JSC Selenginsky Pulp and Paper Mill was established with an annual harvesting volume of 25 -30 thousand m^3. As a result of this and a number of other reasons, many brooks and rivulets in these areas have ceased to exist.

On the territory of the Republic of Buryatia there are many lakes located in the areas of relief lowering: Baikal, Baunt, Busan, Kotokel, Gusinoe, Eravna lakes, Kapylyushi and a large number of small fresh and salt lakes.

The Baikal Basin, a region directly associated with a unique freshwater natural feature, is located in the center of the Asian continent, on its territory meet the boundaries of large natural areas - the South-Siberian mountain tundra-mountain taiga, Baikal-Dzhugdzhur mountain-tundra-mountain taiga and Central Asian mountain dry-steppe. This determines a large variety of natural landscape conditions, from the mountain-tundra and mountain-taiga forest to dry-steppe, which, in turn, provides a wide range of conditions for the growth of forests and their biological diversity in this region. The landscapes of the basin are very diverse: from loaches on the crests of mountain ranges to dry steppes in the basins of Transbaikalia, however, mountain-taiga, forest mountain-valley and mountain-steppe landscapes prevail.

The volume of water mass of Baikal is 23,000 cubic kilometers. Baikal is located above sea level at an altitude of 456 meters, the catchment area is 588,000 square kilometers. Only one river flows out of the reservoir, the Angara. Baikal is characterized by frequent winds reaching hurricane speeds of up to 40 meters per second. There are strong storms on Baikal in October-November. For 7-8 months, Baikal provides shipping service to hard-to-reach areas of its coast. There are 30 islands on Baikal, the largest is Olkhon Island, its area is 722.4 square km. The shores of Lake Baikal are mostly mountainous, with large rocky slopes. The mountains bordering the coast reach a height of 2000 meters and only at the confluence of large rivers Selenga, Upper Angara, Barguzin and others into the lake form wide deltas. The lake is characterized by a peculiar flora and fauna, over 1,800 species of animals and plants live here, a significant part of which (up to 1,100 species) is endemic.

The second largest lake of the Republic of Buryatia is the Goose Lake, located in the Selenga Steppe. The lake is lined with mountains, but the shores are mostly low and sandy, forming beaches accessible to holidaymakers. In addition to these lakes, there are several lake groups on the territory of the Republic of Buryatia. The Eravna group has 17 fairly large lakes, which are round or oval in shape, have sandy shores and are distinguished by a shallow depth (up to 6–7 meters). They are located at an altitude of 950-970 meters above sea level.

The remoteness of the territory of the Republic of Buryatia from the oceans, its location in the center of the vast Eurasian continent and the mountain-valley relief caused a peculiar and in its own unique climate. A specific feature of the climate is its sharp and frequent spatial variability: the presence of mountain ranges of different heights and orientations, the existence of intermontane depressions and intermountain valleys, the shape and degree of spelling severity — all this strongly influences the local circulation of air masses, dramatically changing the main climatic indicators, creates a mosaic - a mixed, heterogeneous climate picture.

In general, the climate is formed under the influence of three contrasting components: the dry and cold climate of the northern regions, the hot and dry climate of the Mongolian deserts, and the humid Pacific exposure. The average temperature in summer is + 18.5°C, in winter –22°C, and the average annual temperature is –1.6 °C. The average annual rainfall is 244 mm. The territory of the Republic of Buryatia is characterized by a sharply continental climate with large annual and daily fluctuations in air temperature and with an uneven distribution of precipitation over the seasons of the year.

This mountainous taiga territory in the central part of the Asian continent is characterized by cold winters and hot summers. Low winter temperatures are fairly easily tolerated by humans due to the dryness of the air. In the summer, the heat is felt only in the middle of the day, the morning and evening hours are cool. Autumn is long and rather warm, until Baikal freezes, air temperature indicators in the Republic of Buryatia at this time of year are often higher than in the European regions of the Russian Federation. Spring begins to be felt at the end of March.

A specific feature of the climate of the Republic of Buryatia is the development in the cold period of the year of the powerful northeast spur of the Siberian anticyclone, which occurs in September – October and disappears in April – May. Therefore, winter in Transbaikalia is characterized by a large number of sunny days and low air temperatures. The lowest temperatures are noted in the river valleys and basins, where stagnation and intense air freezing take place. The average air temperature in January is –20° –30°C, the minimum is from –45° to –55°C. Partly cloudy, little windy or calm weather with the lowest annual rainfall. The thickness of the snow cover is small.

Severe windless winter gives way to late windy and dry spring with night frosts, as the barometric pressure during this period decreases, and cold air flows from the northern regions of Siberia to the territory. This contributes to the return of cold weather and the emergence of long and strong winds.

The summer is short, in the first half of the dry period, with individual dry winds that developed in Mongolia, in the second half (July-August) cyclonic activity gradually increased, resulting in moist air from the Pacific Ocean. Wind speed increases and the bulk of precipitation falls: July and August account for 60–70% of the annual norm. The average July temperature reaches 15° 20°C, and the maximum is

Figure 5. Administrative map of the Republic of Buryatia with the boundaries of districts (Dorzhiev et al., 2017).
The numbers indicate the areas: 1 – Severo-Baikal'sky; 2 - Muysky; 3 - Kurumkansky; 4 - Bauntovsky; 5 - Barguzinsky; 6 - Pribaikalsky; 7 - Khorinsky; 8 - Eravninsky; 9 - Okinsky; 10 - Tunkinsky; 11 - Kabansky; 12 - Ivolginsky; 13 - Tarbagataisky; 14 - Zaigraevsky; 15 - Kizhinginsky; 16 - Zakamensky; 17 - Dzhidinsky; 18 - Selenginsky; 19 - Mukhorshibirsky; 20 - Kyakhtinsky; 21 - Bichursky.

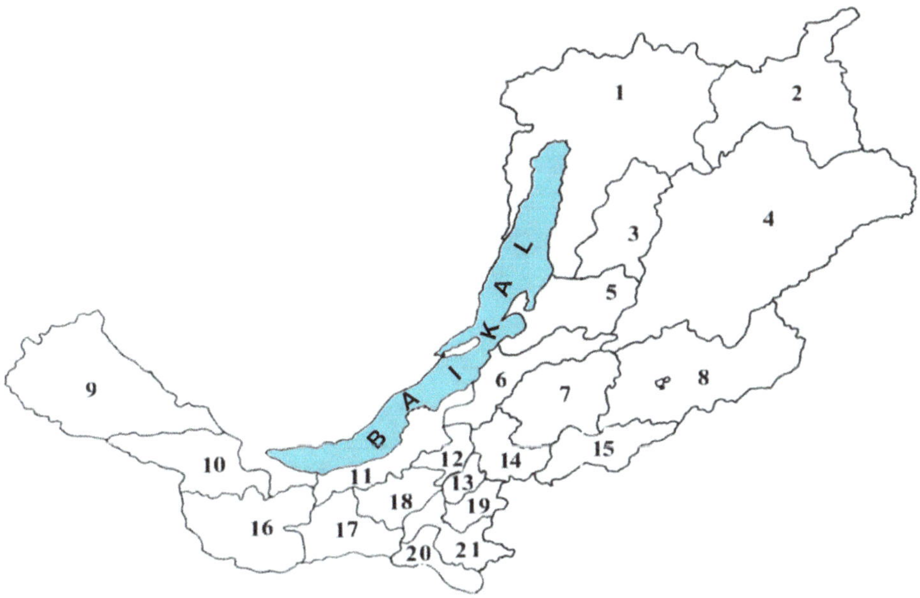

30° – 38°C above zero. Autumn is short and dry, with sharp daily fluctuations in temperature and often with early frosts.

A significant feature of the climate of the Republic of Buryatia is the long duration of sunshine (1900 - 2200 hours per year), and by this indicator the Republic of Buryatia is not inferior to the southern regions of the Russian Federation, and sometimes exceeds.

FORESTRY OF THE REPUBLIC OF BURYATIA

The forest fund of the Republic of Buryatia is divided into 37 forest areas (Russian Federation, 2006, Federal Forestry Agency, 2008) The number of forest areas created objectively reflects the optimal number of territorial units subordinated to the executive authority in the field of organizing the use, conservation, protection and reproduction of forests, the external borders of forest areas coincide with the borders of other Forestiers within the same region, but do not always coincide with the boundaries of the districts of the Republic of Buryatia (Vashchuk, 2014), i.e. one forest area may be located in several districts; at the same time, several forest areas may be located in one area (table). Within the borders of the existing state nature reserves and national parks in the republic five forest areas are created:

Table 1. Location of forest areas in relation to areas of the Republic of Buryatia

Districts	Forestries
Barguzinsky	Barguzinskoe, Ust-Barguzinskoe, Kuytunskoe, Zabaykalsky National Park
Bauntovsky	Vitimskoe, Romanovskoe
Bichursky	Bichurskoe, Buyskoe
Dzhidinsky	Dzhidinskoe, Baikal State Natural Biosphere Reserve
Eravninsky	Eravninskoe, Kondinskoe, Khorinskoe
Zaigraevsky	Zaigraevskoe, Verkhne-Taletskoe, Ulan-Udenskoe, Khandagataiskoe`
Zakamensky	Zakamenskoe
Kabansky	Kabanskoe, Gusinoozerskoe, Babushkinskoe, Baikal State Natural Biosphere Reserve
Kizhinginsky	Kizhinginskoe, Kudunskoe, Khandagataiskoe
Kurumkansky	Kurumkanskoe, Verkhne-Barguzinskoe, State Nature Reserve "Dzherginsky
Kyakhtinsky	Kyakhtinskoe
Mukhorshibirsky	Mukhorshibirskoe
Okinsky	Okinskoe
Pribaikalsky	Kikinskoe, Pribaikalskoe, Baikal'skoe
Severo-Baikal'sky	Angoyanskoe, Severo-Baikal'skoe, Uoyanskoe, State Natural Biosphere Reserve "Barguzinsky"
Selenginsky	Ivolginskoye, Gusinoozerskoye, Selenginskoye, Baikal State Natural Biosphere Reserve
Tunkinsky	Tunkinsky National Park
Tarbagataisky	Zaudinskoe, Mukhorshibirskoe, Khandagataiskoe
Khorinsky	Kurbinskoe, Verkhne-Taletskoe, Khorinskoe
Ivolginsky	Ivolginskoye, Ulan-Udenskoe
Muysky	Muyskoe, Uoyanskoe

(Government of the Republic of Buryatia, 2008)

1. Baikal State Natural Biosphere Reserve.
2. State Natural Biosphere Reserve "Barguzinsky".
3. Dzherginsky State Nature Reserve.
4. Transbaikal National Park.
5. Tunkinsky National Park.

Forest parks in the Republic of Buryatia and Sakha (Yakutia), Trans-Baikal Territory and the Irkutsk Region were not created. The boundaries of the districts of the Republic of Buryatia are shown in Figure 5.

The territorial and administrative affiliation of forest areas is given in the Table 1.

The Republic of Buryatia is characterized by the following types of high conservation value forests:

Figure 6. Charts with the accumulation of the quantity (a) and fire area (b) from 2001 to 2018 for the forest districts of the Republic of Buryatia according to the data of the Federal Forestry Agency

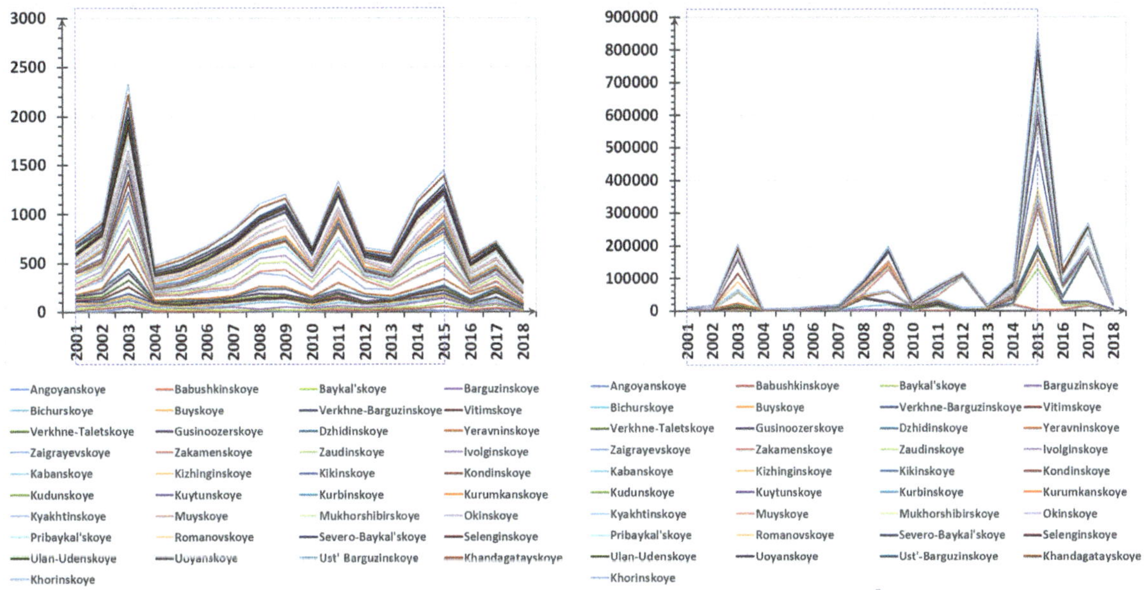

(a) *(b)*

- Forests with a high level of biodiversity that are internationally significant. This type of forest includes forests of protected natural areas that perform numerous functions, one of which is the preservation of biodiversity, as well as forests with the presence of endemics (plant and animal species distributed only in the territory of the Republic of Buryatia).
- Large forest landscapes that are significant at the regional or national levels. These are forests in which ecological processes and ecosystem functioning have not been completely or largely anthropogenic in recent times.
- Rare and endangered ecosystems. Some ecosystems are rare due to climatic conditions. This type of high conservation value forest is dedicated to conserving endangered forest ecosystems, communities or forest types.
- Forest areas that perform natural protective functions (water protection, anti-erosion, and others).
- Forest areas necessary for the preservation of the traditional types of environmental management of the small ethnic groups of the North.

The length of the republic from north to south, the spread of various types - continuous, intermittent and island - permafrost, mountainous nature of the relief, determine in the territory of Buryatia the isolation of a multiple network of ecological niches with sharply contrasting climatic regimes (Badmaev & Bazarov, 2019), especially in conditions of global warming. The impact of climate change on forest fires has been studied globally, but uncertainties about regional and level influences still exist (Strydom & Savage, 2017). The biome-level forest fire potential is related to fire risk indicators provided by synoptic information, but landscape-scale behavior is determined by local weather conditions (Jolly et al., 2015).

Figure 7. Distribution of forest fires (a) and their area, ha, (b) in 2003 and number (c) of forest fires and their area, ha, (d) in 2015, by forestries of Buryatia

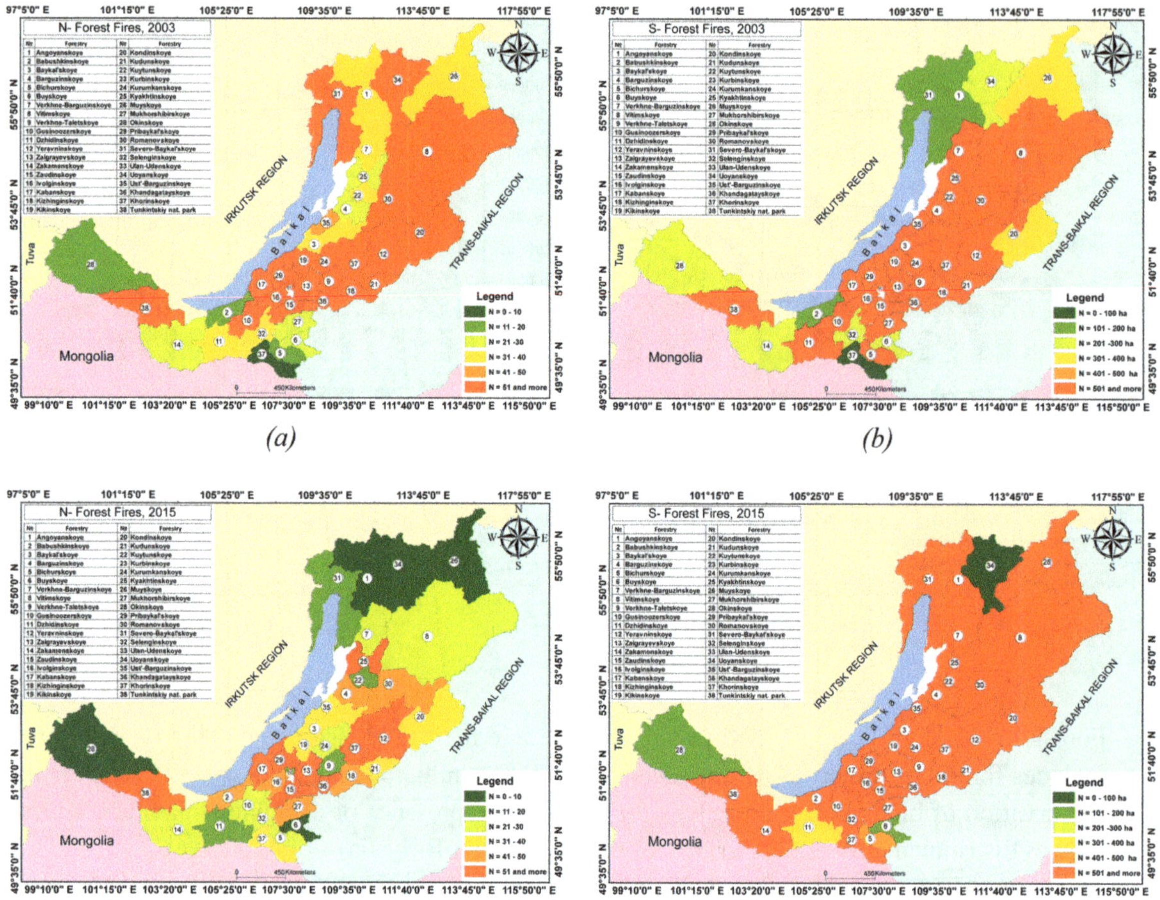

(a) (b) (c) (d)

In the present work, for the first time, the spatial and temporal distribution of the number and area of forest fires by forest areas of the Republic of Buryatia from 2001 to 2018 (Figure 6 a, b)

Dotted rectangles indicate time intervals of data that coincide with those in Figure 4. The progress of the selected areas in the graphs of figures 4 and 5 is the same. It can be seen that the peaks of the number of forest fires accounted for 2003, 2009, 2011, 2015 and 2017, which, in principle, is consistent with the peaks of fire distribution areas in 2003, 2009, 2012, 2015 and 2017. Only 2011 with the peak of fires and 2012 with maximum areas are not agreed. The prevailing number of fires in the territory of Buryatia was registered in 2003, while the maximum area of burnt forests fell in 2015. In Figure 7 a, b, c, d shows the distribution of the number and area of fires in these years in the forest areas of Buryatia is shown.

Figure 8. Distribution of total solar radiation and precipitation in Buryatia

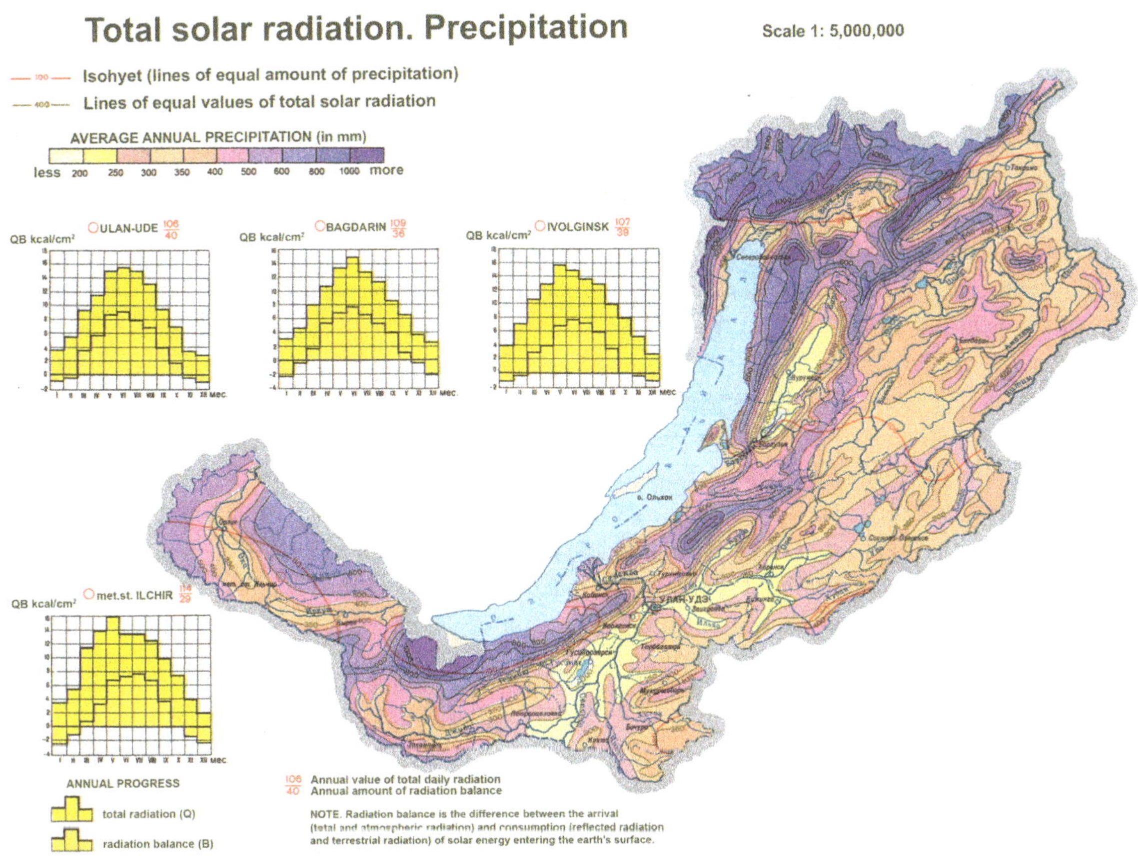

Figure 9. Distribution of days of the fire-dangerous season by weather-fire hazard classes (Sidorov & Sanzhieva, 2018)

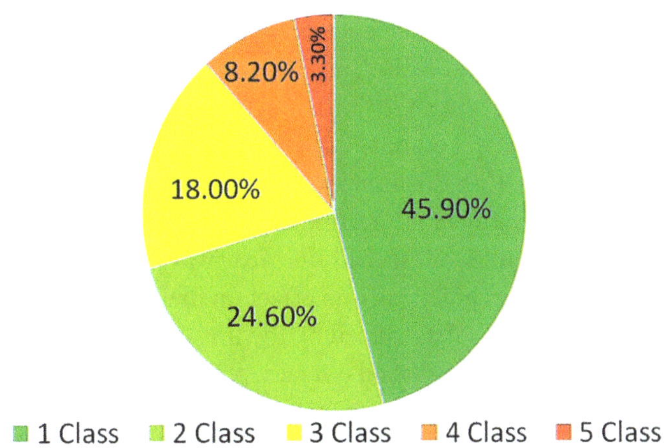

In spatial terms, the largest number and area of fires are recorded in the valleys of major rivers: The Selenga, Uda, Chikoi, Irkut and Barguzin. It is here, along the river valleys and the slopes of the foothills, that sod forest soils are formed under pine forests on sandy and sandy foam sediments. These landscapes have a low water capacity and extremely high flammability, especially in spring and summer (April, May), when the moisture coefficient is 0.1-0.2 and corresponds to the values of semi-desert (Sochava, 1967, Imetkhenov, 2000, Nogina, 1964). The smallest numbers and areas of fires are observed on the ridges of the Eastern Sayans, Khamar-Daban and the Stanovoye Highlands. It is in these mountainous landscapes under dark-coniferous (cedar, spruce, fir) taiga and larch forests on burozem and permafrost taiga soils the maximum amount of precipitation falls (Figure 8).

In the occurrence of forest fires during the year there is a certain frequency. The duration of the fire-dangerous season in Buryatia ranges from 140 to 160 days - from the beginning of May (sometimes from the end of April) to the end of October. According to long-term averages, the number of forest fires in April-June is distributed as follows: April - 13%, May - 36%, June - 24%, July - 13%, August - 7%, September - 6%, October - 1% . In some years, up to 34% of all forest fires occur in April, and up to 30% in September (Antropov, Seredkin, & Schepin, 2013). In Figure 9 shows the distribution of days of fire-hazardous season according to weather-fire hazard classes of the forests of the Republic of Buryatia.

Despite a small percentage of the most flammable days at high air temperatures in June-July (3.3 and 8.2%), forest fires occurring during this period can reach 4th and 5th hazard classes, which corresponds to the scale of emergency and high fire danger, as a result of which huge fires of forest territories can be covered.

DIFFERENCES IN FIRE CONDITIONS IN HUMID AND ARID TYPES OF CLIMATE

In order to assess the quantitative relationship between the frequency and area of fires, a correlation analysis was conducted in contrasting natural conditions of the forest areas of the Sayano-Baikal Stanovoye Highland (Babushkinskoe) and Trans-Baikal middlelands (Ivolginskoye). Analysis of the causal relationships of the prerequisites for the occurrence and area of fire spreading from weather conditions in the territory of the Babushkinsky and Ivolginsky forest areas will be carried out on the basis of continuous time series of a sufficient length of the monthly average temperature of the atmospheric air and the amount of liquid precipitation of the corresponding weather stations. Babushkinsky forest area is located on the southern coast of Lake Baikal and is characterized by a humid type climate. Ivolginsky forest area is located on the Trans-Baikal middle mountains and is typical of dry steppe arid ecosystems. The causes of the occurrence and spread of fires are far from being limited to these meteorological parameters; therefore, the correlation coefficient of more than 0.4 is decisive for the significance and reliability of the closeness of communication. With this in mind, the quantity and area of fires for Babushkinsky forestry was the most significant air temperature from January to August (Figure 10, a, b), whereas for Ivolginsky forestry the temperature in April of the same year was the most decisive for the number of fires (Fig 10 c), and this parameter has practically no effect on the area of fire spread (Figure 10 d).

The effect of precipitation on fires is determined by the inverse correlation, i.e. the more rains fell, the fewer fires, and vice versa. In a humid coastal climate, precipitation neither past nor the same year is statistically significant either for quantity or for the area of forest fires. For Ivolginsky forest area, precipitation in April, July and the average for the year for the amount, and precipitation in April for the fire area for the same year are statistically significant (Figure 11 a, b).

Figure 10. Correlation of quantity (a – Babushkin, c – Ivolginsk) and fire area (b – Babushkin, d – Ivolginsk) versus average monthly temperature of the current year

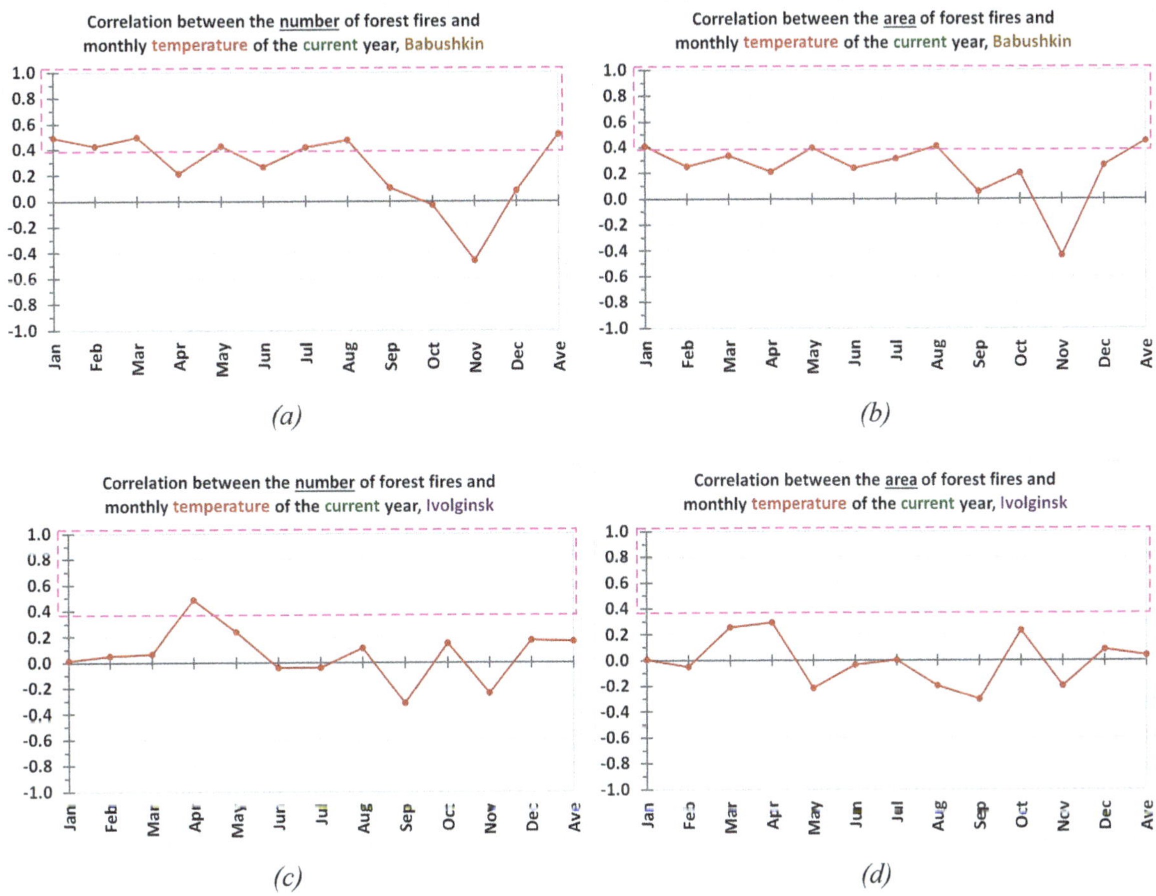

Figure 11. Correlation of quantity (a) and fire area (b) from the monthly precipitation of the current year

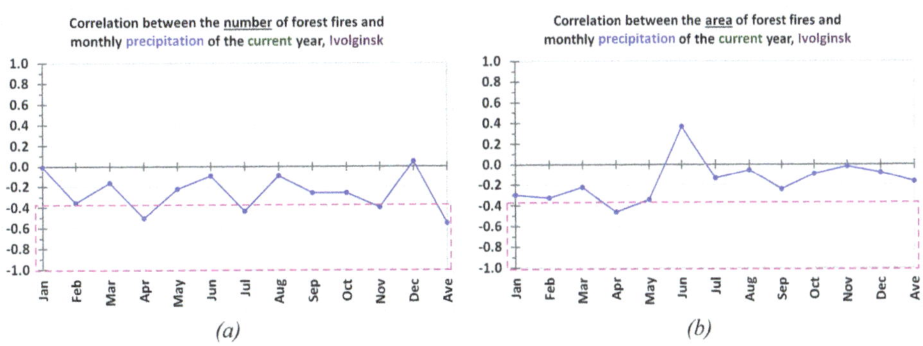

Figure 12. Correlation of quantity (a) and fire area (b) from the monthly precipitation of the previous year

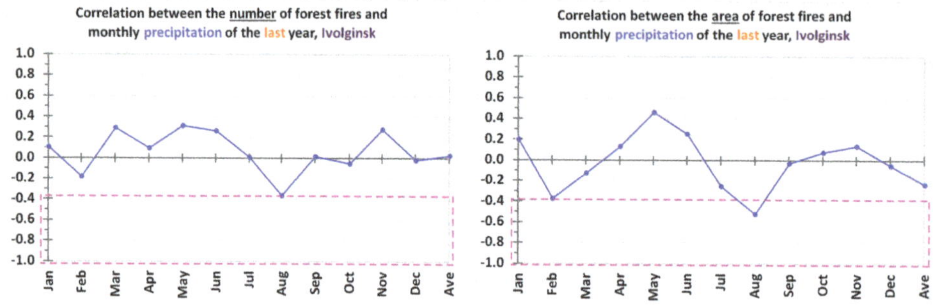

(a) (b)

Also defining both for the amount and for the fire area is precipitation in August of the previous year (Figure 12 a, b).

Thus, the established correlations indicate the influence of meteorological parameters on the occurrence and spread of fires. The coastal Babushkinsky forest area is characterized by a large amount of precipitation, which is often recorded in the winter months, while forest fires occur predominantly in summer, so there is no surprise that there is no correlation between these parameters. In general, the Babushkinsky forest area, both in the number of fires and in area, is at the very bottom of the list.

For the more continental climate of the Transbaikal middlelands, precipitation is one of the main factors determining the potential danger of forest fires. Moreover, the connection between these indicators is characterized by anti-correlation, i.e. if the amount of precipitation decreases, the number of forest fires increases, and vice versa, which is consistent with (Obyazov, 2012).

Thus, the main fact of the occurrence of fires among those considered for a humid climate is elevated temperatures, and for arid it is the absence of rain at the end of the previous year's summer, as well as the absence of rain this spring in combination with simultaneously hot weather. The impact of rain on forest fires is primarily determined by their effect on the moisture content of forest fuel, which also depends on soil moisture (Kislov, Varentsov, & Tarasova, 2015, Curic & Zivanovic, 2013, Schunk, Wastl, Leuchner, Schuster, & Menzel, 2013).

Modern technical means have recently acquired the fantastic ability to deliver the necessary information for research directly to the user's computer. Fairly reliable meteorological information is available on the Internet at relevant sites. However, their time frames are quite limited, and the sets are incomplete. For example, information about the dryness of forest fuel can only be obtained in the field, by means of samples. However, its state can be judged at least indirectly, having on hand the time series of data on soil moisture and its temperature. Moreover, aggregate monitoring of atmospheric and soil meteorological parameters is highly desirable. By atmospheric it is necessary to additionally have information on the direction and speed of wind, air humidity, solar insolation and the depth of snow cover. Such information can provide innovative equipment Atmospheric-soil measuring complex (ASMC).

Figure 13. ASMC scheme

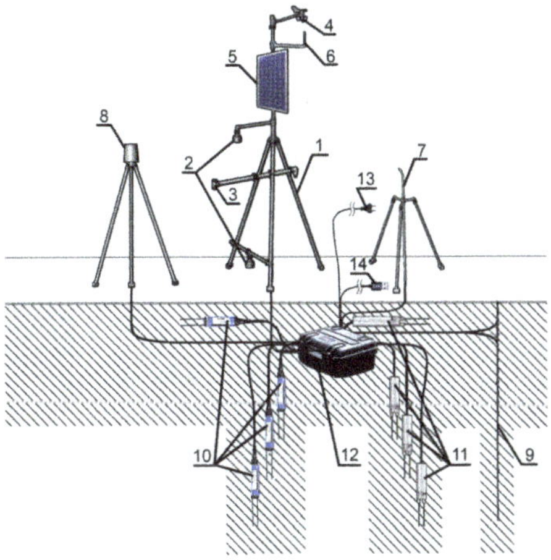

N	Component	Range, error
1	Mast	3 m
2	Air temperature and humidity sensors	-55...+50°C, ±0.3°C 0...100%, ±3.5%
3	Sensor of total solar radiation (2 channels)	Range of 0.2...10 µm, ±10% 0-2000 W/m², -55...+50°C
4	Wind speed and direction sensor	0.9...78 m/s, ±5% 0...360°, ±7%
5	Solar battery	
6	GSM-antenna	
7	Snow depth sensor	0...0.7 m, ±0.05 m
8	Precipitation sensor	0...1000 mm/h, ±5%
9	Temperature profile probe	-55...+50°C, ±0.2°C
10	Soil moisture sensors	0-70% (±2%); -15...+50°C
11	Water conductivity sensors	$10...2\times10^3$ µS/cm; ±20%
12	Case with electronics and atmospheric pressure sensor	500...1150 hPa, ±10 hPa 375...862 mm Hg, 7.5 mm Hg
13	Backup power supply	220 V
14	USB	

Figure 14. Permafrost types and ASMC locations
(Badmaev, Kulikov, & Mironov, 2011)

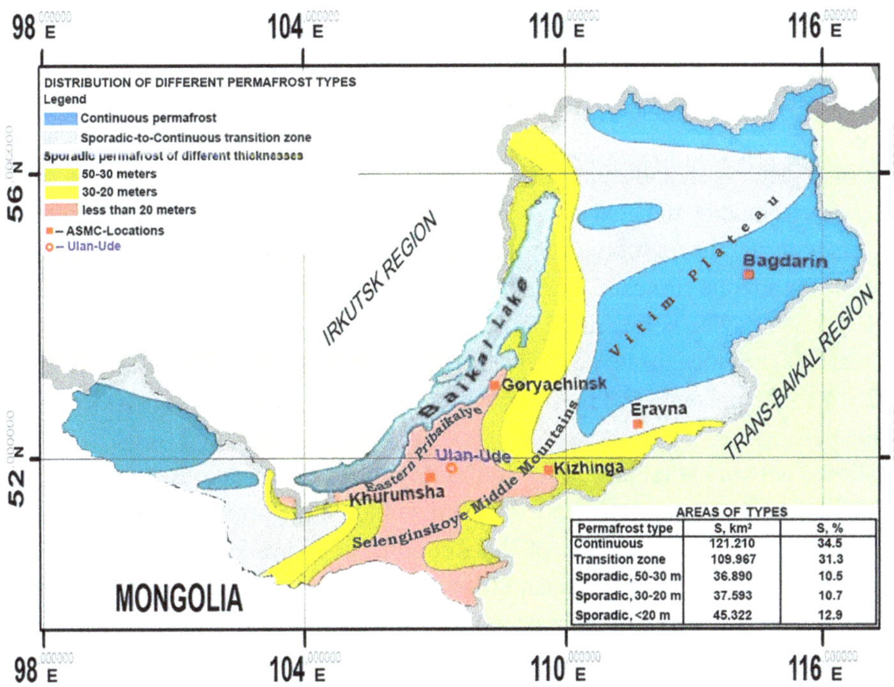

ATMOSPHERIC-SOIL MEASURING COMPLEX

The ASMC developed at the Siberian Branch of the Russian Academy of Sciences (Bazarov, Badmaev, Kurakov, & Gonchikov, 2018) is an automatic weather station with a paired set of soil sensors. It is designed for long-term automatic measurements and recording of the main parameters of the atmosphere and soil, such as:

- temperature and humidity,
- atmospheric pressure
- wind speed and direction,
- amount of liquid precipitation,
- level of snow cover and
- level of solar radiation
- soil temperature profile,
- soil moisture sensors
- water conductivity sensors.

Figure 13 shows the ASMC reference scheme, which embodies all the useful qualities of experimental embodiments, with the exception of unsuccessful options, as a whole. The composition and technical characteristics of its sensors are given in the table. The numbering of items in the table coincides with the numbers of the components in the diagram. Dual temperature and humidity sensors are located at heights of 0.2 and 2 m. Electronic thermometers on a soil temperature probe are placed 2 cm to a depth of 20 cm, which is sufficient for studying the temperature of the soil as fire predictors; further inland, respectively, less.

The network of ASMCs is spread across the territory of Buryatia (Figure 14). The location of the stations is currently determined by the goal of on-line information on the spatial and temporal dynamics of climate parameters in the territories of the continuous, intermittent and island permafrost of the southern boundary of the cryolithozone stretched over almost the entire territory of the Baikal region (Badmaev & Bazarov, 2019). Each device is measured synchronously eight times a day every three hours with a conditional start at 7:00 am local time, i.e. at 7:00; 10:00; 13:00; 4pm; 19:00; 22:00; 01:00; 04:00, etc. from summer 2014 to the present. The measurement results are transmitted via a GSM antenna to the data server of the network.

The observational network, first developed in the Republic of Buryatia, Russian Federation, makes it possible to obtain data on atmospheric and soil climate indicators promptly and at regular intervals.

According to the forest zoning, forests of the Republic of Buryatia belong to the South Siberian mountain zone on an area of 28,224.6 thousand hectares, partly to the taiga zone on an area of 1,413.8 thousand hectares and occupy four forest areas: the Baikal mountain forest - 13,566.8 thousand hectares (45.7%), Transbaikal Mountain-Frozen - 11077.9 thousand hectares (37.4%), Altaesayan Mountain-Taiga - 3579.9 thousand hectares (12.1%) and East Siberian Taiga Frost-frost - 1413, 8 thousand hectares (4.8% of the total area). The assignment of the territory of the Republic of Buryatia to the four forest areas is due to the characteristic features of natural and climatic conditions, complex, heavily dissected mountain-relief relief, differences in terrain heights, geological features and climate influence (Government of the Republic of Buryatia, 2008). The development of the APIC network, even for these areas only, not to mention the equipment of each forest area, should significantly increase the amount of data collected needed for more qualitative quantitative research in this area.

CONCLUSION

In the territory of the Republic of Buryatia, due to global changes, significant transformations of the natural environment are observed. In particular, statistically significant trends of increase in temperature and decrease in precipitation with identification of wet and dry periods have been established. In the past 20 years, due to the abnormally high air temperatures and the increasing aridity of the territory, the frequency and area of fires have increased.

In spatial terms, the largest number and area of fires are recorded in the valleys of major rivers: the Selenga, Uda, Chikoi, Irkut and Barguzin. It is here, along the river valleys and the slopes of the foothills, that sod forest soils are formed under pine forests on sandy and sandy sandy sediments. These landscapes have low water capacity and extremely high burning capacity, especially in the spring and summer period (April, May), when the moisture coefficient is 0.1–0.2 and corresponds to the values of semi-desert. The smallest numbers and areas of fires are observed on the ridges of the Eastern Sayans, Khamar-Daban and the Stanovoye Highlands. It is in these mountain landscapes under dark-coniferous (cedar, spruce, fir) taiga and larch forests on burozem and permafrost taiga soils the maximum amount of precipitation falls.

A correlation analysis was carried out in the contrasting climatic conditions of the forest areas of the Sayano-Baikal Stanovoye Highland and the Zabaikalye Highlands in order to assess the quantitative relationship between meteorological parameters with frequency and area of fires. It has been established that in arid ecosystems of the Transbaikal middlelands, the occurrence of fires is significantly affected by precipitation at the end of summer (August) of the previous year in combination with hot and dry weather in spring (April) of the current year. At the same time, it was found that for fires in the humid coastal climate of the Sayano-Baikal Stanovoye Highlands, air temperature is of decisive importance, and the relationship between precipitation and the number and area of fires is insignificant.

Coordination of work between the executive authorities (the Center for Forest Protection, the General Directorate of the EMERCOM of Russia for the Republic of Buryatia, the Ministry of Natural Resources of the Republic of Buryatia) and scientific organizations, as well as the use of modern methods of remote sensing of the Earth, analysis of satellite imagery, UAVs and APIC situations with forest fires and creating a single database.

The urgent need to use a network of atmospheric and soil measurement systems as monitoring and meteorological equipment to study the occurrence of forest fire situations depending on weather and climatic conditions is convincingly shown. So, for example, precipitation, which fell heavily in August last year, adversely affect the occurrence of fire danger this year exclusively in the form of accumulated soil moisture.

ACKNOWLEDGMENT

The study was supported by the Russian Foundation for Basic Research, grant N 17-29-05093 "Development of methods to monitor forest fire danger caused by human activity in Baikal Lake basin".

REFERENCES

Abatzoglou, J. T., & Kolden, C. A. (2013). Relationships between climate and macroscale area burned in the western United States. *International Journal of Wildland Fire*, *22*(7), 1003–1020. doi:10.1071/WF13019

Adame, J. A., Lope, L., Hidalgo, P. J., Sorribas, M., Gutiérrez-Álvarez, I., Del Águila, A., ... Yela, M. (2018). Study of the exceptional meteorological conditions, trace gases and particulate matter measured during the 2017 forest fire in Donana Natural Park, Spain. *The Science of the Total Environment*, *645*, 710–720. doi:10.1016/j.scitotenv.2018.07.181 PMID:30031329

Antropov, V. F., Seredkin, A. D., & Schepin, A. A. (2013). *Forestry of Buryatia*. Ulan-Ude, Russia: Ecos. (In Russian)

Badmaev, N., & Bazarov, A. (2019). Monitoring network for atmospheric and soil parameters measurements in permafrost area of Buryatia, Russian Federation. *Geosciences*, *9*(1), 6. doi:10.3390/geosciences9010006

Badmaev, N. B., Kulikov, A. I., & Mironov, I. A. (2011). Permafrost. In Buryatia. Encyclopedic Reference Book in 2 Volumes (Vol. 1, pp. 54-55). Ulan-Ude, Russia: LLC Publishing House "Ecos". (In Russian).

Barbero, R., Abatzoglou, J. T., Larkin, N. K., Kolden, C. A., & Stocks, B. (2015). Climate change presents increased potential for very large fires in the contiguous United States. *International Journal of Wildland Fire*, *24*(7), 892–899. doi:10.1071/WF15083

Bazarov, A. V., Badmaev, N. B., Kurakov, S. A., & Gonchikov, B.-M. N. (2018). A mobile measurement system for the coupled monitoring of atmospheric and soil parameters. *Russian Meteorology and Hydrology*, *43*(4), 271–275. doi:10.3103/S106837391804009X

Bedia, J., Herrera, S., & Gutierrez, J. M. (2014). Assessing the predictability of fire occurrence and area burned across phytoclimatic regions in Spain. *Natural Hazards and Earth System Sciences*, *14*(1), 53–66. doi:10.5194/nhess-14-53-2014

Belyakin, A. A., & Volokitina, A. V. (2010). Forest type pyrological characteristics in the southern Pribaikalye. [In Russian]. *Bulletin of the Krasnoyarsk State Agrarian University*, *7*, 91–96.

Curic, M., & Zivanovic, S. (2013). Dependence between deficit and surplus of precipitation and forest fires. *Disaster Advances*, *6*(6), 64–69.

De la Cueva, A. V., Quintana, J. R., & Canellas, I. (2012). Fire activity projections in the SRES A2 and B2 climatic scenarios in peninsular Spain. *International Journal of Wildland Fire*, *21*(6), 653–665. doi:10.1071/WF11013

De Vicente, J., & Crespo, F. (2012). A new wildland fire danger index for a Mediterranean region and some validation aspects. *International Journal of Wildland Fire*, *21*(8), 1030–1041. doi:10.1071/WF11046

Dorzhiev, T. Z., Yuhai, B., Badmaeva, E. N., Batsaihan, V., Urbazaev, C. B., & Yushan, Y. (2017). Forest fires in republic of Buryatia for 2002–2016. [In Russian]. *Nature of Inner Asia*, *3*, 22–37.

Evdokimenko, M. D. (2013). Forest fire causes and distribution in the Baikal Region. [Forestry Journal]. *Lesnoy Zhurnal*, *4*, 33–39.

Federal Forestry Agency. (2008). *On the determination of the number of forest areas on the territory of the Republic of Buryatia and the establishment of their borders (Order of Oct. 16, 2008 No. 310)*. Moscow, Russia: Federal Forestry Agency. (In Russian)

Federal Hydrometeorology and Environmental Monitoring Service. (2008a). Assessment report on climate change and its effects on the territory of the Russian Federation: Vol. I. *Climate change*. Moscow, Russia: Roshydromet. (In Russian)

Federal Hydrometeorology and Environmental Monitoring Service. (2008b). Assessment report on climate change and its effects on the territory of the Russian Federation: Vol. II. *Implications of climate change*. Moscow, Russia: Roshydromet. (In Russian)

Feltman, J. A., Straka, T. J., Post, C. J., & Sperry, S. L. (2012). Geospatial analysis application to forecast wildfire occurrences in South Carolina. *Forests*, *3*(2), 265–282. doi:10.3390/f3020265

Flannigan, M., Cantin, A. S., de Groot, W. J., Wotton, M., Newbery, A., & Gowman, L. M. (2013). Global wildland fire season severity in the 21st century. *Forest Ecology and Management*, *294*, 54–61. doi:10.1016/j.foreco.2012.10.022

Flannigan, M. D., Krawchuk, M. A., de Groot, W. J., Wotton, B. M., & Gowman, L. M. (2009). Implications of changing climate for global wildland fire. *International Journal of Wildland Fire*, *18*(5), 483–507. doi:10.1071/WF08187

Garmaev, E. Z., Tsydypov, B. Z., Dabaeva, D. B., Andreyev S. G., Ayurzhanayev, A. A., & Kulikov A. I. (2017). The lake Baikal level regime: retrospection and current status. *Water Sector of Russia: Problems, Technologies, Management*, 4-18. (In Russian).

Girardin, M. P. (2010). Wildfire risk inferred from tree rings in the Central Laurentians of boreal Quebec, Canada. *Dendrochronologia*, *28*(3), 187–206. doi:10.1016/j.dendro.2009.05.006

Government of the Republic of Buryatia. (2008). *Annex to the Resolution dated Dec. 17, 2013 No. 665 "Forest Plan of the Republic of Buryatia" (approved by the decree of the Government of the Republic of Buryatia dated Dec. 31, 2008 No. 608)*. Ulan-Ude, Russia: Republican Forestry Agency, official site. (In Russian)

Gruza, G. V., & Rankova, E. Y. (2004). Detection of changes in climate state, climate variability, and climate extremity. [In Russian]. *Russian Meteorology and Hydrology*, *4*, 31–43.

Guo, F., Innes, J. L., Wang, G., Ma, X., Sun, L., Hu, H., & Su, Z. (2015). Historic distribution and driving factors of human-caused fires in the Chinese boreal forest between 1972 and 2005. *Journal of Plant Ecology*, *8*(5), 480–490. doi:10.1093/jpe/rtu041

Hansen, J., Ruedy, R., Sato, M., & Lo, K. (2010). Global surface temperature change. *Reviews of Geophysics*, *48*(4), RG4004. doi:10.1029/2010RG000345

Heyer, J. P., Power, M. J., Field, R. D., & van Marle, M. J. E. (2018). The impacts of recent drought on fire, forest loss, and regional smoke emissions in lowland Bolivia. *Biogeosciences*, *15*(14), 4317–4331. doi:10.5194/bg-15-4317-2018

Holsten, A., Dominic, A. R., Costa, L., & Kropp, J. P. (2013). Evaluation of the performance of meteorological forest fire indices for German federal states. *Forest Ecology and Management*, *287*, 123–131. doi:10.1016/j.foreco.2012.08.035

Imetkhenov, A. B. (Ed.). (2000). *Atlas of the Republic of Buryatia*. Moscow, Russia: Federal Service of Geodesy and Cartography of Russia.

Jolly, W. M., Cochrane, M. A., Freeborn, P. H., Holden, Z. A., Brown, T. J., Williamson, G. J., & Bowman, D. M. J. S. (2015). Climate-induced variations in global wildfire danger from 1979 to 2013. *Nature Communications*, *6*(1), 7537. doi:10.1038/ncomms8537 PMID:26172867

Kale, M. P., Ramachandran, R. M., Pardeshi, S., Chavan, M., Joshi, P. K., Pai, D. S., ... Roy, P. S. (2017). Are climate extremities changing forest fire regimes in India? An analysis using MODIS fire locations during 2003-2013 and gridded climate data of India meteorological department. *Proceedings of the National Academy of Sciences India Section A-Physical Sciences*, *87*(4), 827–843. doi:10.100740010-017-0452-8

Kelly, R., Montgomery, W. I., & Reid, N. (2008). Differences in soil chemistry remain following wildfires on temperate heath and blanket bog sites of conservation concern. *Geoderma*, *315*, 20–26. doi:10.1016/j.geoderma.2017.11.016

Kislov, A. V., Varentsov, M. I., & Tarasova, L. L. (2015). Role of spring soil moisture in the formation of large-scale droughts in the East European Plain in 2002 and 2010. *Izvestiya. Atmospheric and Oceanic Physics*, *51*(4), 405–411. doi:10.1134/S0001433815020061

Koutsias, N., Xanthopoulos, G., Founda, D., Xystrakis, F., Nioti, F., Pleniou, M., ... Arianoutsou, M. (2013). On the relationships between forest fires and weather conditions in Greece from long-term national observations (1894-2010). *International Journal of Wildland Fire*, *22*(4), 493–507. doi:10.1071/WF12003

Kulikov, A. I., Kulikov, M. A., & Smirnova, I. I. (2008) Thermal state of the active layer in the permafrost zone of the Baikal region in the context of global warming. In *Proceedings of the International Symposium "Climate Change in Central Asia: Socio-Economic and Environmental Impacts"*. Chita, Russia: Institute of Natural Resources, Ecology, and Cryology of the Siberian Branch of the RAS. (In Russian).

Liu, W., Wang, S., Zhou, Y., Wang, L., Zhu, J., & Wang, F. (2016). Lightning-caused forest fire risk rating assessment based on case-based reasoning: A case study in DaXingAn Mountains of China. *Natural Hazards*, *81*(1), 347–363. doi:10.100711069-015-2083-1

Makarenko, E. L. (2016). Forest fires and their consequences in the central ecological zone of the Baikal natural territory. *Interaktivnaya nauka (Interactive science)*, 5, 9-12. (In Russian).

Matsoukis, A., Kamoutsis, A., & Chronopoulos, K. (2018). Estimation of the meteorological forest fire risk in a mountainous region by using remote air temperature and relative humidity data. *International Letters of Natural Sciences, 67,* 1-8. doi:10.18052/www.scipress.com/ILNS.67.1.

Ministry of Natural Resources of the Republic of Buryatia (2008-2018). *State report "On the state and environmental protection of the Republic of Buryatia"* (Vols. 2007-2017). Ulan-Ude, Russia: The Official Portal of the Republic of Buryatia.

Moritz, M. A., Morais, M. E., Summerell, L. A., Carlson, J. M., & Doyle, J. (2005). Wildfires, complexity, and highly optimized tolerance. *Proceedings of the National Academy of Sciences of the United States of America, 102*(50), 17912–17917. doi:10.1073/pnas.0508985102 PMID:16332964

Nogina, N. A. (1964). *Soils of Transbaikalia.* Moscow, Russia: Nauka. (In Russian)

Obyazov, V. A. (2012). Effects of changes in meteorological conditions on the forest fire conditions in Zabaikal'skii krai. *Russian Meteorology and Hydrology, 37*(6), 373–378. doi:10.3103/S1068373912060039

Papadopoulos, A., Paschalidou, A. K., Kassomenos, P. A., & McGregor, G. (2013). Investigating the relationship of meteorological/climatological conditions and wildfires in Greece. *Theoretical and Applied Climatology, 112*(1-2), 113–126. doi:10.100700704-012-0726-7

Parente, J., Pereira, M. G., Amraoui, M., & Fischer, E. M. (2018). Heat waves in Portugal: Current regime, changes in future climate and impacts on extreme wildfires. *The Science of the Total Environment, 631-632*, 534–549. doi:10.1016/j.scitotenv.2018.03.044 PMID:29529441

Rolinski, T., Capps, S. B., Fovell, R. G., Cao, Y., D'Agostino, B. J., & Vanderburg, S. (2016). The Santa Ana wildfire threat index: Methodology and operational implementation. *Weather and Forecasting, 31*(6), 1881–1897. doi:10.1175/WAF-D-15-0141.1

Russian Federation. (2006). *Article 81 of the Forest Code of the Russian Federation. The powers of the state authorities of the Russian Federation in the field of forest relations* (Forest Code of the Russian Federation of 04.12.2006 N 200-ФЗ). Russian Federation. (In Russian).

San-Miguel-Ayanz, J., Moreno, J. M., & Camia, A. (2013). Analysis of large fires in European Mediterranean landscapes: Lessons learned and perspectives. *Forest Ecology and Management, 294*, 11–22. doi:10.1016/j.foreco.2012.10.050

Schunk, C., Wastl, C., Leuchner, M., Schuster, C., & Menzel, A. (2013). Forest fire danger rating in complex topography - results from a case study in the Bavarian Alps in autumn 2011. *Natural Hazards and Earth System Sciences, 13*(9), 2157–2167. doi:10.5194/nhess-13-2157-2013

Shabbar, A., Skinner, W., & Flannigan, M. D. (2011). Prediction of seasonal forest fire severity in Canada from large-scale climate patterns. *Journal of Applied Meteorology and Climatology, 50*(4), 785–799. doi:10.1175/2010JAMC2547.1

Sidorov, A. A., & Khankhunov, Yu. M. (2015). On the problems with forest fires in the Republic of Buryatia. In *Proceedings of the Scientific-Practical Conference with International Participation.* Ulan-Ude, Russia: East Siberia State University of Technology and Management. (In Russian).

Sidorov, A. A., & Khankhunov, Yu. M. (2016). Forest fires in the Republic of Buryatia: liquidation, prevention and consequences. In *Proceedings of the Scientific-Practical Conference with International Participation*. Ulan-Ude, Russia: East Siberia State University of Technology and Management. (In Russian).

Sidorov, A. A., & Sanzhieva, S. E. (2018). Chronology of forest fires in the Republic of Buryatia. [In Russian]. *Bulletin of the Krasnoyarsk State Agrarian University*, *4*, 204–208.

Sochava, V. B. (Ed.). (1967). *Atlas of Transbaikalia. (Buryat ASSR and Chita region)* [*Maps*]. Moscow, Irkutsk, Russia: Main Department of Geodesy and Cartography. (In Russian)

Strydom, S., & Savage, M. J. (2017). Potential impacts of climate change on wildfire dynamics in the midlands of KwaZulu-Natal, South Africa. *Climatic Change*, *143*(3-4), 385–397. doi:10.100710584-017-2019-8

Teakles, A. D., So, R., Ainslie, B., Nissen, R., Schiller, C., Vingarzan, R., ... Huang, L. (2017). Impacts of the July 2012 Siberian fire plume on air quality in the Pacific Northwest. *Atmospheric Chemistry and Physics*, *17*(4), 2593–2611. doi:10.5194/acp-17-2593-2017

Tulokhonov, A. K., & Puntsukova, S. D. (2016). Forest fires in the Republic of Buryatia in the context of climate change. [In Russian] Society: Politics, Economics, *Law*, *3*, 72–78.

Ukraintsev, A. V., & Plyusnin, A. M. (2015). Forest fires in Zaigraevsky district of the Republic of Buryatia in 2010–2012: Causes of fire and damage. [In Russian]. *Geography and Natural Resources*, *2*, 60–65.

Vashchuk, L. N. (2014). *Irkutsk forest inventory 120 years: the stages of development of forest accounting*. Irkutsk, Russia: Forward LLC.

Wastl, C., Schunk, C., Leuchner, M., Pezzatti, G. B., & Menzel, A. (2012). Recent climate change: Long-term trends in meteorological forest fire danger in the Alps. *Agricultural and Forest Meteorology*, *162*, 1–13. doi:10.1016/j.agrformet.2012.04.001

Wastl, C., Schunk, C., Lüpke, M., Cocca, G., Conedera, M., Valese, E., & Menzel, A. (2013). Large-scale weather types, forest fire danger, and wildfire occurrence in the Alps. *Agricultural and Forest Meteorology*, *168*, 15–25. doi:10.1016/j.agrformet.2012.08.011

Won, M. S., Lee, S. Y., Lee, M. B., & Ohga, S. (2010). Development and application of a forest fire danger rating system in South Korea. *Journal of the Faculty of Agriculture, Kyushu University*, *55*(2), 221–229.

Yao, J., Raffuse, S. M., Brauer, M., Williamson, G. J., Bowman, D. M. J. S., Johnston, F., & Henderson, S. B. (2018). Predicting the minimum height of forest fire smoke within the atmosphere using machine learning and data from the CALIPSO satellite. *Remote Sensing of Environment*, *206*, 98–106. doi:10.1016/j.rse.2017.12.027

Zdereva, M. Y., & Vinogradova, M. V. (2009). Medium-range forecasting of the extent of forest fire hazard from meteorological conditions. *Russian Meteorology and Hydrology*, *34*(1), 8–15. doi:10.3103/S1068373909010026

Zhao, Z., Kooperman, G. J., Pritchard, M. S., Russell, L. M., & Somerville, R. C. J. (2014). Investigating impacts of forest fires in Alaska and western Canada on regional weather over the northeastern United States using CAM5 global simulations to constrain transport to a WRF-Chem regional domain. *Journal of Geophysical Research, D, Atmospheres, 119*(12), 7515–7536. doi:10.1002/2013JD020973

Zink, M., Samaniego, L., Kumar, R., Thober, S., Mai, J., Schäfer, D., & Marx, A. (2016). The German drought monitor. *Environmental Research Letters, 11*(7), Article number 074002. doi:. doi:10.1088/1748-9326/11/7/074002

ADDITIONAL READING

Evin, G., Curt, T., & Eckert, N. (2018). Has fire policy decreased the return period of the largest wildfire events in France? A Bayesian assessment based on extreme value theory. *Natural Hazards and Earth System Sciences, 18*(10), 2641–2651. doi:10.5194/nhess-18-2641-2018

Fares, S., Bajocco, S., Salvati, L., Camarretta, N., Dupuy, J. L., Xanthopoulos, G., ... Corona, P. (2017). Characterizing potential wildland fire fuel in live vegetation in the Mediterranean region. *Annals of Forest Science, 74*(1), 1. doi:10.100713595-016-0599-5

Gabysheva, L. P., & Isaev, A. P. (2015). Forest fires impact on microclimatic and soil conditions in the forests of cryolithic zone (Yakutia, North-Eastern Russia). *Siberian Journal of Forest Science, 6*, 96–111. doi:10.15372/SJFS20150609

Grishin, A. M. (1992). *Mathematical modeling of forest fires and new ways to deal with them.* Novosibirsk, Russia: Nauka. (In Russian)

Grishin, A. M. (2001). Modelling and forecasting ecological catastrophes. [In Russian]. *Ecological Systems and Devices, 2*, 12–21.

Grishin, A. M. (2002). *Modelling and forecasting catastrophes (Basic definitions and concepts of the theory of catastrophes and general regularities of their ignition and development).* Tomsk, Russia: Tomsk State Universityю. (In Russian)

Hiers, J. K., Stauhammer, C. L., O'Brien, J. J., Gholz, H. L., Martin, T. A., Hom, J., & Starr, G. (2019). Fine dead fuel moisture shows complex lagged responses to environmental conditions in a saw palmetto (Serenoa repens) flatwoods. *Agricultural and Forest Meteorology, 266-267*, 20–28. doi:10.1016/j.agrformet.2018.11.038

Littell, J. S., McKenzie, D., Peterson, D. L., & Westerling, A. L. (2009). Climate and wildfire area burned in western U.S. ecoprovinces, 1916–2003. *Ecological Applications: A Publication of the Ecological Society of America, 19*(4), 1003-1021. doi:. doi:10.1890/07-1183.1

Littell, J. S., Peterson, D. L., Riley, K. L., Liu, Y., & Luce, C. H. (2016). A review of the relationships between drought and forest fire in the United States. *Global Change Biology, 22*(7), 2353–2369. doi:10.1111/gcb.13275 PMID:27090489

Matthews, S. (2014). Dead fuel moisture research: 1991-2012. *International Journal of Wildland Fire, 23*(1), 78–92. doi:10.1071/WF13005

Nelson, R. M. Jr. (2000). Prediction of diurnal change in 10-h fuel stick moisture content. *Canadian Journal of Forest Research, 30*(7), 1071–1087. doi:10.1139/x00-032

KEY TERMS AND DEFINITIONS

Arid: Specialized environment very dry and without enough rain for plants.

Arid Climate: A dry climate with high air temperatures, experiencing large daily fluctuations, and a small amount of precipitation (less than 100-150 mm / year) or their complete absence.

Atmosphere and Soil Measurement Complex (ASMC): An autonomous automatic meteorological complex, which simultaneously with atmospheric weather measures soil moisture and temperature profiles.

Baikal: Is the largest freshwater lake by volume in the world, containing 22–23% of the world's fresh surface water. With 23,615.39 km^3 of fresh water, it contains more water than the North American Great Lakes combined. With a maximum depth of 1,642 m, Baikal is the world's deepest lake. It is considered among the world's clearest lakes and is considered the world's oldest lake – at 25–30 million years. It is the seventh-largest lake in the world by surface area.

Global Warming: A gradual increase in world temperatures caused by gases such as carbon dioxide that are collecting in the air around the earth and stopping heat escaping into space.

Humid Climate: A type of climate in areas with excessive moisture, in which the amount of precipitation is more than can evaporate and seep into the soil.

Precipitation: Any product of the condensation of atmospheric water vapor that falls under gravity. The main forms of precipitation include drizzle, rain, sleet, snow, graupel and hail.

Temperature: A physical quantity expressing hot and cold. It is measured with a thermometer calibrated in one or more temperature scales. We use the Celsius scale (denoted °C).

Wildfire or Wildland Fire: An uncontrolled fire in an area of combustible vegetation occurring in rural areas.

Chapter 9
Using Geographic Information Systems for Study of the Natural Fire Risk on the Territory of the Jewish Autonomous Region

Vladimir Aleksandrovich Glagolev
Institute for Complex Analysis of Regional Problems of Far East Branch, Russian Academy of Science, Russia

Anna Mihailovna Zubareva
Institute for Complex Analysis of Regional Problems of Far East Branch, Russian Academy of Sciences, Russia

ABSTRACT

This chapter presents a software complex to assess the fire hazard of a territory. The structure includes four subsystems: (1) collection and processing of ground-based remote monitoring data for environmental objects, including daily actual and forecast data of hydrometeorological stations, data on non-forest vegetation fires, and information on the state of vegetation in different time periods and locations of sources of fire; (2) a forecast of the spread of vegetation fires consisting of their calculated period and place of occurrence and calculated time of probable spread of grass fire to the nearest forest area; (3) a probabilistic forecast of fires to determine the areas of vegetation with high forest fire risks on a given day by using the author's method; and (4) planning strategies for the activities of fire services. The verification of this work was carried out on the territory of the Jewish Autonomous Region.

DOI: 10.4018/978-1-7998-1867-0.ch009

Figure 1. The number of vegetation fires and their total area (per 100 thousand hectares) on the territory of the subjects of the Far Eastern Federal District from 1992 to 2012

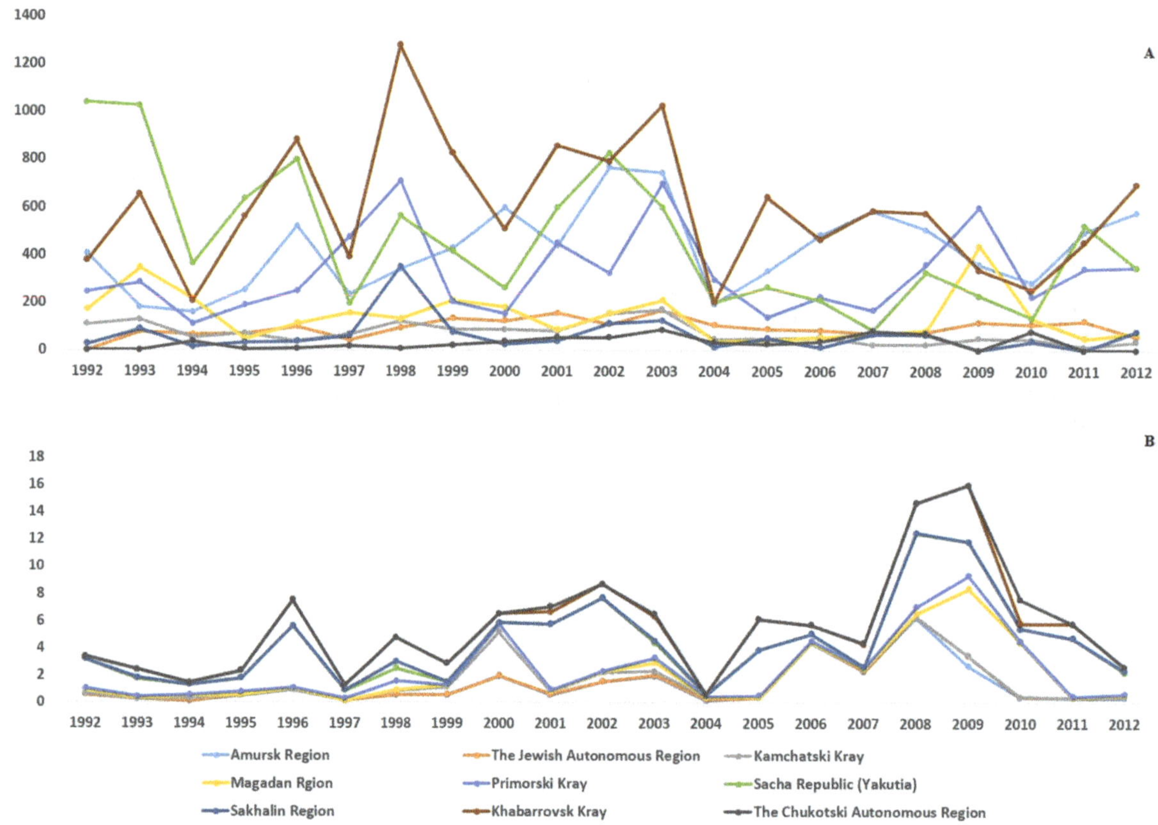

INTRODUCTION

The forest fund of the Far Eastern Federal District is distinguished by high fire hazard and fire hazard, which is due to the climatic, forest, and geomorphological features of the region. The natural fire danger of forests in this area is one of the highest in Russia because more than 80% of its forest fund is among the highest (I-III) fire danger classes.

The highest fireability of forests was studied by the Federal State Statistics Service (http://www.gks.ru, The current state of the forests…2009) for the period of 1992–2012 over the territory of the Far Eastern Federal District. in 1998, 2002 and 2003 (Fig. 1) The number of fires for the selected years 3592, 3282 и 3819 and the corresponding areas covered by fire were both the highest recorded in recent history in 2008 and 2009, (1471 and 1601 thousand hectares, respectively).

According to data analysis, the actual burning rate and the number of fires in the Amursk Region, the Jewish Autonomous Region, and the Primorsky and Khabarovsk Territories have both decreased, but the degree of fireiness in many of these areas has increased.

High fire rates can be attributed to majority of the forest fund being allocated beyond the protected area and the extremely low availability and degree of development of the territory. A comprehensive assessment of the area's pyrological characteristics is therefore necessary (table 1). The assessment

Table 1. The degree of the actual fire rate and fire area (per 100 thousand hectares) in the territory of the Far Eastern Federal District from 1999 to 2012

Number	Far Eastern Federal District								
	Amursk Region	The Jewish Autonomous Region	Kamchatski Kray	Magadan Region	Primorski Kray	Sacha Republic (Yakutia)	Sakhalin Region	Khabarovsk Kray	The Chukotski Autonomous Region
1	363,7	53,6	472,8	461,4	165,9	3103,2	87,1	824,6	721,5
2	422	97	71	141	322	458	68	598	40
3	1,2 lowered	1,8 lowered	0,1 low	0,3 low	1,9 lowered	0,1 low	0,7 lowered	0,7 lowered	0,0 low
4	344,2 high	144,7 increased	60,9 average	116,9 increased	113,8 increased	63,7 average	62,4 average	138,1 increased	19,3 low

Note: 1- the area of the territory of the Far Eastern Federal District; 2 - the average long-term number of fires; 3 and 4 - the degree of actual flammability by the number of fires and by the relative area of the fire, respectively.

can be accomplished by a geographic information system (GIS) that allows stakeholders to visualize the databases of assessment factors, which are collected to determine the pyrological vulnerability of territories. Examples of databases for visualization are the set of electronic maps that can be used to improve fire service activities. In this regard the purpose of our research is - studying of natural fire danger of the territory of the Jewish Autonomous Region with use of geographic information systems

BACKGROUND

In Russia, the first work on the development and design of programs to systematically calculate forest fire protection was performed by laboratory of mathematical methods under the direction of G. N. Korovin (1977). The programs can provide input, control, and adjustment of information files and systematically calculate ground and aviation systems to detect and extinguish forest fires.

In subsequent work, the possibilities of averaging the precipitation data of meteorological stations covering the area were considered (Marchenko, 1993). As a result of the calculation, the weighting coefficients were derived for the meteorological stations and subsequently used to calculate the indicators of fire danger. In this study, a set of programs is proposed. PL/1 The complex system consists of three modules: (1) input, control, and formation of information arrays; (2) calculation of averaging coefficients for precipitation; (3) calculation of current values of fire danger indicators (i.e., complex meteorological, humidity of litter N° 1, and humidity of litter N° 2 indicators) for forests and their class boundaries. The meteorological information is organized and stored as an array of data. Records are arranged in a strict sequence of ascending row keys. Each record key contains a five-digit weather station code and a date code.

The possibility of determining fire hazard periods is considered in the work (Algorithms and programs for calculating…1974). The software package «OREOL» is designed to effectively estimate the intensity of fire-dangerous periods and determine the results of functioning of forest protection systems at the aviation department and linkage levels. The program includes the following modules in PL/1: (1) implementation of the creation of administrative, reference, and meteorological information files; (2) assessment of the intensity of fire hazard periods and the characteristics of fires during their localization;

and (3) determination of the performance and effectiveness of the regional forest conservation system and its elements.

The expansion of the program as a complex GIS was performed on the basis of the work of N. A. Marchenko (Marchenko, 1993). With the assistance of GISs to monitor the fire danger of natural–territorial complexes in Southern and Middle Sikhote-Alin, the extrapolations of entered data according to landscapes by weather stations are carried out, and the increase rate of possible fires according to the parameters is determined. The current fire hazard assessment method developed in the program allows to take into account landscape conditions affecting fire and related intraseasonal dynamics at a certain taxonomic level and subsequently determine the parameters of possible fires.

In the early 1990s, a series of studies on geographic information databases based on the digital images of NOAA were considered (Geographic information systems…1992). Baseline information was used to create maps of fire hazards, their distribution, and their indicators across all image points. As a method of calculation of each point, a technique in which the dew point for interpolation according to the TOVS instrument was developed, and radiometric temperature was measured by using the fifth channel of AVHRR (Advanced Very High Resolution Radiometer).

The best-known systems to assess and predict fire hazards are from the United States and Canada, the development of which began in 1959. Particularly in the United States, the system has been implemented all over the country since 1972. At present, the US National Fire Hazard Rating System (NFDRS) is a combination of computer programs and algorithms that enable the assessment and prediction of the degree of fire danger in forests, particularly by predicting the behavior of forest fires in a given area during a 24-hour period. The forecast of fire behavior is based on meteorological (Sommers, Coloff, Conard, 2012) and satellite data, information about the state of forest combustible materials (FCM) (Frantz, Stellmes, Röder, Hill, 2016), and terrain data (Díaz-Delgado, Lloret, Pons, 2004). NFDRS results in relative assessments of potential growth and behavior of forest fires are determined daily on the basis of the fire hazard class of a certain territory, and fire hazard forecasts are generated monthly for weather conditions (Keetch, Byram, 1988).

The fire hazard rating system of Canada based on the forest data in CFFDRS (Canadian Forest Fire Danger Rating System Advanced Very High Resolution Radiometer) was formed in 1968 (Rogeau, Flannigan, Hawkes, Parisien, Arthur, 2016). A map of the distribution of natural fire hazard classes is compiled daily on the basis of the values of the heat intensity index of the frontal heat edge, which is spread over a section covered with a reference complex of plant combustible materials. Studies addressing territory fireability have been conducted in multiple countries of the Mediterranean: Portugal (Goncalves, Lourenco, 1990, Oliveira, Jose, Pereira, Joao, Carreiras, 2012), Spain (Díaz-Delgado, et al., 2004), France, Italy (Salis Michele, Arca Bachisio, Alcasena Fermin, 2016), as well as the USA (Preisler, Riley, Stonesifer, Calkin, 2016), Canada (Rogeau, et al., 2016), Australia (http://www-as.uwyo.edu/~geerts/cwx/notes/chap16/oz_bush.html, Kumar, Dharssi, 2017), China (Kunpeng Yi, Yulong Bao, Jiquan Zhang, 2016, Xiaorui Tian, Fengjun Zhao, Lifu Shu, Mingyu Wang, 2013), and Finland (Bovio, Nosenzo, Viegas, Ferreira, Sol, 1999, Heikinheimo, 1998). Although people have taken a large number of measures to combat forest fires, fire continues to play an important role in forest ecosystems.

Foreign and Russian colleagues presume by experience that various meteorological elements, regional empirical dependencies of meteodata, and fire danger indicators, as well as the obtained forecasts, can be used to predict the spread of vegetation fires during fire hazard assessments.

In modern conditions of firefighting, high efficiency, accuracy, organization, and temporal and spatial consistency of data about the state of the forest fund are necessary. The GIS is focused on overcoming the above limitations, the aims of which is to contribute to the effective solution of scientific and applied problems associated with assessing fire hazards of a territory (Abushenko, et al. 1998, Sukhinin, Kashkin, Ponomarev, 1999). The southern territory of the Far East Region is the most pyrologically vulnerable in Russia, and the Jewish Autonomous Region is among the top three areas in terms of number and area of fires per unit area of the selected subject in this study. Thus, the formation of a GIS that takes into account all the above characteristics and tasks is highly important for this region.

The author's geographic information system contains meteodata web portals are published in a type of the documents structured by XML. Delineation of the fires by manual entry or vectorization of satellite pictures is made for import of metadata of the fires of vegetation. In the first way by means of instrumental GIS MapInfo Professional 8.5 the vector layer in which the fires it is fixed by polynominal objects is edited, and for each of them attributive information is entered. In the second - the program - the vectorizer Easy Trace, allowing to transfer subjects of the bitmap image of the satellite picture to a vector layer is used.

Visualization of results is executed in the form of the electronic spreadsheets Microsoft Excel 2007, work sets of vector layers of MapInfo Professional 8.5. Owing to what in MapInfo Professional 8.5 and in the subsequent versions does not support storage of attributive data in Oracle MySQL 5 calculation results are exported to a universal vector layer of MIF (Mapinfo Interchange Format).

In developing the conceptual framework for creating the information system called "Fires of the Jewish Autonomous Region," formulating the purpose and identifying the problems are necessary prior developing the system's software requirements, defining the construction principles, and considering the issues related to the organization of storing spatial and attribute data.

METHODOLOGY

The basic parameters of an information system depend on the basic principles of its design. An analysis of existing information systems to solve the problems associated with the assessment and forecast of fire danger of a territory and the formalization of the main tasks can help single out the following basic principles of building the information system:

1. Integration of meteodata and formation of a single information space. This aspect can help with the analysis of fire hazard information based on various criteria, thereby improving the quality of management decision-making for forest protection services and organizations.
2. Integration with other information systems in terms of databases and tools (basic software). The use of databases from other systems can expand the functional limits of the information system.
3. Extensibility of the list of tasks. This principle will ensure the adaptation of information systems across several problems, including the existing one, along with the changing conditions of operation, thereby increasing life cycle and versatility when using various fire hazard assessment techniques.
4. Openness of databases and implemented technologies according to industry and international standards to guarantee the portability of information across different information systems.

5. Use of modern technologies to solve some of the functions of information systems, including geographic information components and database management systems in the form of a databank (Glagolev, 2015).

The implementation of the above-described principles can help create an information system with several advantages relative to existing analogues.

The main features of the information systems developed in this study are as follows:

1. Collection, management, analysis, and presentation of data;
2. Integration of information systems with various forms of data, i.e., storing various data in the system is possible;
3. Presence of mathematical modeling methods and decision-making systems;
4. A large number of attribute and dynamic data;
5. Possibility of expanding the functional content to connect new subsystems, application databases, web portals, and GISs; and
6. Development of substantive parts of the information systems in accordance with system implementation, and development of requests of potential users.

Thus, to assess and predict the fire hazard of vegetation, a comprehensive study is needed, including an assessment of fire hazard according to weather conditions, and the calculation of the probability of occurrence of fires according to natural anthropogenic conditions.

The study of the assessment and prediction of the occurrence of fires over vegetative areas in the territory of the Jewish Autonomous Region (JAR) includes three stages (pig.2) derived from the following studies: "Evaluation and forecast of daily fire danger of vegetation according to weather conditions"

1. Development of methods for the continuous forecasting of fire hazard indicators according to weather conditions; and
2. Daily spatial forecast of fire hazard indicators according to the weather conditions and pyrological state of forest combustible materials (FCM) in the forest areas of JAR.

"Spatial forecast of occurrence of fires of vegetation"

1. Development of a probabilistic model to predict the occurrence of vegetation fires;
2. Daily forecast of fires over fire-hazardous areas in the territory of JAR;
3. Generate recommendations for forest protection services.

"The program complex assessment and prediction of fire danger of vegetation"

1. Designing the software architecture;
2. Creating a database of weather data and fire-of-vegetation data;
3. Development of copyright information systems.

Figure 2. The algorithm forecast of meteorological fire danger

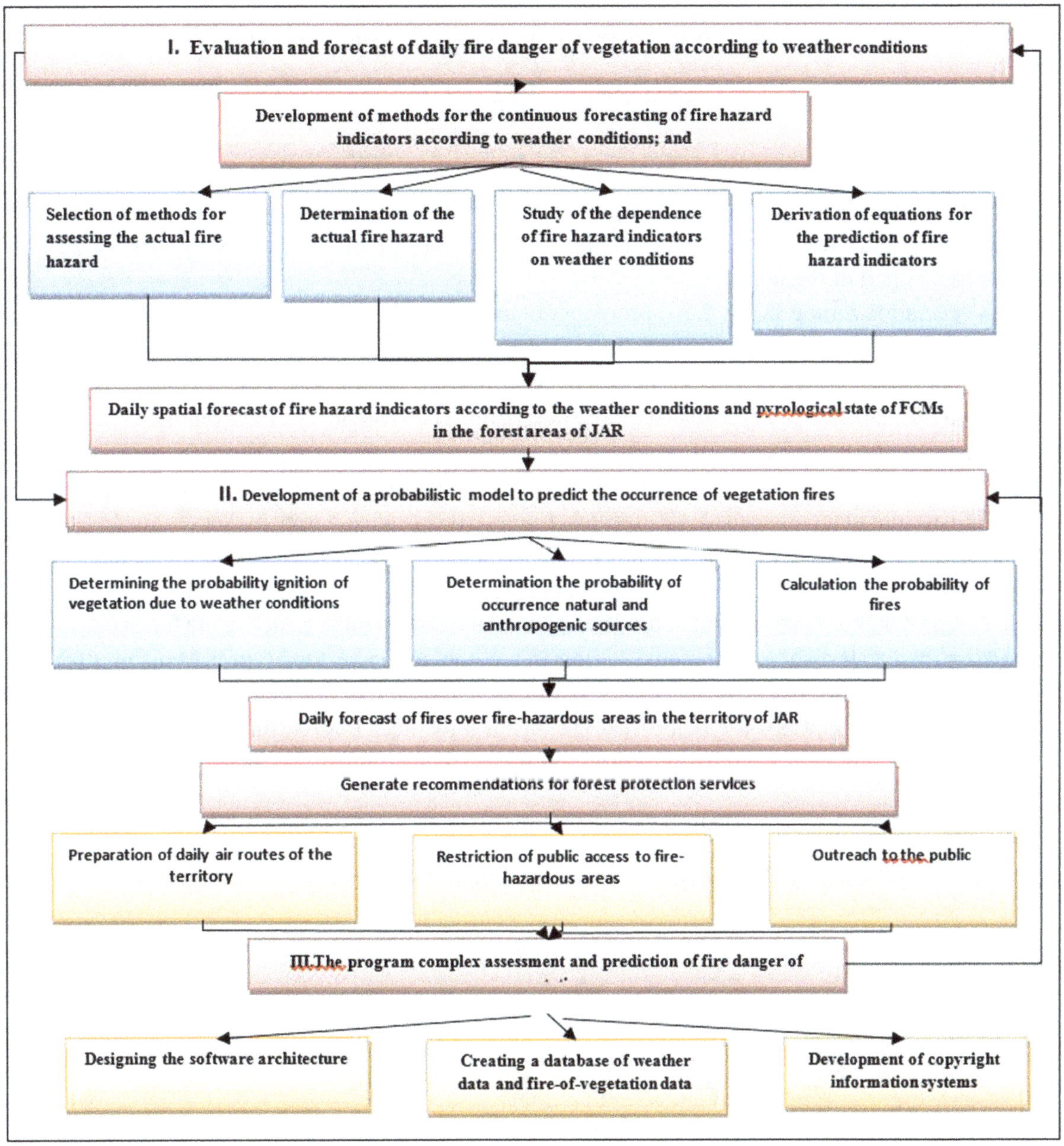

The final action of the algorithm in assessing and predicting the occurrence of vegetation fires is the development of the author's GIS (fig. 2), which is particularly relevant for the territory of the Jewish Autonomous Region, which is mostly dominated by areas with high values of pyrological characteristics.

Pyrological Characteristics of the Study Area

The geographical location of the Jewish Autonomous Region at the junction of the Eurasian continent, with the Pacific Ocean and its outpost within the Khabarovsk Territory (Sea of Okhotsk), and the characteristics of the monsoon climate both determine the diversity of plant world and the richness of plant resources in this part of the northwestern Pacific.

Forest vegetation zoning shows that the territory belongs to the Amur–Primorsky coniferous–deciduous area of the coniferous–deciduous forest zone. The geobotanical classification indicates the existence of southern subzone of taiga and coniferous–deciduous forests (Kolesnikov, 1995).

Autonomy is characterized by a diverse vegetation cover. The diversity can be explained by horizontal and vertical zonality, participation of intrazonal groups, and anthropogenic activity (Fig. 3).

Most of the territory of the Jewish Autonomous Region is occupied by forests (60%). The species richness of vascular plants exceeds 1,400 species. Forest cover is 45%. Timber reserves are 167 million m3. The forests of the third group, depending on national economic significance, location, and ecological functions, comprise 61% of the total forest area of the region. This group has an operational nature and is designed to meet the needs of the timber industry without introducing imbalance in the forest ecosystem.

The flora of the Jewish Autonomous Region is a combination of four types: Manchu, Okhotsk, Daurian, East Siberian.

The main plant formations are beloberezovye, cedar–broadleaf forests, black birch oak and larch forests combined with dwarf birch and willow thickets, larch–beloberezovye woodlands with dwarf

Figure 3. Map diagram of the vegetation of the Jewish Autonomous Region

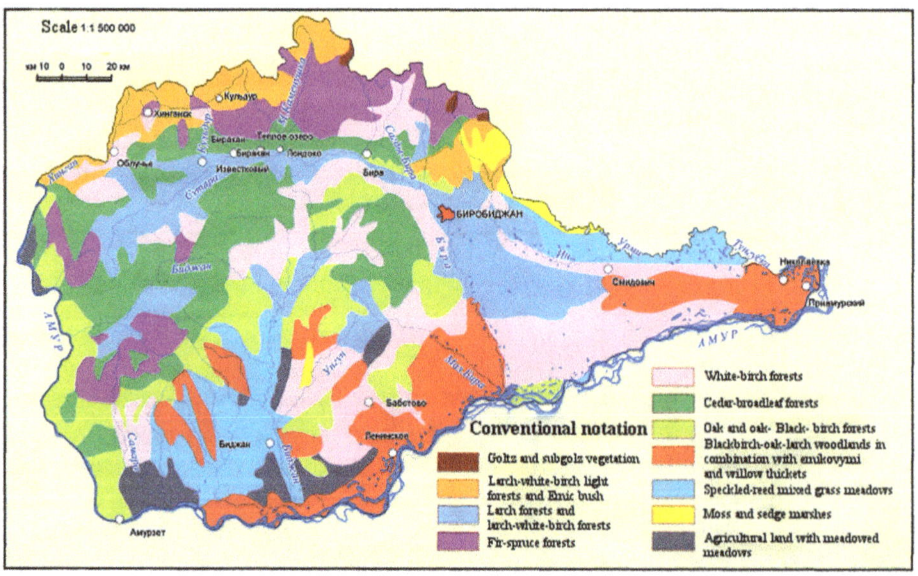

birch thickets, oak, oak, black birch, fir, spruce, larch and larch–beloberezovye forest (Kurentsova, 1967, Natural Resources…, 2004).

The northern mountainous areas, which belong to some of the highest areas of the Small Khingan and Pompeevsky ranges that occupy dark coniferous forests, are greatly influenced by fires and logging. Minor areas are occupied by virgin dark coniferous spruce–fir forests. Cedar–broadleaved forests are located in the middle mountain belt of the Sutarsky, Pompeevsky, Shukhi-Pokta ranges and the southern areas of the Bureinsky range. The main plant formations of the lowland areas, the coast of the river, and the Amur and some flat areas are oak forests. Depending on the degree of moistening and soil types, there exist Amur and Manchurian lindens, Dahurian birch, small-leaved maple in oak forests, two-colored hazel wood, two-colored Lespedean, and small-colored beetle in the underbrush. The south of the autonomy, particularly along the remnant mountains, comprises steppe species (feather grass, Baikal skullcap, and many others). The western part along the coast of the river comprises Amur, and oak trees on the southeastern border of its range that grows pine. Along the river valleys are willow, ash tree, alder, and poplar forests. The eastern part of the autonomy comprises meadow and bog vegetation. White pillars, aspen groves, and oak groves are found on the elevated sites. Significant areas are occupied by larch woodlands, sedge, reed grass, and forb grass meadows. Sedge and moss swamps are located in the wettest areas.

A scale in which vegetation is assigned to one of five hazard classes is used to assess the natural pyrological properties of the vegetation (Starodumov, 1964). The JAR territory has raw reed grass meadow grass bogs, priruslovyh thickets willows, sedge versigrass–reed grass meadows, sedge marshes with residues of larch of marshes assigned to I class coniferous–broad-derived mixed broadleaf–oak forests and woodlands assigned to II class, grass–moss combined with forest creeps, and sedge marshes assigned to V class of natural fire danger (Doroshenko, 2007).

Previous reports have established that plant formations belonging to the class of “very high” fire danger occupy 32.4% of the total autonomy area. Formations with high fire hazard classifications are located in 24.4% of the area. The vegetation with average fire hazard classification accounts for 25% of the territory of the Jewish Autonomous Region. Low-danger larch forests and light forests are located in 17.8% of the autonomy area. Formations with very low fire risk occupy less than 1% of the area of the region. Thus, the plant formations in the autonomy with very high and high classes of fire danger occupy more than 56% of the territory.

The fire hazard of the natural–territorial complexes of the Jewish Autonomous Region is analyzed in consideration of the ball-based assessment of the pyrological indicators of vegetation, climate, and soils but without taking into account the spatial distribution of fire sources and vegetation periods. The findings show that the natural complexes of vegetation is most susceptible to fires and the spread of fires in the 41,7% autonomy area. These areas include lowland natural complexes with meadow vegetation on the Middle Amur lowland, the foothills of the Lesser Khingan, and the valleys of mountain rivers. Natural complexes with average natural fire risks comprise 41.7% of the territory of the Jewish Autonomous Region. These areas are located within the territory of the Small Khingan Range (ridges of Uldura, Chocks, Daur), the isolated mountain ranges of the Middle Amur Lowland, and the valleys of lowland rivers (Amur, Big Bira, Bidzhan and others).

The main object of burning is forest combustible materials (FCMs). The forest phytocenosis, being a complex genetically interconnected dynamic system, consists of woody and under-tree and shrub–grass–moss layers, which includes litter and litter (peat or turf). The role of individual components of phytocenoses in the occurrence and development of forest fires is uneven due to differences in their morphological structures, chemical compositions, and other pyrological properties. In general, these components form a complex FCM that is closely associated with specific forest formations, groups and types of forest, and different land categories (Kolesnikov, 1995).

In the Jewish Autonomous Region, the main types of LGM are grass and grass vegetation with litter (rags), foliage from trees and shrubs, bushy lichens and shrubs, mosses, tree timbers and litter, needles in crowns, and her litter. In assessing the main groups of different types of FCM according to the degree of danger of fires on the lands of the forest fund of the Far East Republic, a scale is developed to determine which fleece grass fires may occur in the grass, lichen, dead cover–coniferous, deciduous–grass groups of FCM. The maximum probability of fire occurrence in each type of FCM can occur if the daily complex indicator of fire danger according to weather conditions is equal to or exceeds the critical value. Depending on the zonal–geographical characteristics of the territory, a certain number of days in a season can be specified for each type of FCM; on these days, the areas are in a state of "fire maturity," which is sufficient for a fire to occur. For example, on the territory of the Jewish Autonomous Region, cereal and motley grass litter, which has the highest pyrological properties, can be in a fire hazard state for 131–147 days or less if the fire hazardous leaf litter from trees and shrubs only occur 30–40 days a year. Consequently, easily combustible FCMs are a resource of "primary" fire, which can change the moisture characteristics of other types of combustible materials, which also indicate a shift to a fire-hazardous state (i.e., occurrence of "secondary" fires).

The level of vegetation in the Far Eastern Federal District has a steady upward trend, especially in abnormally dry seasons. At the same time, no clear patterns can be observed in the frequency of occurring extremely dry seasons, which reduces the possibility of using the physicostatistical and thermodynamic models for the reliable long-term forecasting of weather conditions based on long-term ground and satellite monitoring under the weather mode.

The study area consists of 5 municipal districts of Jewish Autonomous Region, which include 5 hydrometeorological stations. The data on meteorological observations and weather forecasts are used to regulate the activities of the offices of the Far Eastern airbase.

The vegetative state in the far-east part of Russia can be attributed to weather and climatic conditions that are highly conducive to the fire-hazardous "maturation" of plant combustible materials.

The territory of the Jewish Autonomous Region belongs to the monsoon climatic region of mid-latitudes, and it differs significantly in intra-annual course of temperatures, air humidity, precipitation, and prevailing wind directions of the other regions along the temperate latitude in the Northern Hemisphere (Petrov, Novorotsky, Lenshin, 2000). Complicated weather conditions, which superimpose with climatic features, determine the basic patterns of fire-danger formation according to season, i.e., long duration (April-October, Sometimes the beginning of November), presence of three periods of different inflammabilities of plant combustible materials (spring – summer, autumn), and uneven distribution of fires within periods.

The climate over the Jewish Autonomous Region is the warmest among the areas of the Far Eastern Federal District. In the first half of the warm period of the year, atmospheric humidification is moderate or insufficient. In the second half of summer, sea tropical air is carried away with tropical cyclones (typhoons). This area is the only territory in the region where a positive difference exists in annual positive and negative temperatures. For example, the southern regions of the Jewish Autonomous Region have 8–10 days of dry wind each month during April–May. In general, spring accounts for approximately 59% of the total number of dry winds during the growing season, in which the relative humidity of air

Table 2. Weather and climatic characteristics of fire-dangerous seasons municipal districts of the Jewish Autonomous region from 1960 to 2010 years.

N°	Districts	Meteostations	Average Annual						
			Temperature Air's in 1-3 p.m., °C	Temperature Dew Points in 1-3 p.m.	Average Precipitation mm / day	Average Precipitation Days Where Temperature Air's More,			
						0°C	5°C	10°C	15°C
1	Birobidzhansky	Birobidzhan	16,6	7,5	2,8	206	189	162	128
2	Smidovichsky	Smidovich	17,7	8	2,2	208	195	169	136
3	Obluchensky	Obluchye	16,7	6,4	2,5	205	189	160	130
4	Leninsky	Leninskoye	17,9	7,8	2	208	197	171	140
5	Oktyabrsky	Ekaterino-Nikol'skoye	17,8	7,8	1,8	203	191	167	135

Note: Air temperature and dew points, daily precipitation amount are average daily data of the fire hazard season according to long-term data.

decreases by five to six times during the spring months relative to the long-term average. The average temperature of air at 13:00–15:00 during the entire fire-dangerous season (Table 2) varies widely (for example, according to MS, Okhotskoye ÷ 10.7 ° C, Komsomolsk-on-Amur, ÷ 15.1 ° C, Birobidzhan ÷ 16.6 ° C). The duration of the period with daily temperatures exceeding 0 °C varies from 171 days to 208 days (MS “Arch”, “Leninskoe”).

Shape, exposure of slopes, and complex orography have significant effects on temperature regimes. The transition to 10 °C (the beginning of the growing season) in the southern and central lowland regions occurs at the end of the first and second decades of May. In autumn, the transition of the average monthly temperature to 10 °C can be observed in the third decade of September. The average duration of a period with temperatures above 10 °C is 146 days. The period with daily temperatures above 15 °C occurs in the valley of the river and the cupid from Blagoveshchensk to Komsomolsk-on-Amur. Ussuri from Khabarovsk to Vladivostok occurs from the second half of June (sometimes earlier) to the second half of September (Table 2). The temperature transition in autumn to 5 °C (i.e., the end of the growing season) in the Amur region ends in October. The temperature of the soil surface has an average annual value close to air temperature. The frost-free period of the soil surface of the flat areas is somewhat shorter than that of air. In the lowland areas, only seasonal permafrost occurs. The beginning of thawing of soil coincides with the time of descent of snow cover and the transition of average air temperature to 0 °C.

A characteristic climatic feature is the concentration of precipitation in the warm season. More than 60% of the annual amount falls inclusively from July to September. The average long-term annual precipitation varies during the fire hazard season from 450 to 800 mm and more: Khabarovsk - 500-600 mm, Bira - 817 mm, Smidovich - 672 mm, Ekaterino-Nikolskoye - 632 mm. The maximum amount of precipitation falls in July and August, whereas the minimum amount falls in February to January. The total number of days with relative humidity of ≥ 80% is 60–80 days. In spring, the average daily humidity is 40%–50%. During long periods without rain, air becomes even drier: for April-May, there are, on average, 10-15 days, when the humidity of the air falls below 30%.

This territory is characterized by large time periods, in which low soil and air humidity are observed in combination with dry winds against the background of high temperatures. This phenomenon contributes to a highly intensive drying of plant combustible materials relative to the other subjects of the

Far Eastern Federal District and the emergence of several maxima with increased degree of fire danger due to weather conditions during the entire fire-hazardous season, which lasts from April to November. Vegetation fires at this time cause the atmosphere to create smoke from combustion products at different altitudes. Most fires occur as a result of the economic activities of the local population, such as those in the southern regions (Birobidzhansky and Komsomolsky). The main peaks of fires are observed in spring in late April and early May and in autumn in late September and early October. The modern assessment of the pyrological characteristics of the territory involves working with a large volumes of information. Therefore, for its processing and systematization, remote sensing data and GISs are used, which are convenient for spatial analysis and modeling and can assist various operations with cartographic layers and digital elevation models with sets of additional thematic information. These methods enable stakeholders to visualize existing databases and combine them with cartographic materials, summarize the collected information in forms acceptable for comparative and analytical activities, and allow for a comprehensive assessment of fire hazards of a territory.

For the territory of the Jewish Autonomous Region, the presence of vegetation and plant-based combustible materials with high pyrological properties and the weather conditions conducive to their transition to a state of "fire maturity" are the main components considered in this study. In this manner, the presence of fire sources with natural or anthropogenic origin can be used to predict the occurrence of fires.

Results and Discussion

The author's geo-information system "Fire''

On the basis of the high-burning capacity of the study area and the significant prevalence of pyrologically vulnerable areas, the GIS called "Fires of the Jewish Autonomous Region" was developed for the territory of the Jewish Autonomous Region. The structure of the GIS includes four subsystems:

1. Subsystem for collecting and processing data of ground-based remote monitoring of environmental objects.

In rendering spatial predictions of the occurrence of grass fires and the speed of their propagation, daily actual and forecast data of hydrometeorological stations, data on vegetation fires, and information on the state of vegetation in different time periods and locations of fire sources are necessary. These data comprise the following:

- Meteorological elements data representing Internet services: State Hydrometeorological Center of the Russian Federation (http://meteoinfo.ru); Space Research Institute of the Russian Academy of Sciences (http://meteo.infospace.ru); Research and Production Center "Map-Maker" (http://gismeteo.ru), Meteorological elements (Seager et al., 2015, Varol, Ertugˇrul, 2016, Van Dijk, Bruijnzeel, 2001) including daily measurements of daily air temperature and dew point temperature, wind direction and speed at 13–15 hours a day, daily precipitation from 9:00 of the previous day to 9:00 of the current day, and public weather forecasts with lead times of 3 days;
- Information about the location of vegetation fires and their states ("greenness") based on satellite images (spacecraft information) in 36 spectral channels with resolutions of 250, 500, and 1000 m per pixel. The pictures are shared on the site. For image recognition, an algorithm developed by

NASA called MOD14 is used (Thermal Anomalies – Fires and Biomass Burning). The Russian counterpart of the site is shown in the ISDM–Rosleskhoz website, which hosts data from satellite monitoring of vegetation fires across the Russian Federation.

The image data obtained in the red (620–670 нм) and near-infrared (841–876 нм) channels (product MOD09GQK) allow for the calculation of the vegetative index (i.e., NDVI), which is a quantitative indicator of photosynthetically active biomass. The calculation of NDVI is based on the two most stable parts (i.e., independent of other factors) of the spectral reflection curve of vascular plants. The red region of the spectrum shows the maximum absorption of solar radiation by chlorophyll of higher vascular plants, whereas the infrared region shows the region of maximum reflection of the cellular structures of leaf plates. In adjusting the values of the index in the case of fixated areas covered by clouds and shadows from clouds, this study proposes the use of satellite information with 1–6 channels for this unit (Satellite monitoring, 2003). The processing of multichannel images is performed by using GIS GRASS 6.0 and ENVI 4.8, in which the index is calculated and then converted from HDF (Hierarchical Data Format) format to geo-TIFF format.

- Fire statistical data from the Jewish Autonomous Region Forest Management are populated according to a certain state standard and include an attribute database. Each fire type can be described according to the characteristics defining its main parameters (area, start and end dates, day of the week, and detector), location (fire number, local forest range, and quarter number) cause of fire, firefighting costs, damage caused to forest fund, and losses of wood on the vine (Krawchuk, Moritz, 2015).

SUBSYSTEM FIRE PROBABILISTIC FORECAST SUBSYSTEM

In preventing the occurrence of vegetation fires, recommendations should be issued for forest protection institutions to include the following: (1) temporary periods of increased danger to regulate the occurrence of fire hazardous areas of vegetation; (2) installation sites with informational banners and warning stands for publicity and can offer explanations to the population; and (3) air travel routes by taking into account the daily territorial distribution of fire hazardous areas of vegetation.

The following steps must be performed to build an optimal route for air touring:

1. Build a network of operational–territorial units (OTUs) for the territory of forest and non-forest funds;
2. Calculate the actual P_i values and forecast P'_{i+n} the values of the complex indicator of fire danger according to weather conditions in all OTU centers;
3. Calculate in each OTU the probability of occurrence of fires of vegetation denoted by $F_{i,j}(B)$ on the i-th current day and the n-th days of the forecast;
4. Select the OTU, particularly the threshold value of the probability of occurrence of fires at a given time period, and determine if this value is exceeding the standard;
5. Build an air route in which the maximum number of flammable OTU enters the survey area.

As an OTU network, stakeholders can use an irregular quarterly network of forestryor a regular network of non-forest land plots.

The actual values P_{i} are calculated by the VG method. The Nesterov and forecast P'_{i+n} methods are described in the paper (Glagolev, 2015).

The author's methodology is used to determine the areas of vegetation with high forest fire risks on a given day. The method involves a deterministic probabilistic model to predict the occurrence of vegetation fires according to natural anthropogenic conditions, which take into account the origin of the source of fire, unlike in existing models (Baranovskiy, Barakhnin, Yankovich, 2017, Grishin, Filkov, 2011, Evers, Gilbertson-Day, Nicolef, Ryan, Tague, 2013, Parisien 2012, Krueger et al., 2016, Williams et al. 2015). Furthermore, the method introduces the concept of "critical distance" (*Rcr*), which is the minimum distance between the studied OTU and the settlements or railways and roads. In this manner, the lichestvo fires can be significantly reduced.

The probability of occurrence of sources of fire in the *j*-th OTU is determined by the relative frequency of occurrence of lightning discharges and the occurrence of fires from the anthropogenic activities of residents from nearby communities and arrivals along the road network.

$$F_{i,j}(B) = \begin{cases} F_{i,j}(C)\left[(F_{i,j}(N)F_{i,j}(B/N) + F_{i,j}(M)F_{i,j}(B/M)\right] npu\, R_N \leq R_{cr} \\ F_{i,j}(C)\left[(F_{i,j}(D)F_{i,j}(B/D) + F_{i,j}(M)F_{i,j}(B/M)\right] npu\, R_N > R_{cr} \end{cases}, \quad (1)$$

where: i is the day of calculation; j is the OTU number; Fi, j (B) is the probability of occurrence of vegetation fires (event B); Fi, j (C) is the probability of vegetation ignition at a certain value of the complex indicator of fire hazard due to weather conditions (event C); Fi, j (N), Fi, j (D) is the probability of an anthropogenic source of fire in OTU from nearby settlements or railways and roads (events N and D); Fi, j (B / N), Fi, j (B / D) is the probability of ignition due to the appearance of anthropogenic sources of fire; Fi, j (M), Fi, j (B / M) is the probability of occurrence of a natural source (event M) and fire due to its occurrence; RN is the distance from OTE to the nearest populated area.

The probability of a fire $F_{i,j}(C)$ in the *j*-th OTU is equal to 1 if the daily complex indicator P_{i} in the center of the OTU (Walsh, et al., 2017, Taufik, Setiawan, Lanen, 2015) exceeds the value of the indicator *Pcr*. This probability can be calculated by the formula:

$$F_{i,j}(C) = \frac{P_i}{P_{cr.}}, \quad (2)$$

where: Pcr - the minimum value of the complex indicator of fire danger at which the occurrence of fires on this site is possible depending on the degree of natural pyrological fire danger of vegetation in the OTU.

The probability of ignition due to the appearance of lightning discharges from either the population from nearby points or the road network $F_{i,j}(M)$, $F_{i,j}(N)$ and $F_{i,j}(D)$ in the *j*-th OTU is calculated as follows:

$$F_{i,j}(N) = \frac{QN_R}{Q}, F_{i,j}(D) = \frac{QD_R}{Q}, F_{i,j}(M) = \frac{QM_{i,j}}{Q}, \quad (3)$$

where: Q - the number of seasons for the base period; QN_R, QD_R - the number of seasons in which fires were recorded that occurred in the fire hazard zone within radius R (km) of a populated area or road network; $QM_{i,j}$ is the number of seasons in which lightning discharges were observed.

The probability of ignition due to the appearance of lightning discharges from either the population from nearby points or the road network is calculated as follows:

$$F_{i,j}(B / M) = \frac{NMP_{i,j}}{NP}, F_{i,j}(B / N) = \frac{NNP_{i,j}}{NP}, F_{i,j}(B / D) = \frac{NDP_{i,j}}{NP}, \quad (4)$$

where: Fi, j (B / M), Fi, j (B / N) and Fi, j (B / D) - the probability of occurrence of fire from different sources; NMPi, j, NNPi, j, NDPi, j - the number of fires from lightning discharges, anthropogenic activities of residents of settlements or arrivals along the road network; NP is the total number of fires for the base period.

The construction of air travel routes is carried out in the instrumental GIS. A vector layer is created for each day of the fire-hazardous season. The author module, which is implemented in the programming environment, searches for and draws out control points of the flight of the aircraft around fire-hazardous vegetation areas.

Fire prevention publicity and mass advocacy may be regulated according to the values $F_{i,j}(N)$ and $F_{i,j}(D)$ of the probability of the emergence of anthropogenic sources of fire and the seasonal frequency of occurrence of fires in forest and non-forest areas. The dates to restrict public access to hazardous areas and the publicity of this information in the media may be implemented in five-day periods of the fire-dangerous season, during which fires are most likely to occur in certain OTUs. The location of information stands and the definition of the most dangerous populated areas may be recorded according to their locations relative to OTUs with high probabilities of forest and non-forest fires.

Subsystem Forecasting the Spread of Vegetation Fires

The method of calculating the occurrence and spread of grass fires includes the following steps: (1) determination of time periods for agricultural burns; (2) daily determination of the degree of drying of herbal combustible materials during periods of agricultural burns; (3) calculation of daily fire danger according to weather conditions and allocation of days in which the occurrence of grass fires is possible due to meteorological conditions; (4) calculation of the likelihood of grass fires by natural and anthropogenic factors; (5) daily calculation of propagation speed of the edge of grass fires; and (6) timing of the likely spread of grass fire to the nearest forest area.

The season of grass fires can be determined from the actual averaged long-term data or on the basis of the growing season. We propose the use of pre-vegetation, vegetation, and post-vegetation periods, which coincide with the dates of steady transition through threshold temperatures: 0, +5, +10, +15 ^{0}C.

We assume that the whole grass is in a state of rags as a manner of determining the degree of drying of the vegetation at the beginning of the pre- and at the end of the post-growing season. On dry days, in which the daily amount of precipitation in the previous, current, and next day lies in the interval from 0 to less than 3 mm, the amount of dry carbon approaches 100%. In other times, for the daily determination of the state of vegetation of the grass and the degree of its drying, we propose the use of NDVI, which is calculated by the following formula:

The density of vegetation (NDVI) at a certain point in the image is equal to the difference of the intensities of the reflected light in the red and infrared ranges divided by the sum of their intensities (Verhulst, 2010).

$$NDVI = \frac{NIR - RED}{NIR + RED}, \tag{4}$$

where: *NIR* – the reflected light in the infrared ranges; *RED* – the reflected light in the red range.

The MacArthur method is used to calculate the rate of spread of grass fire (*w*, km/h)) in meadow areas (McArthur, 1966, Holgate, van Dijk, Cary, Yebra, 2017):

$$w = 0.26e^{5.01\lg(c+0.001) - 23.6 + 0.028t - 0.226\sqrt{rh} + 0.633\sqrt{v}}, \tag{5}$$

where: *rh (relative humidity)*, % – relative humidity; *c*, % – dry grass content, *v* – wind speed, km/h.

Dry grass content *c* in areas of non-forest land is taken on dry days as 100% during the period of stable temperature transition, i.e., from 0 to +5 °C. In other periods, the dry grass content in the OTU can be determined by the values of the NDVI, which is calculated for each pixel of the satellite image by using formula No. 1. Due to the features of reflection in the infrared and red regions of the spectrum, natural objects not related to vegetation have the following fixed values: –0.5 for the artificial materials (concrete and asphalt); –0.25 for water; –0.05 for snow and ice; 0 for clouds; 0.025 for open soil; 0.5–0.7 for sparse vegetation; and 0.7–1 for dense vegetation. If pixels with values from 0.5 to 0.7 predominate in OTU, then the content of dry grass *c* is 50%, and from 0.7 to 1, it is 0%. In the case of high cloudiness in the OTU (i.e., zero index value), the NDVI value is taken from the previous day, provided that no precipitations with volume exceeding 3 mm/day were observed that day.

Humidity rh_i is calculated on the basis of positive air temperature t_i and dew point τ_i:

$$rh_i = 0.01e^{\tau_i a - \frac{\tau_i t_i a}{b+t_i} - \frac{bat_i}{b+t_i}\frac{1}{b+\tau_i}}, \tag{6}$$

where: *a*, *b* – constants equal 17,27 and 237,7. For the negative air temperatures and dew points at the beginning and end of the fire-dangerous season, a general formula is used.

$$rh_i = 1 - 0.05(t_i - \tau_i). \tag{7}$$

The time needed for a grass fire to reach the nearest forest area is determined on the basis of the speed of propagation of the fire and the distance to the forest.

For the regulation of forest protection measures for federal, municipal, and private organizations, the fire prevention recommendations are implemented in the form of electronic reports (documents, tables, and maps) to include the following: identification of non-forest areas with high fire danger (with likelihood of grass fires from 0.5 to 1); identification of dates with high fire danger, during which the highest probability of non-forest fires (i.e., over 0.6) may occur in certain OTU; determination of the time of probable spread of grass fire to the borders of the forest fund and places of creation of protective strips along the forest fund.

Subsystem Fires Vegetation

The atlas of fires in the territory of the JAR contains electronic maps and their attribute components (tables, database models, and external design of objects).

Electronic maps (general geographic or topographic load). The necessary cartographic material is created in the form of complex electronic maps of the Jewish Autonomous Region, including "Plant Formations," "Relief," "Hydrography," "Road Network." The set of maps forms a topographical base supplemented by the boundaries of the area. Thematic electronic maps are layers of GISs aimed at implementing the fire monitoring of the Jewish Autonomous Region. They typically include the "Quarterly network of forest enterprises," "Pyrological characteristics of vegetation," "Fire danger of orographic characteristics of the territory," "Territory passed by fire," "Anthropogenic fire hazard," and "Pyrological assessment geographic information system provides an opportunity geographically," from which the most visual and readable form related to commands on transmit, investigate, query, and analyze various data can be generated.

The attributive databases for each fire type include 14 characteristics for defining the main fire parameters (area, start and end dates, day of the week, and detector), its location (fire number, forestry, and quarter number), and the adjacent territory (vegetation formation and its fire hazard class, angle, height difference, height of the upper edge, and slope exposure). The database can be supplemented by various information elements depending on the specific requirements of the GISs.

FUTURE RESEARCH DIRECTIONS

The fire-hazardous season of 2010, a typical phenomenon for the territory of the Jewish Autonomous Region, is selected for testing. A total of 116 fires have been registered in the forest fund, and 16.2 thousand hectares were reportedly burned by fire, including 3.2 thousand hectares of forest lands. The analysis of actual flammability by using the 2010 data indicates that a significant part of the fire has occurred in the OTU with a predominance of vegetation of I and III degrees of natural pyrological danger at –33.4% and 38%, respectively, and the lowest rate with vegetation of V level (0,4%).

The sources of information are the actual and forecast data of weather stations located in JAR ("Birobidzhan", "Yekaterino-Nikolskoye", "Leninskoye", "Obluchye" and "Smidovich") and the southern part of Khabarovsk Territory ("Beech", "Kur", "Sutur", "Troitskoe", "Kukan", "Khabarovsk"). The information on the fires of vegetation has been obtained from the FBI "Avialesoohrana," the Department of Forest Protection, and the state control and supervision teams of forests managed by the governments of Khabarovsk Territory and Jewish Autonomous Region.

In the territory of the Jewish Autonomous Region, two networks of OTU were built. The maps of all OTUs and their official air tour networks depend on weather conditions (Fig.4).

The irregular network of districts of the District Forestry comprises the forest management plans of OGKU Lesnichestvo JAR, which is divided into six branches. The network includes 22 district forest districts with 3,977 quarters. A regular network of vegetation areas of non-forest land can be represented by square cells measuring 2.5 km × 2.5 km, and their total number is 2,623 units. Each OTU is assigned sequence numbers and center coordinates. The degree of pyrological stability of the vegetation prevailing in each OTU is determined in accordance with the work of A. M. Starodumova (Starodumov, 1964).

Figure 4. Operational – territorial units of vegetation areas and routes of their air traffic on weather conditions in the territory of the Jewish Autonomous region

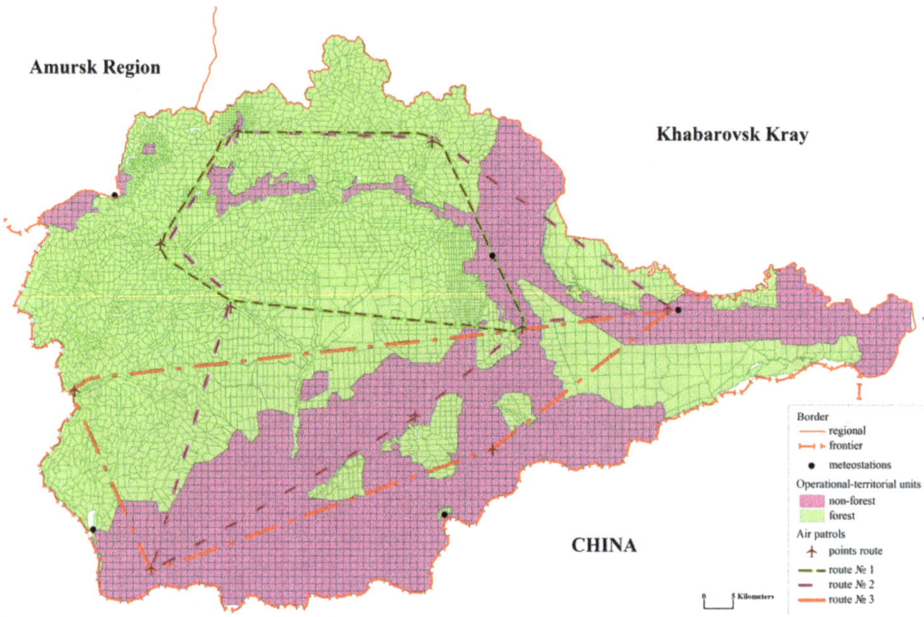

In predicting the probability of occurrence of vegetation fires according to weather conditions $F_{i,j}(C)$ *(equation 2)*, the daily values P_i are calculated according to the method of V.G. Nesterov, and the values *Pcr.* for each plant formation prevailing in each OTU are determined according to the scale of T.V. Kostyrin. The accuracy of the forecast for the current day is 86 ± 6% with a probability of 0.95. The reliability of the short-term forecast decreases slightly from 82 ± 7% to 78 ± 7% but remains fairly high.

In selecting the base period *Q (equation 3)*, an array of forest fire data for the period of 1999–2012 is processed. The array is divided into five periods of 10 years 1999–2008, 2000–2009, 2001–2010, 2002–2011, 2003–2012. In each period, the daily values $F_{i,j}(N)$ and $F_{i,j}(D)$ of the emergence of anthropogenic sources of fire in the three-kilometer zones around the settlements and the road network are determined. The period of 1999–2008, which has the minimum sum of standard deviations $F_{i,j}(N)$ and $F_{i,j}(D)$ from other periods, is chosen as the base (reference).

The likelihood of fires caused by thunderstorms is not calculated due to their practical absence in the area (Petrov, Novorotsky, Lenshin, 2000).

The likelihood of fires $F_{i,j}(B)$ *(equation 1)* is calculated for each current day and every 1–3 days of weather forecasting during the season. The predicted distribution of fires in the entire JAR according to the probability of their occurrence on the current day is determined as follows. The first digit corresponds to the values of $F_{i,j}$ *(B)*, while the second digit corresponds to the number of fires (%). The value of $F_{i,j}$ *(B)* is taken as the threshold value of 0.4 because this can result in the greatest number of fires. For example, in the territory of the branch of the Birobidzhan Forestry JAR, the fires have 72% of occurrence probabilities $F_{i,j}(B)$, which exceeds 0.4.

The check-in of the aerial patrol scheme is carried out by using the example of 35 days of the fire-dangerous season of 2010, the year in which fires of vegetation have been recorded. Ten control points are selected from the official patrol routes of OGKU's "Base of aviation and terrestrial protection of forests

Figure 5. Map of the calculated air route on the territory of the Jewish Autonomous region, 05.05.2010

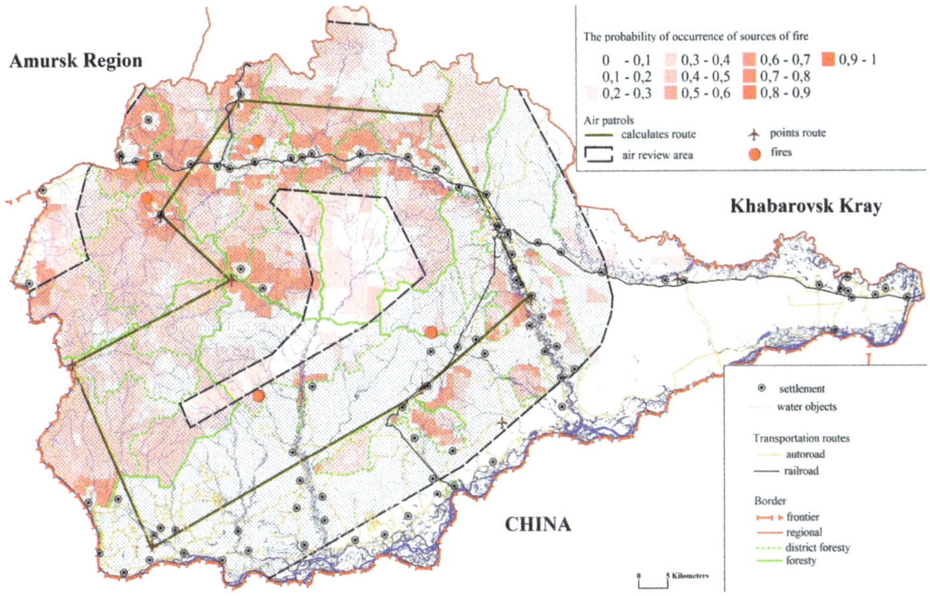

in the Jewish Autonomous Region," which is flown around according to weather conditions depending on the class of fire hazards. The width of the survey aircraft is equal to 30 km. In constructing daily air patrol routes, checkpoints are selected from official routes bordering the largest number of fire-hazard areas of vegetation exceeding the value of 0.4. The obtained routes can cover the territory in which 90% of forest fires have occurred. In the four routes, some fires are located outside the zone of flight, such as in the center of autonomy, which requires further adjustments to the location of control points.

An example of an air route for May 5, 2010, which is built on the basis of the calculated results of the probability of fires in 1,476 OTUs, as shown in Fig. 5.

Periods of restriction of attendance by the population in relation to forest and non-forest funds are determined on the basis of the daily analysis of the probability of the emergence of anthropogenic sources of fire for the base period (1999–2008). Fires are most likely to occur during the periods of April 18–May 20 and October 7–27, during which ensuring the implementation of preventive measures is necessary.

CONCLUSION

Verification of an author's information system showed, a possibility of its application in the forest conservation organizations. Collecting meteodata and the fires of vegetation from 1960- 2017 in the territory of JAR is executed. Forecasts of fire hazard for meteodata during 2015-2016 are carried out and it is established that the maximum of inflammability will be observed in the end of April and the beginning of October in the territory of Leninsky district. According to calculations, the greatest correlation between the vegetation number of fires in day of their detection and values of probability of their emergence amounted during the spring period - 0.74 and during the autumn period - 0.99. Monitoring routes the very dangerous of territories in 2010 are optimized that allowed to capture 90% of ignitions.

The route consisted of 10 control points of flight in the territory of JAR. The made information system promotes identification of settlements where the most active educational fire protection work (fig. 6) is required. 34 settlements in which close the high frequency of emergence of ignitions is observed are set, in this regard these settlements require special attention from forest conservation. Results of a research can be applied during planning of strategy of management of the fires and optimization of the organization of a fire service.

REFERENCES

Abushenko, N. A., Bartalev, S. A., Belyaev, A. I., Ershov, D. V., Zakharov, M. Yu., Korovin, G. N., … Schetinsky, V. E. (1998). Experience and prospects of organizing operational satellite monitoring of the territory of Russia for the purpose of the service of fire protection of forests. Study of Earth from space, 3, 89-95. (in Russia)

Algorithms and programs for calculating operational systems for detecting and extinguishing forest fires on a computer. (1974). LenNIILH. (in Russia)

Baranovskiy, N., Barakhnin, V., & Yankovich, E. (2017). GIS-Technologies and mathematical simulation as tools for lightning-caused forest fire danger prediction. *CEUR Workshop Proceedings*, *1839*, 2–15.

Bovio, G., Nosenzo, A., Xavier Viegas, D., Ferreira, A., & Sol, B. (1999). Comparative study of various methods of fire danger evaluation in southern Europe. *International Journal of Wildland Fire*, *9*(4), 235–24. doi:10.1071/WF00015

Díaz-Delgado, R., Lloret, F., & Pons, X. (2004). Spatial patterns of fire occurrence in Catalonia, NE, Spain. *Landscape Ecology*, *19*(7), 731–745. doi:10.100710980-005-0183-1

Doroshenko, A. M., & Kogan, R. M. (2007). Pyrological vulnerability evaluation of the Middle Amur Region. (By the example of the Jewish Autonomous Region) Fourth International meeting of Regional Northeast Asia wildland fire network, Dec. 16-17, 2007, Harbin, China, Reports. Harbin, 2007. pp. 46-52. (in Russia)

Federal service of statistics. Retrieved from http://www.gks.ru

Oliveira, S. L., Pereira, J. M., & Carreiras, J. M. (2012). Fire frequency analysis in Portugal (1975–2005), using Landsat-based burnt area maps. *International Journal of Wildland Fire*, *21*(1), 48–60. doi:10.1071/WF10131

Frantz, D., Stellmes, M., Röder, A., & Hill, J. (2016). Fire spread from MODIS burned area data: Obtaining fire dynamics information for every single fire. *International Journal of Wildland Fire*, *25*(12), 1228–1237. doi:10.1071/WF16003

Geographic information systems. Survey information. (1992). – Moscow, Russia: TsNIIGAIK. (in Russia)

Glagolev, V. A. (2015). *Estimation and forecast of grassland fire occurrence within Jewish Autonomous Region territory*. Khabarovsk, Russia: Institute for Water and Environmental Problems of Far East Branch of the Russian Academy of Sciences.

Goncalves, Z. J., & Lourenco, L. (1990). Meteorological index of forest fire risk in the Portuguese mainland territory. In *Proceedings of the international conference on forest fire research. Coimbra*. 7, p. 1.

Grishin, A. M., & Filkov, A. I. (2011). Deterministic-probabilistic system for predicting forest fire hazard. *Fire Safety Journal*, *46*(1), 56–62. doi:10.1016/j.firesaf.2010.09.002

Haiganoush, K. (2016). Matthew Jolly Near-term probabilistic forecast of significant wildfire events for the Western United States. *International Journal of Wildland Fire*, *25*(11), 1169–1180. doi:10.1071/WF16038

Heikinheimo, M. (1998). Renewing the system for forest fire risk assessment at the Finnish Meteorological Institute. *International Forest Fire News*, *18*, 65–67.

Holgate, C. M., van Dijk, A. I. J. M., Cary, G. J., & Yebra, M. (2017). Using alternative soil moisture estimates in the McArthur Forest Fire Danger Index. *International Journal of Wildland Fire, 2017*. doi:10.1071/WF16217

Keetch, J., & Byram, G. (1988). A drought index for forest fire control. Res. Paper SE-38. Asheville, NC: U.S. Department of Agriculture, Forest Service, Southeastern Forest Experiment Station, p. 32.

Kolesnikov, B. P. (1995). Sketch of the vegetation of the Far East. Khabarovsk. p. 102 (in Russia)

Korovin, G. N. (1977). Fire hazard assessment in the forest and calculation of parameters of forest fires on a computer: LenNIILH. p. 69 (in Russia)

Kovalcv, A. P. (2009). The current state of the forests of the Russian Far East and the prospects for their use. p. 470 (in Russia)

Krawchuk, M. A., & Moritz, M. A. (2014). Burning issues: Statistical analyses of global fire data to inform assessments of environmental change. *Environmetrics*, *25*(6), 472–481. doi:10.1002/env.2287

Krueger, E. S., Ochsner, T. E., Carlson, J. D., Engle, D. M., Twidwell, D., & Fuhlendorf, S. D. (2016). Concurrent and antecedent soil moisture relate positively or negatively to probability of large wildfires depending on season. *International Journal of Wildland Fire*, *25*(6), 657–668. doi:10.1071/WF15104

Kumar, V., & Dharssi, I. (2017). Evaluation of daily soil moisture deficit used in Australian forest fire danger rating system. *Forest Ecology and Management*, *259*, 685–697.

Kurentsova, G. E. (1967). Sketch of the vegetation of the Jewish Autonomous Region. Vladivostok, Russia: Far Eastern Book Publishing House, p. 64 (in Russia)

LinacreE. (n.d.). Predicting bushfires in Australia. Retrieved from http://www-as.uwyo.edu/~geerts/cwx/notes/chap16/oz_bush.html

Louisa, E., Julie, G.-D., Tessa, N., Kevin, R., & Christina, T. (2013). Research and development supporting risk-based wildfire effects prediction for fuels and fire management. *International Journal of Wildland Fire*, *22*(1), 37–50. doi:10.1071/WF11143

Marchenko, N. A. (1993). Fire hazard assessment of forests by the likelihood of fires with a certain speed of propagation based on the analysis of the state of landscapes. Geography and Natural Resources. 4. pp. 131-136. (in Russia)

McArthur, A. G. (1966). Weather and grassland fire behavior / Leaflet 100. Forestry and Timber Bureau. Commonwealth of Australia, 1966.

Michele, S., Bachisio, A., & Fermin, A. (2016). Predicting wildfire spread and behaviour in Mediterranean landscapes. *International Journal of Wildland Fire*, *25*(10), 1015–1032. doi:10.1071/WF15081

Petrov, E. S., Novorotsky, P. N., & Lenshin, V. T. (2000). The climate of the Khabarovsk Territory and the Jewish Autonomous Region. Khabarovsk, Russia: Dal'nauka. pp. 174 (in Russia)

Research and Production Center. Map-Maker. Retrieved from http://gismeteo.ru

Rogeau, M.-P., Flannigan, M. D., Hawkes, B. C., Parisien, M.-A., & Arthur, R. (2016). Spatial and temporal variations of fire regimes in the Canadian Rocky Mountains and foothills of southern Alberta. *International Journal of Wildland Fire*, *25*(11), 1117–1130. doi:10.1071/WF15120

Satellite monitoring of forest fires in Russia. (2003). Results problems. Perspectives. Novosibirsk, Russia. p. 135. (in Russia)

Seager, R., Hooks, A., Williams, A. P., Cook, B. I., Nakamura, J., & Henderson, N. (2015). Climatology, variability and trends in United States vapor pressure deficit, an important fire-related meteorological quantity. *Journal of Applied Meteorology and Climatology*, *54*(6), 1121–1141. doi:10.1175/JAMC-D-14-0321.1

Sommers, W. T., Coloff, S. G., & Conard, S. G. (2012). Synthesis of knowledge: fire history and climate change. *Final report to the Joint Fire Science Program*, JFSP Project, 09-02-1-09.

Space Research Institute of the Russian Academy of Sciences. Retrieved from http://meteo.infospace.ru

Parisien, M. A., Snetsinger, S., Greenberg, J. A., Nelson, C. R., Schoennagel, T., Dobrowski, S. Z., & Moritz, M. A. (2012). Spatial variability in wildfire probability across the western United States. *International Journal of Wildland Fire*, *21*(4), 313–327. doi:10.1071/WF11044

Starodumov, A. M. (1964). Scale of fire hazard weather for the conditions of the Khabarovsk Territory, Primorye and Amur Region. Collected Works DalNIIHL. 6. pp. 176-185. (in Russia)

State hydrometeorological center of the Russian Federation. Retrieved from http://meteoinfo.ru

Sukhinin, A. I., Kashkin, V. B., & Ponomarev, E. I. (1999). Monitoring forest fire in eastern Siberia from space. *Proceedings of the Society for Photo-Instrumentation Engineers*, *3983*, 206–214. doi:10.1117/12.370494

Taufik, M., Setiawan, B. I., & van Lanen, H. A. J. (2015). Modification of a fire drought index for tropical wetland ecosystems by including water table depth. *Agricultural and Forest Meteorology*, *203*, 1–10. doi:10.1016/j.agrformet.2014.12.006

Tian, X., Zhao, F., Shu, L., & Wang, M. (2013). Distribution characteristics and the influence factors of forest fires in China. *Forest Ecology and Management*, *310*, 460–467. doi:10.1016/j.foreco.2013.08.025

Van Dijk, A. I. J. M., & Bruijnzeel, L. A. (2001). Modelling rainfall interception by vegetation of variable density using an adapted analytical model. Part 1. model description. *Journal of Hydrology (Amsterdam)*, *247*(3-4), 230–238. doi:10.1016/S0022-1694(01)00392-4

Varol, T., & Ertug˘rul, M. (2016). Analysis of the forest fires in the Antalya region of Turkey using the Keetch–Byram drought index. *Journal of Forest Research*, *27*(4), 811–819. doi:10.100711676-016-0235-0

Verhulst, N., & Govaerts, B. (2010). The normalized difference vegetation index (NDVI) GreenSeekerTM handheld sensor: Toward the integrated evaluation of crop management. Part A: Concepts and case studies. Mexico, D.F.; CIMMYT. 2010.

Walsh, S. F., Nyman, P., Sheridan, G. J., Baillie, C. C., Tolhurst, K. G., & Duff, T. J. (2017). Hillslope-scale prediction of terrain and forest canopy effects on temperature and near-surface soil moisture deficit. *International Journal of Wildland Fire*, *26*(3), 191–208. doi:10.1071/WF16106

Williams, A. P., Seager, R., Macalady, A. K., Berkelhammer, M., Crimmins, M. A., Swetnam, T. W., ... Rahn, T. (2015). Correlations between components of the water balance and burned area reveal new insights for predicting forest fire area in the southwest United States. *International Journal of Wildland Fire*, *24*(1), 14–26. doi:10.1071/WF14023

Yi, K., Bao, Y., & Zhang, J. (2016). Spatial distribution and temporal variability of open fire in China. *International Journal of Wildland Fire*, *26*(2), 122–135. doi:10.1071/WF15213

Zhurnist, V. I., Kogan, R. M., Kodyakova, T. E., Rubtsova, T. A., ... (2004). Far East Branch of the Russian Academy of Sciences. Birobidzhan, Russia: ICARP. p. 111 (in Russia)

Section 5

Chapter 10
Analysis of Terrestrial Vegetation Trends and Correlation Between Vegetation Indices and Climatic Factors

Dhanapriya M.
Tamil Nadu Agricultural University, India

Hiren P. Bhatt
https://orcid.org/0000-0001-9980-0291
Space Applications Centre (SAC), Indian Space Research Organisation, India

Vyas S. P.
Space Applications Centre (SAC), Indian Space Research Organisation, India

ABSTRACT

The chapter analyzes terrestrial vegetation trends and correlation of vegetation indices with climatic factors like LST, Net Radiation, and TRMM. The result shows the positive vegetation trend might be due to the increased vegetation growth and productivity. Negative vegetation trend might be due to changes in a crop choice. Correlation of vegetation indices with climatic factors like LST showed weak to moderate positive correlation. Net radiation showed moderate negative correlation and TRMM showed a weak negative or positive correlation for both the Kharif and Rabi seasons.

INTRODUCTION

The robust measure of vegetation activity at the land surface helps to compute vegetation indices from the remotely sensed data based on the optical characteristics of vegetation that are strongly absorptive in the red, but highly reflective in the near-infrared (NIR) regions of the solar spectrum. The optical energy reflected by vegetation is dependent on leaf structure, pigmentation and moisture. MODIS multispectral

DOI: 10.4018/978-1-7998-1867-0.ch010

sensor consisting of several bands of data which provides different information on land properties based on its spectral reflectance through various indices while these features are not available in other sensors. Vegetation index is the indicative factor of vegetation density and growth status, it has positive correlation with vegetation coverage, and, to a certain extent, changes of vegetation coverage. Given the effect of soil background, radiation increases significantly when the vegetation cover is sparse; conversely NIR radiation is reduced to make the relationship between vegetation and soil more sensible. NDVI can be directly used as a substitute of plant photosynthetic activity and helps in detecting biotic responses to climate change. The trend analysis method was followed to detect changes in deforestation, urban expansion or mining activities. The trends show positive and negative trends over land cover class and precipitation changes in the same time period which seems to have an influence on large NDVI trend areas.

BACKGROUND

Roberts *et al.*, (1997) monitored the seasonal changes in atmospheric water vapor, liquid water and surface cover in the Jasper Ridge during 1992 using NDVI showed that green vegetation, NDVI and liquid water were highly correlated. Rasmussen, 1998 found a positive correlation bound NDVI and precipitation. Ahmed *et al.* (2002) recounted that regression between LST, NDVI and Evapotranspiration (ET) in the Gezira Research station during 6th Sep 2001 and 28th Nov 2002 show negative correlation for all land cover types. Herrmann and Anyamba (2005) studied about the dynamic ecosystem that responds to the fluctuations in climate and anthropogenic land use patterns for the West African Sahel from 1982 – 2003. The temporal and spatial patterns of vegetation greenness and rainfall variability were investigated based on NDVI time series dataset. They suggested that rainfall was the important factor for increase in vegetation greenness and as a cause human-induced change may also be super imposed on the climate trend. They also observed that the correlations between the NDVI and rainfall were found to be highly significant throughout the entire Sahel. Peng *et al.* (2007) reported that EVI had strong correlation with temperature on the seasonal fluctuation at Zhejiang province during 2001 – 2004. The sequence of correlation was carried out to extract the most important periods of rainfall responsible for the production of green biomass and for the extent of land cover types using MODIS time series data for North Western Morocco. The correlation between mean daily precipitation and classified area of each land cover type were in the range of medium to high (Hopfner and Scherer, 2011). Zhang *et al.* (2011) reported that a positive correlation between NDVI and temperature was most obvious for coniferous forest and shrubs, that the effect of precipitation on NDVI was not as significant as the effect of temperature and that scrubland NDVI correlated more strongly with precipitation than the NDVI of other vegetation. Chuai *et al.* (2013) performed correlation analyses of NDVI, temperature and precipitation for vegetation types during growing season in Inner Mongolia, China and observed that NDVI correlated differently with temperature and precipitation. Eckert *et al.* (2015) examined whether MODIS NDVI satellite data time series can be used to detect land degradation and regeneration area in Mongolia from 2001 to 2011. The inter-annual and seasonal trends of vegetation cover were analyzed in the Upper Blue Nile using GIMMS NDVI and MODIS NDVI from 1982 to 2011 and found that the GIMMS NDVI had positive trend during 1981 to 2006 and negative trend in MODIS NDVI from 2001 to 2011 (Teferi *et al.*, 2015). The pixel-based linear regression model was used to identify the rate of deforestation and fragmentation at landscape level. The classified result was validated with field data and found that it is an effective method for monitoring forest cover change (Amarnath *et al.*, 2017). Papagiannopoulou *et al.* (2017)

discovered the relationship between climate and vegetation dynamics using satellite data by using Ridge regression, Random forest model and Pearson coefficient. They found that the Random forest model achieved the best results, the Ridge regression performed well in the relation between climatic variables by showing non-linearity in areas were forest gets high R^2 value and Pearson coefficient does not perform well. Comparing MODIS-C5 and C6 vegetation indices products from both Terra and Aqua from 2001 to 2015 proved that Global vegetation showed a remarkable greening trend in the past 15 years and contributes to accurate understanding of vegetation played in the Earth's biogeochemical and climatic systems (Zhang *et al.*, 2017). The analysis of vegetation dynamics using time series MODIS vegetation indices viz., NDVI and EVI was carried out seasonally from 2007 to 2017. The NDVI and EVI profiles was extracted for agriculture and forest ecosystem for the Saurashtra region of Gujarat. It was observed that NDVI and EVI showed a good range of spatio-temporal variation in vegetation condition, but there is little variation in the growth profile over the years except 2009, 2012, 2015 and 2016 (Dhanapriya *et al.*, 2018). With this review of previous study, the aim of this research was carried out to lead a better understanding between the vegetation indices and climatic parameter and also to analyze the decadal seasonal vegetation trends of NDVI and EVI.

MATERIALS USED

The following satellite and ancillary data products were downloaded or procured from various sources for the years 2007 to 2017: MODIS Land Surface Temperature (LST) (MOD11A2) of tile H24V6 at 1 km spatial resolution in 8-day gridded level-6 product in the Sinusoidal projection was downloaded from NASA Earth Observing System (EOS) data gateway (reverb.echo.nasa.gov/reverb, 2017)). MOD11A2 is comprised of day-time and night-time Land Surface Temperature (LST), quality assessment, observation times, view angles, bits of clear-sky days and nights, and emissivity's estimated in Bands 31 and 32 from land cover types. Tropical Rainfall Measurement Mission (TRMM) 3B42 of daily precipitation estimates at 0.25° spatial resolution in NetCDF format was downloaded from NASA Earth Observing System (EOS) data gateway (reverb.echo.nasa.gov/reverb, 2017). TRMM 3B42 depends on inputs from two types of sensors like microwave and infrared (IR). CERES-NetFlux Global net radiation data of 8-day composite with 0.25° spatial resolution were downloaded from NASA Earth Observatory website (https://neo.sci.gsfc.nasa.gov/view.php?datasetId=CERES_NETFLUX_M, 2018). NetFlux data are convolved from Clouds and the Earth's Radiant Energy System (CERES) sensors on NASA's Terra and Aqua satellites. The investigation were carried out in the Saurashtra (60900 sq. km.) region of Gujarat which lies between 68°50' E Longitude, 23°31' N Latitude to 72°26' E Longitude, 20°41' N Latitude which is shown in Figure 1.

METHODOLOGY

The outline of methodology was listed in Figure 2. Preprocessing and processing of satellite data were carried out using MODIS Reprojection Tool (MRT) Release 4.1. and ArcGIS 10.1 software respectively. The time series trend analysis and correlation between vegetation indices and climatic variables were carried out in Python environment using the Arcpy and GDAL libraries. Base layers like state boundary, district boundary, district with taluk boundaries were generated from SOI toposheets of scale 1:50000.

Figure 1. Map showing the study area
Source: Dhanapriya et al., 2018

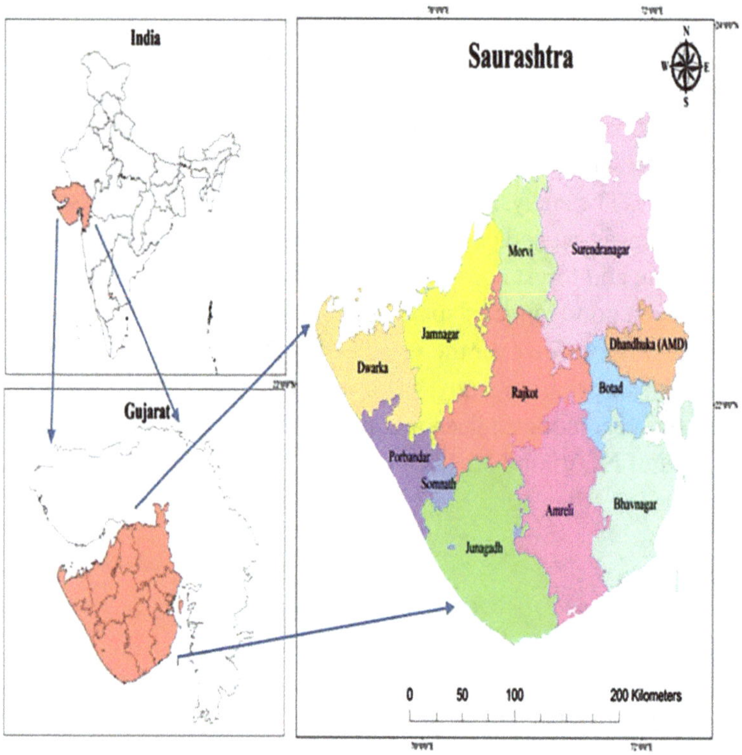

Reprojection was done using MODIS Reprojection Tool (MRT) for projecting from Sinusoidal to Universal Transverse Mercator (UTM 43N). Sub-setting of raster data reduces the time in further processing. Using the base layer, subset was done using ArcGIS Extract by Mask tool. The resampling of LST (1 km), Net Radiation (0.25°) and TRMM (0.25°) are resampled to 250 m spatial resolution to match with MOD13Q1 vegetation products using resampling tools in ArcGIS. By considering the time-lag between MOD13Q1 (Dhanapriya *et al.*, 2018) and other satellite data products, TRMM data were accumulated at 16 days interval to match with the Julian days (16-day composite) of MOD13Q1. Similarly, MOD11A2 and Net radiation data were averaged at 16-day interval. The pixel reliability layer of MODIS products was used to eliminate the cloudy pixels. The highest quality pixels free from cloud cover were only considered and extracted using Flag 0 of pixel reliability layer as mask. A linear regression model was fitted to the time series images of NDVI and EVI both for Kharif and Rabi seasons.

The temporal trends with time series data sets were determined using linear regression model, with time as independent variable and NDVI and EVI as dependent variable. The Linear Regression model was fitted among every pixel of time series data. An output image was generated which indicates the slope of the regression line. Positive slope indicates increasing trend and negative slope indicates decreasing trend. A pixel value nearer to zero indicates no specific trend in those pixels.

A linear regression model adapts the following equation:

$$Y = m X + c$$

Figure 2. Outline of Methodology

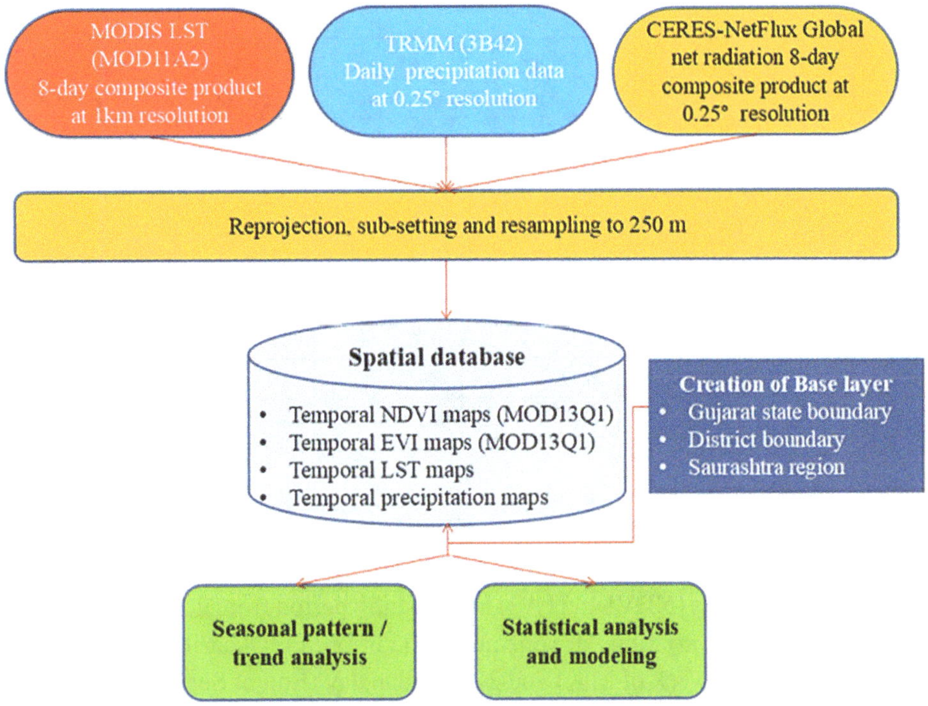

Where

Y is the dependent variable, *viz.*, time series data like NDVI
m is the slope of the regression line
X is the time variable
c is the y-intercept

Pearson's correlation analysis was conducted to understand the correlation between NDVI and EVI with LST, Rainfall and Net Radiation for a temporal period between 2007 and 2017. An output image was generated which indicates the pixel by pixel correlation coefficient. The correlation coefficient is used to represent the strength of the correspondence between NDVI with LST, Rainfall and Net Radiation and EVI with LST, Rainfall and Net Radiation.

RESULTS AND DISCUSSION

The results and research findings are discussed in the following sections: Decadal seasonal and annual trend of vegetation indices and Correlation between vegetation indices and climatic factors.

Table 1. Decadal Seasonal trend of NDVI under different land use classes for Saurashtra region (2007-2017)

Land Use	Slope			Percentage of Pixels Showing	
	Minimum	Maximum	Mean	Positive Trend	Negative Trend
Kharif Season					
Cropland	-40.41	35.00	5.64	82.8	17.2
Deciduous forest	-22.92	33.41	5.80	80.3	19.7
Scrub/Degraded forest	-24.40	30.48	5.09	78.0	22.0
Rabi Season					
Cropland	-45.18	47.70	-0.02	48.5	51.5
Deciduous forest	-21.81	26.79	-2.41	21.4	78.6
Scrub/Degraded forest	-31.56	36.73	-0.89	43.1	56.9

Decadal Seasonal and Annual Trend of Vegetation Indices

The MODIS vegetation products (NDVI and EVI) for the study period from 2007 to 2017 were subjected to trend analysis to study the variations in vegetation condition. For Kharif season, the satellite data between 145th Julian day (Last week of May) and 273rd Julian day (End of September) was used to study the decadal seasonal Kharif trend. The satellite data from 289th Julian day (October) to 049th Julian day (February) were considered to derive the decadal seasonal Rabi trend. Linear regression analysis was performed for NDVI and EVI separately for Kharif and Rabi seasons. A decadal annual trend of NDVI and EVI were also carried out considering the January to December data as a whole for agricultural and forest areas. The following sub sections explain the results obtained pertaining to pixel based linear trend analysis of vegetation indices *viz.,* NDVI and EVI for decadal seasonal Kharif, decadal seasonal Rabi and decadal annual for Saurashtra region.

Decadal Seasonal Trend of NDVI

The decadal seasonal trend of NDVI under different land use classes for Saurashtra region are presented in the Table 1. The results of trend analysis for the various land use classes showed both negative and positive trends. The slope values ranged from -40.41 to 35.00 in cropped area where Kharif crops are grown. The minimum and maximum slope value for Rabi season cropland ranged between -45.18 and 47.70, respectively. Among the cropland class, the Rabi season shows a maximum deviation when compared to Kharif season. The slope value of deciduous forest class during Kharif season ranged from -22.92 to 33.41 while during Rabi the values ranged between -21.81 and 26.79. The NDVI slope value for Kharif season under scrub/degraded forest varied from -24.40 to 30.48 and for Rabi season the slope values varied between -31.56 to 36.73. Irrespective of seasons, both the forest class has lesser deviation in slope values when compared to cropland class.

During Kharif season, more than 78 per cent of total pixels in all the land cover classes showed a positive trend and around 22 per cent of the pixels showed negative trends in NDVI. For Rabi season, 78.6 per cent of total pixels in deciduous forest showed a negative trend in NDVI followed by scrub/

Figure 3. Decadal Seasonal (Kharif and Rabi) trends of NDVI for cropland and forest of Saurashtra region

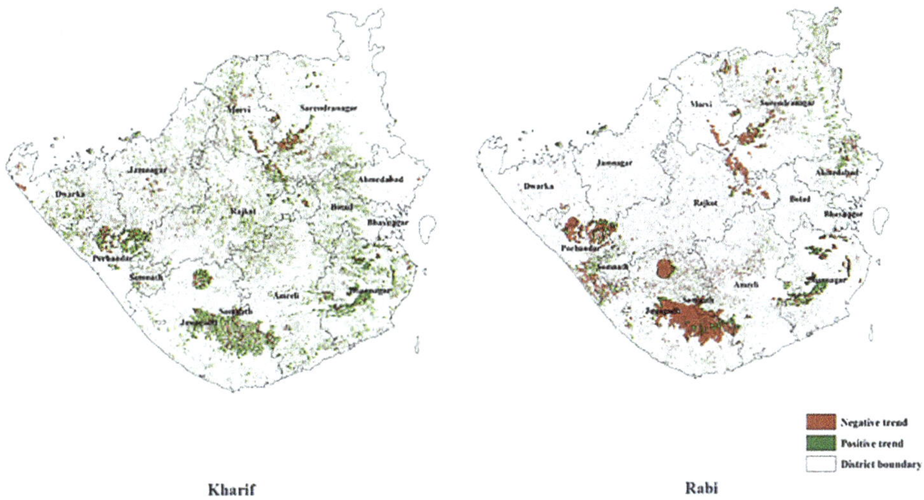

degraded forest (56.9 per cent) and cropland (51.5 per cent) (Table 1; Figure 3). In general, the Kharif season showed a greening trend whereas in Rabi season its browning in nature. The positive greening trend during the Kharif season might be due to the increased vegetation growth and productivity during the period (Teferi *et al.,* 2015, Nemani *et al.,* 2003; Slayback *et al.,* 2003; Zhou *et al.,* 2001).

The negative browning trend in Rabi season for deciduous forest and scrub/degraded forest could be caused by leaf fall followed by regrowth or the dry deciduous nature of the tree species or the already existing vegetation decline or might have strengthened by a period of constantly poor rains (Teferi *et al.,* 2015).

Decadal Seasonal Trend of EVI

The decadal seasonal trend for EVI during Kharif and Rabi are presented in the Table 2. The minimum and maximum slope value for Kharif cropland ranged between -29.62 to 27.93, respectively. Under land cover class deciduous forest, the slope value ranges from -20.09 to 19.97 and -16.45 to 18.88 during Kharif and Rabi seasons, respectively. The slope value of scrub/degraded forest class during Kharif season ranged from -30.19 to 22.52 while during Rabi the values ranged between -22.11 and 28.94. Irrespective of land cover class, Rabi season EVI found to have more deviation when compared with Kharif season.

During Kharif season, more than 70 per cent of total pixels in cropland and scrub/degraded forest class showed a positive trend except in deciduous forest the positive trend was around 57.5 percent. For Rabi season, 73.3 per cent of total pixels in deciduous forest showed a negative trend in NDVI followed by scrub/degraded forest (53.7 per cent) and cropland (45.4 per cent) (Table 2 and Figure 4). EVI shows a similar trend in both Kharif and Rabi as that of NDVI. EVI follows the same pattern as NDVI, the Kharif season showed a greening trend whereas in Rabi season it is browning in nature. The declining vegetation (browning) trend in Rabi season might be due to changes in a crop choice or disruption of cropping, which is in line with the result obtained by Dubovyk *et al.* (2012).

Table 2. Decadal Seasonal trend of EVI under different land use classes for Saurashtra region (2007-2017)

Land Use	Slope			Percentage of Pixels Showing	
	Minimum	Maximum	Mean	Positive Trend	Negative Trend
Kharif Season					
Cropland	-29.62	27.93	3.39	76.7	23.3
Deciduous forest	-20.09	19.97	0.96	57.5	42.5
Scrub/Degraded forest	-30.19	22.52	2.45	70.2	29.8
Rabi Season					
Cropland	-34.03	35.80	0.61	54.6	45.4
Deciduous forest	-16.45	18.88	-1.50	26.7	73.3
Scrub/Degraded forest	-22.11	28.94	-0.37	46.3	53.7

Decadal Annual Trend of Vegetation Indices

The decadal annual trend on NDVI and EVI for different land cover classes are given in Table 3. The NDVI slope value for cropland varies between -13.37 to 15.67 and for EVI the slope value ranges from -10.07 to 10.33. Irrespective of the vegetation indices, the slope value for deciduous forest and scrub/degraded forest found to be narrow when compared to cropland class. The deciduous forest has slope value ranges from -5.29 to 7.85 and from -3.49 to 5.59 for NDVI and EVI, respectively. The minimum NDVI slope value for scrub/degraded forest is -7.99 and maximum is 8.18 and the EVI slope value ranges from -6.79 to 6.64. Irrespective of land cover classes, it is observed that about 70 per cent of pixels are showing positive trend for decadal annual NDVI and EVI (Table 3 and Figure 5). Both NDVI and EVI image showed negative trend in some parts of Rajkot, Dwarka, Porbandar, Amreli and Jamnagar districts. The browning (negative) trend might be associated with the degraded saline and sodic soil fields or those abandoned from cultivation or dominated by coastal sandy soils with insecure supply of irrigation water which in turn, causes decline in vegetation growth and cultivated land degradation. Similar results are reported by Sonnenschein *et al.* (2011) and Dubovyk *et al.* (2012).

CORRELATION BETWEEN VEGETATION INDICES AND CLIMATIC FACTORS

In order to have a better understanding on the behavior of climatic factors under different land cover classes in Saurashtra region, time series plot on LST, net radiation and TRMM rainfall were generated and presented in Figures 5 to 7 respectively.

Figure 4. Decadal Seasonal (Kharif and Rabi) trends of EVI for cropland and forest of Saurashtra region

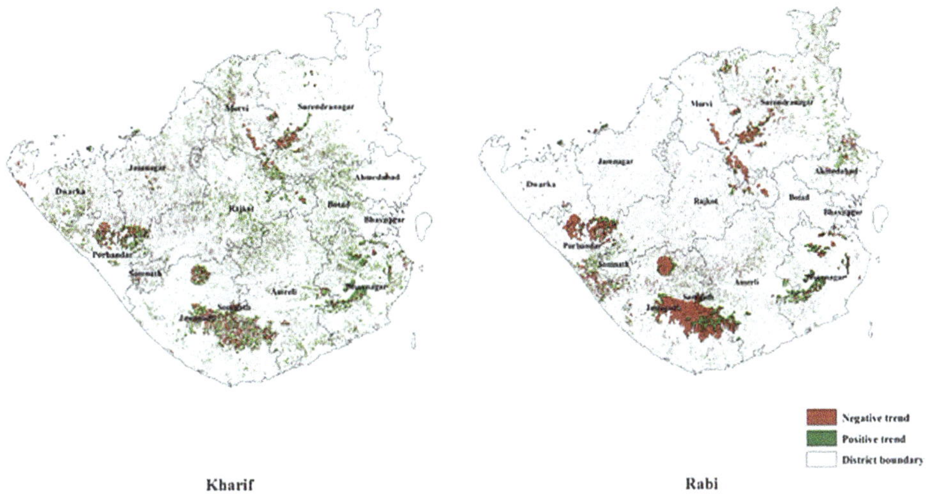

Table 3. Decadal Annual trends of vegetation indices for cropland and forest (2007-2017)

Land Use	Slope			Percentage of Pixels Showing	
	Minimum	Maximum	Mean	Positive Trend	Negative Trend
NDVI					
Cropland	-13.37	15.65	0.93	72.5	27.5
Deciduous forest	-5.29	7.85	1.60	91.0	9.0
Scrub/Degraded forest	-7.99	8.18	1.21	80.7	19.3
EVI					
Cropland	-10.07	10.33	0.67	71.7	28.3
Deciduous forest	-3.49	5.59	0.71	80.8	19.2
Scrub/Degraded forest	-6.79	6.64	0.69	78.7	21.3

Time Series Plot of Land Surface Temperature (LST)

The Land Surface Temperature (LST) is the radiative skin temperature of the land surface, as measured in the direction of the remote sensor. LST is the mixture of bare soil and vegetation temperature and the estimation of LST depends on the albedo, soil moisture and vegetation covers. The land surface temperature derived from MODIS sensor data for the study region for two seasons *viz.,* Kharif and Rabi under different land cover classes are presented in Table 4 and Figure 6. Irrespective of the land cover classes, LST values were found to be higher during the March of every year (81st Julian day) (Figure 6). The zonal maximum averaged LST during Kharif season in cropland shows variation over years, the value ranged from 32.84°C to 36.24°C, whereas during Rabi season the LST ranged from 31.62°C to 36.07°C. During Kharif season, deciduous forest shows a minimum LST value of 30.61°C and maximum

Figure 5. Decadal Annual trends of vegetation indices for all croplands and forest of Saurashtra region

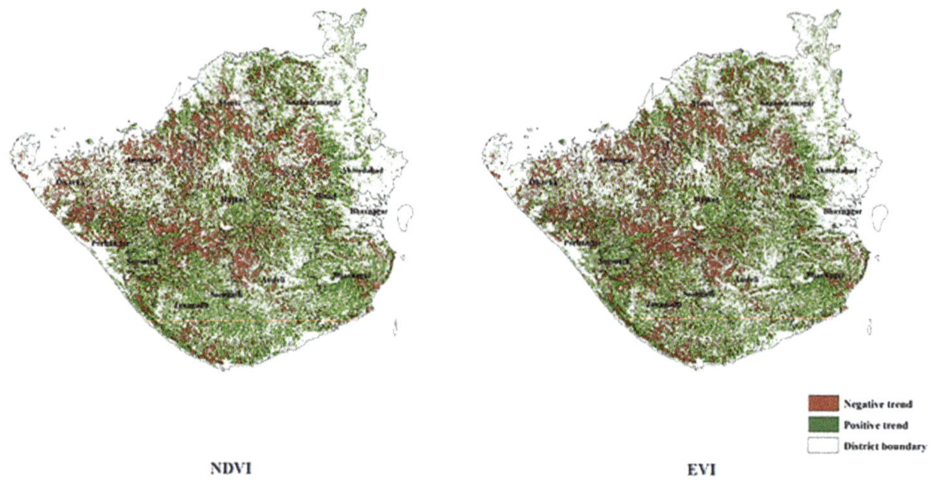

Table 4. Zonal maximum averaged LST (° C) during Kharif and Rabi season under cropland and forest for Saurashtra region (2007-2017)

Year	Cropland		Deciduous Forest		Scrub/Degraded Forest	
	Kharif	Rabi	Kharif	Rabi	Kharif	Rabi
2007	34.62	32.63	31.96	30.21	35.27	32.45
2008	33.82	34.36	30.61	32.21	33.76	34.27
2009	36.24	35.73	32.73	33.03	36.52	35.19
2010	33.47	31.79	31.46	29.68	34.27	31.78
2011	32.84	32.81	31.14	31.16	33.60	33.00
2012	35.98	36.07	31.87	32.98	35.23	34.88
2013	33.08	31.62	31.24	29.97	33.39	31.82
2014	36.06	33.93	32.27	31.30	35.68	33.55
2015	35.80	34.88	32.55	32.32	36.03	34.43
2016	35.59	33.75	32.35	31.26	35.61	33.74
2017	33.85	33.64	31.02	31.28	33.92	33.47
Mean	**34.67**	**33.75**	**31.75**	**31.40**	**34.84**	**33.51**
Standard Deviation	**1.30**	**1.47**	**0.70**	**1.15**	**1.09**	**1.16**
Coefficient of variation	**3.76**	**4.35**	**2.19**	**3.66**	**3.13**	**3.46**

Figure 6. Zonal Averaged Maximum LST temporal profile under different land use classes for Saurashtra region (2007-2017)

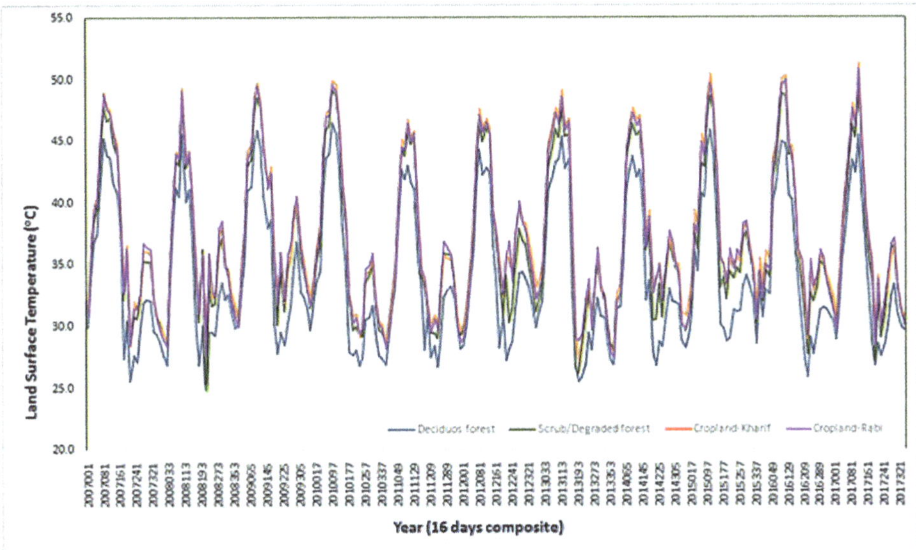

of 32.73°C whereas for scrub/degraded forest the values ranges between 33.39°C and 36.52°C. The LST values ranges from 29.68°C to 33.03°C and 31.78°C to 35.19°C during Rabi season for deciduous forest and scrub/degraded forest respectively. The mean LST value during Kharif season found to be higher for scrub/degraded forest (34.84°C) followed by cropland (34.67°C) and deciduous forest (31.75°C) whereas during Rabi season the mean LST found to be higher in cropland (33.75°C) followed by scrub/degraded forest (33.51°C) and deciduous forest (31.40°C). Irrespective of land cover classes and season, the coefficient of variation found to be less than 5 per cent, which indicates the very less inter-annual variation in LST value for the past decade (Table 4).

Time Series Plot of Net Radiation

Net radiation is the balance between incoming and outgoing energy at the top of atmosphere, sometimes called as Net flux. Net radiation will be positive or negative, positive values indicates more of incoming radiation than outgoing radiation. The unit of Net radiation will be given W/m^2. The net radiation data derived from Clouds and the Earth's Radiant Energy System (CERES) sensors on NASA's Terra and Aqua satellites for the study region are presented in Table 5 and Figure 7. Irrespective of land cover classes, the zonal averaged net radiation found to be positive during Kharif season and the value ranges between 50.84 W/m^2 (cropland) and 72.25 W/m^2 (Deciduous forest) whereas during Rabi season the net radiation value found to be negative with value ranging from -38.96 W/m^2 (cropland) to -23.89 W/m^2 (deciduous forest). During Kharif season, the mean positive net radiation values found to be higher in deciduous forest (60.48 W/m^2) followed by scrub/degraded forest (58.49 W/m^2) and cropland (58.10 W/m^2). Similar pattern was found during Rabi season but in the form of negative net radiation. Irrespective of land cover classes, the inter-annual variation for net radiation during Kharif and Rabi season recorded less coefficient of variation and the value ranged from 8.67 to 10.35 per cent.

Table 5. Zonal averaged Net Radiation (W/m²) during Kharif and Rabi season under cropland and forest for Saurashtra region (2007-2017)

Year	Cropland		Deciduous Forest		Scrub/Degraded forest	
	Kharif	Rabi	Kharif	Rabi	Kharif	Rabi
2007	68.60	-31.08	72.25	-23.89	69.06	-28.72
2008	58.53	-32.35	61.03	-26.10	58.48	-31.06
2009	60.01	-33.75	64.31	-26.58	60.92	-32.27
2010	67.22	-33.33	67.83	-26.26	67.20	-31.32
2011	56.23	-31.46	57.37	-25.92	55.90	-30.04
2012	57.28	-32.77	60.79	-26.55	58.01	-32.38
2013	58.01	-37.53	59.02	-31.58	58.05	-36.29
2014	54.21	-33.31	57.45	-26.13	54.97	-31.27
2015	53.00	-35.46	55.74	-27.90	53.67	-33.44
2016	50.84	-39.77	51.46	-33.23	51.26	-38.13
2017	55.17	-38.96	58.04	-30.73	55.86	-36.61
Mean	**58.10**	**-34.52**	**60.48**	**-27.72**	**58.49**	**-32.87**
Standard Deviation	**5.51**	**2.99**	**5.82**	**2.87**	**5.44**	**2.96**
Coefficient of variation	**9.49**	**-8.67**	**9.62**	**-10.35**	**9.29**	**-9.01**

Figure 7. Zonal averaged Net radiation temporal profile under different land use classes for Saurashtra region (2007-2017)

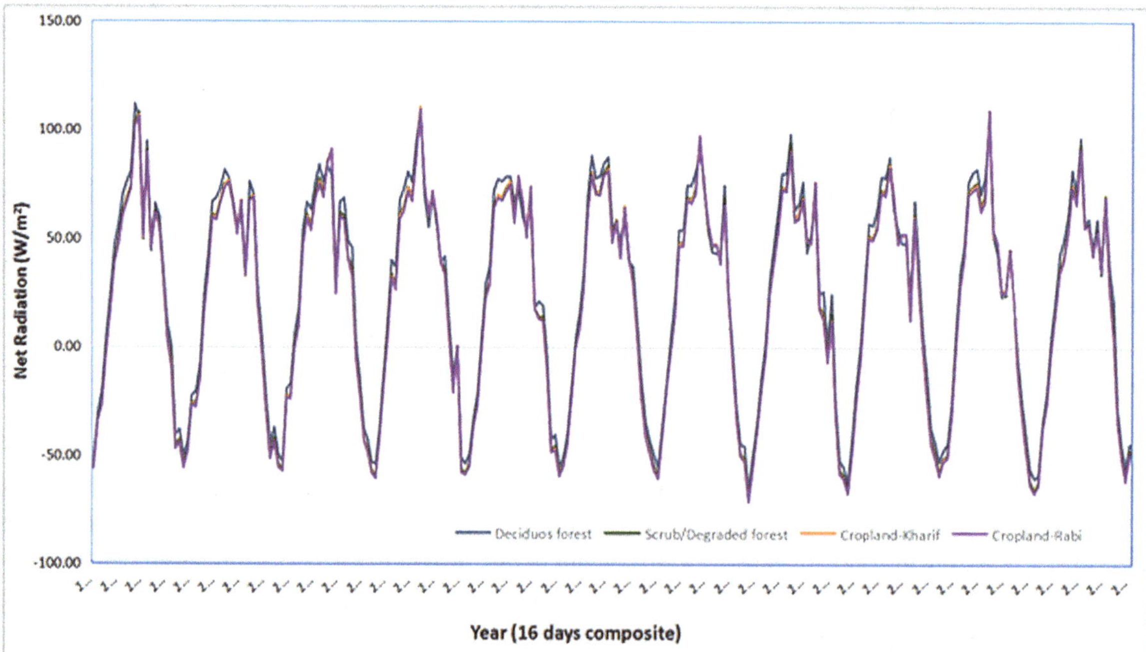

Table 6. Zonal total averaged TRMM rainfall (mm) during Kharif and Rabi season under cropland and forest for Saurashtra region (2007-2017)

Year	Cropland		Deciduous Forest		Scrub/Degraded Forest	
	Kharif	Rabi	Kharif	Rabi	Kharif	Rabi
2007	335.17	223.43	425.30	216.73	392.95	211.48
2008	202.41	249.25	300.21	275.60	285.61	240.61
2009	171.18	260.07	223.62	313.38	232.62	274.05
2010	246.17	192.08	262.27	204.96	296.67	202.60
2011	252.11	224.13	287.43	310.61	293.91	264.13
2012	299.55	145.89	322.92	152.59	336.30	144.03
2013	847.92	14.57	927.17	8.08	950.14	7.53
2014	631.93	37.14	780.92	34.65	679.50	25.57
2015	729.03	11.41	766.07	4.70	797.48	4.17
2016	571.60	16.07	665.89	12.70	631.52	12.82
2017	813.98	10.74	878.06	7.36	934.89	9.11
Mean	**463.73**	**125.89**	**530.90**	**140.12**	**530.15**	**126.92**
Standard Deviation	**258.85**	**107.64**	**273.33**	**129.78**	**275.56**	**115.35**
Coefficient of variation	**55.82**	**85.50**	**51.48**	**92.62**	**51.98**	**90.89**

Time Series Plot of TRMM Rainfall

Rainfall is a key factor that controls active variations in vegetation development in arid and semi arid regions as vegetation cover in those regions are particularly vulnerable to the quantity, duration and frequency of rainfall (Fravolini *et al.*, 2005; Liu *et al.*, 2012; Gamon *et al.*, 2013; Dutta *et al.*, 2015; Kundu *et al.*, 2015). The total rainfall (mm) derived from Tropical Rainfall Measuring Mission (TRMM) for the study region (2007 to 2017) are given in Table 6 and Figure 8. In general, total rainfall received during Kharif season was higher than Rabi season, however, the rainfall received from 2013 to 2017 during Rabi season was very minimum compared to the other Rabi season years. The rainfall received during the Kharif season of years 2013 to 2017 was abrupt when compared to the normal rainfall pattern (Figure 8). Irrespective of land cover classes, the Saurashtra region showed higher variability in the rainfall pattern in both the seasons. The coefficient of variation during Kharif season varied from 51.48 per cent to 55.82 per cent whereas in Rabi season the CV found to vary between 85.50 per cent to 92.62 per cent (Table 6). In general, the total rainfall received during Kharif season varied from 171.18 mm to 950.14 mm whereas in Rabi season it varied between 4.17 mm and 313.38 mm. The mean total rainfall received during Kharif season found to be higher in deciduous forest (530.90 mm) followed by scrub/degraded forest (530.15 mm) and cropland (463.73 mm). Similar pattern was found to be in Rabi season also. Vegetation condition is highly influenced by the climatic conditions of an area. In order to understand the influence of climatic parameters on the vegetation condition, NDVI and EVI for Kharif, Rabi and annual times were subjected for correlation analysis with satellite derived Land Surface Temperature (LST), Net Radiation and TRMM rainfall parameters. Correlation coefficient values ranges between -1 and +1, where, 0 indicates no linear relationship, -1 denotes the perfect linear negative relationship and

Figure 8. Zonal Averaged Total Rainfall temporal profile under different land use classes for Saurashtra region (2007-2017)

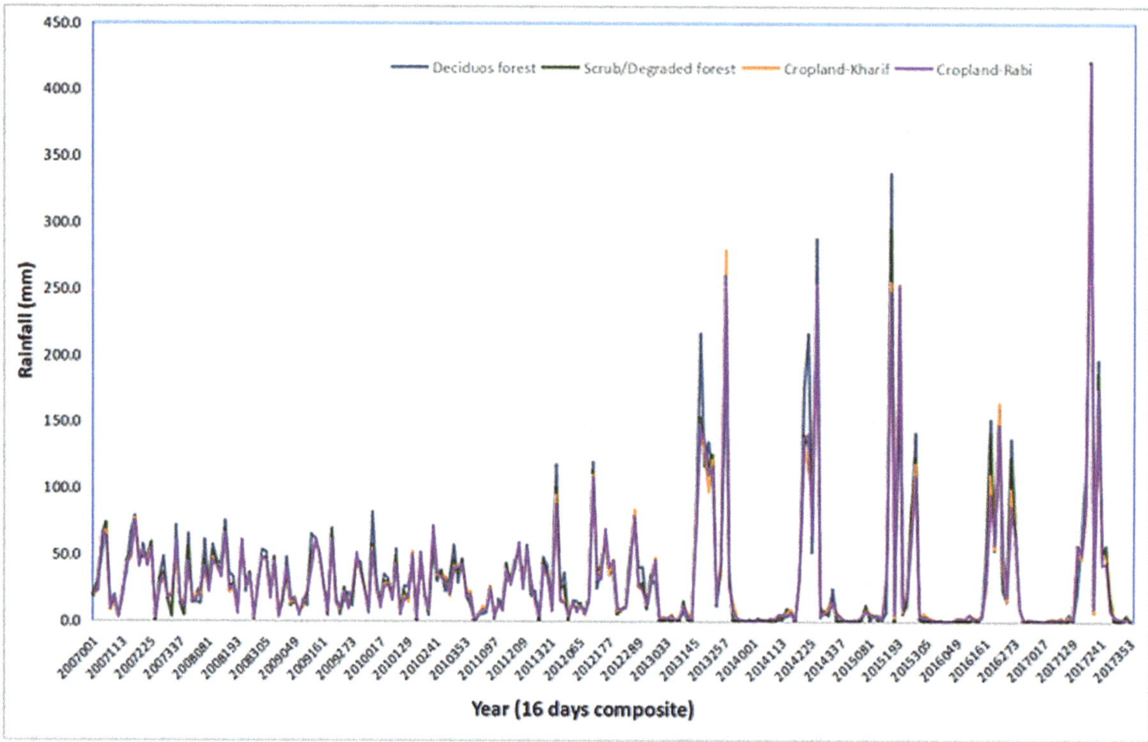

+1 denotes the perfect linear positive relationship with the variables. To have a better understanding on the derived output correlation images, the images are grouped into -1.0 to -0.6, -0.6 to -0.2, -0.2 to 0.2, 0.2 to 0.6 and 0.6 to 1.0 which indicates strong negative correlation, moderate negative correlation, weak negative or positive correlation, moderate positive correlation and strong positive correlation respectively. The results of the pixel-based correlation analysis between NDVI and EVI with LST, net radiation and rainfall data for Saurashtra region is depicted in the following sub-sections.

Pixel-Based Correlation of NDVI and EVI With Land Surface Temperature

Image of pixel wise correlation between NDVI and LST during Kharif, Rabi and annual times for Saurashtra region was generated (Figure 9). In general, correlation coefficient values ranges between -0.84 to 0.65 within the cropland, deciduous forest and scrub/degraded forest area. It has been noted that the correlation between NDVI and LST during Kharif season are coming under weak negative or positive correlation and moderate positive correlation class. Among the land cover classes, it is observed that 66.20 per cent of total pixels in deciduous forest class showed a moderate positive correlation during Kharif season. It could be observed that 12.79 per cent of total pixels in crop land category showed a strong negative correlation during Rabi season. Irrespective of land cover class during Rabi season, most pixels fall in between moderate negative correlation and weak negative or positive correlation class. The annual correlation between NDVI and LST showed that about 75 per cent of pixels in cropland (consid-

Table 7. Pixel by pixel correlation of NDVI with Land surface temperature

Land Use	Correlation Coefficient			Percentage of Pixels Under Correlation Coefficient Class				
	Minimum	Maximum	Mean	-1.0 to -0.6	-0.6 to -0.2	-0.2 to 0.2	0.2 to 0.6	0.6 to 1.0
Kharif Season								
Cropland	-0.27	0.65	0.20	-	0.01	47.48	52.51	-
Deciduous forest	-0.30	0.63	0.25	-	0.03	33.73	66.20	0.04
Scrub/Degraded forest	-0.29	0.61	0.14	-	0.21	70.53	29.25	0.01
Rabi Season								
Cropland	-0.84	0.57	-0.26	12.79	59.77	26.34	1.09	-
Deciduous forest	-0.77	0.50	-0.11	0.13	39.25	54.17	6.45	-
Scrub/Degraded forest	-0.82	0.52	0.005	0.82	12.38	71.86	14.94	-
Annual								
Cropland	-0.39	0.64	0.09	-	0.28	75.30	24.41	0.01
Deciduous forest	-0.45	0.66	0.09	-	1.99	82.36	15.60	0.05
Scrub/Degraded forest	-0.45	0.55	-0.04	-	12.32	84.07	3.62	-

ered all the croplands from Kharif, Rabi and Zaid season), deciduous forest and scrub/degraded forest fall in weak negative or positive correlation category (Table 7). The pixel wise correlation between EVI and LST for Kharif season, Rabi season and annual times are depicted in Figure 9 and land cover wise correlation data are presented in Table 8. It could be observed that more than 80 per cent total pixels under cropland, deciduous forest and scrub/degraded forest during Kharif season are coming under the weak negative or positive correlation category. EVI follows the similar pattern of NDVI during Rabi season, most pixels from all three land cover classes falls in between moderate negative correlation and weak negative or positive correlation class. In cropland category during Rabi season, it is observed that 12.54 per cent of total pixels showed a strong negative correlation (Table 8). The weak to strong negative correlation re-establishing that cropland and forest class decrease surface temperature during Rabi season (Sun and Kafatos 2007 and Ramachandra *et al.,* 2016). It could be observed that 87.75 per cent of total pixels in cropland category for decadal annual correlation have weak negative or positive correlation. Whereas in deciduous forest and scrub/degraded forest, 39.76 and 61.77 per cent of total pixels showed a moderate negative correlation for annual duration respectively. The moderate negative correlation was spatially spread over in Junagadh, Porbandar, Morvi and Surendranagar districts. During Kharif season and annual duration, the weak to moderate positive correlation in all the land cover classes and vegetation indices may have been caused due to high temperature even with dense vegetation which enhances photosynthesis and respiration for plant growth (Mao *et al.,* 2011 and Ramachandra *et al.,* 2016).

Table 8. Pixel by pixel correlation of EVI with Land surface temperature

Land Use	Correlation Coefficient			Percentage of Pixels Under Correlation Coefficient Class				
	Minimum	Maximum	Mean	-1.0 to -0.6	-0.6 to -0.2	-0.2 to 0.2	0.2 to 0.6	0.6 to 1.0
Kharif Season								
Cropland	-0.44	0.53	0.11	-	0.41	81.89	17.70	-
Deciduous forest	-0.45	0.51	0.09	-	1.23	82.95	15.83	-
Scrub/Degraded forest	-0.50	0.46	-0.01	-	8.45	87.26	4.29	-
Rabi Season								
Cropland	-0.83	0.67	-0.22	12.54	53.76	31.39	2.31	-
Deciduous forest	-0.77	0.56	-0.03	0.12	16.40	68.85	14.63	-
Scrub/Degraded forest	-0.83	0.58	0.06	0.87	7.75	68.95	22.44	-
Annual								
Cropland	-0.57	0.46	-0.03	-	8.80	87.75	3.45	-
Deciduous forest	-0.59	0.47	-0.13	-	39.76	59.43	0.81	-
Scrub/Degraded forest	-0.63	0.29	-0.22	-	61.77	37.78	0.44	-

Figure 9. Correlation coefficient class for decadal seasonal and decadal annual vegetation indices with Land surface temperature

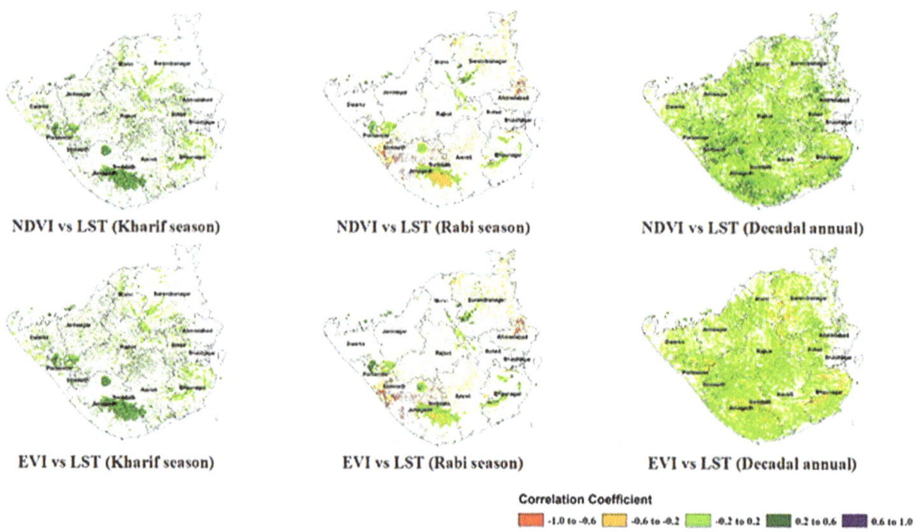

Table 9. Pixel by pixel correlation of NDVI with Net Radiation

Land Use	Correlation Coefficient			Percentage of Pixels Under Correlation Coefficient Class				
	Minimum	Maximum	Mean	-1.0 to -0.6	-0.6 to -0.2	-0.2 to 0.2	0.2 to 0.6	0.6 to 1.0
Kharif Season								
Cropland	-0.67	0.69	-0.40	1.02	77.31	21.65	0.01	-
Deciduous forest	-0.64	0.69	-0.29	0.03	67.73	32.21	0.03	0.01
Scrub/Degraded forest	-0.65	0.69	-0.33	0.09	72.66	27.22	0.02	0.01
Rabi Season								
Cropland	-0.78	0.70	0.01	0.43	15.79	76.29	7.47	0.02
Deciduous forest	-0.72	0.59	0.05	-	6.80	76.44	16.76	-
Scrub/Degraded forest	-0.70	0.47	0.05	0.09	5.97	79.87	14.07	-
Annual								
Cropland	-0.86	0.49	-0.12	1.63	24.35	73.94	0.09	-
Deciduous forest	-0.57	0.69	-0.11	-	18.45	81.53	0.01	0.01
Scrub/Degraded forest	-0.62	0.69	-0.05	-	7.30	92.67	0.02	0.01

Table 10. Pixel by pixel correlation of EVI with Net Radiation

Land Use	Correlation Coefficient			Percentage of Pixels Under Correlation Coefficient Class				
	Minimum	Maximum	Mean	-1.0 to -0.6	-0.6 to -0.2	-0.2 to 0.2	0.2 to 0.6	0.6 to 1.0
Kharif Season								
Cropland	-0.69	0.70	-0.39	1.14	76.80	22.04	0.01	-
Deciduous forest	-0.63	0.70	-0.28	0.05	66.57	33.31	0.07	0.01
Scrub/Degraded forest	-0.65	0.70	-0.30	0.09	69.99	29.86	0.04	0.01
Rabi Season								
Cropland	-0.77	0.70	0.03	0.25	13.40	76.13	10.19	0.02
Deciduous forest	-0.71	0.63	0.10	-	7.11	47.86	45.02	-
Scrub/Degraded forest	-0.70	0.55	0.07	0.08	6.94	66.14	26.84	-
Annual								
Cropland	-0.81	0.49	-0.07	0.72	15.60	83.44	0.24	-
Deciduous forest	-0.46	0.70	-0.02	-	1.14	98.77	0.08	0.01
Scrub/Degraded forest	-0.53	0.70	0.00	-	1.25	97.47	1.27	0.01

Figure 10. Correlation coefficient class for decadal seasonal and decadal annual vegetation indices with Net Radiation

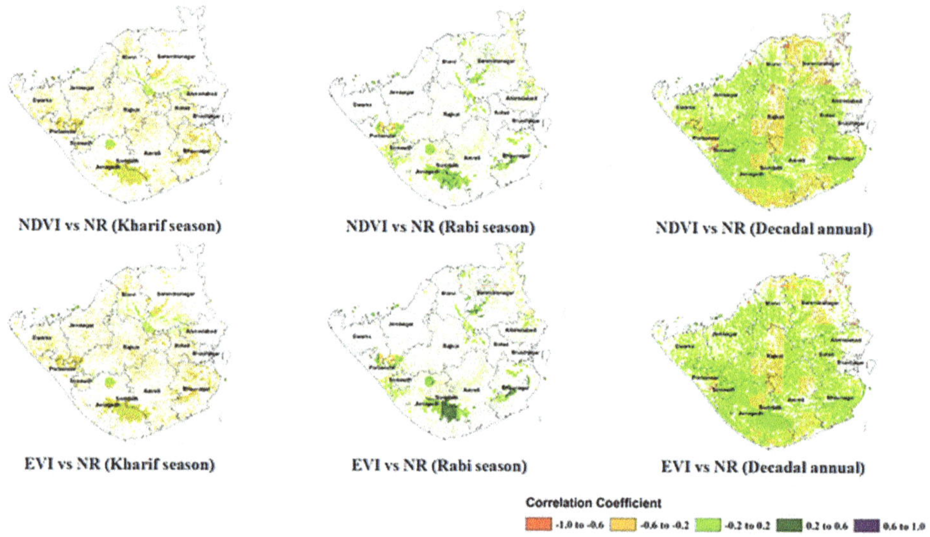

Pixel-Based Correlation of NDVI and EVI with Net Radiation

The correlation image between NDVI and Net radiation for Kharif, Rabi and annual times for the study region are presented in Figure 10. Irrespective of season, the correlation coefficient between NDVI and net radiation values ranges between -0.86 to 0.70 in cropland, deciduous forest and scrub/degraded forest areas (Table 9). It is observed that about 70 per cent of pixels in cropland, deciduous forest and scrub/degraded forest during Kharif season reported to have a moderate negative correlation. In Rabi season and annual duration, the most of the pixels observed to have weak negative or positive correlation and the pixel per cent varies between 73.94 to 92.67. During Rabi season, 16.76 per cent in deciduous forest and 14.07 per cent of total pixels in scrub/degraded forest falls in moderate positive correlation class (Table 9). The pixel wise correlation between EVI and net radiation for Kharif season, Rabi season and annual duration are depicted in Figure 9. In general, the correlation coefficient values ranges between -0.81 and 0.70 in the land cover areas of cropland, deciduous forest and scrub/degraded forest (Table 10). Similar to NDVI, it could be observed that about 70 per cent of total pixels in all the land cover classes during Kharif season has moderate negative correlation. In Rabi season, the weak negative or positive correlation and moderately positive correlation class occupies major percentage of area (Table 10). In Kharif season, outgoing radiation is higher than the incoming radiation, hence the vegetation indices are negatively correlated with vegetation growth whereas in Rabi season, outgoing radiation is very less where we could observe the weak to moderate positive correlation (Wu *et al.,* 2014). The correlation coefficient values for annual duration clearly depicts the weak negative or positive correlation class and the value ranges between 83.44 per cent and 98.77 per cent of the total area.

Table 11. Pixel by pixel correlation of NDVI with TRMM rainfall

Land Use	Correlation Coefficient			Percentage of Pixels Under Correlation Coefficient Class				
	Minimum	Maximum	Mean	-1.0 to -0.6	-0.6 to -0.2	-0.2 to 0.2	0.2 to 0.6	0.6 to 1.0
Kharif Season								
Cropland	-0.45	0.62	-0.11	-	16.10	82.07	1.83	-
Deciduous forest	-0.41	0.59	-0.07	-	8.93	86.48	4.59	-
Scrub/Degraded forest	-0.41	0.57	-0.09	-	13.00	84.42	2.58	-
Rabi Season								
Cropland	-0.67	0.56	0.03	0.06	17.28	78.58	4.08	-
Deciduous forest	-0.46	0.40	0.07	-	0.30	97.12	2.58	-
Scrub/Degraded forest	-0.52	0.58	0.04	-	2.60	91.07	6.32	-
Annual								
Cropland	-0.46	0.45	-0.06	-	8.77	91.17	0.06	-
Deciduous forest	-0.27	0.19	-0.01	-	0.31	99.69	-	-
Scrub/Degraded forest	-0.27	0.38	0.02	-	0.20	99.53	0.27	-

Table 12. Pixel by pixel correlation of EVI with TRMM rainfall

Land Use	Correlation Coefficient			Percentage of Pixels Under Correlation Coefficient Class				
	Minimum	Maximum	Mean	-1.0 to -0.6	-0.6 to -0.2	-0.2 to 0.2	0.2 to 0.6	0.6 to 1.0
Kharif Season								
Cropland	-0.47	0.61	-0.11	-	16.94	81.21	1.85	-
Deciduous forest	-0.47	0.60	-0.08	-	13.95	81.36	4.69	-
Scrub/Degraded forest	-0.47	0.57	-0.08	-	11.30	86.12	2.58	-
Rabi Season								
Cropland	-0.65	0.56	0.03	0.06	18.09	77.53	4.32	-
Deciduous forest	-0.49	0.40	0.09	-	0.36	94.07	5.57	-
Scrub/Degraded forest	-0.56	0.53	0.04	-	2.71	90.25	7.04	-
Annual								
Cropland	-0.44	0.40	-0.02	-	5.61	94.06	0.33	-
Deciduous forest	-0.20	0.24	0.07	-	-	99.50	0.50	-
Scrub/Degraded forest	-0.20	0.35	0.08	-	-	95.60	4.39	-

Figure 11. Correlation coefficient class for decadal seasonal and decadal annual vegetation indices with TRMM Rainfall

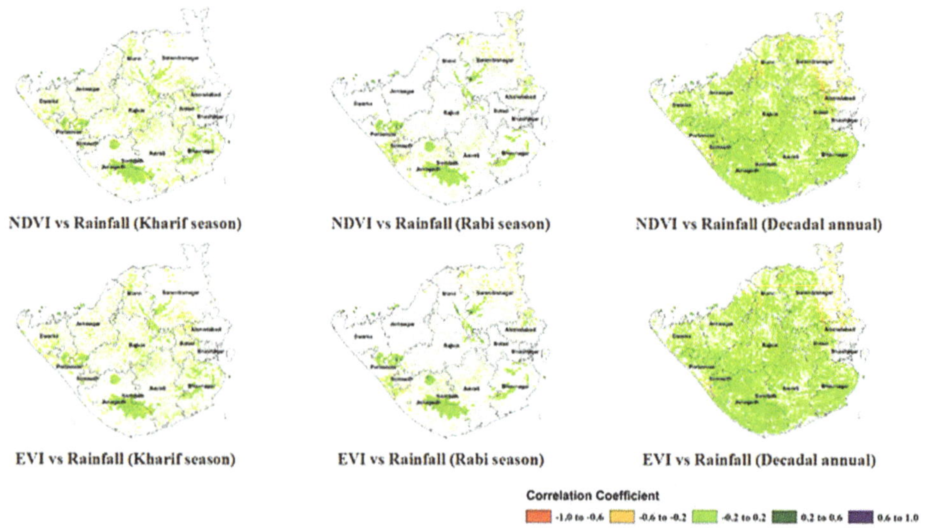

Pixel-Based Correlation of NDVI and EVI With TRMM Rainfall

The pixel-based correlation between NDVI and EVI with TRMM rainfall was generated for the Saurashtra region and presented in Figure 11. Irrespective of land cover classes and season, the correlation between NDVI and rainfall varies between -0.67 to 0.62. A weak negative or positive correlation (-0.2 to 0.2) is observed in all the land cover class and seasons (Table 11). The Saurashtra region is characterized by dry hot climate with shallow fine textured soils, in despite of high rainfall during Kharif season, the weak correlation exists between NDVI and rainfall, which might be due to high surface run-off (Kundu *et al.,* 2018). It is observed that 16.10 per cent of total pixels during Kharif season and 17.28 per cent of total pixels during Rabi season under cropland class has moderate negative correlation. This shows that the increase in rainfall leads to decline in vegetation health which might be due to abrupt change in rainfall pattern during Kharif and Rabi seasons (Table 6 and Figure 8). Several authors reported similar results (Wang *et al.,* 2003 and Mingjun *et al.,* 2007). In general, the correlation coefficient values between EVI and TRMM rainfall ranges between -0.65 and 0.61 in the land cover areas of cropland, deciduous forest and scrub/degraded forest (Table 12). Irrespective of land cover class and season, more than 80 per cent pixels reported weak negative or positive correlation. It clearly indicates that the rainfall is not a key factor for vegetation growth in Saurashtra region. Mingjun *et al.* (2007) reported that precipitation has weak correlation with NDVI for some parts of Tibetan Plateau. The Saurashtra region receives very less rainfall when compared to the northern parts of Gujarat. The annual correlation between NDVI and rainfall showed a weak negative or positive correlation, more than 95 per cent of total pixels in all the land cover types falls in that category. This might be due to the poor rainfall or varied rainfall pattern which doesn't support the vegetation growth. Similar results were also obtained by Revadekar *et al.* (2011) for some parts of Indian region. From this, it could be inferred that, the correlation analysis shows significant impact of inter-annual variability of rainfall on the NDVI and EVI. Similar conclusions were drawn by Revadekar *et al.* (2012). The weak negative or positive correlation exists for both seasonal and annual duration which might also be due to the well irrigation structures established in the study region or the type and method of crops cultivated or low rainfall or the study region covered by coastal belt which leads to variability in climate or due to the shallow sandy/ clayey problematic soils. In Kharif season, irrigated rice is dominated than other rainfed crops whereas in Rabi season irrigated wheat is dominated than the rainfed gram. This shows that stable establishment of crop through continuous irrigation which could be the reason for weak negative or positive correlation between rainfall and NDVI/EVI (Chuai *et al.,* 2013). The reason for weak correlation between vegetation indices and rainfall may be due to scanty rainfall during winter (Rabi) season. Many other researchers have also reported the variability of vegetation and its relation to climatological data over the Indian region (Saikia, 2009 and Prasad *et al.,* 2008)

CONCLUSION

Decadal Seasonal and Annual Trend

The results of decadal seasonal and annual trend analysis for NDVI and EVI under various land cover classes showed both negative and positive trends. Irrespective of land cover classes, NDVI and EVI during Kharif season showed a greening (positive) trend in vegetation growth whereas during Rabi season it is observed to have browning (negative) trend. The positive greening trend during the Kharif season might be due to the increased vegetation growth and productivity during the period. The declining vegetation (browning) trend in Rabi season might be due to changes in a crop choice or disruption of cropping lands. Decadal annual NDVI and EVI witnessed positive greening trend except in some parts of Rajkot, Dwarka, Porbandar, Amreli and Jamnagar districts, where it is browning in nature.

Correlation Between Vegetation Indices and Climate Factors

The mean LST value during Kharif season found to be higher for scrub/degraded forest (34.84 °C) followed by cropland (34.67 °C) and deciduous forest (31.75 °C) whereas during Rabi season the higher mean LST found to be higher in cropland (33.75 °C) followed by scrub/degraded forest (33.51 °C) and deciduous forest (31.40 °C). Irrespective of land cover classes and season, the coefficient of variation found to less than 5 per cent, which indicates low inter-annual variation in LST value during the study period. During Kharif season, the mean positive net radiation values found to be higher in deciduous forest (60.48 W/m^2) followed by scrub/degraded forest (58.49 W/m^2) and cropland (58.10 W/m^2). Similar pattern was found during Rabi season but in the form of negative net radiation. It can be concluded that during Kharif season more of outgoing radiation is observed than incoming radiation while in Rabi season it was *vice–versa.* In general, total rainfall received during Kharif season was higher than Rabi season, however, the rainfall received from 2013 to 2017 showed unexpected increase during Kharif season and very minimum rainfall during Rabi season. The total rainfall received during Kharif season varied from 171.18 mm to 950.14 mm whereas in Rabi season it varied between 4.17 mm and 313.38 mm in the study area. The mean total rainfall received during Kharif season found to be higher in deciduous forest (530.90 mm) followed by scrub/degraded forest (530.15 mm) and cropland (463.73 mm). Similar pattern was found to be in Rabi season also. During decadal Kharif and annual times, weak to moderate positive correlation was noticed between LST and vegetation indices whereas during Rabi season it showed strong to weak negative correlation. From the correlation study between vegetation indices and net radiation, it could be observed that about 70 per cent of total pixels in all the land cover classes during Kharif season has moderate negative correlation. In Rabi season, the weak negative or positive correlation and moderately positive correlation classes occupies the major percentage of area. The correlation study between TRMM rainfall and vegetation indices showed a weak negative or positive correlation for both the seasons and annual times which indicates that the rainfall is not a key factor for the vegetation growth in Saurashtra region. In general, the rainfall and vegetation indices are reported to have strong positive relationship but some authors have also reported negative correlation between them. The weak negative or positive correlation between rainfall and vegetation indices might be due to the well irrigation structures established in the Saurashtra region or the type and method of crops cultivated. The low rainfall of the study region covered by coastal belt which leads to variability in climate or due to the shallow sandy / clayey problematic soils or may be due to the domination of irrigated crops which has stable vegetation coverage.

ACKNOWLEDGMENT

I deem it great pleasure to place on record my deep sense of gratitude and profound thanks to Dr. R. Kumaraperumal, Dr. Balaji Kannan, Mr. Hiren Bhatt, Dr S P Vyas, Dr. Markand Oja, Sashikant sharma, Dr. Palazhanivelan and Dr. Jagadeeshwaran for meticulous guidance, suggestion, valuable counseling and constant encouraging throughout the course of the investigation. The dataset of Land use/Land cover used for this research was received from Space Applications Centre, ISRO, Ahmedabad. This research received no specific grant from any funding agency in the public, commercial, or not-for-profit sectors.

REFERENCES

Ahmed, B. M., Hata, T., Tanakamarua, H., Abdelhadi, A., & Tadac, A. (2002). The spatial analysis of surface temperature and evapotranspiration for some land use/ cover types in the Gezira Area, Sudan. Proceedings of Joint Symposium URBAN – VRS 2005 held at Temple, Arizona, March 14-16, 2005. (Eds.) M. Moeller, & E. Wenth, ISPRS Archieves, Volume XXXVI- 8/W27, p. 5.

Amarnath, G., Babar, S., & Murthy, M. S. R. (2017). Evaluating MODIS-vegetation continuous field products to assess tree cover change and forest fragmentation in India–A multi-scale satellite remote sensing approach. *The Egyptian Journal of Remote Sensing and Space Sciences*, *20*(2), 157–168. doi:10.1016/j.ejrs.2017.05.004

Chuai, X., Huang, X., Wang, W., & Bao, G. (2013). NDVI, temperature and precipitation changes and their relationships with different vegetation types during 1998–2007 in Inner Mongolia, China. *International Journal of Climatology*, *33*(7), 1696–1706. doi:10.1002/joc.3543

Dhanapriya, M., Kumaraperumal, R., Kannan, B., & Hiren Bhatt, P. (2018). Spatio-temporal analysis of vegetation dynamics for Saurashtra region of Gujarat. AgricINTERNATIONAL, 5(2), 43-56.

Dubovyk, O., Menz, G., & Khamzina, A. (2012). Trend analysis of MODIS time-series using different vegetation indices for monitoring of cropland degradation and abandonment in Central Asia. *Proceedings of IEEE International Geoscience and Remote Sensing Symposium (IGARSS)*, Munich, Germany, July 22-27, 2012, pp. 6589-6592. 10.1109/IGARSS.2012.6352089

Dutta, D., Kundu, A., Patel, N. R., Saha, S. K., & Siddiqui, S. R. (2015). Assessment of agricultural drought in Rajasthan (India) using remote sensing derived Vegetation Condition Index (VCI) and Standardized Precipitation Index (SPI). *The Egyptian Journal of Remote Sensing and Space Sciences*, *18*(1), 53–63. doi:10.1016/j.ejrs.2015.03.006

Eckert, S., Hüsler, F., Liniger, H., & Hodel, E. (2015). Trend analysis of MODIS NDVI time-series for detecting land degradation and regeneration in Mongolia. *Journal of Arid Environments*, *113*, 16–28. doi:10.1016/j.jaridenv.2014.09.001

Fravolini, A., Hultine, K. R., Brugnoli, E., Gazal, R., English, N. B., & Williams, D. G. (2005). Precipitation pulse use by an invasive woody legume: The role of soil texture and pulse size. *Oecologia*, *144*(4), 618–627. doi:10.100700442-005-0078-4 PMID:15891829

Gamon, J. A., Huemmrich, K. F., Stone, R. S., & Tweedie, C. E. (2013). Spatial and temporal variation in primary productivity (NDVI) of coastal Alaskan tundra: Decreased vegetation growth following earlier snowmelt. *Remote Sensing of Environment*, *129*, 144–153. doi:10.1016/j.rse.2012.10.030

Herrmann, S. M., & Anyamba, A. (2005). Exploring relationship between rainfall and vegetation dynamics in the Sahel using coarse resolution satellite data. *Proceedings of the 31st International Symposium on Remote Sensing of Environment*, 118.

Kundu, A., Denis, D., Patel, N. D., & Dutta, D. (2018). A Geo-spatial study for analysing temporal responses of NDVI to rainfall. *Singapore Journal of Tropical Geography*, *39*(1), 107–116. doi:10.1111jtg.12217

Kundu, A., Patel, N., Saha, S., & Patel, N. D. (2015). Monitoring the extent of desertification processes in western Rajasthan (India) using geo-information science. *Arabian Journal of Geosciences*, *8*(8), 5727–5737. doi:10.100712517-014-1645-y

Liu, B., Zhao, W., & Wen, Z. (2012). Photosynthetic response of two shrubs to rainfall pulses in desert regions of northwestern China. *Photosynthetica*, *50*(1), 109–119. doi:10.100711099-012-0015-9

Mao, D., Wang, Z., Luo, L., & Ren, C. (2011). Integrating AVHRR and MODIS data to monitor NDVI changes and their relationships with climatic parameters in Northeast China. *International Journal of Applied Earth Observation and Geoinformation, 18*(2012), 528-536.

Mingjun, D., Yili, Z., Linshan, L., Wei, Z., Zhaofeng, W., & Wanqi, B. (2007). The relationship between NDVI and precipitation on the Tibetan Plateau. *Journal of Geographical Sciences*, *17*(3), 259–268. doi:10.100711442-007-0259-7

Nemani, R. R., Keeling, C. D., Hashimoto, H., Jolly, W. M., Piper, S. C., Tucker, C. J., ... Running, S. (2003). Climate-driven increases in global terrestrial net primary production from 1982 to 1999. *Science*, *300*(5625), 1560–1563. doi:10.1126cience.1082750 PMID:12791990

Papagiannopoulou, C., Miralles, D. G., Decubber, S., Demuzere, M., Verhoest, N. E., Dorigo, W. A., & Waegeman, W. (2017). A non-linear Granger-causality framework to investigate climate-vegetation dynamics. *Geoscientific Model Development, 10*(5), 1945–1960. doi:10.5194/gmd-10-1945-2017

Peng, D., Huang, J., & Wang, X. (2007). Correlation analysis of regional vegetation seasonal fluctuation and climate factors based on MODIS-EVI. *Journal of Applied Ecology, 18*(5), 983–989. PMID:17650845

Prasad, V. K., Badarinath, K. S., & Eaturu, A. (2008). Effects of precipitation, temperature and topographic parameters on evergreen vegetation greenery in the Western Ghats, India. *International Journal of Climatology, 28*(13), 1807–1819. doi:10.1002/joc.1662

Ramachandra, T. V., Kumar, U., & Dasgupta, A. (2016). *Analysis of land surface temperature and rainfall with landscape dynamics in Western Ghats, India. Technical report* (p. 181). Bangalore, India: Centre for Ecological Sciences, Indian Institute of Science.

Revadekar, J. V., Patwardhan, S. K., & Kumar, K. (2011). Characteristic features of precipitation extremes over India in the warming scenarios. *Advances in Meteorology, 2011*, 1–11. doi:10.1155/2011/138425

Revadekar, J. V., Tiwari, Y. K., & Kumar, K. (2012). Impact of climate variability on NDVI over the Indian region during 1981–2010. *International Journal of Remote Sensing, 33*(22), 7132–7150. doi:10.1080/01431161.2012.697642

Roberts, D., Green, R., & Adams, J. (1997). Temporal and spatial patterns in vegetation and atmospheric properties from AVIRIS. *Remote Sensing of Environment, 62*(3), 223–240. doi:10.1016/S0034-4257(97)00092-8

Saikia, A. (2009). NDVI variability in North East India. *Scottish Geographical Journal, 125*(2), 195–213. doi:10.1080/14702540903071113

Slayback, D. A., Pinzon, J. E., Los, S. O., & Tucker, C. J. (2003). Northern hemisphere photosynthetic trends 1982–1999. *Global Change Biology, 9*(1), 1–15. doi:10.1046/j.1365-2486.2003.00507.x

Sonnenschein, R., Kuemmerle, T., Udelhoven, T., Stellmes, M., & Hostert, P. (2011). Differences in Landsat-based trend analyses in drylands due to the choice of vegetation estimate. *Remote Sensing of Environment, 115*(6), 1408–1420. doi:10.1016/j.rse.2011.01.021

Sun, D., & Kafatos, M. (2007). Note on the NDVI-LST relationship and the use of temperature-related drought indices over North America. *Geophysical Research Letters, 34*(24), 4. doi:10.1029/2007GL031485

Teferi, E., Uhlenbrook, S., & Bewket, W. (2015). Inter-annual and seasonal trends of vegetation condition in the Upper Blue Nile (Abay) Basin: Dual-scale time-series analysis. *Earth System Dynamics, 6*(2), 617–636. doi:10.5194/esd-6-617-2015

Wang, J., Rich, P. M., & Price, K. (2003). Temporal responses of NDVI to precipitation and temperature in the central Great Plains, USA. *International Journal of Remote Sensing*, *24*(11), 2345–2364. doi:10.1080/01431160210154812

Wu, D., Zhao, X., Zhao, W., Tang, B., & Xu, W. (2014). Response of vegetation to temperature, precipitation and solar radiation time scales: A case study over mainland Australia. *Proceedings of 2014 IEEE International Geoscience and Remote Sensing Symposium (IGARSS)*, Quebec city, QC. pp. 855-858.

Yagoub, Y. E., Li, Z., Musa, O. S., Anjum, M. N., Wang, F., Xu, C., & Bo, Z. (2017). Investigation of vegetation cover change in Sudan by using MODIS data. *Journal of Geographic Information System*, *9*(3), 279–292. doi:10.4236/jgis.2017.93017

Zhang, Y., Song, C., Band, L. E., Sun, G., & Li, J. (2017). Reanalysis of global terrestrial vegetation trends from MODIS products: Browning or Greening. *Remote Sensing of Environment*, *191*, 145–155. doi:10.1016/j.rse.2016.12.018

Zhang, Y., Zhang, X., & Liu, S. (2011). Correlation analysis on Normalized Difference Vegetation Index (NDVI) of different vegetation and climatic factors in Southwest China. *Journal of Applied Ecology*, *22*(2), 323–330. PMID:21608242

Zhou, L., Tucker, C. J., Kaufmann, R. K., Slayback, D., Shabanov, N. V., & Myneni, R. (2001). Variations in northern vegetation activity inferred from satellite data of vegetation index during 1981 to 1999. *Journal of Geophysical Research, D, Atmospheres*, *106*(17), 20069–20083. doi:10.1029/2000JD000115

Chapter 11
Classification of Territory on Forest Fire Danger Level Using GIS and Remote Sensing

Elena Petrovna Yankovich
https://orcid.org/0000-0001-5873-8444
Tomsk Polytechnic University, Russia

Ksenia S. Yankovich
University of Paris 7 (Paris Diderot University), France

ABSTRACT

The vegetation cover is the most important factor in forest fires, because it reflects the presence of forest fuels. The study of the variability of the vegetation cover, as well as observation of its condition, allows estimating the level of fire danger of the forest quarter. The work presents a geo-information system containing a set of tools to determine the level of fire danger of the forest quarter. The system is able to predict (determine the probability) and classify forest quarters according to the level of fire danger. The assessment of forest fire danger of Tomsk forestry of Tomsk region has been carried out. Fire probability maps of forest quarters were created based on remote sensing data and ArcGIS software.

INTRODUCTION

The assessment of the forest fire danger probability is of great importance in creating an effective forest fire management system. Identifying forest areas that are highly likely to cause burning of forest fuel is an important component of fire management planning. The development of geo-information technologies greatly facilitates this process, allowing you to create maps and analyze the factors causing the occurrence of fires in forested areas.

In this work, geographic information systems (GIS) technologies are used in combination with remote sensing data to solve problems of forest fire management. The use of satellite imagery materials is more applicable to remote monitoring of forests, as well as updating land-based forest inventory data.

DOI: 10.4018/978-1-7998-1867-0.ch011

Medium resolution remote sensing data remain essential for many tasks. The obvious advantage of medium spatial resolution surveys is their spectral characteristics - the ability to obtain data not only in the visible and near, but also in the middle and far infrared parts of the spectrum. An important circumstance is also the fact that accumulated long-term archival materials of remote sensing.

Free-access materials from multispectral space surveys of the Landsat 7 ETM + system were used in the work. In this work, various geo-information tools were used to combine the parameters necessary for calculating the level of forest fire danger and creating maps of fire danger of forest quarters. Processing input data, calculation of intermediate and final results is carried out in the ArcGIS environment. Forest fire danger assessment was carried out for Tomsk forestry of Tomsk region.

BACKGROUND

The ability to determine the place where a fire is probably to occur is a prerequisite for planning fire management. The vegetation cover is the most important factor in forest fires, because it reflects the presence of forest fuels (Roy, 2003). The use of remote sensing data in the classification and mapping of vegetation becomes the main method of fuel assessment. Existing methodologies for determining the type of vegetation include the traditional classification with and without training (Hansen et al., 2000 ; Churches et al., 2014), the method based on phenology (Yan et al., 2015), and the object classification (Mitchell et al., 2016).

Vegetation type maps are necessary for the spatial calculation of fire danger and for estimation of fire risks by using them in models that simulate the growth and intensity of a fire in a landscape. Forest fuel maps are used in various widespread systems for the forest fire danger forecasting. For example, such models are used in the North American models National Fire Danger Rating System (NFDRS), Fire behavior (BEHAVE), Fire Area Simulator (FARSITE), National Fire Management Analysis System (NFMAS) (Deeming et al., 1978 ; Finney, 1998 ; Lundgren et al., 1995 ; Arroyo et al., 2008). The McArthur Forest Fire Danger Rating System and the McArthur Grassland Fire Danger Rating System (McArthur, 1967) are widely used in Australia. The Canadian Forest Fire Danger Rating System is used in Canada, which consists of two main subsystems: the Fire Weather Index (FWI) and Fire Behavior Prediction System (Arroyo et al., 2008; van Wagner, 1987). The Forest Fire Satellite Monitoring Information System of Russian Federal Forestry Agency (SMIS-Rosleshoz) is used in the Russian Federation and is based on the Nesterov index. The ground-based observations, aerial photography, modeling and remote sensing data (Arroyo et al., 2008; Lasaponara and Lanorte, 2007; Keane et al., 2001; Wang et al., 2004; Ustin et al., 2004) are used for mapping. In general, data on the greenness of vegetation, meteorological data, data on wetting the surface and moisture content of forest fuel are necessary to forecast and to monitor forest fires (Kononov and Ka, 2008).

In addition, numerous studies showed the feasibility of geographic information system and remote sensing data (Chowdhury and Hassan, 2015; Hernandez-Leal et al., 2008). The GIS is a widely used tool for processing spatial data and displaying results. A model of a potential forest fire using satellite data and the GIS was developed in China to identify areas with a high probability of forest fire (Butt et al., 2015). In (Sheriza et al., 2010) five fire danger classifications were identified in developed forest fuel maps and the degree of fire danger in peat-bog forests was assessed. Vegetation types in the study area were analyzed using digital classification systems, namely, two vegetation indices: an extended vegetation index (AVI) and a Tasseled Cap (TC) conversion.

MATERIALS AND METHODS

According to the structure of the Forestry of the Russian Federation, the minimum forest taxation unit is a site. In general, local forestry consists of quarters, and each quarter consists of sites. As a rule, several local forestry are combined into forestry.

It is convenient to use a quarter as a unit of mapping for spatial analysis of a territory, since its borders and location are precisely known. Information about the composition of the vegetation is stored in the books of forest descriptions of and tied to the stands. Therefore, in order to divide quarters by fire risk levels based on the species composition of the forest area and the age of the stand, it is necessary to determine the number of fire hazardous sites in each quarter.

Tomsk forestry of the Tomsk region is located in the central part of the Tomsk region in the territory of the Tomsk municipal district. The territory of forestry is located in the taiga zone, the West-Siberian southern taiga lowland area. According to official data, the total forest area is 53 611 ha.

Free-access materials from multispectral satellite imagery of the Landsat ETM + system (spatial resolution 30 m, 7 spectral ranges from visible to thermal) were used in the work. This resolution allows you to identify areas of 0.2 ha.

The images available for download are a set of files for each spectral range in TIFF format, which requires preliminary preparation. To obtain a single multispectral image, it is necessary to merge the individual layers into one file. This procedure is performed in an ERDAS environment using the Image Interpreter module (Layer Stack operation).

Prepared rasters must be classified. Classification - the process of sorting (distributing by classes) of image elements (pixels) into a finite number of classes based on the values of their attributes (DN - digital numbers). If a pixel satisfies a certain classification condition, it belongs to a certain class that corresponds to this condition. In this work, two methods of classification were considered: uncontrolled and classification with training.

The use of uncontrolled classification is advisable if the source data belongs to known classes. During the processing of the input raster, the program independently identifies clusters of similar data, combining them into classes.

Classification by standards allows the program to be "trained" by identifying the region and its properties in manual mode. The created standards are stored in the corresponding data set. Usually they are determined during field surveys, noting the coordinates of objects of a particular class. The program processes the source raster, assigning the area containing the reference pixels to the parameters of the corresponding class. After creating a sufficient number of standards 25-35, classification is carried out using the formed set. Next is the processing of raster information by the method of principal components. Then the raster that passed the classification must be converted to a vector view. Vectorization was performed in the ArcGIS software environment (Raster to Polygon operation).

In addition to the materials of space imagery, the work uses:

- maps for the Tomsk forest area
- regulations of Tomsk forestry
- data from the state forest registry

As a common research platform, ArcGIS software was used, which includes an integrated set of GIS applications. For storing data in GIS, a geodatabase format was chosen. The choice of this model was determined by the tasks to be solved.

The algorithms of the geographical information system for forest fire danger quantitative assessment are implemented using the Python language built into ArcGIS.

The main task of the GIS is to quantify the probability of a forest fire from forest descriptions of forest areas, statistical information using remote sensing data. Criteria for assessing forest fire danger are determined on the basis of the theory of probability (Kuznetsov and Baranovskiy, 2013).

System components:

- Information entry and storage subsystem - file geodatabase, which includes feature classes, satellite imagery, stand-alone tables.
- Visualization and editing subsystem.
- Analysis subsystem.

Input data:

- Polygonal feature class containing information about the territorial location of forest quarters;
- Tables in MS Excel format, with forest descriptions of forest stands;
- Layer template file.

The output is an autonomous table with an estimate of the probability of fire danger of the forest quarter, maps illustrating the levels of fire danger.

GIS system allows you to:

- Determine the fire hazard of a forest area by forest-based descriptions.
- Conduct a quantitative assessment of the forest fire danger of the quarter due to the action of focused solar radiation (as well as anthropogenic load and thunderstorm activity).
- Classify forest tracts according to the level of fire danger under conditions of exposure to forest fuel combustible material of concentrated solar energy flows.
- Create thematic maps on the levels of forest fire danger of the quarter.

RESULTS AND DISCUSSION

The resulting images are a set of files for each spectral range in TIFF format. They were pre-prepared. Individual files (layers) were combined into a single file to obtain a multispectral image. Multichannel raster was used for classification (figure 1).

The combination of channels 1, 4 and 8 was chosen for visual interpretation. This made it possible to better distinguish different types of surface.

Two types of classification were used with and without training. The use of autonomous classification is advisable if the source data belongs to known classes (*Figure 2*). During the processing of the input raster, the program independently identifies clusters of similar data, combining them into classes.

Figure 1. The combination of channels in natural colors, cropped by area of interest

Figure 2. The result of the unsupervised classification (20 classes)

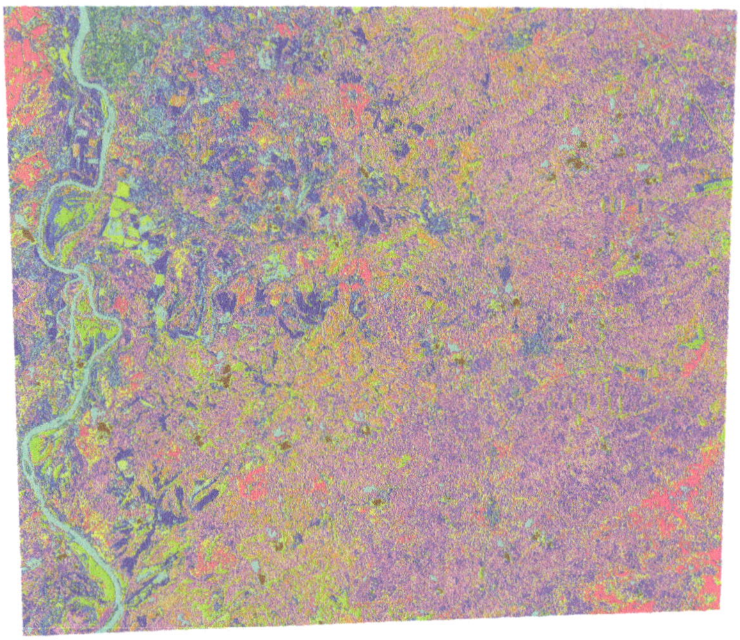

Figure 3. The result of the supervised classification

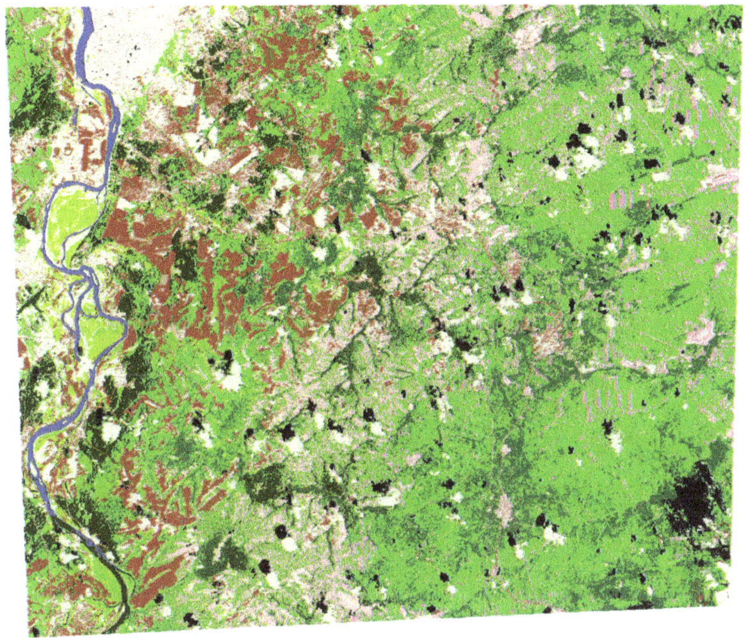

The next type of classification is a classification by standards. This type of classification allows the program to "learn" by identifying the region and its properties in manual mode. The created standards are stored in the corresponding data set. Usually they are determined during field surveys, noting the coordinates of objects of a particular class. The program processes the source raster, assigning the area containing the reference pixels to the parameters of the corresponding class.

To perform the classification, it is necessary to create a database of standards. The task was carried out using the maps of the State Forest Inventory 2013, containing information on the location of individual parts of forestry and divided into areas by tree species. Since map files do not contain coordinates, you must first bind. The most appropriate method of binding in this case is the binding by characteristic points (turning points of the border of the region, the riverbed, and roads). ArcGIS software was used to complete this task.

The set of standards is created based on ERDAS IMAGINE software using the standards editor (Signature Editor). The program provides the possibility of improving the image for perception using a set of operations. To improve the base of standards, you can use various checks provided by the program.

After creating a sufficient number of standards, a classification is carried out using the generated set. The number of classes corresponds to the number of classes created when selecting standards (Figure 3).

The use of the vegetation index did not show significant changes in the classification results. And the pretreatment using the principal component method allowed us to obtain more accurate results. The rasters vectorized using ArcGIS tools were processed to determine the level of fire danger.

As a result of raster processing, intermediate characteristics of classes of 7 components were obtained. The first component made it possible to confidently divide forest areas by species (pine, cedar, birch, aspen, fir, flood plain vegetation). The second component includes settlements and agricultural land.

Figure 4. Categories of land in the forestry and adjacent area

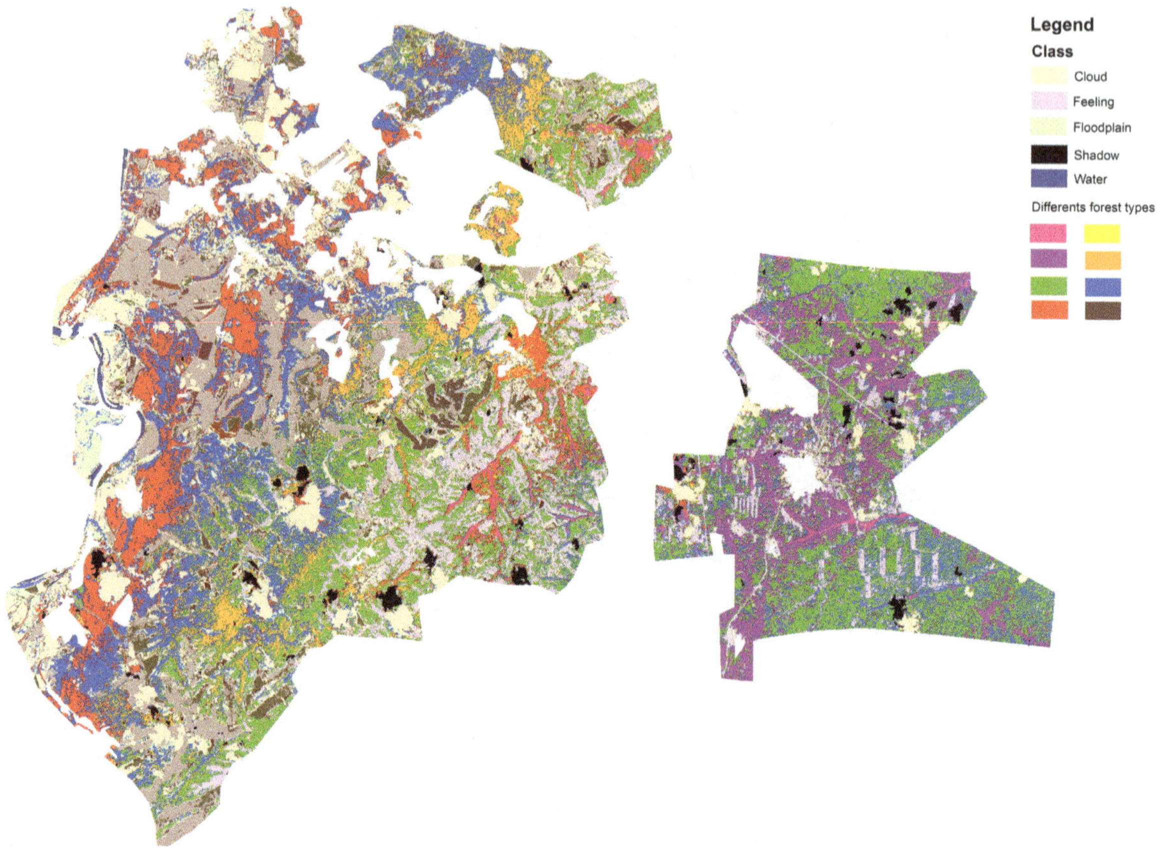

The classified raster is converted into a vector species and the square of clearings, groves, various forest plantations (coniferous and deciduous trees) are determined.

With the help of the developed tools, the system classified the quarters according to the level of fire danger.

CONCLUSION

Based on the developed model, which includes additional tools for assessing the probability of forest fire danger due to focused solar radiation, thunderstorm activity and anthropogenic load, maps of probability of forest fire danger are created. The system has a user-friendly interface and allows the user to change parameters as needed when determining the fire danger of a forest quarter.

The system can be used by the relevant authorities in planning the control of forest fires and minimizing their consequences, as it allows assessing the probability of a forest fire and mapping the most likely forest areas that may be affected by a fire.

ACKNOWLEDGMENT

This work was implemented with the financial support of the "Russian Foundation for Basic Research". [Project N 17-29-05093].

REFERENCES

Arroyo, L. A., Pascual, C., & Manzanera, J. A. (2008). Fire models and methods to map fuel types: The role of remote sensing. *Forest Ecology and Management*, *256*(6), 1239–1252. doi:10.1016/j.foreco.2008.06.048

Butt, A., Shabbir, R., Ahmad, S. S., & Aziz, N. (2015). Land use change mapping and analysis using Remote Sensing and GIS: A case study of Simly watershed, Islamabad, Pakistan. *The Egyptian Journal of Remote Sensing and Space Sciences*, *18*(2), 251–259. doi:10.1016/j.ejrs.2015.07.003

Chowdhury, E. H., & Hassan, Q. K. (2015). Operational perspective of remote sensing-based forest fire danger forecasting systems. *ISPRS Journal of Photogrammetry and Remote Sensing*, *104*, 224–236. doi:10.1016/j.isprsjprs.2014.03.011

Churches, C., Wampler, P., Sun, W., & Smith, A. (2014). Evaluation of forest cover estimates for Haiti using supervised classification of Landsat data. *Int. J. Appl. Earth Obs.*, *30*, 203–216. doi:10.1016/j.jag.2014.01.020

Deeming, J. E., Burgan, R. E., & Cohen, J. D. (1978). The National Fire-Danger Rating System — 1978. General technical report INT-39. Intermountain Forest and Range Experiment Station, Ogden, UT: USDA Forest Service.

Finney, M. A. (1998). FARSITE: Fire area simulator — Model development and evaluation. USDA Forest Service, Rocky Mountain Research Station, Research Paper RMRS-RP-4 (Fort Collins, CO).

Hansen, M., Defries, R., Townshend, J., & Sohlberg, R. (2000). Global land cover classification at 1 km spatial resolution using a classification tree approach. *International Journal of Remote Sensing*, *21*(6-7), 1331–1364. doi:10.1080/014311600210209

Hernandez-Leal, P. A., Gonzalez-Calvo, A., Arbelo, M., Barreto, A., & Alonso-Benito, A. (2008, December). Synergy of GIS and remote sensing data in forest fire danger modeling. *IEEE Journal of Selected Topics in Applied Earth Observations and Remote Sensing*, *1*(4), 240–247. doi:10.1109/JSTARS.2008.2009043

Keane, R. E., Burgan, R. E., & van Wagtendonk, J. (2001). Mapping wildland fuel for fire management across multiple scales: Integrating remote sensing, GIS, and biophysical modeling. *International Journal of Wildland Fire*, *10*(4), 301–319. doi:10.1071/WF01028

Kononov, A. A., & Ka, M. H. (2008). Model-associated forest parameter retrieval using VHFSAR data at the individual tree level. *IEEE Transactions on Geoscience and Remote Sensing*, *46*(1), 69–84. doi:10.1109/TGRS.2007.907107

Kuznetsov, G. V., & Baranovskiy, N. V. (2013). Focused sun's rays and forest fire danger: new concept / *Proceedings of SPIE. Vol. 8890*, paper 889011; 10.1117/12.2033929

Lasaponara, R., & Lanorte, A. (2007a). On the capability of satellite VHR QuickBird data for fuel type characterization in fragmented landscape. *Ecological Modelling*, *204*(1-2), 79–84. doi:10.1016/j.ecolmodel.2006.12.022

Lundgren, S., Mitchell, W., & Wallace, M. (1995). A status report on NFMAS—an interagency system update project. *Fire Management Notes*, *55*, 11–12.

McArthur, A. G. (1967). Fire behaviour in eucalypt forests. Canberra, Australian Capital Territory: Commonwealth of Australia Forest and Timber Bureau, 25.

Mitchell, M., Wilson, R., Twedt, D., Mini, A., & James, J. (2016). Object-based forest classification to facilitate landscape-scale conservation in the Mississippi Alluvial Valley. *Remote Sens. Appl. Soc. Environ.*, *4*, 55–60.

Roy, P. S. (2003) Forest fire and degradation assessment using satellite remote sensing and geographic information system. Available at http://www.wamis.org/agm/pubs/agm8/Paper-18.pdf

Sheriza, M. (2010, May 7). Razali, Ainuddin Nuruddin, Ismail A. Malek, and Norizan A. Patah Forest fire hazard rating assessment in peat swamp forest using Landsat thematic mapper image. *Journal of Applied Remote Sensing*, *4*(1), 043531. doi:10.1117/1.3430040

Ustin, S. L., Roberts, D. A., Gamon, J. A., Asner, G. P., & Green, R. O. (2004). Using imaging spectroscopy to study ecosystem processes and properties. *Bioscience*, *54*(6), 523–534. doi:10.1641/0006-3568(2004)054[0523:UISTSE]2.0.CO;2

van Wagner, C. E. (1987). *Development and structure of the Canadian forest fire weather index system. Rep. No. Forest Technology Report 35*. Ottawa, Canada: Canadian Forestry Service.

Wang, L., Sousa, W. P., Gong, P., & Biging, G. (2004). Comparison of IKONOS and QuickBird images for mapping mangrove species on the Caribbean coast of Panama. *Remote Sensing of Environment*, *91*(3-4), 432–440. doi:10.1016/j.rse.2004.04.005

Yan, E., Wang, G., Lin, H., Xia, C., & Sun, H. (2015). Phenology-based classification of vegetation cover types in Northeast China using MODIS NDVI and EVI time series. *International Journal of Remote Sensing*, *36*(2), 489–512. doi:10.1080/01431161.2014.999167

ADDITIONAL READING

ArcGIS Desktop. https://desktop.arcgis.com/

Jensen, J. R. (2000). *Remote Sensing of the Environment: An Earth Resource Perspective* (p. 544). NJ: Prentice Hall.

Jwan Al-doski. (2011). Shattri B. Mansor and Helmi Zulhaidi Mohd Shafri (2013) Image Classification in Remote Sensing. Journal of Environment and Earth Science Vol. 3, No.10, Zhang JH, Yao FM, Liu C, Yang LM, Boken VK. Detection, emission estimation and risk prediction of forest fires in China using satellite sensors and simulation models in the past three decades--an overview. *International Journal of Environmental Research and Public Health*, *8*(8), 3156–3178. PMID:21909297

KEY TERMS AND DEFINITIONS

Image Classification: A process of grouping pixels into several classes of land use/land cover (LULC) based on the application of statistical decision rules in the multispectral domain or logical decision rules in the spatial domain

Landsat 7: The seventh satellite of the Landsat program. Launched on April 15, 1999, Landsat 7's primary goal is to refresh the global archive of satellite photos, providing up-to-date and cloud-free images.

Python: An interpreted, high-level, general-purpose programming language. Created by Guido van Rossum and first released in 1991, Python's design philosophy emphasizes code readability with its notable use of significant whitespace. Its language constructs and object-oriented approach aim to help programmers write clear, logical code for small and large-scale projects.

Remote Sensing: The collecting of information about the earth using aircraft and satellites Geographic information systems: a system designed to capture, store, manipulate, analyze, manage, and present spatial or geographic data.

Supervised Classification: Uses the spectral signatures obtained from training samples to classify an image.

Unsupervised Classification: Finds spectral classes (or clusters) in a multiband image without the analyst's intervention.

Section 6

Chapter 12
Influence of Climatic Conditions on Western Siberian Forest Fires

Valentina Petrovna Gorbatenko
Tomsk State University, Russia

Marina Alexandrovna Volkova
Tomsk State University, Russia

Olga Vladimirovna Nosyreva
https://orcid.org/0000-0002-3722-6834
Tomsk State University, Russia

George Georgievich Zhuravlev
Tomsk State University, Russia

Irina Valerievna Kuzhevskaia
Tomsk State University, Russia

ABSTRACT

Current climate changes in Russia are attended by the increase in frequency of dangerous weather events. This chapter researches long-term variations of the dangerous weather's events on Western Siberia and to reveal general regularity, which can be associated with forest fires. The researches have been carried out for the territories of southeast of Western Siberia. The duration of the fire season increases due to climate change. This is due both to the earlier snowfall and the onset of the phenological spring, and to the increase in the duration of the thunderstorm period. Thunderstorms in Siberia are a much more frequent cause of forest fires (28%) than in other territories. Wildfire frequency is correlated with air temperature and drought anomalies.

DOI: 10.4018/978-1-7998-1867-0.ch012

INTRODUCTION

In this chapter, we consider the influence of climate on the fireiness of forests on the example of the taiga of Western Siberia.

Forest fires are a serious and growing hazard over territory of the Russia and over Siberia too. One of the features of global warming, observed from the middle of 70s XX century, is its spatial inhomogeneity. The global climate change issue has raised serious questions about changes in repetition of weather extremes, such as thunderstorm, heavy rain, strong wind, duration of high-low temperature periods. All these characteristics are important for initiating or preventing forest fires. Besides, extreme temperature conditions affect the rigidity and strength of structures, performance capabilities, the ecological situation of the area (fire hazards, the level of pollution), housing and utilities infrastructure and other branches of economy. Every year many billions of dollars are spent on fire management and fire suppression.

Forest fires are frequent in the Siberian taiga and are predicted to increase in frequency as a result of increased fire risk under drought conditions, and prolonged fire seasons caused by climate change.

The goal of the present park work is presentation of long-term variations of the dangerous weather events above Western Siberia and to reveal general regularity, which can be associated with forest fires.

BACKGROUND

The global climate is warming and this may have a profound and immediate impact on wildland fire activity (Flannigan at al., 2013). Some suggest that wildland fire activity has already increased due to climate change. Hansen and colleagues (Hansen, Ruedy, Sato, & Lo, 2010) showed in their work that warming for the territory of Siberia has more rapid growth, than for Northern Hemisphere, in general. The positive trend of annual averaged surface air temperature is 0.36°C per every 10 years with higher values in some months. It was found that the rise in temperature occurs in February, March, October and May basically (Intergovernmental Panel on Climate Change [IPCC], 2013, 2014). Note that in May in Siberia most often the fire-dangerous season begins. Increases in average annual temperatures create conditions that dramatically elevate the risk and severity of forest fires.

The authors Ponomarev and Kharuk (2016) noted an increase of fire frequency and burned area in Altai-Sayan region of southern Siberia. The important issue of climate impact on the wildfires and burns dynamics in Siberia was discussed only in a few papers. The authors of Flannigan at al. (2013); Gorbatenko, Gromnitskaya, Konstantinova, Ershova, & Nechepurenko (2015); Ponomarev, Kharuk, & Ranson, (2016); Polyakov, Barashkova, & Kuzhevskaya (2014) discuss it in their research. It was also shown that the occurrence of extreme fire events in Central Siberia and the Trans-Baikal region were related to soil moisture and precipitation anomalies (Ponomarev et al., 2016). Lightning frequency, air temperature and moisture, wind speed have an impact on the fire regime too. Under scenarios of climate change many predictions show an intensification of the fire regime, i.e. more frequent and/or more intense fires (IPCC, 2014). It is known, that the increase in mid-annual values of temperature of air even on 1 degree provides increase in storm activity at least on 10 per cent (Gorbatenko et al., 2015). Therefore, will increase not only quantity of thunderstorms, but also quantity of downpours, squalls and other phenomena that accompany them. Weather conditions conducive to forest fires in Western Siberia:

- The increase in the duration of the fire season: when early warm weather sets in, it leads to premature snow melt and runoff. In recent decades, snow cover in Siberia is melting one to three weeks earlier than 30 years ago and the duration of the fire hazard season is increasing (Konstantinova, Gorbatenko, & Polyakov, 2017).
- Dry conditions: if in some areas the amount of precipitation has decreased, the probability and duration of forest fires naturally increases. At higher temperatures, the soil dries faster and the relative humidity is less (Ponomarev et al., 2016).
- Lightning: One major impact of climate change is increasingly severe storms. Severe storms carry a lot of lightning, and are often caused by forest fires (Konstantinova, Gorbatenko, & Ershova, 2010).

The structure of long-term fluctuations of meteorological elements such as air temperature, precipitation, thunderstorm activity, has been studied in detail and is used in the practice of long-term weather forecasts. Knowledge of the trends in the perennial variability of these characteristics will also be useful for predicting the fire danger of the forests of Western Siberia. Consequently, the systematic analysis of dangerous weather events characteristics is necessary for activity management aimed at decreasing either the predicted or incurred losses associated with climatic risks.

Let us consider in more detail the dynamics of forest fires in the territory of one of the regions of Western Siberia.

FOREST FIRES IN TOMSK REGION

How has the number of forest fires changed in the long run? It is known that the Siberian taiga is expected to become more susceptible to forest fires. This is discussed by the authors (Bunn, Goetz, Kimball, & Zhang, 2007; Flannigan et al., 2013; Gorbatenko et al., 2015; Ponomarev et al. 2016), in their studies. And their assumptions according to our research are justified.

The influence of climate trends on the threat of an increase in the number and area of forest fires is illustrated by the example of Tomsk Region. The studies were carried out for a part of the territory of Western Siberia (between the coordinates of 55-61 N and 75-90 E) - Tomsk region (Figure 1a,b).

The spatial distribution of forest fires (data retrieved from National Aeronautics and Space Administration [NASA]) in the territory of the Tomsk region (their sum over 10 years) is shown in Figure 2. Registration of forest fires was carried out using a MODIS (*MODerate resolution Imaging Spectroradiometer*) spectroradiometer installed on EOS AM-1 (Terra) and EOS PM-1 space platforms. Spectroradiometers can detect fires ranging from fractions of a hectare to several tens of hectares, depending on the intensity of combustion and the state of the atmosphere. They use data from the satellites Terra and Aqua and Suomi NPP, the flight of which is carried out up to sixes times a day. The thermal channels MODIS of the satellites Terra and Aqua have a resolution of 1 km (picture with pixels 1 km wide) (Kovacs, Ranson, Sun, & Kharuk, 2004).

Figure 3 shows the ratio of the causes of fires in the study area. It is obvious that forest fires most often occur near settlements and along major rivers of the region. Most of the fires occur in the southern part of the territory, where the highest density of population and areas sown by crops. In Western Siberia, the inhabitants traditionally have been using controlled ground fires (grassland fires) for preparing agricultural fields for planting season.

Figure 1. Territory of Western Siberia (A) and meteorological stations of Tomsk region (B) (Location maps – schemes)

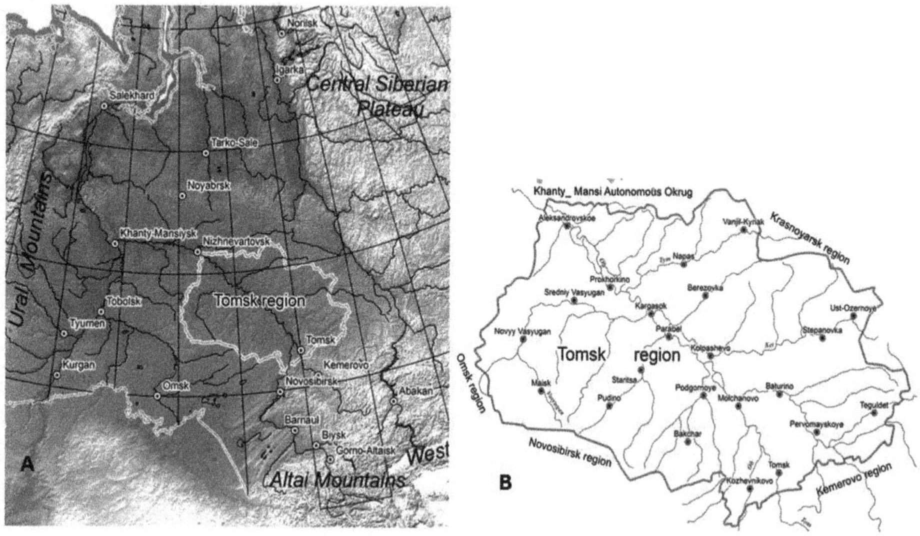

Figure 2. Fires registered in the territory of the Tomsk region by satellite over a ten-year period

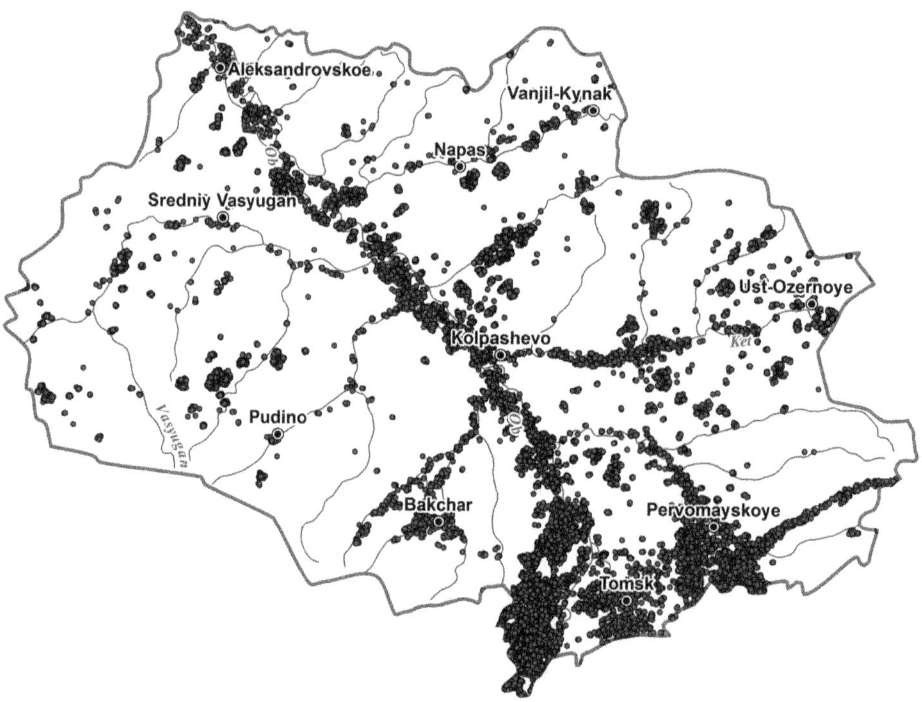

Figure 3. Causes of forest fires: human factor (agricultural burns (1) and other human factor (3)), fires from lightning (2), the cause is not established (4)

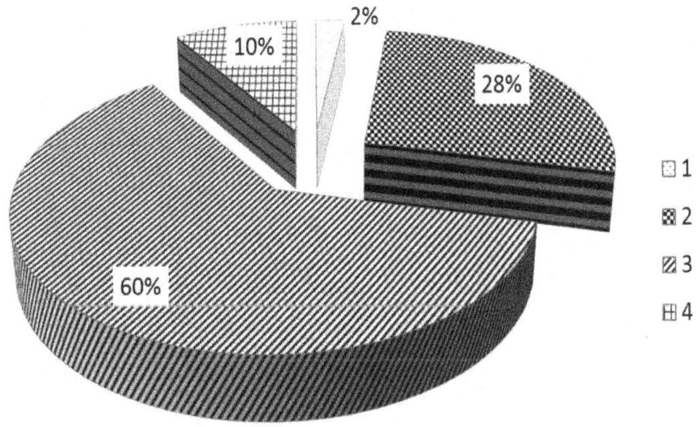

It is estimated that about 62% of wildfires are started by humans (about 2% of these are agricultural crops); the cause of fire in 10% of cases could not be established. The other 28% are started by lightning. Heat is usually supplied by a lightning strike to a tree or dry grass. An areas of wildfires, with are started by lightning, are the largest. People also can provide heat by carelessly starting a fire or leaving a fire unattended.

The ratio of causes of ignition varies somewhat in different years and is highly dependent on weather conditions (Gorbatenko et al., 2015). On the Figure 4 the perennial variability of the number of forest fires per year (n) and total annual forest area (ha) affected by fires (S) is presented. The correlation coefficient between these characteristics is high and statistically significant. Both the annual number of forest fires and the area of burnt forests demonstrate the absence of directed tendencies to increase them. This happens despite the fact that the average annual air temperature values continue to increase.

Most often forest fires occur in May and June (Figure 5). Although the season of fires in some years begins in April.

Note that in May forest fires begin in Siberia very often (Ponomarev et al., 2016). High values of air temperature and low rainfall this month are the most important factors for the occurrence and development of forest fires. Thunderstorms pose a great threat to forests in May too. Despite the rarity of the May thunderstorms in these latitudes, their number has increased in recent decades.

As a rule, fires in separate parts of a forest occur with a periodic periodicity (Gorbatenko et al., 2015). Such a quasi-periodicity on the territory of the Tomsk region can be seen in Figure 4. It is due, most likely, to the repetition of synoptic processes that ensure a high inflammability of forests.

Conclusion: Over the past 30 years there are no tendencies to increase the number of forest fires and burnt-out areas on the territory of the Tomsk Region. This result can be explained by the close attention of the forest protection and measures to prevent them. Consequently, in those regions where forest fires are regularly monitored, it is possible to prevent their wide spread.

Figure 4. The number of forest fires per year (n) and the total annual forest area (thousand hectares) affected by fires (S)

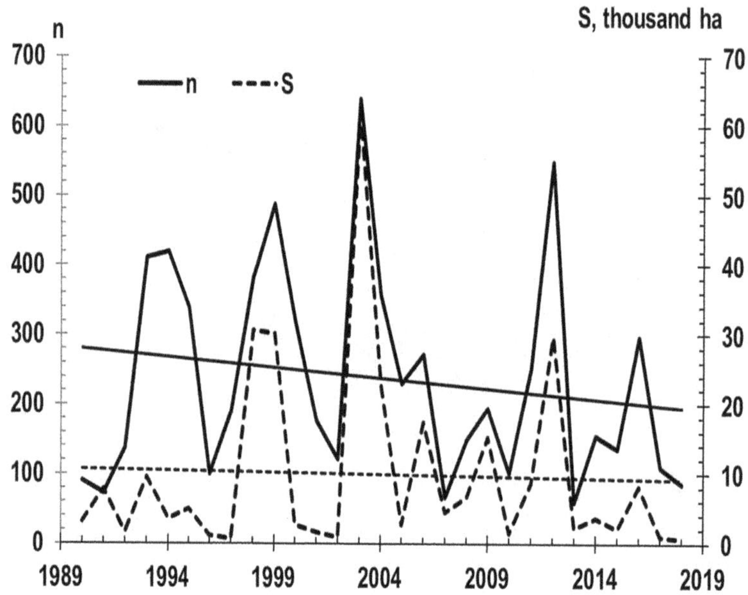

Figure 5. The proportion of forest fires (%) in different months

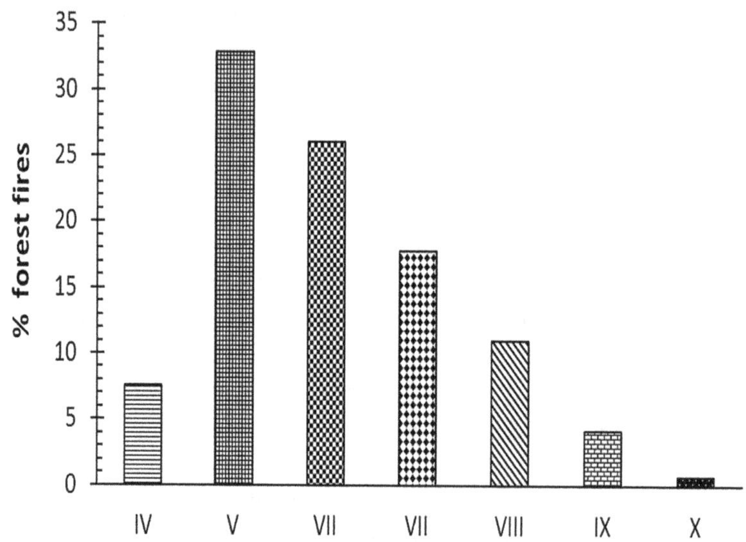

The question remains: how will the frequency of forest fires change if the trend for climate warming continues?

It is very difficult to control taiga fires throughout Siberia. Nevertheless, attracting satellite monitoring results will help to take early measures to prevent the spread of fire to large areas.

THUNDERSTORM ACTIVITY ON SOUTHEAST WESTERN SIBERIA

How often do thunderstorms cause a forest fire? What is the thunderstorm activity in this northern region and how does it change over time?

Western Siberia in general and the territory of the Tomsk region are characterised by temperate thunderstorm activity approves researches (Gorbatenko and Konstantinova, 2011; Konstantinova, 2017). Thunderstorm activity over Western Siberia increases from north to south and is maximum over the Altai Mountains - 35 thunderstorm days per year (Figure.6).

In Tomsk region there are on an average 42 thunderstorm hours and 22 thunderstorm days per year; maximum about 65 thunderstorm hours and 28 thunderstorm days and minimum are 23 thunderstorm hours and 17 thunderstorm days. There are areas of particular risk of wildfires inside regions, because the spatial distribution of thunderstorm activity as in other regions appeared to be spotty. Lighting discharge density is estimated on the satellite data. For the territory of Tomsk Region the NASA's data containing the information on quantity of lightning discharges for each trapezium of a net with an interval in 1° of latitude and a longitude, was recalculated in values of lightning discharges density on 1 km^2 per a year (Figure 6b). The daily data of Optical Transient Detector (OTD) from the "MicroLab-1" satellite were used for mapping of the lightning discharges (NASA b, 2019).

For Tomsk region the lightning discharges density per year changes from 2.2 st.$km^{-2}yr^{-1}$ to 0.8 st.$km^{-2}yr^{-1}$, with a mean value 1.3 st.$km^{-2}yr^{-1}$. The increased lightning discharges density is marked in two regions: the first is between rivers Ob and Chulym; the second is in a valley of Ob River, from north to the southeast (Figure 6a). The analysis of spatial distribution of characteristics of thunderstorm activity has allowed establishing, that mesoscale inhomogeneity of values thunderstorm days and durations of thunderstorms and density of lightning discharges to the ground are essential and statistically significant.

Different possibilities for the realization of thunderstorm situation are more often provided (formed) by the peculiarities of an underlying surface – not only orography, but also its temperature and humidity.

Relationship between peculiarity of the thunderstorm activity and number and areas of forest fires demonstrates a comparison of Figures 2 and 6. The coincidence is, of course, not absolute, but it is interesting that lightning, like fires, also occurs more frequently in areas with a higher population density.

The next aim of this part of work is the presentation of the periodic structure and trends of time series of number of days with thunderstorm on the Western Siberia (Figure.7). The analysis of long-term data on visual thunderstorm observations has no alternative in Siberia today when solving a task of monitoring thunderstorm activity. The common measure of thunderstorm activity is a number of thunderstorm days per year observed visually. Under thunderstorm day we mean day, during which a thunderclap has been heard at least once. Then the data are summarized over a month and then over a year. More detailed information on the density of lightning discharges to the ground can be obtained with the help of lightning multistation system. However, these measurements are short and do not permit us to judge about the time variations of this characteristic. The presence of time dependence of density of lightning discharges to the ground on the number of thunderstorm days for every point of thunderstorm observations is undoubted

Figure 6. Average values number of thunderstorm days per year in Siberia (A) and of lighting discharge density on Tomsk region (B)

Figure 7. Number of thunderstorm days (T) in Tomsk region

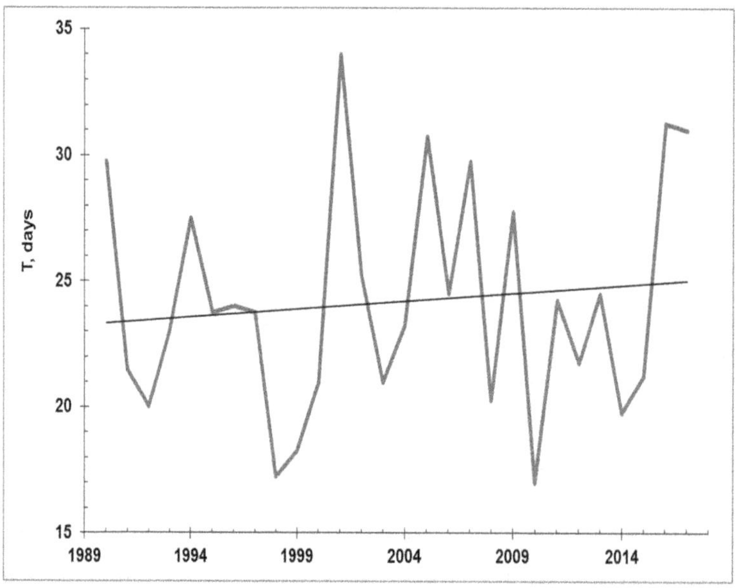

(Konstantinova et al., 2017). So, analyzing long-term variations of thunderstorm day's number, it is possible to assess the expected range of time changes of density of lightning discharges to the ground.

Temporal ranges of thunderstorm activity characteristics are characterized by the compound statistical structure (Figure 7). The comparison of the works, carried out for the different regions, permits to affirm, that above the different territories there are cycles of different periodicity (Konstantinova et al., 2017).

The time changes of characteristics of thunderstorm activity are discussed from time to time in literature (Gorbatenko & Konstantinova, 2011). Like other meteorological quantities, thunderstorm activity characteristics undergo considerable variations of annual mean values that are caused by the changeableness of the circulation processes in the atmosphere. Analyzing the asynchronous variations of the thunderstorm activity in observation data, climatologists try to explain them either by natural fluctuations or by observation errors or by anthropogenic impact on characteristics of atmospheric electricity (Konstantinova et al., 2010).

We have analyzed the historical thunder-day data collected at 12 state meteorological stations of Tomsk region. The differences of the annual number of thunderstorms in stations are not very evident. There is considerable changes interannual variability of thunderstorm characteristics (Figure 4). Not many stations maximal of thunderstorms days were observed last years. Nevertheless statistically significant trends in analyzed rows of thunderstorm it was not revealed.

Like the other meteorological elements, thunderstorm activity characteristics are defined by a considerable variation of average annual values that is caused by the changeableness of the circulating processes in atmosphere. Mostly this goes parallel with the variability of synoptic flow conditions favorable for the development of thunderstorms. The data of thunderstorm days was examined for time fluctuations. After elimination of nonstationarity, in order to find periodical oscillations in the time series under study, the assessments of autocorrelation and spectral functions were calculated. We cannot interpret unambiguously the regularities, nevertheless we can assert that:

- In recent decades, there has been an increase in the thunderstorm season (the number of days between the first and last thunderstorm of the year). Consequently, the duration of the fire hazard season also increases.
- There is a slight trend to increase the number of thunderstorm days. The increase is due to the spring months be attributed to and may be due the manifestation of global warming.
- In the time series of thunderstorm activity, one can distinguish cycles with short (2-5 years), middle (6-12 years) and long periods (15-23 years), though the duration of this cycles above the different stations is not quite similar. Therefore, they are difficult to take into account in the forecast of forest fires.
- As to interannual variability of thunderstorm days in Western Siberia it was found that the absolute maximum for days with thunderstorm take place in July (Konstantinova et al., 2017).

Table 1. The list and the criteria for hazardous temperature events in the warm period for the territory of the south-east of Western Siberia

Hazardous Weather Events	HWE Characteristics
Heat wave	The value of warm extreme temperature is +35 °C and above during 3 days and more for May-August period
Anomalously hot weather	The value of warm extreme temperature is +30 °C and above during 3 days and more for April-September period
High temperatures	The value of maximal air temperature is +30°C and above of any duration

(Federal State Budgetary Institution, n.d.)

Table 2. The absolute maximum of temperature for the period 1961-2017

Station of Observations	The Absolute Maximum	
	Tabs, °C	Date
Alexandrovskoe (northern part of the region)	34.9	02.07.1969
Kolpashevo (center part of the region)	34.6	03.07.1969
Pervomayskoe (south-east part of the region)	35.3	11.07.1974
Sredny Vasugan (west part of the region)	37.0	01.07.1969
Tomsk (south part of the region)	35.6	12.07.2014
Ust-Ozyornoe (center part of the region)	36.2	02.07.1969

HAZARDOUS TEMPERATRUE EVENTS IN FIRE RISK PERIOD

Current climate changes are attended by the increase in frequency of hazardous weather events (HWE) (IPCC, 2013). In 1996–2012, in Russia the number of hazardous events increased, including those that caused considerable damage to the economy and humans (Frolov et al., 2014). On average, the losses associated with climatic risks accounts for 80 per cent of that caused by natural disasters. According to observations of Federal Service for Hydrometeorology and Environmental Monitoring (2005), the loss makes up 30-60 billion rubles (0.4-0.85 billion €) per year on the background of steady increase of HWE ranging about 6.3 per cent per year. It should be noted that, apart from the direct loss, HWE can enhance other negative factors of the environment and different natural disasters. Thus, the increase in duration of high temperature periods exerts a pernicious influence directly on people's health and indirectly, through the quality of water, on food and air pollution, forest fires. Consequently, the systematic analysis of hazardous weather events characteristics is necessary for activity management aimed at decreasing either the predicted or incurred losses associated with climatic risks.

This part of chapter discusses extreme temperature events in a fire hazard period over the Tomsk Region.

For each hazardous weather events (table 1) around the stations located in in different areas of the Tomsk region the following characteristics were calculated: the total number of HWE days (1961–2017),

Table 3. Hazardous temperature events characteristics around Tomsk region

HWE	Statistical Characteristics	Station of Observations					
		Alexandrovskoe	**Kolpashevo**	**Pervomayskoe**	**Sredny Vasugan**	**Tomsk**	**Ust-Ozyornoe**
Anomalously warm weather (AWW)	The total number of AWW days (1961–2017)	48	44	81	**102**	60	77
	Average continuous duration of AWW periods (days)	5.8	6.5	5.8	**6.7**	6.9	8.3
	Maximal continuous duration of AWW periods (days)	**7**	9	9	12	11	**14**
	The frequency of AWW (%) per the May- August period	0.7	0.6	1.2	**1.5**	0.6	1.1
High temperature periods (HTP)	The total number of HTP days (1961–2017)	231	255	**428**	425	319	362
	Average continuous duration of HTP periods (days)	1.1	**2.9**	1.5	1.5	1.1	1.6
	Maximal continuous duration of HTP periods (days)	12	12	14	**19**	14	14
	The frequency of HTP (%) per the May- August period	3.4	3.7	**6.2**	**6.2**	4.6	5.3

average and maximal continuous duration of HWE periods, the frequency of days with HWE per the May- August period, the duration of periods with HWE every 30, 50, 100 years.

The location of Tomsk region in the center of the Eurasian Continent provides highly continental climate which expresses itself in forming low temperature conditions in the cold season and high temperature conditions in the warm season. The absolute maximum of seasonal extreme warm temperature in Tomsk region was observed on July (table 2). The range of recorded absolute maxima is 34.6 – 37.0. Such a small range shows the uniformity of the temperature conditions of the warm period. Consequently, the whole area is equally exposed to fire danger.

Table 3 shows the statistical characteristics of Anomalously warm weather and High temperature periods.

The "heat wave" HWE is not typical of Tomsk region as no period of this kind was recorded for the 1961-2017 period.

Anomalous warmth (AWW) was observed more often in the western part of the territory (the Sredny Vasugan station), the total number of AWW days (1961–2017) being 102 days with a maximum continu-

Table 4. Duration of extreme temperature periods (days) around Tomsk region stations, expectative once every k years

Stations	Hazardous Weather Events					
	Anomalously Warm Weather			High Temperature Periods		
	k, years			k, years		
	30	50	100	30	50	100
Alexandrovskoe	9	10	12	10	11	14
Kolpashevo	10	12	14	11	13	15
Pervomayskoe	10	11	13	12	15	18
Sredny Vasugan	11	13	17	12	16	19
Tomsk	11	12	16	10	13	17
Ust-Ozyornoe	11	14	18	15	15	17

ous duration of AWW periods 12 days (table 3). The frequency of AWW was recorded 1.5% for 4 months (May-August period). The increased number of days with abnormally warm weather in the south (Tomsk station) and south-east (Pervomayskoe station) of the territory corresponds to areas with high fire risk.

The total number of high temperature days for the period under study ranges from 231 days in the Alexandrovskoe station to 428 days in the Pervomayskoe station. So the southern part of the Tomsk region is characterized by a higher frequency of high temperature days.

Consequently, temperature conditions in the southern half of the Tomsk region contribute to increased fire risk during the absence of precipitation. According (Volkova, Cheredko, & Titovskaya, 2019, p. 64) the frequency of occurrence of high temperatures period in 1998–2016 "became higher than in the period of explicit warming" (1976–1997).

The events associated with the warmth extremes are formed more often in June and July over the Tomsk region area.

In order to get the climatic parameters of various reliability the model based on simple Markov chains was used. The calculations of maximal duration of the periods with the relevant HWE predicted for the predetermined number of years. The specified time periods of 30, 50, 100 years were considered in this chapter.

The values of period duration with HWE under study in each station are given in Table 4. Thus for example, once every 100 years in Tomsk anomalous warmth can be observed during 16-17 days. The period with high temperatures may continue for 14-19 days every 100 years for the study area.

In order to prove model validity the correlation between calculated and actual probabilities of occurrence periods under study was estimated. The correlation was found to be rather high (Pearson coefficient is 0.98). Consequently, the model data match the actual ones. Furthermore, use was made of Pearson test to estimate the correlation between the actual and the calculated distribution. In every respect, the condition was satisfied, that is, the hypothesis on model relevance can be significant.

The results of these calculations show a homogeneous distribution over the territory of the expectative duration of HWE period. Consequently, these periods with a precipitation deficit can significantly increase the fire hazard of the entire region equally.

Figure 8. Spatial distribution of fires in 2012 from May to September

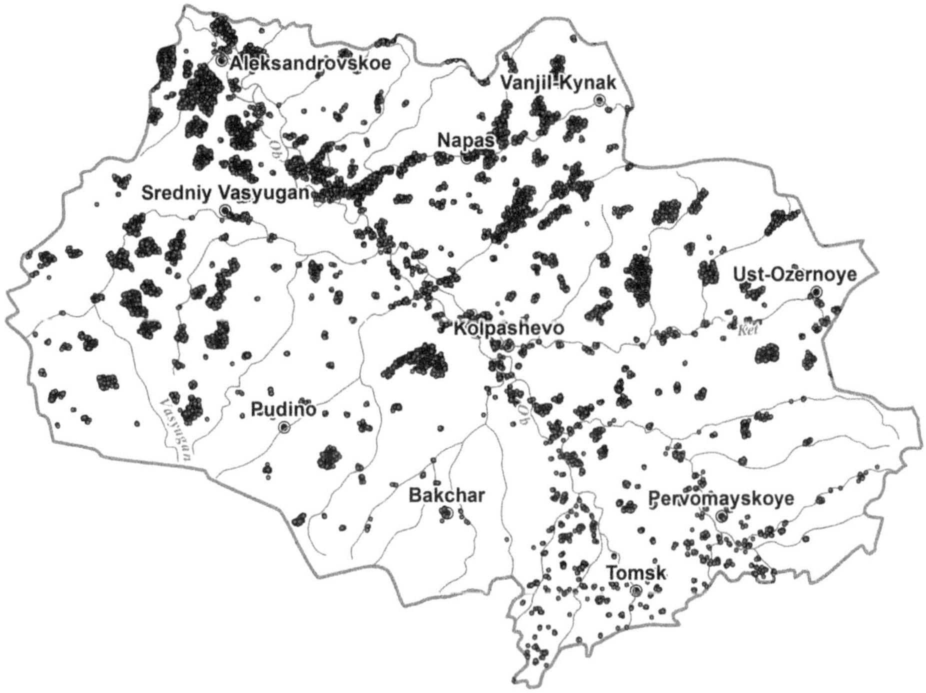

The increased number of days with abnormally warm weather in the south (Tomsk station) and south-east of the territory (Pervomayskoe station) corresponds to areas with high fire risk.

ANOMALOUS WEATHER AND FOREST FIRES IN 2012

In the summer of 2012, a large number of forest fires were observed and the forest area covered by fires was also large (Figure 4). This was caused by the anomalous weather established in the region. According to the data of the Tomsk Center for Hydrometeorology and Environmental Monitoring (Federal State Budgetary Institution "West-Siberian Administration for Hydrometeorology and Environmental Monitoring") in the period from May 30 to August 2, 2012 in Tomsk region the abnormally hot and dry summer was observed having the first rank in recent 60 years. In the Siberian Federal District, according to official data, 162 days (from May to October) were recorded with the weather providing the highest class of fire danger. From June to September, the emergency class of fire danger remained in the Siberian taiga daily. Spatial localization of fires is shown in Figure 8, data retrieved from NASA. Forest and peat fires were registered practically on the whole territory of Tomsk region and especially difficult conditions were formed in the northern parts of the region. In all, 551 forest fires with the total area of 102120.3 ha were registered during the period of emergency situation. Forest fires in Tomsk region and neighboring regions in the summer of 2012 formed a high level of air pollution with combustion products. Smoke trails that are well identified on satellite images, spread to the settlements including Tomsk and were accompanied by haze with the decrease in vertical and horizontal visibility to 80–100 m and with the smell of combustion products.

The objective of the present part is to analyze meteorological conditions and forming them synoptic processes in Tomsk region in summer 2012. The data from the Roshydromet meteorological network on the territory of Tomsk region were used as the initial data. The data on wildfires are taken from the website of the Ministry for Emergency Situations of Russia.

The air temperature at the height of 2 m and the precipitation of diurnal resolution are considered and analyzed using the normal (for the period of 1971-2000) accepted by the World Meteorological Organization (WMO). Hydrothermal conditions during the period of the hot summer were assessed using the anomalies of air temperature and amount of precipitation. Hydrothermal conditions during the period of the hot summer were assessed using the hydrothermal moistening coefficient (HTC) on base the anomalies of air temperature and amount of precipitation (Gringof & Kleshchenko, 2011). For more information on the HTC coefficient, see "Atmospheric humidity regime during fire hazardous seasons" of this chapter.

Synoptic Conditions

During the most arid month, July, the low-gradient anticyclonic field was registered on the territory of the region near the surface, and isobars gained the cyclonic curvature only occasion ally under the polar front coming to the southern areas of the region. At the level of 500 hPa, the anticyclone was formed on July 1 as a multi-core pressure formation, and the migration of these cores inside the main closed formation was observed on the next days which al ternated with its periodical transformation into the pressure ridge. The upper-level trough was located over Tomsk region from July 3 to 11 but mainly anticyclonic conditions were registered since July 12. The pressure ridge axis passed along 55° E, moved slowly to the east, reached 85° E by July 21, and spread north wards up to the Arc tic coast. The main anticyclonic core de fining the weather in West and East Siberia was situated in the south west of the territory of the Sakha (Yakutia) Republic.

By July 25, the territory under study was located in the narrow zone of high pressure limited by the quasistationary upper-level trough from the west which meridionally oriented axis passed along the east ern Urals. By late July the anticyclone genesis became more intensive over the European territory and became a dominant process in this part of Russia. A trend towards the latitudinal orientation started being manifested in the Siberian high. By July 31, the cyclonic curvature of isobars over Tomsk region determined the formation of clouds, and occasional rains locally reaching the "heavy" gradation were registered during a month starting from August 2. Thus, the observed anticyclonic regime can be characterized as a durable blocking process.

In summer 2012, weather over Siberia was determined by two types of blocking synoptic processes (Berezhnaya, Golubev, & Parshina, 2012). The first scheme determined the weather in early summer. The separation of the in dependent anticyclone from the ridge of the Azores high over the Scandinavian Peninsula and its movement to the north of the Urals and Siberia. The second scheme of the process was implemented at the end of the summer and includes the formation of anticyclones in the south of the European part of Russia. These synoptic objects developed and moved to the northeast direction to Siberia.

As a result of such development of processes, one long-wave upper-level ridge is formed and the blocking anticyclone is generated in its area over Siberia. The observed blocking process had the signs of "meridional" type of the blocking during the central phase (Berezhnaya et al., 2012).

Results: Such a synoptic situation contributes to the formation of drought and a high level of fire danger in Siberia. If the frequency of the synoptic blocking processes increases, it will also increase the fire danger in the region too.

Surface Air Temperature

Significant positive anomalies of monthly mean air temperature *(ΔT)* on the territory of Tomsk region varied from 1.3 to 7.2 °C (for June-July) and were observed for the first time in recent 60 years. For ex ample, $\boldsymbol{\Delta T}$ in Tomsk in June reached 5.3 °C and amounted to 5.7 °C for the territory on average. The maximum $\boldsymbol{\Delta T}$ in June was observed in the north of the region at Aleksandrovskoe station and was equal to 7.2 °C. In July, at the certain weakening of heat, the monthly mean anomaly of air temperature averaged for the territory under study was 2.3°C (Fig. 9a).

The factor of hot weather combined with the air and soil aridity and with the overheating of the soil surface (50-55 °C) caused the spontaneous ignition of vegetation in many areas of the region.

At the end of the first ten-day period of August, the blocking anticyclone started collapsing. Due to the intensification of cyclonic activity on the territory under study, the heat weakened dramatically that affected the monthly mean air temperature in August, and it turned out to be within normal limits at al stations.

Precipitation

The extreme aridity of summer 2012 was favored by the winter period with variable regime of precipitation (the deficit amounted to 11% of the normal in February) and by the abnormally warm pre-spring period: the anomaly of temperature in Tomsk was equal to 5.3 °C in March and 3.9 °C in April. Weather conditions determined weak snow accumulation that favored the deep freezing of the soil (the soil freezing depth was larger than the normal by 20–30 cm). The amount of precipitation in winter by the beginning of the snow melting turned out to be below the normal by 59%; the similar situation was registered in 1967 only.

The earlier melting of the snow cover (April 8) as compared to the date registered before (April 12) took place; this is 22 days earlier than the normal (April 30). Such weather conditions were formed due to the fact that the territory of the region was mainly influenced by anticyclonic formations from January to March 2012 which occasion ally gave way to short cyclonic intrusions.

At the same time, the anticyclonic weather in June and July caused by the durable blocking of the zonal transport and by the absence of active fronts favored the dehydration of air masses and the precipitation absence. Since the middle of June, no rains at all were registered in some areas (1% of the normal at Aleksandrovskoe station and 3% of the normal at Napas station). The amount of precipitation observed on the territory of the region on average made up 32% of the normal in June and 20% of the normal in July (Fig. 9c, 9d).

In late June and early July, moderate and heavy rains were registered in some areas of the region. During this period, the southwestern part of the region turned out to be influenced by the polar front coming from the southwest and by the associated centers of convective instability that caused occasional rains being heavy in some places. For example, at Napas station, no rain was registered since June 15, and 46 mm of precipitation or 58% of the normal fell on July 7.

Figure 9. Schematic maps of distribution of meteorological parameters on the territory of Tomsk region in (a, c, e) June and (b, d, f) July 2012: (a, b) anomalies of average daily air temperature (°C) at the height of 2 m, (c, d) amount of precipitation (% of the normal), and (e, f) mean hydrothermal coefficient. A zone of "severe drought" is grey-colored

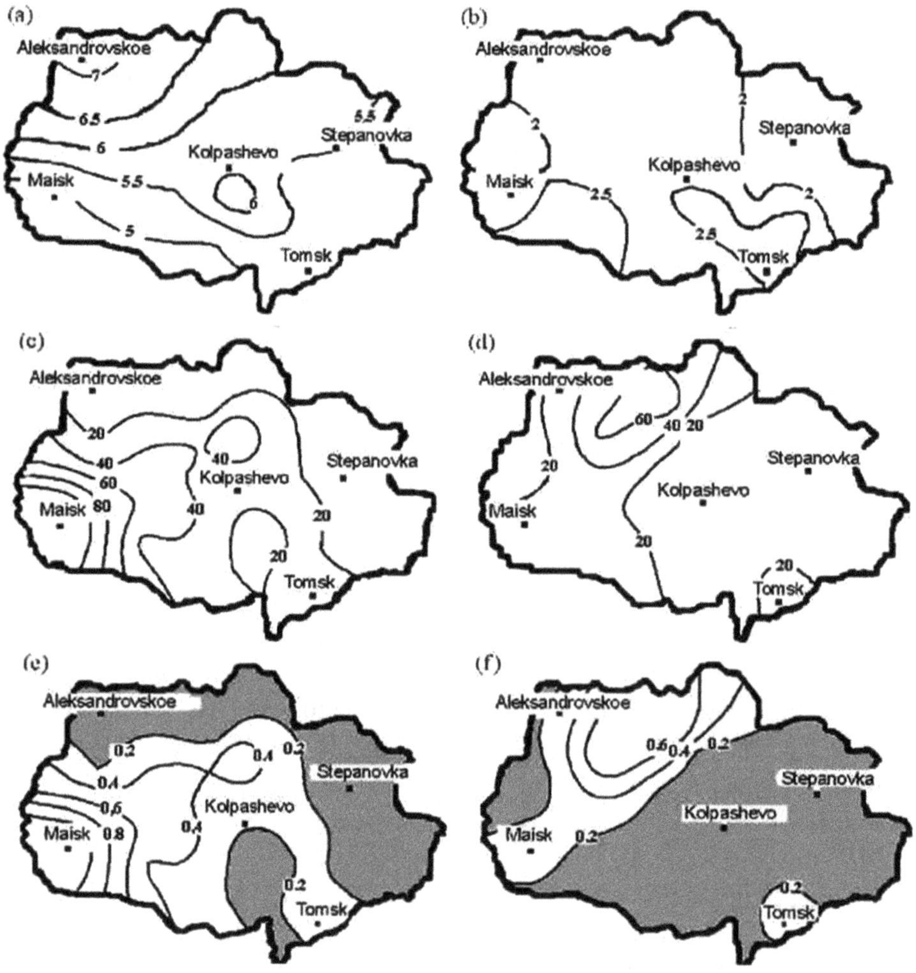

Since the second half of July, synoptic conditions were determined by the blocking anticyclone again (only the dislocation of its center varied from June to July). The precipitation was absent practically everywhere till the end of the first ten-day period of August and was registered only in the form of light occasional rains. The total amount of precipitation in July amounted to 33.4% at Tomsk station. The amount of precipitation above the normal (up to 199% at Pudino station) was registered in August on the territory of southwestern and southeastern areas of the region, and the deficit maintained in the northern areas.

Hydrothermal Conditions

The estimation of HTC in June 2012 and the comparison of its values with the scale from (Gringof & Kleshchenko, 2011) corroborated that in June in the most part of the territory the atmospheric drought of three gradations was formed: severe, heavy, and medium (Fig. 9e, 9f). Only the south west of the region was characterized by optimum heat and moisture conditions.

The heavy atmospheric drought in combination with the soil drought against a back ground of high temperatures in June and July was a key factor to increase fire hazard.

The hydrothermal moistening coefficient was compared in the summer of 2012 with hydrothermal conditions in previous years. It is shown that drought in 2012 in combination with a high precipitation deficit has had no analogs in recent 50 years.

Conclusion

The analysis of the actual data on meteorological and synoptic conditions demonstrated that the extreme regime of summer 2012 on the territory of Tomsk region is rare from the point of view of the combination of unfavorable weather conditions and their duration. Droughts are an integral part of the climate of West Siberia, and the organizators of economic complexes should consider more thoroughly official long-term weather forecasts for planning their activities, in particular in the field of forest fires protection.

ATMOSPHERIC HUMIDITY REGIME DURING FIRE HAZARDOUS SEASONS

The variability of weather during the warm period of the year has increased significantly in recent decades. This leads to an alternation of heavy rains and long warm and dry periods, often with abnormal heat. Such weather contributes to the occurrence and spread of forest fires.

Wildfires regime is usually identified by weather/climate, quantity and condition of combustible materials and ehe presence of the fire source. Fire danger increases with the temperature rise and precipitation reduction.

It is important to take into account climatic factors not only in summer, but also in other seasons of the year. How the characteristics of the winter period change over Western Siberia is described in the publications (Garmaeva & Nosyreva, 2016; Nosyreva, Barashkova, & Kizhner, 2018; Gorbatenko, Sevastyanov, Konstantinova, & Nosyreva, 2019). In the cited works, the features of the temperature regime and snow cover in spring and autumn are analyzed.

Rainfall also plays an important role – forest fires occur when the daily rainfall is little or equal to zero. The growth of fire danger can be significant even with a small increase in surface temperature, if there are long periods without precipitation.

In addition to the total amount of precipitation, a significant factor is the non-uniformity of their precipitation. In hot weather, the absence of precipitation for several days leads to the drying up of vegetation and an increase in the probability of forest fires. Despite the fact that an increase in precipitation is observed in a considerable part of the territory of the Russian Federation, an increase in the duration of periods without precipitation is observed.

Figure 10. The hydrothermal coefficient in Tomsk region, % (100% is the average long-term value of a month)

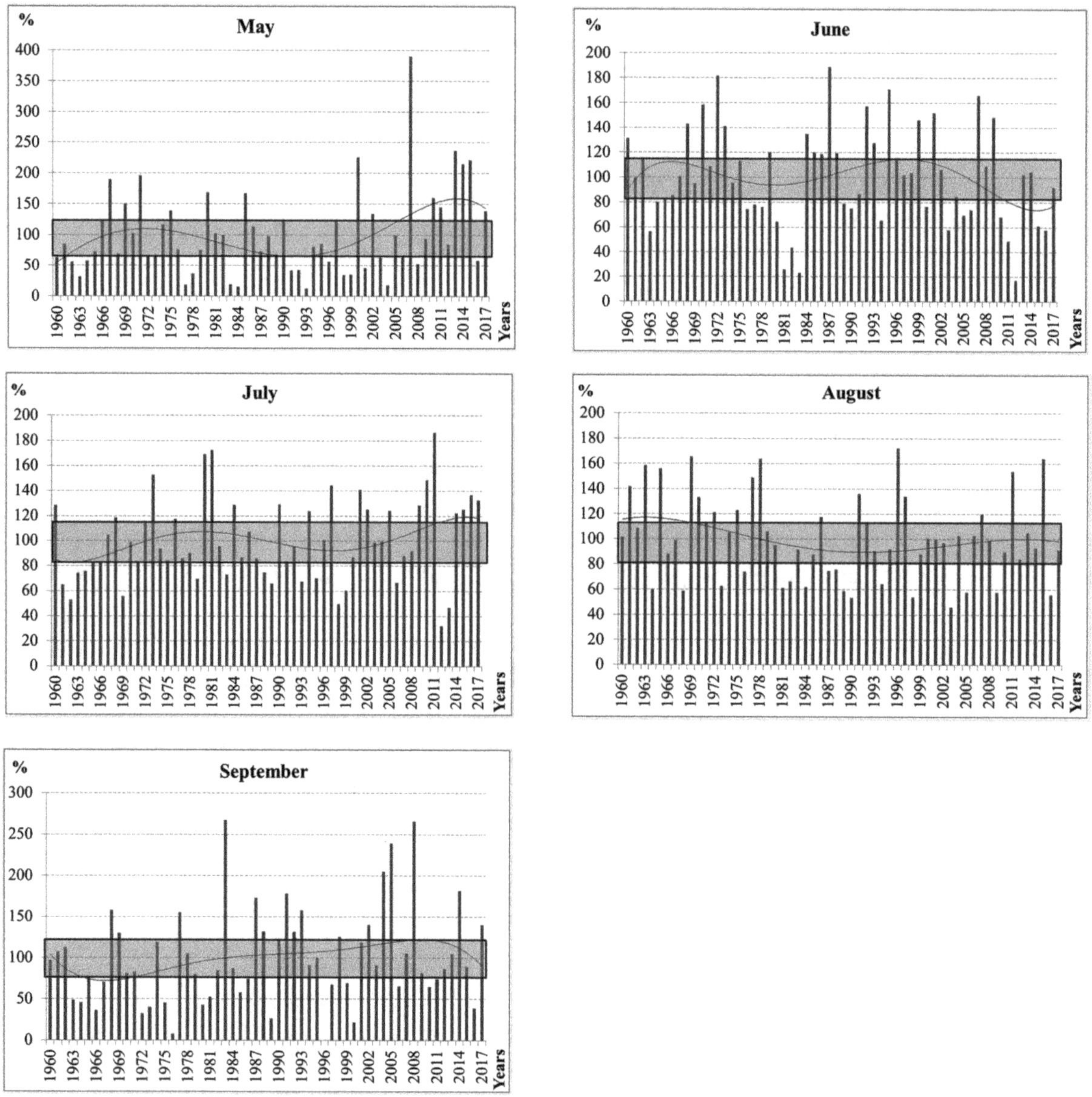

The fire hazard period is directly related to the warm period of the year, when the average daily air temperature becomes above 10 °C.

The purpose of this part of work is to study the moisture regime during the fire hazard period, the analysis of the frequency of droughts and their effect on the fire hazard assessment in the Tomsk region. For the study, data were used for the 58-year period from 1960 to 2017.

The fire danger of forests is largely determined by the level of drought, representing a balance of moisture and drying factors. It is determined from the beginning of the fire hazard season and predicts the fire hazard of vegetation.

Table 5. Drought repeatability in Tomsk region, %

Station	Drought Type	May	June	July	August	September	The Whole Period
Napas	very strong drought	18	58	64	55	18	43
	severe drought	1	5	5	6	1	4
Sredniy Vasyugan	very strong drought	25	60	64	54	24	46
	severe drought	1	6	6	7	2	4
Ust'-Ozeornoye	very strong drought	21	62	67	59	19	45
	severe drought	1	5	5	6	1	4
Kolpashevo	very strong drought	25	62	67	57	23	47
	severe drought	2	5	7	6	2	4
Pudino	very strong drought	29	63	63	58	24	47
	severe drought	2	4	6	6	2	4
Pervomaiskoye	very strong drought	30	64	68	59	27	50
	severe drought	2	5	6	6	3	4
Backchar	very strong drought	31	63	67	60	25	49
	severe drought	2	5	6	7	2	4
Tomsk	very strong drought	33	63	65	60	29	50
	severe drought	2	5	6	7	2	4

For the fire hazard assessment in Tomsk region for the period from 1960 to 2017 for the months of the fire hazard period, according to the data on the average daily air temperature and precipitation amount, the hydrothermal moistening coefficient (HTC) was calculated:

$$HTC = \frac{\Sigma R}{0.1\Sigma T_{\geq 10^{\circ}C}}$$

where ΣR is the total precipitation (mm) for the period with $\Sigma T_{\geq 10\,°C}$; $\Sigma T_{\geq 10\,°C}$ is the accumulated average daily values of air temperature of more than 10 °C (accumulated effective temperatures).

Heat and moisture supply of the territory is determined by the index values (Gringof & Kleshchenko, 2011):

- <0,4 – Dry area,
- 0,41-0,7 –Very arid area,
- 0,71-1,0 – Arid area,
- 1,1-1,3 – Insufficiently humid area,
- 1,31-1,6 – Sufficiently humid area,
- >1,6 – Wet area

Figure 11. Drought repeatability in fire hazard period in the Tomsk region

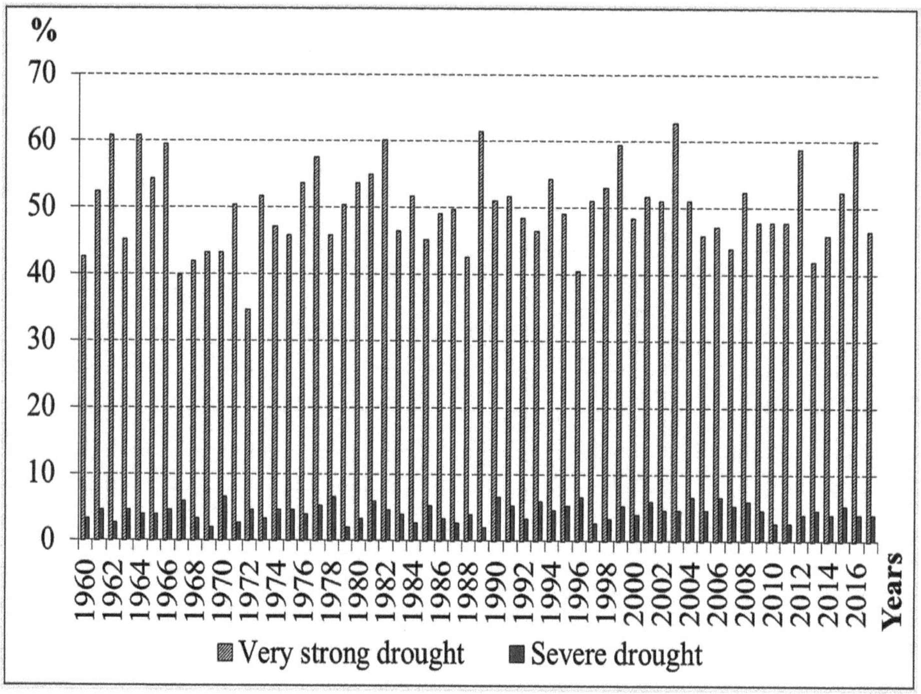

According to the obtained values of HTC, graphs of the distribution of the magnitude were plotted depending on the average long-term value of a certain month, taken as 100%, for the territory of Tomsk Region (Figure 10). In order to determine the period of the "norm" in the distribution of the series, the standard deviation σ was used: $Xi \pm 0,5\sigma$. In general, during the study period, the majority of changes in HTC occur within the normal range, but in some years, there are significant deviations from it.

For example, in 2007 a positive deviation from the normal range by 255% was noted (HTC was equal to 3.7 which corresponded to excessively wet conditions throughout the region). In June and July 2012 in the Tomsk region the lowest value of HTC was fixed (0.3 and 0.41 respectively), which is typical for dry and very arid conditions. In addition, low values were observed in June 2003-2006 and in July 1998-1999. There is a periodicity with a duration of about 20-30 years in the dynamics of HTC. In recent years, an increase in HTC values has been observed in June and July; in May, August and September - a slight decrease, which was formed due to the low value of HTC in 2016 against the background of neighboring years with a higher HTC.

In this paper, the frequency of very strong and severe droughts was calculated for the territory for each month and for the fire hazard period (table 5).

Throughout the Tomsk region and in all months, the frequency of repeatability with "very strong drought" is observed more than with "severe drought". The studied characteristic has a latitudinal distribution over the territory, a parameter change from north to south (most pronounced in May and September). It was found that over the entire study period, the frequency of very strong droughts in the territory varies from 43% in Napas (north) to 50% in Tomsk (south). The highest frequency of droughts is observed in July, which reached values of 67% (very strong drought), 7% (severe drought). In the dynamics of droughts in the territory of the Tomsk region, for the entire vegetation period, fluctuations

Figure 12. Precipitation amount in fire hazard period in the Tomsk region

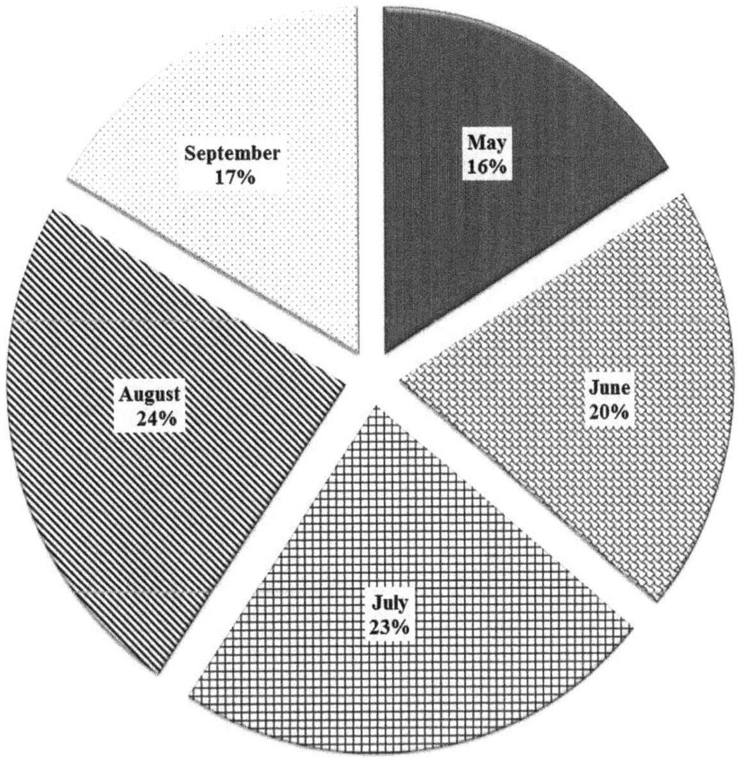

with a period of 11–20 years can be traced (Figure 11). In the last eleven-year period, the frequency of droughts is mostly within the normal range, with the exception of 1998–1999, 2003, 2012, and 2016, when values were around 60% over the period.

Further, precipitation for the fire hazard period and for each month for the stations of the Tomsk region was calculated. Figure 12 shows the distribution of precipitation amount by months relative to the total precipitation for the entire period. July and August (23 and 24%, corresponding to 70 and 73 mm respectively) make the largest contribution to the moisture regime of the fire hazard period, the amount of precipitation on average was 302 mm throughout the Tomsk Region.

Figure 13 shows the deviation of precipitation amounts from the average long-term value for each month as a percentage. As well as in the dynamics of HTC, periodic fluctuations can be traced. It shows that 1–2 waves can be distinguished. In general, variations in the characteristics for June and July are about 20–30 years; for May, August and September, 17–40 years.

In may the average long-term precipitation amount is 49 mm, which corresponds to 100%. We can distinguish the presence of some waves in the fluctuations of parameter. In May, the periods have the following form: from 1960 to 1975 there was an increase in the quantity studied, from 1976 to 2000 there was a decrease in the characteristic, from 2001, and to the present, an increase in the amount of precipitation is observed. In most cases, fluctuations in the studied characteristics are observed within the normal range (the range of the "norm" is from 84 to 116%). A slight decrease in size in recent years can be explained by a significant deviation of precipitation amounts in 2016 towards more arid conditions (31 mm).

Figure 13. Precipitation amount in the Tomsk region, % (100% is the average long-term value of a month)

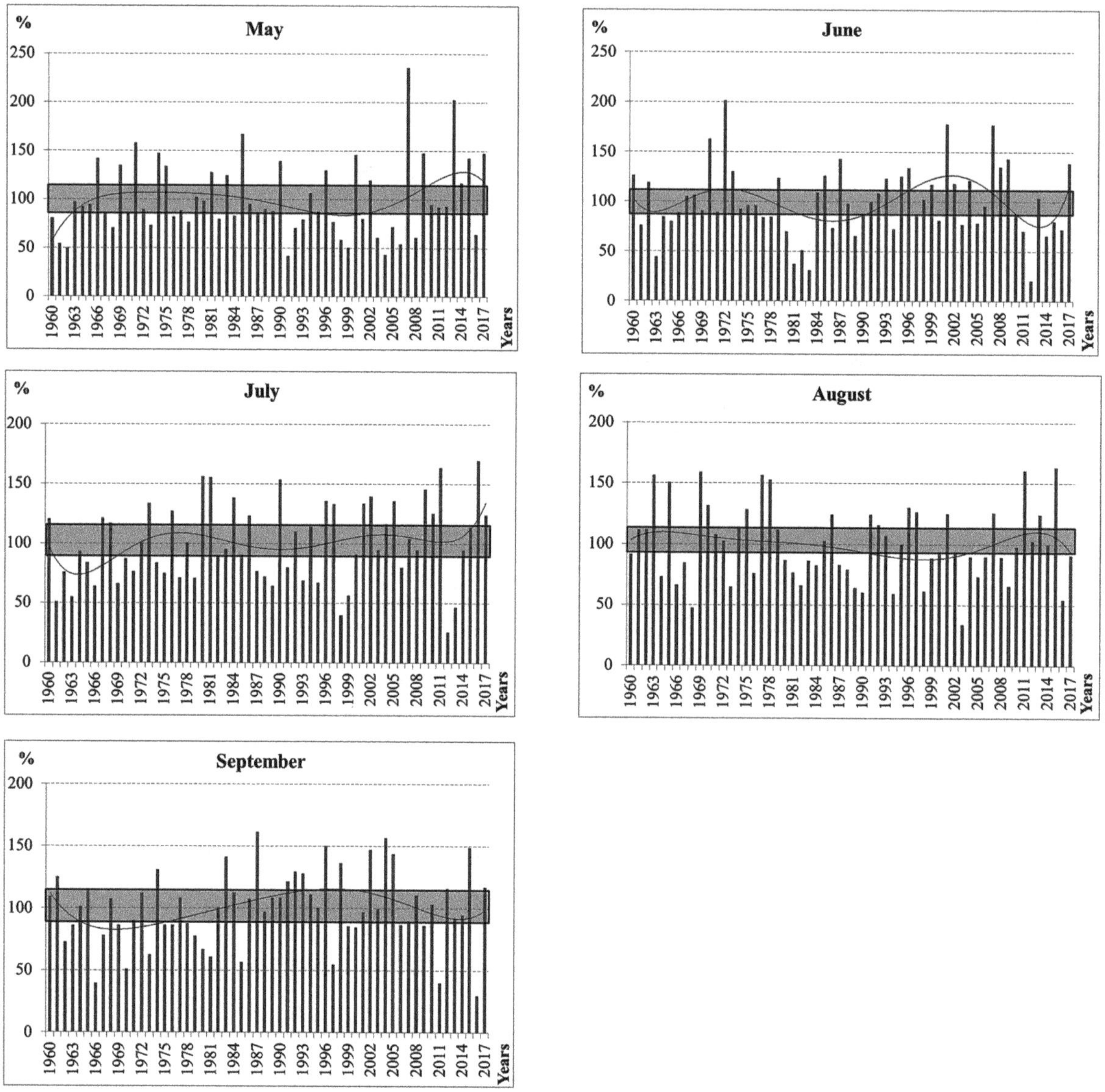

In the last decade, there is excessive moisture throughout the Tomsk region. In most cases, the characteristic values are within the normal range, except for some individual years. In August, the highest perennial mean value of precipitation is observed over the entire study period in the Tomsk Region (the total precipitation is 73 mm, which corresponds to excessively wet conditions throughout the region).

In recent years, there has been a tendency for precipitation to increase in June, July and September. However, in 2016 in May and in August, in 2012 in June and July, the largest negative deviations from the norm were observed (from 40 to 60%), which corresponded to very dry conditions.

As can be seen from the analysis of the presented characteristics, conditions with insufficient moisture were observed in 1998–1999, 2003, 2012–2013, 2016. In the same periods, the maximum numbers of fires were noted (Figure 4). Consequently, the repetition of such periods will lead to an increase in the number of fires and their area.

Thus, in the conditions of modern climatic changes, one of the main factors determining the fire hazard of the territory is the hydrothermal characteristics of the warm period of the year, including the distribution of moisture during the fire hazard period.

CONCLUSION

- Climate-induced fire frequency and burned area are increasing within Siberian forests. However, in those regions where forest fires are regularly monitored, it is possible to prevent their widespread occurrence. It is very difficult to control taiga fires throughout Siberia areas. Only the results of satellite monitoring will help to take early measures to prevent the spread of fire over large areas.
- The duration of the fire season increases due to climate change. This is due both to the earlier snowfall and the onset of the phenological spring, and to the increase in the duration of the thunderstorm period.
- Thunderstorms in Siberia are a much more frequent cause of forest fires (28%) than in other territories.
- Wildfire frequency is correlated with air temperature and drought anomalies. This result supports the hypothesis of climate-driven increase of fire frequency in boreal forests.
- An increase in the number of days with abnormally warm weather in the south and southeast of the territory will result in a magnified fire hazard of forest areas.
- If the frequency of the synoptic blocking processes increases, it will also increase the fire danger in the region too.
- An increase in the number and duration of May droughts will contribute to an increase in the fire danger of forests.

ACKNOWLEDGMENT

This research was supported by the Russian Foundation for Basic Research [grant N° 18-45-700010 p_a]

REFERENCES

Berezhnaya, T. V., Golubev, A. D., & Parshina, L. N. (2012). Anomalous hydrometeorological phenomena on the Russian Federation Territory in July 2012. [in Russian]. *Meteorology and Hydrology, 11*, 107–118.

Bunn, A. G., Goetz, S. J., Kimball, J. S., & Zhang, K. (2007). Northern high-latitude ecosystems respond to climate change. *Eos (Washington, D.C.), 88*(34), 333–335. doi:10.1029/2007EO340001

Federal State Budgetary Institution. West-Siberian Administration for Hydrometeorology and Environmental Monitoring. *The list and the criteria for severe weather events for the service territory of the West Siberian Administration of Hydrometeorology.* Retrieved from http://www.meteo-nso.ru/pages/115 (in Russian)

Field, C. B., Barros, V. R., Dokken, D. J., Mach, K. J., Mastrandrea, M. D., Bilir, T. E., ... White, L. L. (Eds.). (2014). *IPCC 2014: Climate Change 2014: Impacts, Adaptation, and Vulnerability. Summaries, frequently asked questions, and cross-chapter boxes. A contribution of Working Group II to the Fifth Assessment Report of the Intergovernmental Panel on Climate Change.* New York, NY: Cambridge University Press.

Flannigan, M., Cantin, A. S., de Groot, W. J., Wotton, M., Newbery, A., & Gowman, L. M. (2013). Global wildland fire season severity in the 21st century. *Forest Ecology and Management, 294*, 54–61. doi:10.1016/j.foreco.2012.10.022

Frolov, A. V., Blinov, V. G., Golitsyn, G. S., Dymnikov, V. P., Izrael, Yu. A., & Kattsov, V. M. … Semenov, S. M. (2014). The second Roshydromet assessment report on climate change and its consequences in the Russian Federation. General summary. Moscow, Russia: Federal Service for Hydrometeorology and Environmental Monitoring (Roshydromet)

Garmaeva, S. D., & Nosyreva, O. V. (2016). Assessment of the normative depth of seasonally frozen soil layer in Western Siberia using GIS. In Proceedings of 16th International Multidisciplinary Scientific GeoConference SGEM 2016. Sofia, Bulgaria: STEF 92 Technology Ltd. doi:10.5593/SGEM2016/B23/S11.015

Gorbatenko V. P., Gromnitskaya A. A., Konstantinova D. A., Ershova T. V., & Nechepurenko O. E. (2015). Assessment of the role of climatic factors in the occurrence and spread of forest fires in the Tomsk region. *Bulletin of Tomsk State University, 395*, 233–240. (in Russian). doi:10.17223/15617793/395/38

Gorbatenko, V. P., & Konstantinova, D. A. (2011). Mesoscale convection and dangerous weather phenomena in southeast of Western Siberia. In *Proceedings of 7th Asia-Pacific International Conference on Lightning* (pp. 160–164). Chengdu, China. Beijing, China: Tsinghua University 10.1109/APL.2011.6111094

Gorbatenko, V. P., Sevastyanov, V. V., Konstantinova, D. A., & Nosyreva, O. V. (2019). Characteristic of the snow cover for the Western Siberia territory. *IOP Conference Series: Earth and Environmental Science, 232*, 1–5. 10.1088/1755-1315/232/1/012003

Gringof, I. G., & Kleshchenko, A. D. (2011). Basics of agricultural meteorology. Obninsk, Russia: Federal Service of Russia for Hydrometeorology and Monitoring Environment. All Russian Research Institute of Hydrometeorological Information – World Data Center. (in Russian)

Hansen, J., Ruedy, R., Sato, M., & Lo, K. (2010). Global surface temperature change. *Reviews of Geophysics*, *48*(4), RG4004. doi:10.1029/2010RG000345

Konstantinova, D. A., Gorbatenko, V. P., & Ershova, T. V. (2010). Spatial distribution of ligtning discharges density on Western Siberia. In *Proceedings of 30th International Conference on Lightning Protection (ICLP)*. Cagliari, Italy: Electrical and Electronics Engineers. 10.1109/ICLP.2010.7845826

Konstantinova, D. A., Gorbatenko, V. P., & Polyakov, D. V. (2017). Spatial distribution of the thunderstorm activity characteristics for the territory of Western Siberia. In *Proceedings of SPIE: 23rd International Symposium on Atmospheric and Ocean Optics: Atmospheric Physics*. doi:10.1117/12.2285994

Kovacs, K., Ranson, K. J., Sun, G., & Kharuk, V. I. (2004). The relationship of the Terra MODIS fire product and anthropogenic features in the Central Siberian landscape. *Earth Interactions*, *8*(18), 1–25. doi:10.1175/1087-3562(2004)8<1:TROTTM>2.0.CO;2

National Aeronautics and Space Administration. (NASA). *Optical Transient Detector (OTD) data sets*. Retrieved from https://ghrc.nsstc.nasa.gov/lightning/data/data_otd.html

National Aeronautics and Space Administration. (NASA). *Fire information for resource management system*. Retrieved from https://firms.modaps.eosdis.nasa.gov/map/

Nosyreva, O. V., Barashkova, N. K., & Kizhner, L. I. (2018). Determining the duration of the heating season in Tomsk. *IOP Conference Series: Earth and Environmental Science*, *211*, 1–9. doi:10.1088/1755-1315/211/1/012056

Polyakov, D. V., Barashkova, N. K., & Kuzhevskaya, I. V. (2014). Weather and climate description of anomalous summer 2012 in Tomsk region. *Russian Meteorology and Hydrology*, *1*(1), 22–28. doi:10.3103/S106837391401004X

Ponomarev, E. I., & Kharuk, V. I. (2016). Wildfire occurrence in forests of the Altai-Sayan Region under current climate changes. *Contemporary Problems of Ecology*, *9*(1), 29–36. doi:10.1134/S199542551601011X

Ponomarev, E. I., Kharuk, V. I., & Ranson, K. J. (2016). Wildfires dynamics in Siberian Larch Forests. *Forests*, *7*(6), 125. doi:10.3390/f7060125

Stocker, T. F., Qin, D., Plattner, G.-K., Tignor, M., Allen, S. K., Boschung, J., ... Midgley, P. M. (Eds.). (2013). *IPCC 2013: Climate change 2013: The physical science basis. Contribution of Working Group I to the Fifth Assessment Report of the Intergovernmental Panel on Climate Change*. New York, NY: Cambridge University Press.

Volkova, M. A., Cheredko, N. N., & Titovskaya, A. A. (2019). Spatio-temporal distribution of periods with low and high temperatures in Western Siberia for 1961-2016. In *Proceedings International Young Scientists School and Conference on Computational Information Technologies for Environmental Sciences CITES 2019*. Moscow, Russia: IMCES.

Chapter 13
Influence of Environmental Conditions on the Susceptibility of the Territories to the Occurence of Forest Fires:
Forest Fire Danger

Olga Anatolievna Pasko
Tomsk Polytechnic University, Russia

Vasily Fedorovich Kovyazin
St. Petersburg Mining University, Russia

Nadezhda Anatolyevna Lebedeva
https://orcid.org/0000-0003-4095-2631
Kherson State Agrarian University, Ukraine

ABSTRACT

The chapter describes the conditions of predisposition of territories to the emergence of forest fires on different continents of the world. Information on the types of forest fires and the characteristics of burning materials is given. Human and economic losses from the forest fires and other emergencies (earthquake, flood) are compared. The causes of forest fires and their dynamics in Europe, USA, Canada, Southeast Asia, and Russia are given in this chapter. An analysis of the factors of their occurrence is given. Forest fires in the Russian Federation, where they annually cover large areas, have been studied in detail. The dynamics of the burning of Russian forests in the regions and administrative districts of the Tomsk region is considered. The causes of fire emergence is revealed. The forecast of forest fires is given and zoning of forest areas of the region as for fire danger is carried out. The research identifies the role of natural conditions in the occurrence of forest fires at various territorial levels (continents, countries, regions, areas).

DOI: 10.4018/978-1-7998-1867-0.ch013

INTRODUCTION

Forests are to be an important element of a country's natural resource potential and play a diverse role. The Russian Federation accounts for over a quarter of all world timber reserves (82 billion m3), more than two thirds of its territory is covered by forests of various composition.

Annually, forest fires occur in various regions of the Federation, posing a danger to people, settlements, animals and birds. Fires cover large areas, often hundreds of thousands of hectares, leading to the burning of millions of hectares of forest, including those covered with valuable coniferous species. In the massive fires there is a large-scale smoke, which disrupts the operation of air, water and road transport. Smoky clouds slow the ripening of crops. There is a burnout of settlements. Appealing to ambulance is increasing, and the mortality rate rises sharply. All this causes enormous economic, environmental and social damage to regions and countries.

Several thousand people from the Ministry of Emergency Situations and Forest Protection and hundreds of different vehicles, including aviation, are involved in the fight against fire. With the transfer of the forest fund to tenants, the involvement of forest areas into economic circulation, the increasing of number of loggers, expeditions, tourists and collectors of forest products there is an increase in fire danger in all federal districts of Russia. Therefore, the task of identifying the main factors of predisposition of the territory to the occurrence of forest fires and the development of new methods of prevention are particularly relevant for the foreseeable future. The reduction of ecosystems' burning becomes a major contribution of researchers to the preservation of forest resources and the protection of nature of the Russian Federation.

The aim of the research is to identify the common global and regional patterns of the forest fires' occurrence and course.

To achieve this goal, the following tasks are solved to:

- Determine the number and types of forest fires, the burned-out area at different territorial levels (country, region, administrative region)
- Reveal the role of the regional environmental conditions on the predisposition of the territory to forest fires
- Differentiate the regional and seasonal pyrogenic situation of the region, its temperature, the amount of precipitation and wind strength
- Assess the environmental consequences and financial losses from fires
- Propose some modern preventive measures for the burning forests reduce, worsen the state of forest stands, increasing their productivity, sustainability and diversity

The object of our research is the statistical material as for the forest fires, presented on the official websites of different countries. Information on the forest fund and fires that have been obtained from forestry's inventory materials of all the 85 subjects of the Russian Federation. The forest fund of the Russian Federation, which amounts to 1172 million ha, is characterized by a timber stock of 81.3 billion m3 and an average of annual growth, that is about 931 million m3 (Kovyazin, Romanchikov, Pasko, 2015). Information as for the forest management gives an idea of the country's forest complex, it is updated every 10 years and is placed in special collections (Moved from below).

Table 1. Brief specification of GIS technologies on the state of forest resources

Subject of development	Information and expert system
The nature of use	operational and analytical reference
The level of application	Federal, regional, local
Typical assignment	Organization and analysis of heterogeneous materials
Object of analysis	State of forest ecosystems
Specialization uses	Inventory valuation and expert monitoring procedures

(Farber, 2002)

Research methods: collection of statistical information on forest fires from the official websites of countries and regions of Russia; structural and comparative analysis of statistical data; literary sources and media materials review; studying the reports of the Ministry of Natural Resources and Ecology, the Ministry of Emergency Situations, the Ministry of Defense and the Russian Guard of Russia and the main forestry's documents as for the fires in Tomsk Region - "Books of forest fires".

For each fire in the administrative district, the date and the reason of its occurrence and liquidation, the burned-out area and type of fire were specified. Actual data on the fire was used from the sites of the Central Aviation Forest Protection Aviation Forest Base, All-Russian Research Institute for Civil Defense and Emergencies, the space information reception and processing laboratory, the space monitoring center of the Institute of Solar-Terrestrial Physics, Baikal Information Center and Engineering-technological center "ScanEx"; information system FORUS (Forest Resources Information System), which displays the distribution of forests by country. Forest fire obtained data using the MODIS (Moderate Resolution Imaging Spectra radiometer) radiometer for Earth exploration from space using the EOS (Earth Observing System) program, the Forest Fire Book and reports from the Ministry of Emergency Situations and the media are systematized.

Various programs were used for data processing: the Microsoft Office package, the Photoshop graphic editor with which the scanned forest fund maps and fires on it were processed, MapInfo Professional. The main content of the GIS technology specifications for data processing is given in Table 1.

BACKGROUND

For the larger half of forest zone of the Russian Federation, forest fire is one of the most dangerous natural phenomena leading to significant economic losses and negative environmental consequences. Only in the forest-tundra and northern sparse taiga, in particular on the permafrost, ground fires constitute a natural mechanism for preventing forest degradation and their transformation into grass-brush thickets and swamps. The liquidation of fire in this region leads to the accumulation of a thick surface organic material layer, which reduces the depth of the active soil layer and causes the permafrost to rise. There is a worsening of forest conditions, a decrease in the forests' productivity, shrubbing and water logging.

However, even in such conditions frequent repetition of fires leads to a deterioration of the state of forests, their productivity and stability. Fires are the main cause of the treeless belt 100–250 km wide on the border of taiga and tundra, which gradually expands to the south. They have a negative impact on biodiversity at the species, ecosystem and landscape levels.

Figure 1. World fire map
Source: Global forest watch fires, 2019

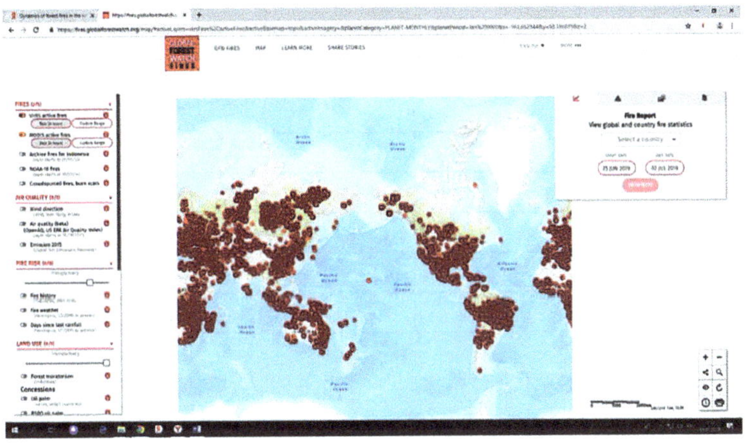

General Information About Forest Fires

In the scientific literature there are various definitions of the term "forest fire". For example, this is "undesirable combustion of forests and other natural objects" (Tedim @ Leone, 2015), "burning of tropical, temperate and boreal forests either by natural fire or artificial fire in connection with clearing land and deforestation" (Yadav @ Devi, 2018); "A natural phenomenon that occurs during long periods of drought caused by climatic events" (Göltenboth @ Widmann, 2006); spontaneous and "uncontrollable spread of fire across forest areas" (Scherbov, Lazareva @ Zhurkova, 2015), etc.

During natural forest fires, unplanned forest burning occurs, for example, due to lightning. An anthropogenic forest fire is the result of the unauthorized burning of forests to release land from dried herbs for crops.

Forest Fires in the World

Annually there are from 300 thousand to three million fires occur almost all continents (Fig. 1). Often they are very long, widespread and have catastrophic consequences.

Fire, like earthquakes or volcanic eruptions, is perceived as a risk that can be avoided, and huge amounts of money are spent in developed countries (Donovan @ Brown, 2007). Fire extinguishing is carried out in order to prevent a more powerful fire in the areas most exposed to them, primarily covered with forests (North, Stephens et al., 2015; Stephens et al.2014).

Despite on the low proportion of boreal forests, the scale of their burning in the world has increased 10 times in recent years (Yadav @ Devi, 2018). The resulting ash is carried away by the winds and has a devastating effect on air quality. It contributes to the development of "greenhouse effect" (Sapkota, Symons, Kleissl, Wang, Parlange, Ondov, Breysse, Eggleston, Buckley, 2005). Similar fires constantly occur in the savannas of Africa, Australia, steppes and grasslands of North America, Europe and Asia.

The growth of cities and population density, changes in land use patterns increase the risk of man-made fires, reduce social and environmental protection of the environment (Tedim @ Leone, 2015) and lead to an increase in the number and extent of forest fires. Despite the fact that the number of injured,

Table 2. Global comparison of human and economic losses caused by forest fires, earthquakes and floods between 1901 and 2014

Indicator	Forest Fires	Earthquakes	Floods
Dead people	3753	2 574 627	6 947 908
Wounded man	6812	2 614 875	1 329 923
Affected people, million	6	190	3604
Risk of death, %	0,06	1.4	0.02
Total direct damage (US $ million)	54 828	774 771	681 427
Cost per event (US $ million)	142	600	152
Cost per person ((US $)	9138	4078	189

(Guha-Sapir et al, 2019).

wounded and killed people in forest fires is much lower than in earthquakes and floods, the cost of measures for eliminating this natural disaster per person is extremely high (Table 2).

According to statistics, over the past 5 years, a number of regions most affected by fires have been identified. The largest forest fire in Canada was registered on 504.4 thousand hectares and 1921 buildings were destroyed. The largest forest fires in the world are typical for the following regions:

1. Russia.
2. North America - USA, Canada.
3. Western Europe - Portugal, France, Spain.
4. Australia.

Forest Fires in Europe

According to 24 countries data (with the exception of Russia), large-scale (> 40 hectares) fires affect an average of about 300 thousand hectares of European territory. 72.008 hectares (29% of fumes) are in specially defended natural areas protected under the "Natura 2000" network (Evans, 2012). In these territories, post-burn succession occurs spontaneously (Berry @ Arsenault, 2015). The percentage of the area of specially protected natural areas affected by fires varies from 0.002% in Germany (one fire affected 133 hectares) to 2.137% in Portugal (100 fires affected 40 837 hectares). For 31 years (from 1980–2011), 20,609 fires were registered in Portugal, which resulted in the burning of 135,934 ha (ICNF, 2012). There is a tendency of fire risk growth and burnout area (Fig. 2).

Forest fires are the major threat of the productivity and sustainability of European forests. As a result of socio-economic changes, depopulation and aging of the population, abandonment of land, changes in the use of forest products, the spread of flammable species (spruce and eucalyptus) has been occurred. The increase in the accumulation of fuel and the combustibility of ecosystems, the expansion of areas of specially protected natural areas also increases the potential damage from fires. The picture is complemented by the urbanization of natural areas and the rejection of agricultural production. In general, these causes allow forests to approach settlements and increase their pyrogenic danger. Most of the fires in Europe have their anthropogenic origin. They induce the emergence of new fires, lead to a

Figure 2. Forest fire occurrence and burned area between 1980 and 2011.
Source: ICNF, 2012.

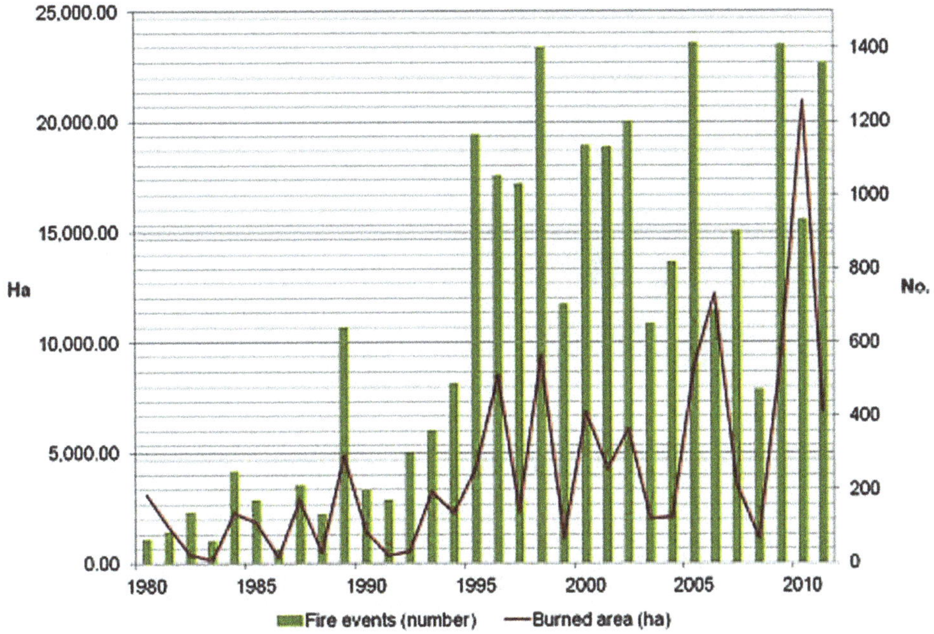

Figure 3. The occurrence of a forest fire (a) and the corresponding area burned (b) in the European Mediterranean region during the period of 1980–2010.
Source: San-Miguel-Ayanz, Moreno & Camia, 2013.

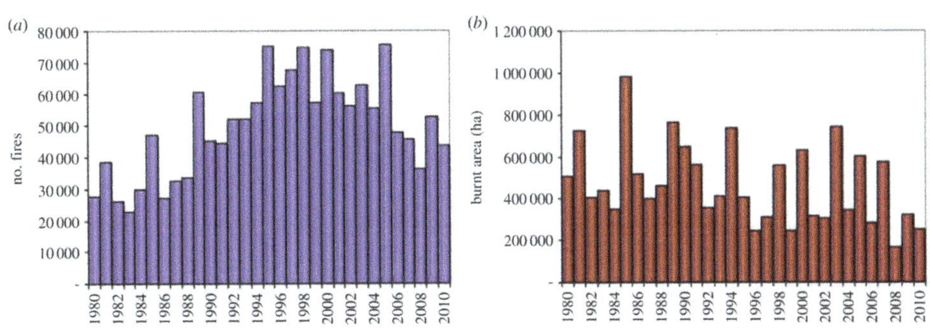

Figure 4. The fire in Oklahoma.
A photo from space, 2019

change in their number and area (Fig. 3) and affect their features such as, for example, the flammability of forests, the extinguishing technique, people's readiness for fires.

For example in Czech Republic, which area is 78.866 km2, more than 33.9% of the territory is covered by forest. The annual forest fire area is 296 ± 136 hectares (European Forest Fire Information System, 2019). In the Czech Republic they are considered socially and economically undesirable. Burnt areas are planted on a mandatory basis within two years from the formation of burns. Partially or completely burnt trees are uprooted, and seedlings are quickly planted in their place (Berry and Arsenault, 2015).

In Ukraine and Belarus, forest fires are also considered to be a problem of national importance. Annual statistics of forest fires in Ukraine is about 1,250 cases on an area of 1,250 hectares; in the Republic of Belarus it is about 320 on an area of 25 hectares. Employees of the Ministry of Emergency Situations, forestry, armed forces, population and volunteers are engaged in fire fighting. It should be noted that forest and peat fires in these countries are eliminated very quickly. This is caused both by a small area of ignition and by a high level of prevention of forest fires - from the installation of protective ditches and lanes, to monitoring the state of the forest during a fire-hazard period.

Forest Fires in the USA and Canada

There are different points of view as for the importance of forest fires in the life of human society. According to traditional approach, they are perceived as some natural hazard (Bleiege Bird, Bird & Codding, 2016; Mistry, Bilbao, Berardi, 2016). At the same time, a lot of communities that are neighbors near the fire perceive it as a valuable natural factor.

Information about the change in the area of fires in North America is controversial. According to data, its growth amounts to 0.1% of the forest area per year (Giglio, Randerson & Van Der Werf, 2013); according to another data such change is substantially (Rocca, Miniat, & Mitchell, Dennison, Brewer, Arnold, Moritz 2014) and is considered to be stable, but A.L.R. Westerling (2016) notes reduction and growth of the forest fires' areas. There is information that the average annual burning area in 1984-2009 is only 14% of the area of 1500-1850 years (before the European settlement) (Mallek, Safford, Viers, Miller, 2013).

Figure 5. Fire statistics in Russia for 7 years
Source: Authors' development

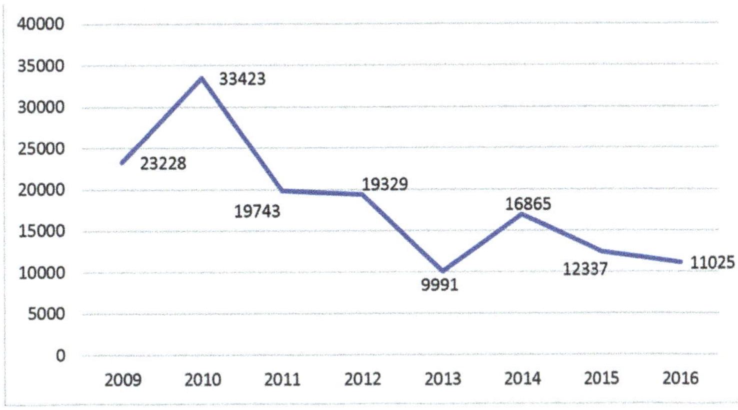

A sharp reduction of the number of fires in Europe and North America is contributed by their rapid detection due to satellite monitoring (Fig. 4) (Fujioka 2009). Automatic processing of remote sensing data has been observed in Europe and North America and the use of effective fire extinguishing equipment (Kudryavtsev, Lukin, Malinetskii, Mitin, Naumenko, Podlazov, Rumyantsev, Toropygina, 2008).

In the European part of Russia, the number of fires, on the contrary, multiply increased, possibly due to the reorganization of forestry, the abolition of the federal government, the reduction of the state forest service staff.

Forest Fires in South-East Asia

Burning-out of forests usually occurs in dry periods under the influence of uncontrolled logging and slash-and-burn agriculture (*Göltenboth & Widmann, 2006)* with the expansion of the plantation areas, pepper and vanilla usually produce forest burning, as a result of which massive wildfires develop. There are cases when the entire region was covered with haze and smoke for several months, and air traffic was stopped.

Forest Fires in Russia

In Russia, on average, between ten thousand and about 34 thousand fires occur annually (Fig. 5), and in some areas even a state of emergency is introduced. The fire destroys forests on an area of more than 18 million hectares, of which 0.5-2.1 million hectares are forest land and 12-36 thousand hectares are those, which protected areas by the forest fund. Large financial and labor resources are needed to restore forests in Russia.

Over the years, vast forest areas, both in the European part of the country and in Siberia, are subjected to the destructive action of fire (Kalinin, Kovyazin, et al., 1993). The most fire-dangerous regions of the country are:

Table 3. Characteristics of forest fires in Russia from 2012 to 2016

Indicators	Years				
	2012	2013	2014	2015	2016
Registered fires, PCs	18 010	10 249	17 058	11 400	11 025
Passed by fire, million hectares	2,3	1,4	3,7	2,5	2,4

(The combustibility of the forests of Russia, 2019)

Figure 6. Fire in the Far East (A) and u in the South of Italy (B)

- **The Far East:** Especially Khabarovsk and Primorsky Krai
- **Siberia:** Khanty-Mansiysk Autonomous Okrug, Trans-Baikal Territory, Krasnoyarsk Territory and Tomsk Region
- **Volga and Ural:** Sverdlovsk region.

The most widespread in terms of territory coverage and damage in Russia are forest fires in the taiga. The huge areas occupied by the forest, the remoteness of large settlements, difficult access roads allow the fire to gain its strength quickly. Fire extinguishing is often carried out with the help of aviation.

The European part of Russia, due to the high level of land development, is not a region of high fire risk. In Moscow Region, peat fires are the most common, carrying smoke to the capital, worsening the unfavorable ecological situation.

One of the lowest levels of fires is noted in the Karelia Republic. About 150 fires of about 320 hectares occur here annually, which is 13 times lower than the national average. Statistics of forest fires in Russia by year are presented in table 3.

As can be seen from table 3, the number of fires in 2015–2016 decreased to 30–33% compared with 2014. This may be related to the monitoring of forest resources (Afonin & Belov, 2001; Pasko, Tokareva, Alshaibi, Chernikova, Cabral, 2019), carries out technical improvement of the fire service, and the scientific selection of the best means of counteracting fire, depending on its type, nature and speed of forest fires spread (Fig. 6). At the same time, in Russia, regular fire monitoring is carried out only on 2/3 of the total forest fund area. In areas of Siberia and the Far East, covering 1/3 of the forest fund, there is practically no active fight against fire and accounting for fires. Fires extinguishing in the Siberian taiga is often carried out using aviation. The number of fires per 1 million hectares of the Russian forest fund is smaller, but the average area of one fire is larger than in Europe and North America. This circumstance, as well as the presence of large unprotected forest lands, indicates an insufficient level of forest fire protection in our country.

The damage from forest fires amounts to billions of dollars. Direct damage is connected with the loss of wood and wildlife and the cost of reforestation, there is some indirect damage in the form of emissions of various combustion products such as aerosols, greenhouse gases and carcinogens (Gorev, 2004).

According to the statistical reporting of Russian regions, annually in a fume category the area of fires goes from 10 to 20% of that is covered by fire, and the rest of this area is damaged by forest stands of varying degrees (Korovin @ Isaev 1998). With such a ratio of the areas covered by fire and dead from the fire of forest stands, the annual area of forest fires will average at least 5-6 million hectares, including in actively protected areas. The burning capacity of forests on unprotected territory is estimated indirectly, through the accumulated area of burns and dead stands.

According to the state forest accounting, the accumulated area is 28.4 million hectares. At the same time, about 56% fall on the actively protected territory, and 44% on the unguarded one. The same or close to it ratio should also take place for the areas of forest fires in the protected and unguarded territory. With an average annual area of forest fires in the zone of active forest protection of about 1.2 million hectares, the area covered by fire in the unguarded territory of the forest fund is about 0.8 million hectares per year, and in Russia as a whole - about 2.0 million ha per year (The Flammability of Russian Forests, 2019).

In the absence of instrumental methods for monitoring fire areas and the long-standing practice of underestimating the damage caused by fire, the obtained estimates of forest burning should be considered as minimal. It is known that the share of forest fires accounts for about 60% of all tree stands, annually dying from the negative impact of the whole complex of anthropogenic and natural factors. This means that out of 28.4 million hectares of fires and dead stands in the forest fund of Russia about 17 million hectares should be burning (Forest fires - one of the most serious problems of Russian forests, 2019). Their area is twice more than the accumulated one of unforested clearings.

The annual areas of forest stands dying from fire should be 3.0-4.0 million hectares with an annual area of clear cutting from 1.5 to 2.0 million hectares and the same rate of reforestation on burns and cuttings. Taking into account forest cultivation works and measures to promote natural regeneration, as well as the presence of fumes' areas in the north of the country with worse reforestation conditions, the actual rates of forest regeneration processes on them in 2-3 times lower than on logging. Annual areas of forest stands dying from fire should be at least 1.0 million hectares.

In 2019 according to Rosleskhoz, the largest burned-out area of forests is amounted to more than 3.2 million hectares in Siberia, and much less in the European part, but only in the Pskov region 380 hectares of ripe coniferous forest stand have been burned. A special fire regime was in force this summer in 44 regions of Russia, more than 2 thousand people were involved in fighting fires, more than 200 pieces of equipment and about 20 aircraft took part in putting out the fire. Federal Aviation Protection involved in extinguishing more than 150 paratroopers. In Russia, on average, from ten thousand to 34 thousand fires occur annually, and in some areas of the country even an emergency regime is introduced. Fire destroys forests on the area of more than 18 million hectares, of which 0.5 - 2.1 million hectares - forest land and 12-36 thousand hectares - protected areas of the forest fund. Significant financial and labor resources are required to extinguish fires and restore forests after the fire.

Among the causes of forest fires' origin there are anthropogenic and natural reasons. The anthropogenic ones include the failure to comply with fire safety measures when carrying out certain types of work in the immediate vicinity of forests, whether it to be logging operations, mining operations or agricultural work. Responsibility for forest fires according to the law has to be borne by persons directly or indirectly influencing their occurrence and spread. Violators face a fine from 1,500 to 400,000 rubles or correctional labor for up to 2 years. Natural causes of forest fires are usually associated with thunderstorms, which can be a source of forest vegetation ignition. The reasons of forest fires have territorial specificity, which is expressed in different ratios of a number of common indicators.

Characteristic features of the spatial and temporal structure of forest burning, which have principal importance for their protective organization, are sharp variations in the number and area of forest fires by region of the country and periods of fire hazardous seasons. It has been established that from 50 to 90% of the forest area annually covered by fire falls on 3-4 regions of the country with extreme weather conditions. The area of emergency zones (IV-V classes), where a significant part of the fires gets out of the control of the regional authorities and takes on the character of a natural disaster, makes up only a few percent of the forest fund lands annually. Moreover, up to 95% of the total area covered by fire is accounted for by large forest fires, the number of which does not exceed 5% of the total number of sunburns in forests.

CONDITIONS OF FOREST FIRES OCCURRENCE IN THE TOMSK REGION

Natural Conditions

The Tomsk region is located in the Asian part of the Russian Federation. Its forest reserves are estimated at 2.6 billion cubic meters. Annual calculated cutting area exceeds 41 million cubic meters. In terms of total forest reserves, this region takes the 3rd place among the others of the Siberian Federal District and the 8th place in the Russian Federation (Evseeva, 2001). In the Tomsk region, about 380 forest fires are fixed annually, affecting 22.5 thousand hectares of forest area (Forest complex, 2019). Forest fires have a negative impact on the development of the logging and timber processing sectors of the region; damage to flora and fauna is done (Fig. 8); economy and budget of the region are getting worse; Significant funds are required to eliminate the consequences of fires (Gorev, 2004; Environmental movement instrument, 2001; Grishin, 2001).

The forests of the Tomsk region are divided into two categories according to their intended purpose: protective and operational (Fig. 10). The operational forests occupy 26954.2 thousand hectares (94% of the forest fund area), protective forests occupy 1,720.0 thousand hectares - 6% of the forest fund area (the names of the categories of protective forests are given in accordance with forest management materials and forestry regulations for forest areas (Doroshenko & Bondarev, 2001)

The total area of the forest fund of the Tomsk Region is 28,674.2 thousand hectares, of which 68.2% of the forest lands are covered by forests, 67% of which are forested, 1.1% of which are artificial stands. Land which is not covered with forest occupies 1% of the forest area, including burning and dead plantings - 0.3%, cuttings - 0.4%, open lands, vacant lots and clearing - 0.2%. Unbroken forest cultures and nurseries make up less than 01% of the forest fund area. Non-forested lands occupy 31.8% of the forest fund area, including swamps - 30.2%, rivers and lakes - 1%, arable land, hayfields and pastures - 0.2%, roads, glades and homesteads - 0.2%, sand and other lands - 0.2%. Coniferous plantations make up 53.7% of the forest area, or 10,315.6 thousand hectares. Of these are: pine - 29.0%, cedar - 19.0%, spruce and fir - 6%. Birch stands, occupy 36% of the area or 6912.3 thousand hectares. The total wood stock is 2848.8 million m3, including coniferous - 1603.1 million m3, softwood - 1244.5 million m3

Figure 7. The satellite images forest fires in the Tomsk region

Figure 8. A forest fire in the Tomsk region

The annual wood increment is equal to 33.02 million m3 or in terms of one hectare of 1.7 m3. The average stock of wood is determined at 148 m3 / ha, and in conifer stands - 155 m3 / ha. The annual allowable amount of wood removal in the territory of the Tomsk Region is 4,1,166.0 thousand m3, including 14,848.9 thousand m2 for coniferous wood, and 26,317.1 thousand m3 for softwood. The average age of plantations of the main forest-forming species is more than half of the age of logging. The highest average age for cedar forests (184 years old). Its high value indicates the presence of forest plantations that are promising for harvesting timber by clear cutting.

Figure 9. General characteristics of the forest Fund of Tomsk region

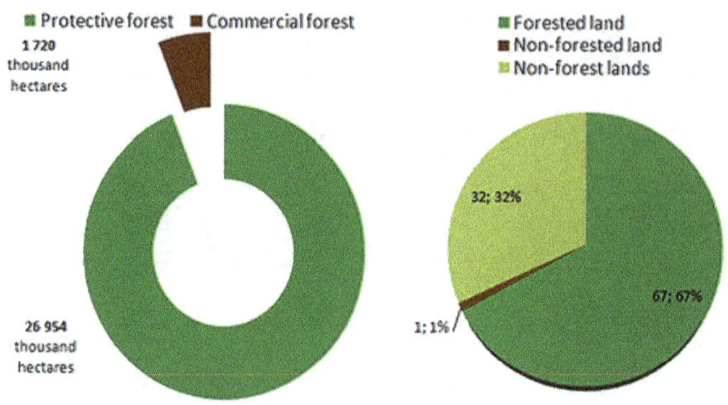

Table 4. Causes of forest fires in Tomsk region (numerator – the total number of the denominator – %)

Causes of Fires	Number of Fires	Area, ha
Agricultural burning	2/1	112/3
Because of lightning	102/23	7162/47
The fault of the population	186/70	3056/33
For unclear reasons	20/6	1923/17
Total	310/100	12253/100

Forest conditions, according to I.S. Melekhova, are one of the signs of the allocation of classes of fire danger according to natural conditions such as

1. The share of peat bogs in the study area.
2. The average number of forest fires, estimated on a five-point scale.
3. The average and maximum density values of lightning discharges into the ground.
4. Average precipitation over the summer period.
5. The value of the average height of the selected area above sea level.
6. Average air temperature over the summer period.
7. Fire danger in points according to the type of vegetation prevailing in this area of the territory (Melekhov, 1965).

Forest cover of the Tomsk region is 75% and is represented by taiga massifs and slopes with steppe vegetation. In the transition zone from dark coniferous taiga and pine forests to birch forests and forest meadows. The region is rich in upland and lowland swamps, rivers and lakes. Soils are podzolic on sandy loam.

Table 5. The area of wildfires in the forestry district from storms, ha

Forestry	2008	2009	2011	2012	2013	2014	2015	2016
Sulzatskoe				23,00		17		1,00
Suiginskoe				172,00		14	119	61,50
Molchanovskoe			137,0					
Klukvinskoe	3,50	20,10		8,5	13	32,30		1,00
Eltarevskoe	1,00				6,35	15,00	8,50	276,00
Aleksandrovskoe	0,10	218,5		967,84	10		0,1	
Strezhevoy				90,00		40		
...	...	...	...	...	...	...	...	...
Kozhevnikovskoe								4,00
TOTAL	4,78	285,6	137	5307,59	109,06	408,81	354,6	1626,41

The causes of forest fires in the Tomsk region are quite diverse (table 4). There are such as local people, a lightning storm, loggers, singosari, high air temperature, the locomotives, which remain of fires, power lines, burning of oil spill, the exhaust gases in the air, the objects of forestry, the fall of the rocket stage expedition. Accordingly, the causes of fires are both natural and anthropogenic factors. Every year, 64 to 527 fires occur in the Tomsk region. They lead to deforestation, reduce the cost of land and forest resources and require additional land management activities.

The high humidity of forest combustible materials keeps Tomsk region forests in a non-flame state for the most of fire-dangerous season. However, upon the onset of a prolonged drought, the moss-vegetation cover and forest litter dry out and turn into the main source of fire danger. Due to this, 98.5% of all fires in the north of the Tomsk Region are grassroots, 1.4% - mounted and 0.1% are underground. A related cause is a significant proportion of coniferous (pine) forests. The maximum of fires (up to 70%) is typical for pine of grass forest types.

Because of the combustible material presence in the vegetation cover therefore, 98.5% of all fires occur on the ground, on the top - 1.4%, and the rarest underground fires - 0.1%. The susceptibility of forests to fire is influenced by natural conditions and climatic factors that affect the drying of combustible materials. The drying time of wood materials depends on the type of forest. Sustainable burning of forest materials is possible only at a certain moisture conditions. The pine forests ripen most quickly on the tops of sandy hills and ridges, pine stands on elevated positions on sandy loams. The forests of hardwood even on drained soils rarely burn.

Other factors that determine the repetition of extreme fire danger seasons are dry periods in late spring and early summer, high summer temperatures, low relative humidity of air and wind, carrying dry and warm air from North Asia, Mongolia and East Siberia, lightning strikes during thunderstorms. These climatic features contribute to the transition of almost all forest areas to mid-summer in a state of fire maturity. The intervals between fires and their frequency determine the dynamics and state of the forest cover and vice versa, so their study should be the basis of fire management. Each forest cover has its own fire regime (type, intensity, size, fire frequency interval) and post-fire dynamics of reforestation. Reconstruction of fires allows to identify the typical fire regimes for this landscape and to predict their development trends for forest ecosystems.

To identify the dependence of the number of fires on meteorological factors, we've done the analysis of the fires' occurrence from thunderstorms in forestry (table 4).

There were no fires from thunderstorms in 2010, according to SE "Tomsk Center for Hydrometeorology and Environmental Monitoring", and in 2012 their number was the maximum (as well as the number of fires). Average monthly air temperatures exceeded the climatic norm in June - by 5-7 degrees, in July - by 2-3 degrees. Hot dry season combined with high thunderstorm activity led to a significant increase in fire danger on forest lands.

Anthropogenic Factors Causing Forest Fires

Following factors should be considered as the cause of forest fires in the territory of the north of the Tomsk region: an oil spill burning, exhausting gases into the air, rocket runoff, expedition activities, the local population, accidents at power lines, etc. To rank these reasons is not possible at the moment, because of the lack of factual material.

Forest fires can threaten settlements, camps, hamper production activities, destroy large tracts of forest resources and lead to the economic losses. On the other hand, they are to be some natural part of the forest ecosystem and important as a valuable self-management tool for improving the environment and eliminating excess fuel. Forecasting the development of a fire situation, monitoring the development of the situation with the use of video surveillance methods, geographic information systems, improvement of fire extinguishing equipment for land management activities (fire barriers, roads and water) can reduce the number of fires and risks of their spread to the Tomsk region oil and gas complex.

It should be noted that a reliable stable change in the number of fires and their area, in the Tomsk region as a whole and in its north, that is similar to the European one, has not been established. Therefore, the assumption of its negative impact on the reorganization of forest service, expressed by Russian scientists, has not yet been confirmed, as the introduction of new effective means for detecting and extinguishing fires, which is to be the characteristic for Europe and North America, has not been reflected.

Development of production and infrastructure improvement, including transport accessibility, on the one hand, contributes to the growth of forest fires, on the other hand, facilitates their detection and accelerates extinguishing (Forest fire and environmental management, 2003). Fires form a danger to oil and gas facilities and disrupt productive forests that could be developed. Their potential occurrence is determined by the oil burning, associated gas and other substances that are used or obtained in the production process. Technological reasons include oil spilled in open areas, air pollution, malfunctioning electrical equipment, non-imported wood, uprooted trees and various waste products (Pasko, 2013).

The situation is getting worth by the fact that in the north of the Tomsk region, where the objects of oil and gas complex are concentrated, mostly pine forests are spread, which are the most fire-dangerous (the growing range of birch forests is limited to 57 degrees north latitude).

The fire danger of the forests of the Tomsk region is determined by the presence of a significant proportion of coniferous forests, developed ground cover and hot dry summer. Especially favorable conditions are created for the occurrence of forest fires in areas with a continental climate. The forests are heavily littered with deadwood and logging residues, so the burning of organic matter lingers for a long time in a thick layer of forest litter and other combustible materials, creating a constant threat of renewed fires in strong winds (Kalinin, Kovyazin et al., 1993). Depending on the weather conditions in the forests of the region, three peaks of seasonal burning are expressed: the spring wave of fires, the summer steady fires and the autumn ones (figure 11).

Table 6. Comparative analysis of distribution local forest fires in hectares from agricultural burning and from lightning

The Origin of the Fires	Years							
	2008	2009	2010	2011	2012	2013	2014	2015
From the storm	4,78	285,6	137	5307,59	109,06	408,81	354,6	1626,41
From agricultural fires	252,8	270	69,3	80,11	175,5	5	27	175,7
The ratio of the magnitudes of thunderstorms to agricultural burning	0,018	1,057	1,976	66,25	0,621	81,762	13,13	9,26

Figure 10. Dynamics of forest fires in the forests of Tomsk region
Source: Pasko & Baranova, 2015

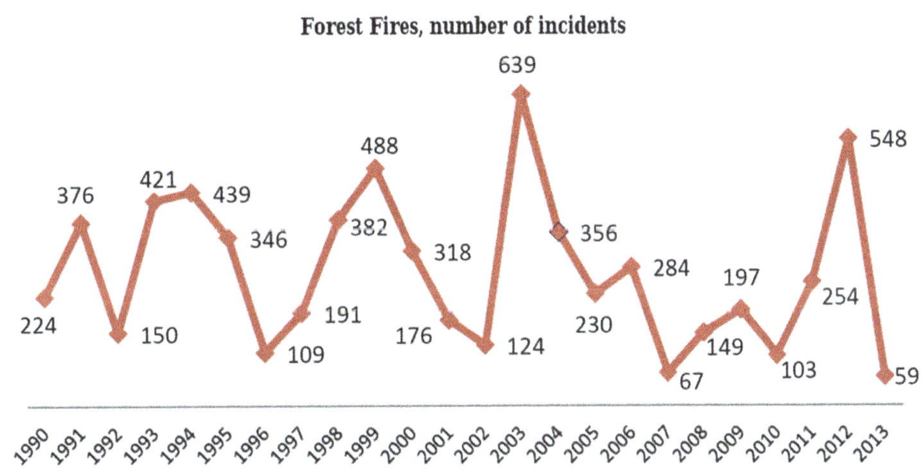

It has been established in the course of the statistic material analysis that the areas with the greatest fire risk are: Kargasoksky, Verkhneketsky and Tomsk; from the middle - Pervomaysky, Kolpashevsky, Molchanovsky with low - Asinovsky, Bakcharsky, Zyryansky, Parabelsky, Teguldetsky districts. Among the most favorable are Aleksandrovsky, Kozhevnikovsky, Krivosheinsky, Chainsky and Shegarsky districts. It should be noted that the districts of Kargasoksky, Verkhneketsky and Tomsk, as well as Parabelsky, Molchanovsky are areas of active exploration and mining. Oil and gas fields are located deep in the forests and, in case of an emergency, involve the transfer of flames to an array of forests. Each company has its own rules for safety, but, despite their existence, accidents happen every year. The situation of fires might be monitored on the website of the Department of Natural Resources and Environmental Protection of the Tomsk Region (Department of natural resources, 2019).

Figure 11. Seasonal fire activity on the territory of the Tomsk region by districts
Source: Pasko &, Baranova, 2014

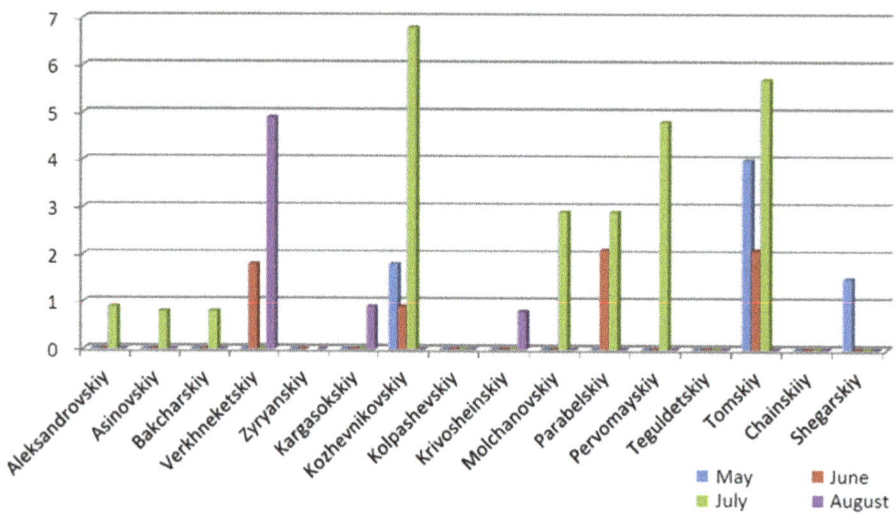

In order to identify the strength of the anthropogenic factor influence on the occurrence of fires, we estimated the frequency of fires from agricultural scorches and thunderstorms in the certain forest districts (table 6). High varied variability of these indicators was revealed. The difference between the maximum and minimum values of the ratio of the thunderstorms' number to the number of fallen reaches 454 units.

Compared with thunderstorms, fewer fires occur as agricultural and grass burning. The most active year in terms of the number of fires is 2009, with the lowest number of fires in 2013.

Statistics shows that the proportion of fires arising from anthropogenic causes is stable, and fires from natural factors (high temperatures) and phenomena (thunderstorms) ratio varies cyclically. Mass thunderstorms are replaced by more calm.

Characteristics of Forest Fires in the Tomsk Region

The number of forest fires varies from year to year considerably. The danger change process is cyclical (Fig. 10). For the period of 1990-2013 the average number of forest fires was 276; the frequency of fire activity — maxima (over 400 fires per year) — about six years; minima (less than 200) — about four years. Intermediate activity peaks (more than 200 fires) are grouped into a 4-6 year cycle. The first phase of each cycle is characterized by an increase in the number of fires and lasts 1-2 years. This is followed by a peak, and the second phase begins ("reduction to a minimum"), lasting 1-3 years. There are short-lived transitions from minimum to maximum and no period for stabilizing the number of fires.

In general, there are three peaks of seasonal inflammability (spring wave of fires, summer sustained fires and autumn fires) in the region, but in three areas in July are predominate (Fig. 11). The territories of these areas belong to the second and the third classes of fire danger with a high probability of low-level fires during the entire fire danger season.

A low frequency of fires has been established, combined with large areas of burnout, which can be explained by the weak development of territory and the difficulty of detecting fire sources. The duration of actual fire season does not exceed 87 days.

Table 7. Comparative analysis of districts of Tomsk region on fire danger

Municipal Districts	Number of Fires, pieces'	Area, ha	Fire Hazard	Grade
Kargasokskoe	107	19066		17
Berhneketskoe	87	50040,29		16
Timiryazev (village))	61	1749,98		15
Pervomaiskoe	47	2537,5		14
Kolpashevskoe	43	4546,39		13
Aleksandrovskoe	41	149210,7		12
Molchanovskoe	31	12445,6		11
Zyryansky	16	291,85		10
Asinovskoe	16	436		9
Tomskoe	13	1528,2		8
Bakchar	12	4683,34		7
Parabelskoe	12	1254		6
Teguldetskoe	11	688,8		5
Shegarskoe	10	75,4		4
Krivosheinsky	10	6864,8		3
Kozhevnikovskoe	9	94,5		2
Kedrovskoe (city)	3	649		1
TOTAL		256163,1		

Figure 12. Dynamics of changes in forest fire areas in Parabel, Kargasok and Alexander districts of Tomsk region in 2008-2013

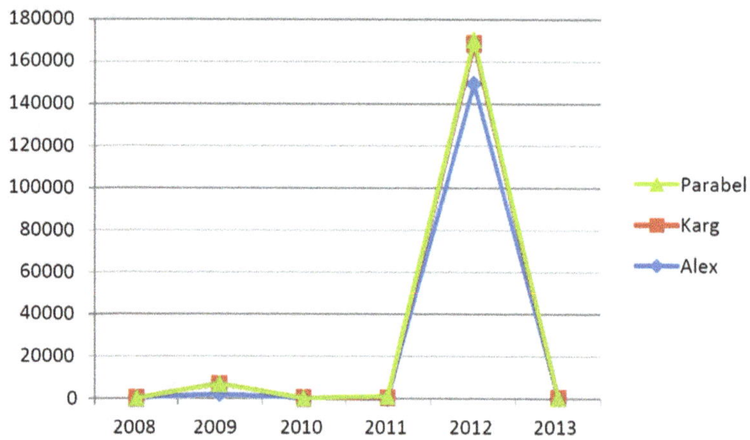

Summarizing the studies compiled in this section, we can say that the degree of fire hazard in the forests in the Tomsk Region is determined primarily by the seasonal activity of the local population and the features of the vegetation cover. The highest incidence of fire is specific for the Tomsk region (more than 35 fires per 100 thousand ha). 85% of fires occur at a distance of 10 km from Tomsk. This is due to the high density of the rural population, the proximity of the regional center and the richness of mushroom and berry lands. Verkhneketsky, Pervomaisky and Kolpashevsky (more than 10 forest fire incidents per 100 thousand hectares) are the next highest incidence regions. The forests of Parabel, Pervomaisk and Kargasoksky regions (less populated) are characterized by a low frequency of fires, but a significant burn-up area. This is due to poor development of the territory and the difficulty of detecting fires. The number of fires from thunderstorms, on the contrary, increases sharply with distance from residential villages: more than 90% of fires occur at a distance of more than 50 km from the settlements.

FOREST FIRE FORECASTING

Knowledge of the basic laws of the occurrence of forest fires allows their forecasting (Gusev. Erican V.A. & Erican E.V., 2011). For the first time it was implemented in 1920 and since then has been rapidly developing in various countries of the world. Forecasts of forest fires takes into account weather factors, terrain, humidity and types of flammable objects and sources of ignition (Nikishchenko, 2011), temperature maximum, temperature change, accumulated non-precipitation period and drought index (Chang, Chaur & Tzuhn, 2015). They allow to analyze and assess the risks of burning flammable objects in the forest based on the logistic regression model. Also, with the help of GIS-systems, spatio-temporal assessment of hydrothermal conditions is carried out.

We conducted a fire hazard prediction according to the Nesterov's method and its comparison with the real situation. The calculations were made for 2013, the northern regions of the Tomsk Region, the forest fire season began on May 10. A total of 59 forest fires were recorded. The area covered by fires amounted to 2211.0 ha, including forest - 2164.3 ha (Fig. 12). Compared to the same period of the pre-

Figure 13. Reliability chart of the forecast for 2013
Source: Source: Authors' development

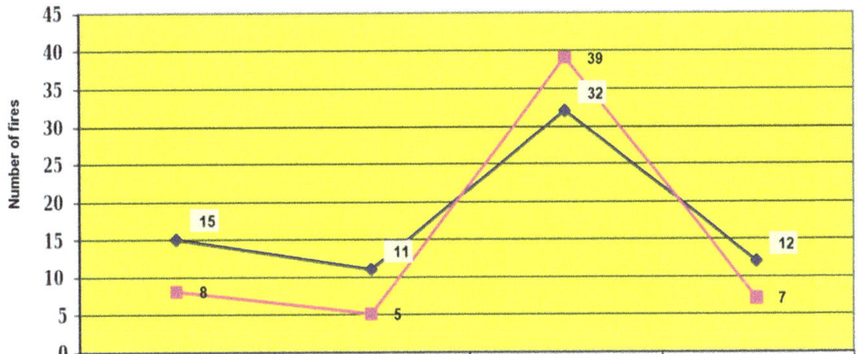

	May	June	June	July
Prediction	15	11	32	12
Fact	8	5	39	7

Prediction Fact

vious year, the number of fires by 548 units of fires and a reduction in the area of fires covered by fires by 295,586.0 ha was noted.

We have carried out a fire dander prediction using the formula:

$$G = \sum_{1}^{n} \left(t - \omega\right) t,$$

where G is the indicator of the ground cover burning, °C; n is the number of dry days (days with precipitation less than 2.5 mm are considered dry); t is the air temperature for 12 hours, °C; ω is dew point, °C (Khodakov &, Zharikova, 2011).

The high convergence of the predicted fire danger with the real one is established (Fig. 13).

Analysis of the statistical material on forest fires collected in the Tomsk Region and the Russian Federation indicates that their main causes are: arson of the local population (64.8%), lightning discharges (16,0%), burning dry grass on agricultural land (7.3%), burning logging residues (2.9%), various types of expeditions (0.9%), activities of other enterprises (5.0%) and not established causes (3.1%). Forest fires occur within a radius of 5-10 km from settlements.

SOLUTIONS AND RECOMMENDATIONS

To prevent the spread of forest fires, it is necessary to carry out timely forestry activities: sanitary logging, cleaning of logging sites from felling residues, cleaning of dead wood, as well as fire-fighting device of the forest fund. It includes: lying of mineralized strips around coniferous young growth, the creation of fire breaks, the construction of forest roads, the device of artificial and maintenance in good condition of reservoirs on forest lands and entrances to them, the creation of barriers of hardwood.

Measures for timely detection of forest fires are also important, such as the construction of a network of fire observation towers and masts and their equipment with video cameras, improvement of communication means of forestry and local authorities and employees of the Ministry of emergency situations. To predict forest fires, organize a meteorological service in the forest areas to monitor the weather and calculate a daily comprehensive indicator of fire danger in the forest, equip the air protection with modern aircraft and technical means of extinguishing. Create landing sites for helicopters in the forest fund of forestry.

Especially relevant at the present time equipment and technology of fire extinguishing. In remote regions, the forest service is often forced to extinguish fires, including large ones, by manual means, i.e. without the use of powerful technical means. However, these methods of extinguishing the fire do not give the expected results in the fight against fire. We need modern seaplanes, helicopters, artificial precipitation, which is especially important for the underdeveloped territories of the Russian Federation.

FUTURE RESEARCH DIRECTIONS

The consequences of catastrophic forest fires affect the development of natural ecosystems for a long time and often lead to irreversible results. Over the past decade, the frequency of catastrophic fires has been increasing both in the Russian Federation and in other countries. Therefore, further research is planned to focus on the environmental problem – the impact of forest fires on human settlements and components of the forest ecosystem.

CONCLUSION

According to the results of the study, the following conclusion has been made.

Forest fires, both natural and man-controlled forest Fund process, i.e. serious risks such as earthquakes and floods, human and environment. The number of human fires ranks third, and the number of people killed in it, after the flood and earthquake. But the price of eliminating the consequences of the fire is 2 times higher than the consequences of the earthquake.

The fire occurs on all continents where there are forests. In their occurrence and distribution affect the natural conditions of the territory: temperature and lack of moisture, wind speed, type and humidity of forest combustible materials, structure and composition of forests. Analyzing forest fires on the planet, the tendency is to increase the area covered by the fire. The cause of forest fires on all continents is the human factor.

In the Russian Federation, there is a tendency to reduce the number of forest fires, but over the past 5 years the burned area of the forest fund constitutes about 2.5 million hectares. Reduction of fires due to its growth in a large forest, and the order of forest fund observations of Russia. But fire fighting is an outdated technology, not in fire fighting modern technology.

Due to the natural conditions in the slope of forest fires is shown by the example of the Tomsk region, located in the Asian part of Russia. This zone is characterized by the average data of forest fires in Russia. In the forest Fund of Tomsk region-all categories of forest land. The wooded land is dominated by coniferous (pine) trees. The number of fires in the forest Fund of the Tomsk region varies from year to year. The greatest number of them was noted in the dry years -2003 and 2012. In other years of observation (from 1990 to 2013), an average of 200 to 450 forest fires occur. Varies in the number of forest area fires and burned it out of the zone. This is what has been said in various natural conditions in the forest because the area covers a large area and forestry is in all its parts.

The human factor is the cause of forest fire in the Tomsk region: burning of felling residues on the cutting area in the cabin of the main and intermediate use, agricultural logs of dry grass and set fire to the forest in order to obtain a wooden house and storage. Coniferous stands and the amount of combustible material to promote the rapid spread of fire in the forest Fund. Sometimes there are forest fires in the area, and lightning.

ACKNOWLEDGMENT

The research is carried out at Tomsk Polytechnic University within the framework of Tomsk Polytechnic University Competitiveness Enhancement Program.

REFERENCES

A forest fire in the South of Italy. (n.d.). Retrieved from https://tipolog.livejournal.com/305437.html

A forest fire in the Tomsk region. (n.d.). Retrieved from https://yandex.ru/images/search?text=%D0%BB%D0%B5%D1%81%D0%BD%D1%8B%D0%B5%20%D0%BF%D0%BE%D0%B6%D0%B0%D1%80%D1%8B%20%D0%BA%D0%BE%D1%81%D0%BC%D0%BE%D1%81%20%D0%A2%D0%BE%D0%BC%D1%81%D0%BA%D0%B0%D1%8F%20%D0%BE%D0%B1%D0%BB%D0%B0%D1%81%D1%82%D1%8C&pos=168&p=5&img_url=https%3A%2F%2Fpbs.twimg.com%2Fprofile_banners%2F1254679254%2F1383113680%2F1500x500&rpt=simage

Afonin, S. V., & Belov, V. V. (2001). System of space monitoring of forest fires in the Tomsk region, *Optics of the atmosphere and ocean, 8,* 14, 692-696.

Berry, L. E., & Arsenault, A. (2015). Regional case studies. The ecological importance of mixed-severity fires. Retrieved from https://www.sciencedirect.com/book/9780128027493/the-ecological-importance-of-mixed-severity-fires

Bliege Bird, R., Bird, D. W., & Codding, B. F. (2016). People, El Niño southern oscillation and fire in Australia: Fire regimes and climate controls in hummock grasslands. *Philosophical Transactions of the Royal Society*, *371*(1696), 20150343. doi:10.1098/rstb.2015.0343 PMID:27216513

Chang, J., Chaur, C., & Tzuhn, C. (2015). Discourse on Taiwanese forest fires, *Wildfire Hazards, Risks and Disasters.* 145-166. Retrieved from https://www.sciencedirect.com/science/article/pii/B9780124104341000087

Dennison, P. E., Brewer, S. C., Arnold, J. D., & Moritz, M. A. (2014). Large wildfire trends in the western United States, 1984–2011. *Geophysical Research Letters*, *41*(8), 2928–2933. doi:10.1002/2014GL059576

Department of natural resources and environmental protection of Tomsk region. (n.d.). Retrieved from http://green.tsu.ru/fires

Doerr, S. H., & Santín, C. (2016). Global trends in wildfire and its impacts: perceptions versus realities in a changing world, *Philosophical Transactions of the Royal Society B: Biological Sciences, 5,* 371(1696), 20150345. doi:. doi:10.1098/rstb.2015.0345

Donovan, G. H., & Brown, T. C. (2007). Be careful what you wish for: The legacy of Smokey Bear. *Front. Ecol.*, *5*(2), 73–79. doi:10.1890/1540-9295(2007)5[73:BCWYWF]2.0.CO;2

Doroshenko, S. G., & Bondarev, D. A. (2001). Monitoring of forest fires according to the space survey. Remote sensing of the earth's surface and atmosphere by aerospace means, Krasnoyarsk, 221-225.

Environmental movement instrument of legal regulation of forest resources use in Tomsk region: Materials of the round table. (2001). *Tomsk: OSU Oblkompriroda*, 30. European Forest Fire Information System. Retrieved from http://effis.jrc.ec.europa.eu

EOS (Earth Observing System). (n.d.). Retrieved from https://eos.com/platform

Evseeva, N. C. (2001). Geography of Tomsk region. Natural conditions and resources, *Tomsk: Publishing house TSU*, 223.

Fire in the Far East. (n.d.). Retrieved from https://www.sibreal.org/a/29352746.html

Forest complex of Tomsk region. (n.d.). Retrieved from http://tomsk.gov.ru/ru/ekonomika/lesnoy-kompleks/

Forest fire and environmental management. A technical report on forest fire as an ecological factor (2003). *University of Oslo,* 266.

Forest fire in the Far East. (n.d.). Retrieved from https://www.sibreal.org/a/29352746.html

Forest fires – one of the most serious problems of Russian forests. (n.d.). Retrieved from https://helpiks.org/6-68946.html

Forest Resources Information System. (n.d.). Retrieved from http://www.fao.org/forestry/4239-0f7d-5c1ac66ee0fe944546751e8d85c85.pdf

Giglio, L., Randerson, J. T., & Van Der Werf, G. R. (2013). Analysis of daily, monthly, and annual burned area using the fourth-generation global fire emissions database (GFED4). *Journal of Geophysical Research. Biogeosciences*, *118*(1), 317–328. doi:10.1002/jgrg.20042

Göltenboth, F., & Widmann, P. (2006). Tropical lowland evergreen rainforest, Ecology of Insular Southeast Asia. (n.d.). Retrieved from https://www.sciencedirect.com/science/article/pii/B9780444527394500176

Gorev, G. G. (2004). Assessment of the climatic predisposition of the territory to forest fires (on the example of the Tomsk region). (*Thesis abstract), Tomsk, TSU: Publishing house*, 24.

Grishin, A. M. (2001). Prediction and simulation of forest fires. Forest fires: occurrence, distribution, extinguishing and environmental consequences. Tomsk: TSU publishing house, 35-46.

Guha-Sapir, R., & Hoyois, P. (2019). EM-DAT: International Disaster Database. Brussels, Belgium: University Cathol. Louvain. Retrieved from http://www.emdat.be

Gusev V. G., Erican V. A., & Erican, E. V. (2011). Forecasting of parameters of forest fires and resources to deal with them. SPb, *SPbTU*, 217.

ICNF. (2012). (n.d.). Retrieved from http://www2.icnf.pt/portal/florestas/dfci/relat/rel-if/copy_of_2012

Kalinin, L. B., Kovyazin V. F., & Belov, S. V. (1993). Forest fire science. St. Petersburg, Russia: St. Petersburg State Forestry Academy, 76.

Khodakov, V., & Zharikova, M. (2011). Forest fires and methods of their research. *Kherson, FLP Green D. S.*, 470.

Korovin, G. N., & Isaev, A. S. (1998). Forest fire protection as the most important element of national security of Russia, *Forest Bulletin,* 8-9. Retrieved from http://old.forest.ru/rus/problems/fires/080908.html ahhh.

Kovyazin, V. F., Romanchikov, A. N., & Pasko, O. A. (2015). Comparative analysis of forest lands cadastral appraisal estimated with regards to wood and food resources. *IOP Conference Series: Earth and Environmental Science.* doi:10.1088/1755-1315/27/1/012039

Kudryavtsev, M., Lukin, V., Malinetskii, G., Mitin, N., Naumenko, S., Podlazov, A., Rumyantsev, A., & Toropygina, S. (2008). Forest conflagration extinguishing problems on Russian Federation territory, *Preprint, Inst. Appl. Math., the Russian Academy of Science*, 40.

Mallek, C. M., Safford, H., Viers, J., & Miller, J. (2013). Modern departures in fire severity and area vary by forest type, Sierra Nevada and Southern Cascades, California. *Ecosphere*, *4*(12), 1–28. doi:10.1890/ES13-00217.1

McCaffrey, S., Toman, E., Stidham, M., & Shindler, B. (2015). Social science findings in the United States. In D. Paton & J. F. Shroder (Eds.), *Wildfire hazards, risks, and disasters* (pp. 15–34). Amsterdam, The Netherlands: Elsevier; doi:10.1016/B978-0-12-410434-1.00002-6

Melekhov, I. S. (1965). Forest pyrology and its tasks. Modern issues of forest protection from fires and control, Moscow, Russia: Forest industry, 5-25.

Mistry, J., Bilbao, B. A., & Berardi, A. (2016). Community owned solutions for fire management in tropical ecosystems: Case studies from indigenous communities of South America. *Philosophical Transactions of the Royal Society*, *371*(1696), 20150174. doi:10.1098/rstb.2015.0174 PMID:27216507

MODIS (Moderate Resolution Imaging Spectroradiometer). (n.d.). Retrieved from https://terra.nasa.gov/about/terra-instruments/modis

Nikishchenko, N. (2011). Comprehensive assessment of natural and anthropogenic factors of fires on the lands of the forest Fund of the Voronezh region thesis abstract, *Voronezh, VSU*, 22.

North, B. M. P., Stephens, S. L., Collins, B. M., Agee, J. K., Aplet, G., Franklin, J. F., & Fulé, P. Z. (2015). Reform forest fire management. *Science*, *349*(6254), 1280–1281. doi:10.1126cience.aab2356 PMID:26383934

Pasko, O. A. (2013). *The forest resource of Tomsk Oblast and its use* (pp. 102–110). St. Louis, MO: Applied and Fundamental Studies, Ltd.

Pasko, O. A., & Baranova, A. V. (2015). Forest fire situation analysis over forest reserve land in Tomsk petroleum province. *IOP Conference Series: Earth and Environmental Science*. Retrieved from 10.1088/1755-1315/24/1/012022

Pasko, O. A., Tokareva, O. S., Alshaibi, A. J. A., Chernikova, T. Y., & Cabral, P. (2019). Assessment of state of cedar forests in Tomsk region using remote sensing data of the earth, *Bulletin of the Tomsk Polytechnic University. Geo Assets Engineering*, *330*(1), 89–101. doi:10.18799/24131830/2019/1

Reports of the Ministry of emergency situations and the media. (n.d.). Retrieved from https://www.mchs.gov.ru/activities/results

Rocca, M. E., Miniat, C. F., & Mitchell, R. J. (2014). Introduction to the regional assessments: Climate change, wildfire, and forest ecosystem services in the USA. *Forest Ecology and Management*, *327*, 265–268. doi:10.1016/j.foreco.2014.06.007

San-Miguel-Ayanz, J., Moreno, J. M., & Camia, A. (2013). Analysis of large fires in European Mediterranean landscapes: Lessons learned and perspectives. *Forest Ecology and Management*, *294*, 11–22. doi:10.1016/j.foreco.2012.10.050

Sapkota, A., Symons, J. M., Kleissl, J., Wang, L., Parlange, M. B., Ondov, J., Breysse, P. N., Diette, G. B., Eggleston, P. A., & Buckley, T. J. (2005). Impact of the 2002 Canadian forest fires on particulate matter air quality in Baltimore city. *Environmental Science & Technology*, 1, 39(1), 24-32.

ScanEx. (n.d.). Retrieved from http://new.scanex.ru

Scherbov, B. L., Lazareva, E. V., & Zhurkova, I. S. (2015). Forest fires and their consequences on the example of Siberian objects, Novosibirsk, Russia: Geo, 154.

Stephens, S. L., Burrows, N., Buyantuyev, A., Gray, R. W., Keane, R. E., Kubian, R., ... van Wagtendonk, J. W. (2014). Temperate and boreal forest mega-fires: Characteristics and challenges. *Frontiers in Ecology and the Environment*, *12*(2), 115–122. doi:10.1890/120332

Tedim, F., Xanthopoulos, G., & Leone, V. (2015). Forest fires in Europe: Facts and challenges wildfire. Hazards, risks, and disasters, 1, 5. Elsevier: Douglas Paton, 77-99.

The book of account of forest fires. (n.d.). Retrieved from https://lektsii.com/1-157112.html

The combustibility of the forests of Russia. Retrieved from https://vuzlit.ru/1100313/gorimost_lesov_rossii

The fire in Oklahoma. A photo from space. (n.d.). Retrieved from https://yandex.ru/images/search?text=forest%20fires%20%D0%BA%D0%BE%D1%81%D0%BC%D0%BE%D1%81&pos=12&img_url=https%3A%2F%2Fstatic2.businessinsider.com%2Fimage%2F5ad8e5e238e2fcb0628b4818-2048%2F26623968147a3e3b45f4ak-2.jpg&rpt=simage

The satellite images forest fires in the Tomsk region. (n.d.). Retrieved from https://i.mr-7.ru/photos/2012/07/640x480_84nZNFB0S4QswU4Nis0X.jpg

World fire map. (2019). (n.d.). Retrieved from https://fires.globalforestwatch.org/home

World fire map. (n.d.). Retrieved from https://fires.globalforestwatch.org/map/#activeLayers=viirsFires%2CactiveFires&activeBasemap=topo&activeImagery=&planetCategory=PLANET-MONTHLY&planetPeriod=Jan%200000&x=-163.652344&y=50.196019&z=2

Yadav, I. C., & Devi, N. L. (2018). Biomass burning, regional air quality, and climate change Earth Systems and Environmental Sciences. USA: Encyclopedia of Environmental Science and Engineering.

ADDITIONAL READING

Bachmann, A., & Allgower, B. (2000). *The need for a consistent wildfire risk terminology, Crossing the Millennium: Integrating Spatial Technologies and Ecological Principles for a New Age in Fire Management* (pp. 67–77). Moscow, ID: The University of Idaho and the International Association of Wildland Fire.

Bytnerowicz, A., Arbaugh, M. J., Riebau, A.R., & Andersen, C. (2009). Wildland fires and air pollution, *Elsevie*r, 643.

Fawcett, J. E. (2012). Fuel Dynamics in Southern Pine Beetle-Killed Stands and Their Implication to Fire. Retrieved September 4, 2019, from https://www.researchgate.net/publication/299800173_Fuel_Dynamics_in_Southern_Pine_Beetle-Killed_Stands_and_Their_Implication_to_Fire_Behavior

Fitting, M. (2000). Databases and Higher Types, Computational Logic. *Springer Lecture Notes in Artificial Intelligence, 1861*, 41–52.

Fujioka, F. (2009). Fire danger and fire behavior modeling system in Austrelia Europe and North America Developments. *Environmental Sciences, 8*, 471–497.

Gamache, I., & Payette, S. (2005). Latitudinal response of subarctic tree lines to recent climate change in eastern Canada. *Journal of Biogeography, 32*(5), 849–862. doi:10.1111/j.1365-2699.2004.01182.x

Göltenboth, F., Langenberger, G., & Widmann, P. (2006) Ecology of Insular Southeast Asia, *The Indonesian Archipelago,* 401-413.

Kasischke, E.S. & Turetsky, M.R. (2006). Recent changes in the fire regime across the North American boreal region – Spatial and temporal patterns of burning across Canada and Alaska, *Geophysical Research Letters, 33*: 9, 5.

Lloyd, A. H. & Bunn, A. G. (2007). Responses of the circumpolar boreal forest to 20th century climate variability. *Environmental research letters, 2,* 4, 13.

Orozko, S. J. (2008) Forest fire risk model for Michoacan, Mexico. *Thesis submitted to the International Institute for Geo-information Science and Earth Observation in partial fulfillment of the requirements for the degree of Master of Science in Geo-information Science and Earth Observation, Specialisation: Integrated Landscape Management*, 88.

Sofronov, M. A., Goldammer, J.A., Volokitina, A.V., Sofronova, T. M., & Sofronov, M.A. (2005). Wildland fire danger, *Krasnoyarsk: IL SO RAN, Max Planck Institute for Chemistry, SibGTU*. 330.

KEY WORDS AND DEFINITIONS

Emergency Forest Fire: The situation in the certain territory, developed as a result of the source of natural emergencies — forest fire (forest fires), which can entail or entailed human victims, the damage to human health and the environment, considerable material losses and violation of conditions of activity of people.

Fire Monitoring: A system of observation and control of fire danger in the forest according to weather conditions, the condition of forest combustible materials, fire sources and forest fires in order to timely develop and implement measures to prevent forest fires and (or) reduce damage from them.

Fire Suppression: Actions of participants of fire extinguishing aimed at the final cessation of combustion, as well as to exclude the possibility of its re-occurrence Gorenje.

Forest Fire: A spontaneous, uncontrolled spread of fire over forest areas.

Forest Fund: A set of forest and non-forest land categories of administrative economic territories.

Forest Protection: A special service organized for the protection of forests under the jurisdiction of the state forestry bodies and control over the protection of collective farm forests, urban forests, forest reserves and fixed forests.

Section 7

Chapter 14
The Influence of Smoke From Forest Fires on the Meteorological and Electrical Characteristics of the Atmosphere

Petr Mikhailovich Nagorskiy
Institute of Monitoring of Climatic and Ecological Systems of the Siberian Branch, Russian Academy of Sciences, Russia

Mikhail Vsevolodovich Kabanov
Institute of Monitoring of Climatic and Ecological Systems of the Siberian Branch, Russian Academy of Sciences, Russia

Konstantin Nikolaevich Pustovalov
Institute of Monitoring of Climatic and Ecological Systems of the Siberian Branch, Russian Academy of Sciences, Russia

ABSTRACT

The impact of smoke from forest fires in western Siberia on meteorological, atmospheric electric, and aerological variables has been analyzed. The anomalous distribution of water vapor in the atmosphere associated with the peculiarities of the evaporation regime and the absence of advective moisture transfer over the southern regions of Western Siberia during the fires. With an increase in the height of the homogeneous surface smoke layer with an unchanged aerosol optical thickness, the cooling of the earth's surface and heating of the atmosphere was weakened. The smoke plume spreads predominantly in the middle of the troposphere, creating aerosol layers elevated above the ground, the lower part of which had a negative volume charge. The effect of diurnal variations in the electrical field in the near-surface layer, differs from the known similar effects.

DOI: 10.4018/978-1-7998-1867-0.ch014

INTRODUCTION

This chapter presents the effects of forest fires on the meteorological variables and electric field variations of the atmosphere. The electric state of the surface layer of the atmosphere with intense smoke is the least studied among other meteorological variables, so the most attention has been paid to the study of this factor.

The complex interrelation between the temporal variations in this field on the one hand and the physical, chemical, and biological processes that form the environmental state on the other hand determines the interest in monitoring atmospheric electricity as both an affecting factor and an indicator of the observed environmental changes. Among the found regularities of temporal variations of electric field strength (***E***), the most well-known ones belong to the "fair weather". These regularities are the following (Chalmers, 1967; Sedunov, 1991; Kupovykh, Morozov, & Shvarts, 1998; Anisimov & Mareev, 2008; Donchenko, Kabanov, Kaul, Nagorskiy & Samokhvalov, 2015; Mareev et al. 2016): diurnal unitary variations caused by change in the Earth's charge as a whole; the electrode effect related to accumulation of voluminous positive charges near the Earth's surface; electrooptical correlation in the optical hazes expressed as the dependence of field intensity on aerosol particle concentration; and some others.

Yet the least known are processes of electrical charge accumulation and the *E* variations under disturbed "fair weather" (thunderstorm, volcanic eruptions and dust storm, natural and technogenic smoke, etc.). Since these atmospheric phenomena are episodic and hardly predictable in time, they complicate field studies in the sense of interpretation of their results. Nevertheless, some of the recently discovered effects related to the dynamics of atmospheric electricity indicate the necessity of taking these effects into account during complex monitoring and simulation of natural climatic processes (Ippolitov, Kabanov, & Smirnov, 2011; Ippolitov, Kabanov, Nagorskiy, Pkhalagov, & Smirnov, 2013; Donchenko et al., 2015). For example, low-frequency oscillations in electric field have been detected in the atmosphere prior to thunderstorms; records of these phenomena can provide control over development of mesoscale convection systems and dynamics of acoustical and gravity waves emerging during such a development (Ippolitov, Kabanov, Nagorskiy, & Smirnov, 2012). The studied effects of selective electrical charge accumulation of aerosol particles depending on their size and composition in the cases of cloud formation (Smirnov, 2010; Mareev et al., 2016), volcanic eruptions (Leblanc et al., 2008; Adushkin, Soloviev, & Spivak, 2018; Firstov et al., 2019) and dust storms (Sow et al., 2011; Gorchakova, Mokhov, & Rublev, 2015; Adushkin et al., 2018; Katz et al., 2018) also indicate the essential differences of variations in electrical field under these conditions compared to "clear weather" ones.

Among the insufficiently studied atmospheric processes, there are those implemented in the smoke from forest fires (Kondrat'ev & Grigor'ev, 2004; Pkhalagov, Uzhegov, Panchenko, & Ippolitov, 2006; Elanskii et al., 2011; Van Donkelaar et al., 2011; Ippolitov et al., 2013; Donchenko et al., 2015). In particular, the size and composition of smoke aerosol particles was changed due to the photochemical transformation of organic compounds (Seinfeld, & Pandis, 1998; Abel, Haywood, Highwood, Li, & Buseck, 2003; Vakkari et al., 2014; Konovalov et al. 2015). It was found (Pkhalagov et al., 2006) during the period of forest fires in Tomsk region in 2004 that as smoke intensified (the concentration of smoke particles increased), electric field potential gradient in the near-surface layer did not increase with respect to the electrooptical ratio, but reduced in contrast (from 200 to 30–60 V/m). With the great spatial extent of smoking during the vegetation period taken into consideration, studies of the meteorological and electric field variations under these conditions are topical for monitoring and modeling of both meteorological and ecological processes.

Table 1. Aerologic stations data of which were used to analyze the vertical structure of humidity

Nº	Station	Index Stations	Coordinates		Height Above Sea Level, m
			Latitude	Longitude	
1	Kolpashevo	29231	58°3'	82°9'	75
2	Novosibirsk	29634	55°02'	82°55'	150

THE INFLUENCE OF SMOKE FROM FOREST FIRES ON THE VERTICAL DISTRIBUTION OF THE TEMPERATURE AND HUMIDITY

In order to study the temperature and moisture reserves of air under smokiness conditions over the territory of south-western Siberia, the data of two-standard standard aerologic observations contained on the site ("University of Wyoming") and the synoptic material taken from the site ("Weather in Tomsk") were used. The studies were conducted based on the examples of two stations (Table 1).

Two situations were compared when studying the content of water vapor in the atmosphere during forest fires in 2012: a) smog (in the bulletin graph, the current weather reported from the meteorological station is recorded as "Visibility is degraded by smoke, for example: like from steppe or forest fires, emissions from industrial enterprises or volcanic ash"); b) clear weather (thunderstorm, dust storm, natural and technogenic smoke, etc.).

For each synoptic situation ("smokiness", "clear"), periods with the maximum duration of conditions (in days) were chosen, presented in Table 2. Synoptic material was used. We selected the data from the vertical sounding of the atmosphere for each situation. This data was used to plot the distribution of the dew point deficit, pressure, temperature, wind speed and absolute humidity over altitude.

Based on the data gathered from the aerologic sounding, the absolute humidity was calculated using the following formula:

$$a = \frac{q \cdot P}{622 \cdot (t + 273{,}15)}, \quad (1)$$

where q is the specific humidity, P is the atmospheric pressure, t is the air temperature.

The values of the selected meteorological quantities were interpolated using a height with a distance of 100 m to a height of 1 km and with a distance of 250 m starting from a height of 1 km to a height of 10 km. Furthermore, using the values at fixed heights, the distribution of the studied parameters and their statistical characteristics were obtained: mean, median, range, and standard deviation.

Table 2. Dates with selected synoptic situations

Station	"Smokiness"	"Clear"
Kolpashevo	12–26.07.2012	12–26.07.2013
Novosibirsk	25–31.07.2012	25–31.07.2006

Figure 1. The vertical averaged profiles of the distribution of absolute humidity at Novosibirsk (left) and at Kolpashevo (right) stations at 19:00 – shown in the upper graphs, at 7:00 – shown in the lower graphs

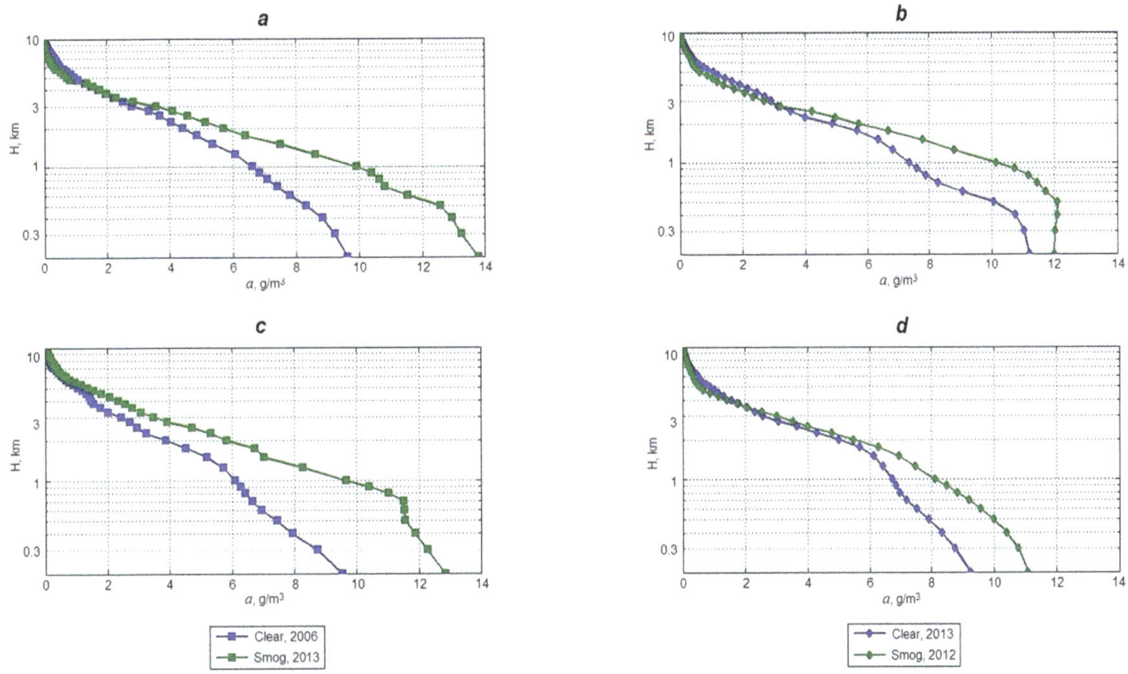

During the next stage of processing, the differences and relationships between the values of the parameters in a "smokiness" and "clear" atmosphere were calculated. Since the sample was insufficient in terms of size, only the arithmetic averages were analyzed from statistical characteristics. Averaging was carried out for fifteen days in order to gather the data obtained in Kolpashevo (15 morning and 15 evening terms), and for seven days in Novosibirsk (7 morning and 7 evening terms). After that, the average values of the studied quantities were compared with each other for each individual situation.

The vertical averaged profiles of the distribution of absolute humidity at the stations of Novosibirsk and Kolpashevo are shown in Figure 1 based on the conditions for a "smokiness" and "clear" atmosphere.

From Figure 1 it follows that the values of absolute humidity during fires in "clear" weather exceed the values when compared to values of the "smokiness" atmosphere throughout the entire altitude range utilized for the study (200 m - 10 km). The magnitude of the excess humidity in these two conditions is shown in Figure 2.

Radiative forcing was an important factor in the presence of excess values of *a* in the vertical distribution. It is associated with the strong heating of the air in the troposphere, the corresponding "cooling" of the underlying surface during the daytime hours, as well as the delay of long-wave radiation at night. These factors led to the fact that during the day high air temperature was preserved, which increased evaporation from the underlying surface. This served as an additional influx of water vapor into the atmosphere. As a result of these factors, water vapor accumulated and was retained within the smog.

The extreme magnitudes of anomalies of absolute humidity reached 13.8 g / m^3 in (Novosibirsk) and 12 g / m^3 (Kolpashevo), which is on average ~ 38 and ~ 15 percent higher than the values of ***a*** in "clear weather".

Figure 2. The vertical profiles of the difference in absolute humidity between "smoke" and "clear" at Novosibirsk (left) and Kolpashevo (right) stations at 19:00 – shown in the upper graphs, at 7:00 – shown in the lower graphs

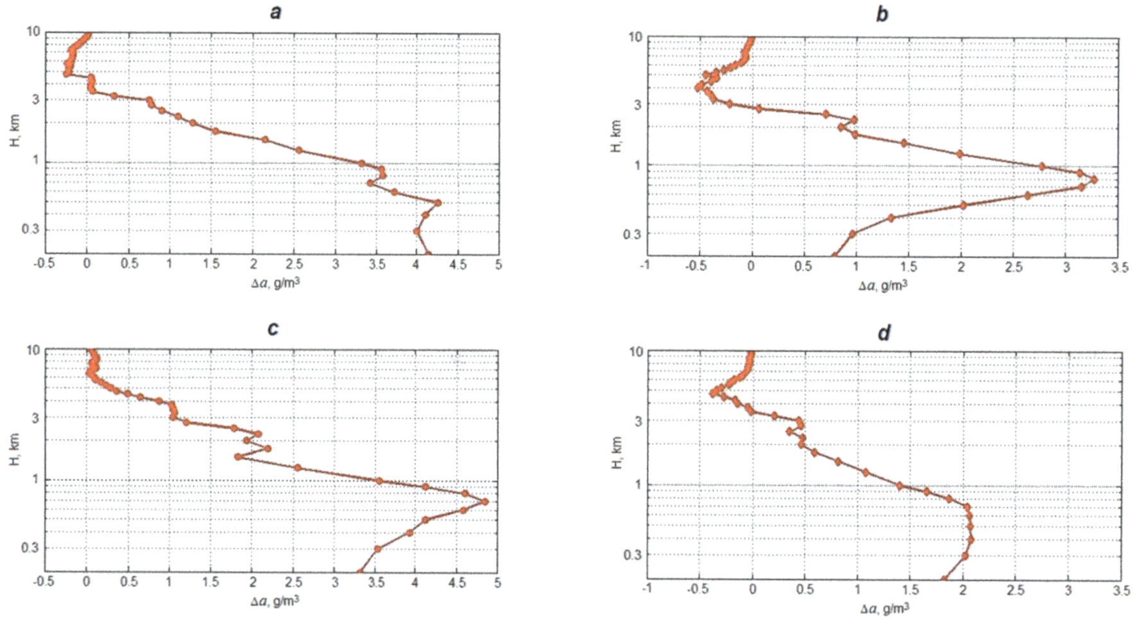

The anomalous distribution of water vapor in the atmosphere was associated with the peculiarities of the evaporation regime and the absence of advective moisture transfer over the southern regions of Western Siberia during the fires, with an additional influx of water vapor from the burning of biomass. The absence of horizontal water vapor transfer in the 0.2-3 km layer played a significant role in the formation of spatial anomalies in the total moisture content. The strengthening (or weakening) of anticyclonic circulation was accompanied respectively by a corresponding increase (or decrease) of advection of water vapor over the center of Western Siberia along the western and northern periphery of the anticyclone. This led to an increase (or decrease) in the average regional moisture content.

Thus, it is possible to conclude that the absolute humidity of the air is substantially exceeded in the presence of aerosol smoke relative to fair weather.

Let us now consider the effect of smoke from forest fires on temperature stratification in the troposphere.

The amount of descending solar radiation in a smokiness atmosphere decreases as a result of an increase in its absorption by the layer of aerosol smoke, which grows larger with the increasing aerosol optical thickness. Moreover, in these conditions, the amount of upward scattered radiation increases. As a result, both the atmospheric surface layer and the earth's surface became cooler (Gorchakova & Mokhov, 2012).

The vertical profiles of the temperature distribution at the Kolpashevo and Novosibirsk stations for the two considered states are shown in Figure 3.

The results of a comparative analysis showed that, in our case, for the atmosphere above station Kolpashevo, the thickness of the "underheated layer" is 500 m at 19:00 and 1 km at 7:00, respectively (Figure 3). Above these layers, the temperature difference between the two states, in general, begins to expand, this is due to the absorption of short-wave radiation by the aerosol smoke.

Figure 3. The vertical averaged profiles of the temperature distribution at Novosibirsk (left) and Kolpashevo (right) stations at 19:00 –shown in the upper graphs, at 7:00 – shown in the lower graphs

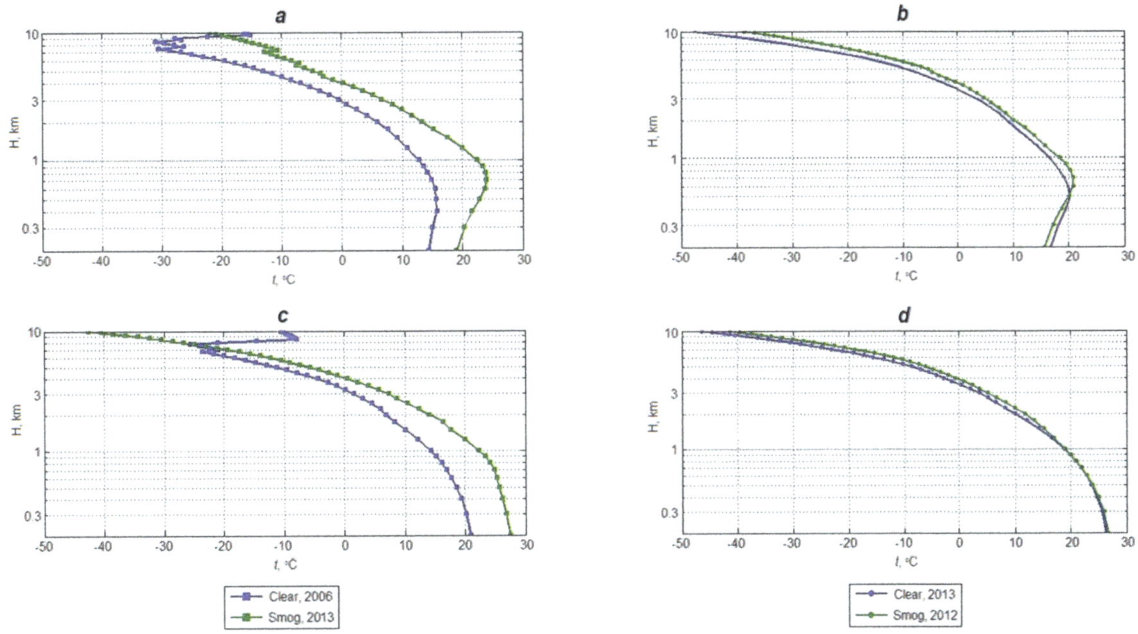

This is confirmed by dust storm simulation results: dust aerosol cools the earth's surface. The rate of cooling of the 10 m near surface atmospheric layer changes in the interval of –(4-21) °C/day without the influence of large aerosol particles over 15 μm in size on solar radiation transfer taken into account; if this influence is taken into account, the rate is –(6-36) °C/day (Gorchakova, et al., 2015).

The layer with a lower temperature is at 200 m for the Novosibirsk station at 19:00 (Figure 4) with a total temperature rise during smokiness conditions. Above it, the temperature difference, just like in Kolpashevo, begins to grow. There was no significant difference in temperature between the "smokiness" and "clear" atmospheres up to a height of 500 m at 7:00. It was revealed in the analysis of the synoptic situation in July 2012, that no continuous smoke was observed over Novosibirsk, as opposed in Kolpashevo. Nevertheless, it can be assumed, according to the authors (Gorchakova & Mokhov, 2012; Gorchakova et al., 2015) and Figure 4, that even a small amount of aerosol smoke can influence the temperature distribution in the layer ranging from 200 m to 1 km.

In the European territory of the Russian Federation during the period of fires in 2010, the overall level of cooling of the near surface-layer of the atmosphere as a result of the weakened solar radiation during the days in which the highest amount of aerosol was expunged was, according to (Gorchakova & Mokhov, 2012; Gorchakova et al., 2015), on average, about 6° C in 12 hours of light during the day. Since in our case the first sounding point was at an altitude of 200 m, the cooling value was calculated for this altitude. The results were 0.7° C for Kolpashevo and 1.9° C for Novosibirsk for 12 hours of daylight.

With an increase in the height of the homogeneous surface smoke layer with an unchanged aerosol optical thickness, the cooling of the earth's surface and heating of the atmosphere was weakened, and the surface-atmosphere system as a whole is cooled more strongly. At the same time, the cooling of the

Figure 4. The vertical profiles of the temperature difference between "smoke" and "clear" at Novosibirsk (left) and Kolpashevo (right) stations at 19:00 – shown in the upper graphs, at 7:00 – shown in the lower graphs

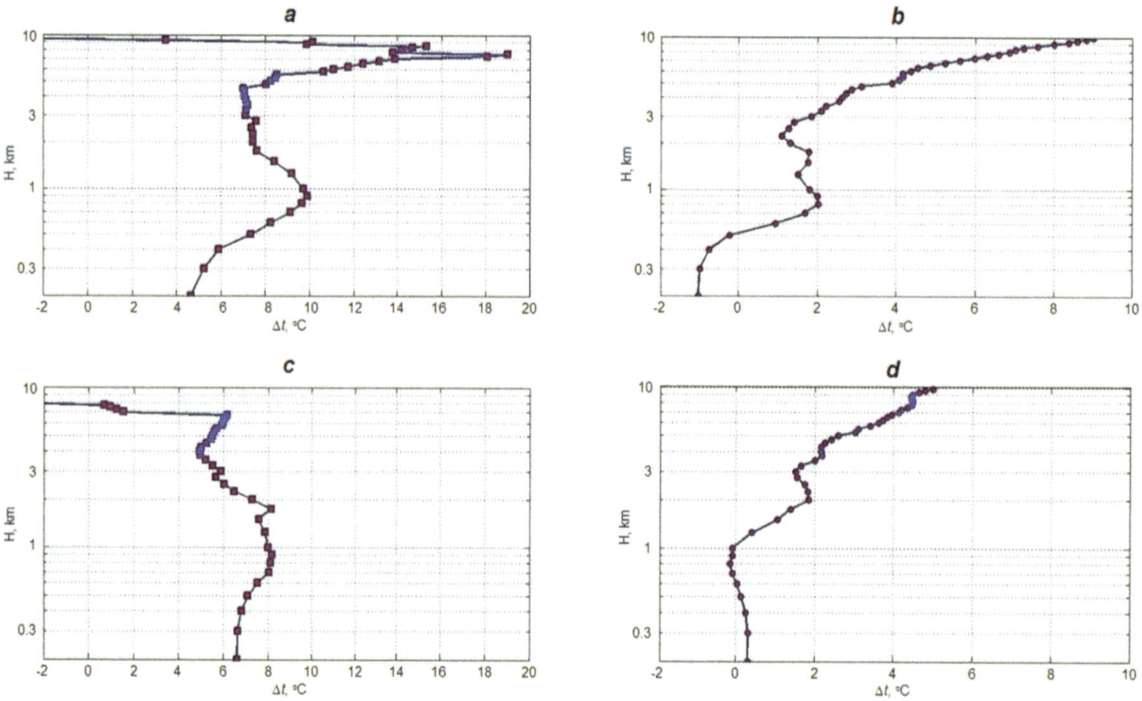

near-surface layer increases in a bright day (Gorchakova & Mokhov, 2012; Gorchakova, et al., 2015; Gorchakova, Mokhov, Anikin, & Emilenko, 2018; Semoutnikova et al., 2018).

Thus, the air temperature in the middle troposphere and the upper part of the lower troposphere increase caused by a layer of smoke aerosol is confirmed, when compared to the case of "clear weather".

THE INFLUENCE OF SMOKE FROM FOREST FIRES ON THE HORIZONTAL DISTRIBUTION OF THE TEMPERATURE AND HUMIDITY

The expedition "Obskaya Guba – Tomsk" was conducted from July 5 to July 23, 2016 by the employees of the NR TPU, POI FEB RAS, IMCES SB RAS and other organizations. Expeditionary measurements were performed on the OM-341 motor ship moving along the Ob River along the route: Salekhard - Obskaya Guba - Salekhard - Tomsk (Figure 5). The period of the expedition coincided with the period of intense forest fires in the Krasnoyarsk region. Anticyclonic circulation over the region facilitated the transfer of smoke to the south-east of Western Siberia (Figure 6) and further to the European territory of Russia (ETR). Smoke aerosol mass during its maximum growth period from 22 July to 26 July 2016 over Siberia (50°-70°, 60°-120 °E) was equal to 2 mln tons approximately, the smoke coverage over ETR with an aerosol optical thickness > 0.3 on July 25, 2016, reached to 5 million km^2, and the total

Figure 5. The route map of the Obskaya Guba – Tomsk expedition

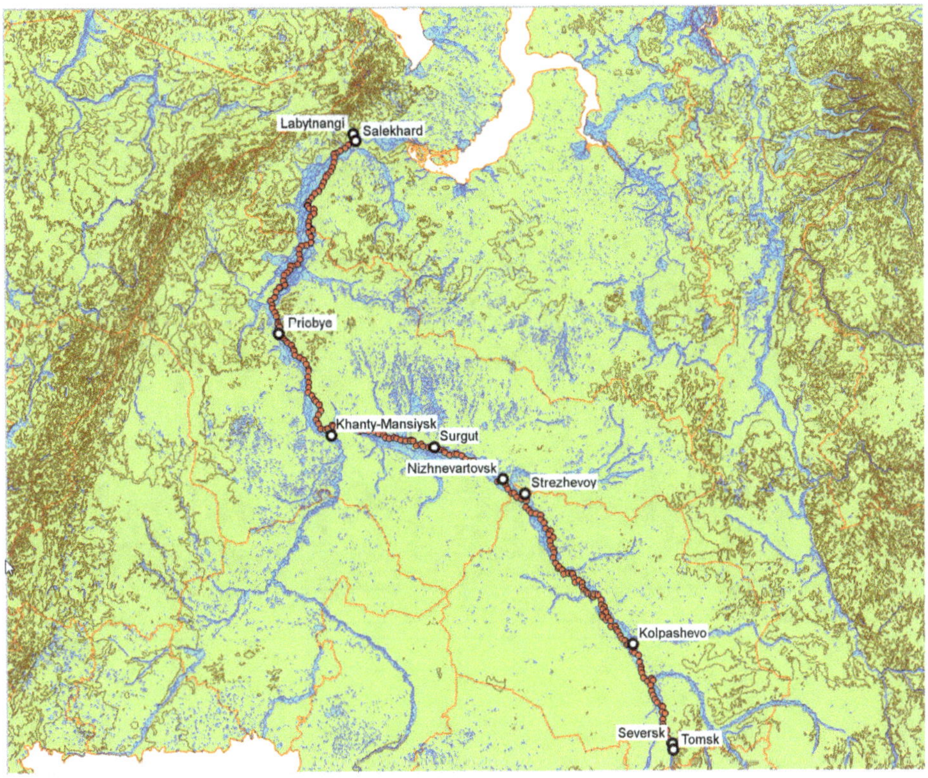

smoke mass was ~1.2 million tons (Sitnov, Mokhov, Gorchakov, & Dzhola, 2017; Semoutnikova et al., 2018; Gorchakov et al., 2019).

The average speed of the vessel was 20 km / h. As part of the measuring complex, the AMK-03 automatic meteorological station (Azbukin et al. 2006), which measured and calculated a large set of meteorological parameters and turbulent parameters of the atmosphere, and the CS110 field mill, which measured the electric field strength, were used. The devices were located on the mast installed on the cabin roof of the motor ship (~ 8 m above the river level).

In the period from July 18 to July 23, at the final section of the route Nizhnevartovsk – Tomsk, the steamboat passed under a smoke plume the source of which were remote forest fires (Figure 7). Thus, the horizontal profiles of the meteorological quantities were obtained at the intersection of the smoke plume almost at a right angle.

Let us consider in detail the dynamics of the meteorological and atmospheric-electrical quantities at the “intersection” of the smoke plume. From July 18 to July 21, an increase in the specific humidity of the air was observed as the ship approached the center of the smoke plume, the maximum value, which was recorded at ~ 22 h on July 20 and was more than 20 g / kg (Figure 8). At the same time, the air temperature almost did not fall below 20° C and changed a little during the course of the day, and the wind speed decreased. These results are in good agreement with the data of aerological sounding given above (fires in Siberia in 2012).

Figure 6. The image of the smoke plume in the visible part of the spectrum (upper graph) and the distribution of the aerosol index (lower graph) according to the data of the VIIRS radiometer from the Suomi NPP satellite ("EOSDIS Worldview") at 07:35 UTC 07/20/2016

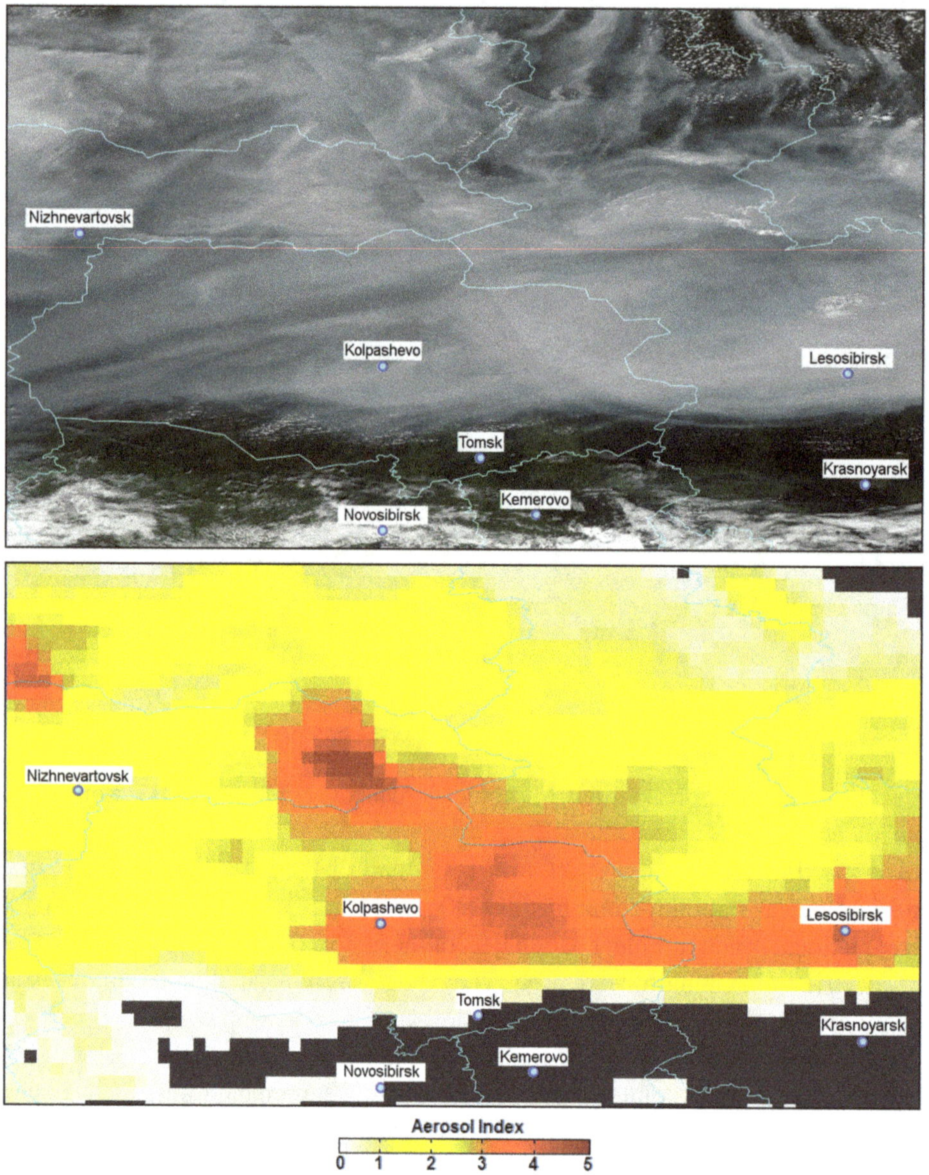

Figure 7. A diagram of the last leg of the route intersecting the smoke plume from the forest fires superimposed on a pseudo-real image, constructed according to MODIS Aqua data ("EOSDIS Worldview") at 07:36 UTC July 21, 2016)

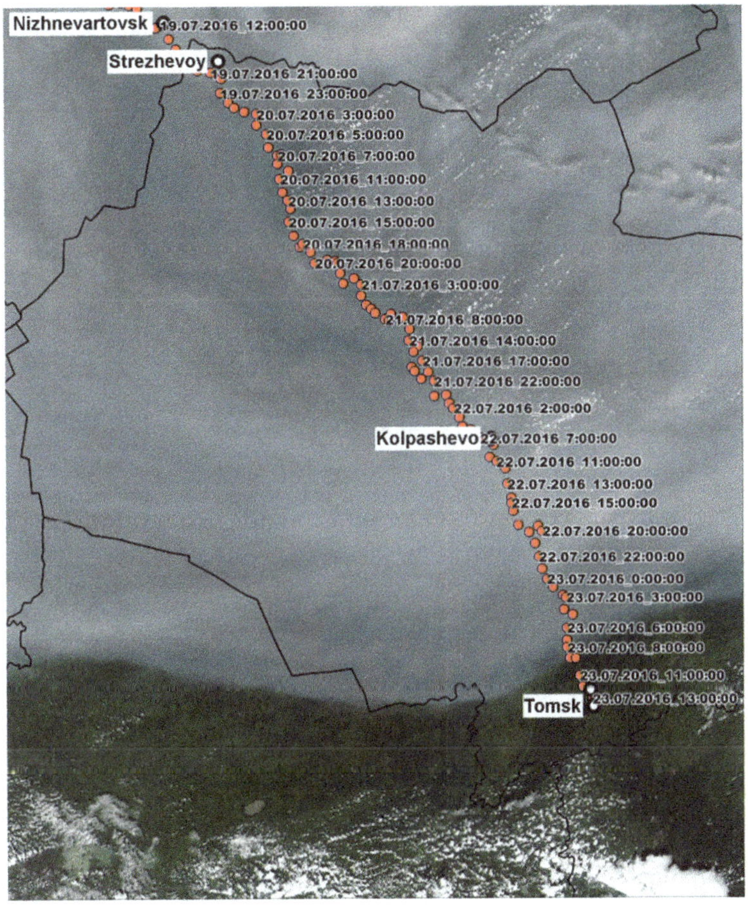

THE MODIFICATION OF THE SURFACE ELECTRICAL FIELD UNDER THE INFLUENCE OF SMOKE FROM FOREST FIRES

It is known that during the passage of clouds and various atmospheric phenomena, there is a distortion of the electrical field in the surface layer of the atmosphere, which is a characteristic of fair weather (Bennett & Harrison, 2007; Cherneva & Firstov, 2018; Adushkin et al., 2018; Pustovalov & Nagorskiy, 2018). According to (Pkhalagov et al., 2006; Ippolitov et al., 2013; Donchenko et al., 2015), the smoke from the forest fires also has a significant effect on the magnitude of the electrical field and its dynamics. However, the conducted studies are based on isolated cases and do not provide a good representation of the quantitative estimates of the effect of this factor on the electrical field of the surface layer. Therefore, the purpose of this part of the study is to obtain, on the basis of a large number of cases, quantitative estimates of the effect of smoke plumes from intense forest fires in Siberia on the electric field.

Figure 8. The dynamics of air temperature (t) and dew point temperature (td), average (V) and maximum wind speed (V_{max}), and specific humidity (q) from the 18th to 23rd July 2016

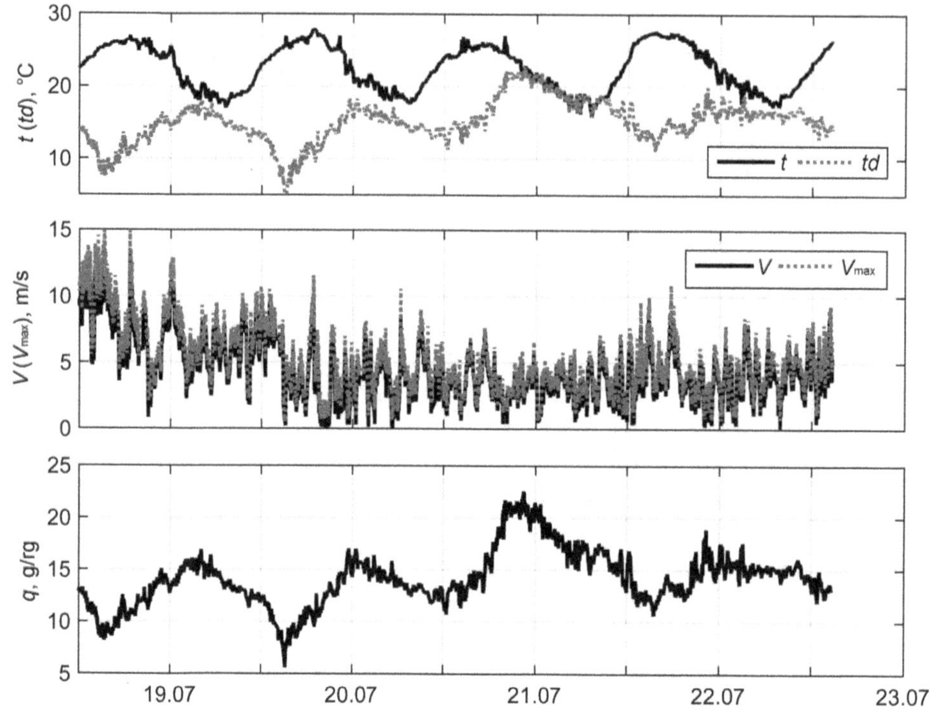

To achieve this goal, the mid-minute data of the electrical field potential gradient ($\boldsymbol{E}(t) = -\nabla\varphi$) for 2006–2016, obtained from the Geophysical Observatory of Institute for Monitoring of Climatic and Ecological Systems (GO IMCES SB RAS, Tomsk) using the Pole-2 field mill, as well as the data from the MODIS spectroradiometer (Terra, Aqua satellites), VIIRS (Suomy NPP satellite) ("EOSDIS Worldview") and the meteorological observations in standard time at the Tomsk meteorological station ("All Russian Research Institute of Hydrometeorological Information") located ~ 6 km from the GO IMCES SB RAS for the corresponding period.

On the basis of the satellite images obtained during the daytime flights of meteorological satellites (12: 00–14: 30), cases of smoke plumes from forest fires passing over the city of Tomsk were selected. The selection took into account the cases of the passage of the smoke plume both in the form of a dense jet and in the form of a thin veil (Figure 9). On the basis of satellite data, 77 cases of passages of smoke plumes over the city of Tomsk were identified for the period from 2006 to 2016. Sources of forest fires were located at a great distance from the research point in the overwhelming majority of cases. During the next stage, on the basis of the Tomsk meteorological station data, the cases in the time-frame after the satellite flights, during which the clouds of *Cb*, *Ns*, St, *Sc*, *Cu* and *As* passed, as well as atmospheric phenomena occurred, were excluded. As a result of filtering, 62 cases were selected.

Variations of the electrical field potential gradient were identified and their statistical analysis were carried out for the selected cases in time intervals corresponding to ± 30 minutes relative to the satellite flight times. In addition, the dynamics of meteorological variables at the research point were analyzed, including the horizontal visibility range.

Figure 9. RGB images of smoke plumes in the form of a jet and veil over the city of Tomsk according to Aqua data at 06:55 UTC on July 25, 2012 (a) and for Suomy NPP at 06:35 UTC 07/23/2016 (b)

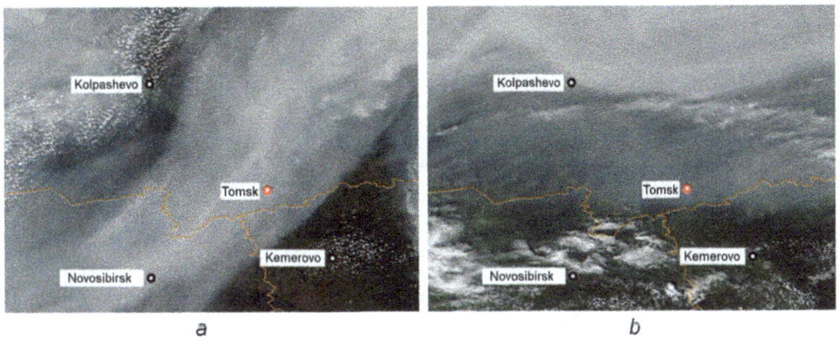

Figure 10. The distribution of the electrical field potential gradient $\nabla\varphi$ values during the passage of the smoke plume (a) and under unperturbed atmospheric conditions (b) in Tomsk for the period of 2006–2016

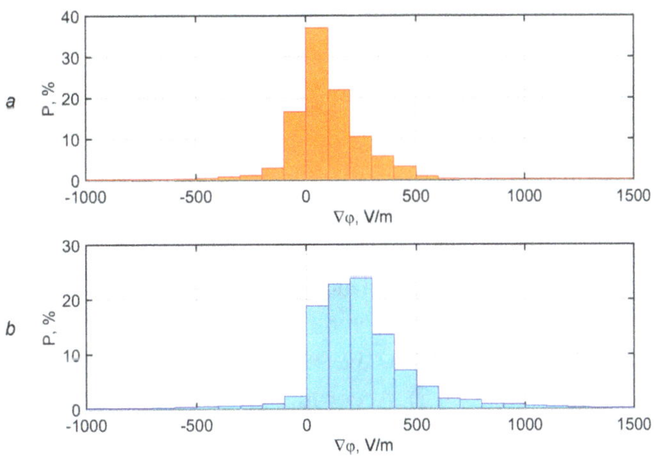

Table 3. Median (Me), interquartile range (IQR), as well as the minimum and maximum values with α = 0.05 (q (0.025) and q (0.975)) $\nabla\varphi$ (V / m) during the passing of the smoke plumes and under undisturbed atmospheric conditions

Characteristic	*Me*	*IQR*	$q_{(0,025)}$	$q_{(0,975)}$
Passage of the smoke plume	70	170	-100	390
Unperturbed conditions	210	230	-140	800

An analysis of the results showed the following: the distribution of the electrical field potential gradient values during the passage of the smoke plume differs from the analogous distribution under unperturbed atmospheric conditions, of which the absence of clouds and atmospheric phenomena are key characteristics. Thus, the distribution mode is shifted toward near-zero values, and its base is narrower (Figure 10).

Figure 11. The recurrence of various values of the meteorological visibility range during the studied events according to the data of the Tomsk meteorological station

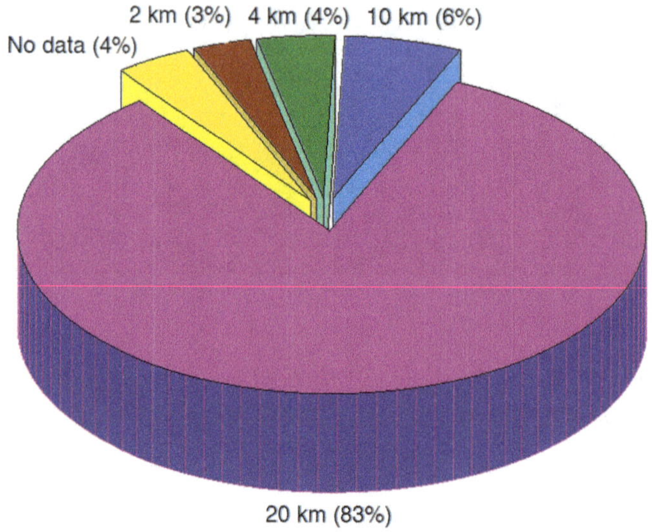

The median value of the electrical field potential gradient during the passage of the smoke plume, in which $\nabla\varphi$ becomes more than 3 times less than a similar value which is present in the absence of cloudiness and atmospheric phenomena. At the same time, the full and interquartile range of the electrical field potential gradient values under the influence of smoke from forest fires is 1.9 and 1.4 times less than under undisturbed atmospheric conditions.

The contradiction of the obtained results to the known dependencies of $\nabla\varphi$ on the aerosol can be explained as follows (Donchenko et al., 2015). According to the Tomsk meteorological station data ("All Russian Research Institute of Hydrometeorological Information"), the horizontal visibility range in 89% of the cases was 10–20 km, only in 11% of cases it fall down to 2–4 km, which means that an occurrence of haze was observed (Figure 11).

An analysis of the aerologic sounding data ("College of DuPage Next Generation Weather Lab [NEXLAB]") and the cloud LIDAR data CALIOP ("CALIPSO–Products–Lidar Browse Images of Productions Release") showed that the elevated inversion at altitudes of 1–5 km corresponded to smoke plumes (Figures 12, 13).

Thus, the smoke plume spreads predominantly in the middle of the troposphere, creating aerosol layers elevated above the ground, the lower part of which, apparently, had a negative volume charge, which caused a decrease in the electrical field in the surface layer.

Based on the presented information, it can be concluded that the smoke from remote forest fires leads to a strong decrease (~ 3 times) in the electrical field in the atmospheric surface layer, as well as a significant decrease in its dispersion relative to unperturbed conditions. The consideration of this factor in the modern models of the Global Electric Circuit seems to be very important because it can influence the electrical state of the surface layer simultaneously over large areas, as well show a correlation with a significant increase in forest fires in northern Eurasia, which has been observed at the beginning of the XXI century.

Figure 12. Example of an aerologic diagram (black curve – air temperature (t), green – dew point temperature (td), H – altitude, P – atmospheric pressure, Wind – wind speed and direction) with marked positions of the lower boundaries of the inversion layers (red line), built according to the aerological Kolpashevo station data from 00 UTC 07/21/2016 ("NEXLAB")

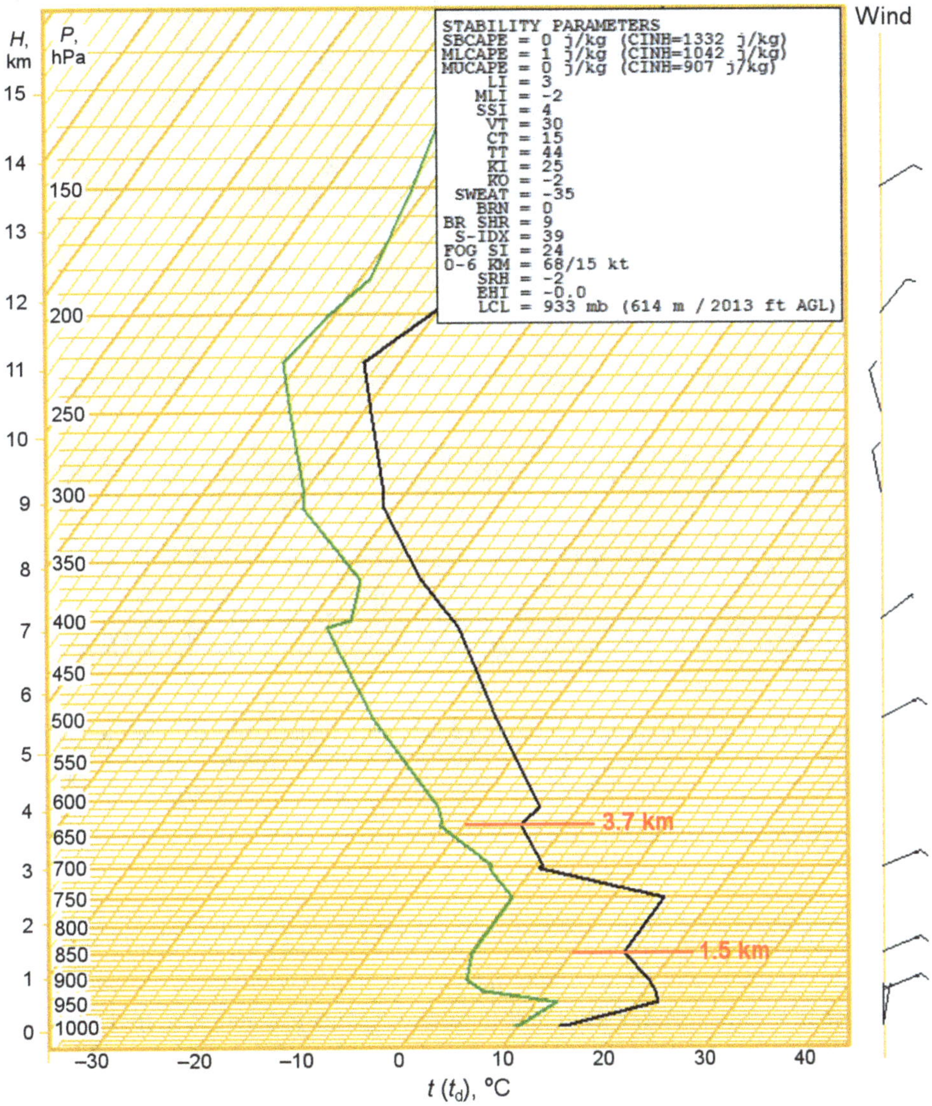

THE INFLUENCE OF SMOKE FROM FOREST FIRES ON THE DAILY VARIATIONS OF THE SURFACE ELECTRICAL FIELD AND THE POLAR ELECTRICAL CONDUCTIVITY

With the great spatial extent of smoking during the vegetation period taken into consideration, studies of the electric field variations under these conditions are topical for monitoring and modeling of both meteorological and ecological processes. The opportunity to carry out such a study was during the long-term period of forest fires in Siberia in summer 2012.

Figure 13. An example of elevated aerosol layers corresponding to the smoke plume, according to CALIOP ("CALIPSO–Products–Lidar Browse Images of Productions Release"), 07:29–07:43 UTC on July 21, 2016, when the CALIPSO satellite flew near Kolpashevo

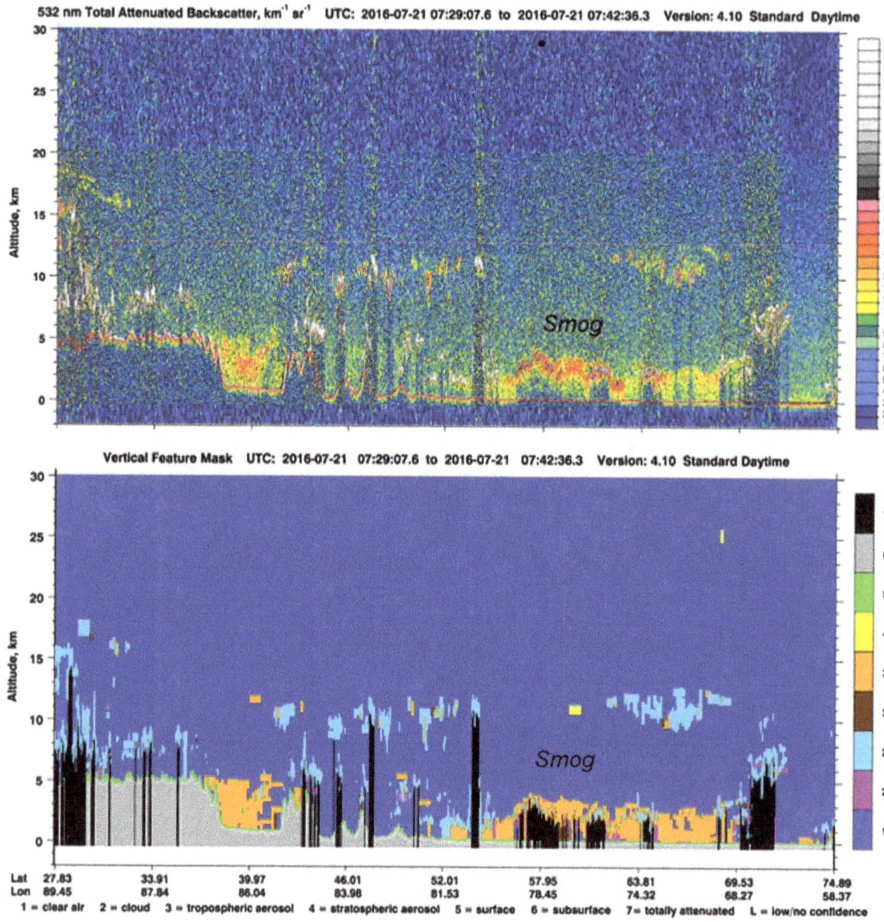

The results of these studies will be presented below, and the revealed effect of diurnal variations in electrical field in the near surface layer from 300 to in the daytime to –300 V/m in the nighttime, under continuous intensive smoking conditions, will be described (Ippolitov et al., 2013; Donchenko et al., 2015)

The experiment devoted to study of variations in the electrical field in the near surface layer of the smoke plume from forest fires was carried out at the GO IMCES SB RAS, in the framework of comprehensive monitoring of meteorological and actinometric parameters, and ultraviolet radiation and natural radioactivity. The electric field potential gradient ($\nabla\varphi = -E$) was measured using a Pole-2 stationary electrical fluxmeter; polar conductivity (negative and positive) was measured by an Elektroprovodnost-2 instrument placed on the roof of the adjacent building (20 m height). The special meteorological observations were implemented using an AMK-03 ultrasonic thermoanemometer (Azbukin et al., 2006) that provided recording of the primary meteorological characteristics (temperature, pressure, humidity) and turbulence characteristics in the near-surface layer at 60 Hz. Among the other recorded characteristics, we can mention the incoming solar radiation in the visible and ultraviolet ranges and the natural

Figure 14. Smoke plume above Tomsk on July 27, 2012 (NPP satellite data, visible band ("Remote sensing satellite data receiving ground station V.E. Zuev Institute of Atmospheric Optics")

radioactivity level (β- and γ-radiation) at the heights of 5, 10, and 25 m in the atmosphere and at the depths of 0.5 and 1 m in the soil. The meteorological optical range was determined by the data of the Tomsk meteorological station, Roshydromet, which is located 1.2 km from the Geophysical Observatory ("Weather in Tomsk").

Forest fires in summer 2012 in Siberia, with variable degrees of smoke in the town of Tomsk, take place from June until August; however, stable conditions favorable to carry out the experiment emerged only in the third ten day period of July, when a smoke plume from a distant forest fire source stably dominated above the territory of the town and, therefore, in the vicinity of the observatory located in the eastern part of town (Figure 14).

The meteorological situation above all of eastern West Siberia in this period was caused by a small gradient pressure field. The air temperature changed from 15 °C at nighttime to 30°C in the daytime; wind was northerly (from 300° to 60°) and weak (1–3 m/s); precipitation was absent, and the meteorological optical range decreased to 500 m or less. In general, meteorological situation in this period can be characterized as stable anticyclonic. There was no significant deviation in variations in the background β- and γ-radiation during this period.

Figure 15. Variations in atmospheric electrical and meteorological parameters during intensive smoke: (1) meteorological optical range (S_m) and electric field potential gradient $\nabla\varphi$); (2) deficit of dewpoint temperature (Δt) and negative electrical conductivity (λ_-); (3) absolute humidity (q) and air density (ρ). Arrows denote the moments of sunrise and sunset

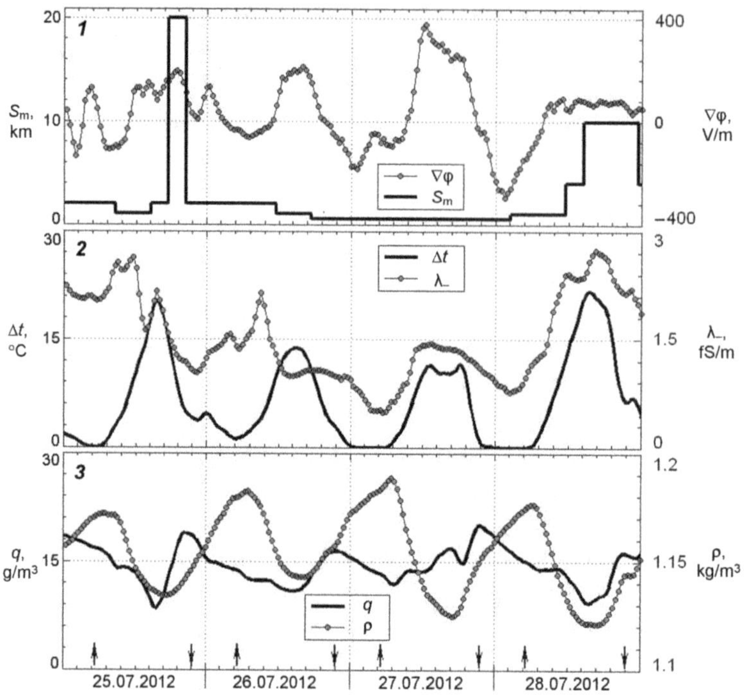

The most stable in terms of meteorological conditions were the days of stable smoke (July 25–28, 2012), when the meteorological optical range S_m was ~500 m and $\nabla\varphi$ variations in the near-surface layer were maximal. It is for these four days, chosen for the experiment, that the most scrutinized processing of data was made, with hourly averaging of series of every half-minute records. The results of such processing are given in Figure 15 (Ippolitov et al., 2013; Donchenko et al., 2015).

As is seen, electric field potential gradient in this period systematically changed from positive values in the daytime to negative ones at nighttime (panel 1). The revealed diurnal the $\nabla\varphi$ variations are principally different from those identified above the oceans (Carnegie curve (Chalmers, 1967; Sedunov, 1991)) and the averaged $\nabla\varphi$ variations, recorded during summer in the middle latitudes with a half-day interval (Anisimov et al., 2008). In the first case, the revealed variations advance the Carnegie curve in shape by approximately π; in the second case, the periods of variations differ about twofold. The absence of similarity based on the diurnal changes of electric field potential gradient is naturally explained by inhomogeneity of the smoke plume and verified by the observed changes in the meteorological optical range (concentration of aerosol particles).

Spatial and temporal variations in the electrical field intensity in the Earth's atmosphere, having abroad range and characterized by change of polarity sign, refer to known phenomena observed from many natural and technogenic effects (Chalmers, 1967; Sedunov, 1991; Anisimov & Mareev, 2008; Ippolitov et al., 2012; Ippolitov et al., 2013; Donchenko et al., 2015; Adushkin et al., 2018). The effect of diurnal variations in the electrical field (from 300 to in the daytime to –300 V/m at nighttime) in the near-surface layer, revealed by us for the first time, differs from the known similar effects by its record during smoke cover from forest fires that often affect vast territories of many regions. Forest fires are sources of aerosol particles, which have electrical charge accumulation owing to thermoemission and friction in powerful convection flows. These particles affect the $\nabla\varphi$ variations in the near-surface atmosphere above the areas covered by the smoke plume.

The more regular variations in this period were the diurnal variations in the atmospheric temperature t_a and the difference between the atmospheric temperature and the dew point $\Delta t = t - t_d$ (panel 2), accompanied by variations in the atmospheric density ρ and water vapor q (panel 3). Variations in negative polar electrical conductivity λ_- (panel 2) and vertical heat flow A_T were of less value, but still discernible and nearly coeval to the variations in the electrical field intensity.

The interrelation between the diurnal variations in the electrical field intensity and the observed variations in meteorological characteristics presented in Fig. 11 suggests the following interpretation of the revealed phenomenon. After midday, with reduced insolation in the homogeneous air mass, temperature t starts to drop and simultaneously q and ρ begin to grow at almost unchanged atmospheric pressure P. The decrease in t and the increase in ρ continue until sunrise. However, growth of water vapor density q lasts only until sunset. After this, at $\Delta t \rightarrow 0$, q starts to drop and lasts until sunrise. In the same time interval, electrical conductivity (amount of light ions) reduces.

After sunset, at atmospheric temperature close to the dew point, smoke aerosol particles in the near-surface layer begin to accumulate water vapor in themselves. Simultaneously, in the interaction between light ions of both positive and positive charges and aerosol particles, the bigger particles become negatively charged and, due to sinking under the effect of gravity, form a negative voluminous charge near the surface. It is this layer of particles with negative voluminous charge that leads to inversion of the direction of electric field potential gradient at nighttime. An analogous effect has been observed when studying the dynamics of charged particles during dust storms and volcanic eruptions (Leblanc et al., 2008; Sow et al., 2011; Adushkin et al., 2018; Firstov et al., 2019).After sunrise, both the water vapor density and the electrical conductivity start to grow. Under the effect of air convective flows that carry smoke aerosol particles upward, and under the rapid dehydration of these particles at $t > t_d$, the negatively charged layer formed near the surface at nighttime dissipates and electric field potential gradient becomes of positive sign again. The insignificant deviations of the diurnal variations from the strictly harmonic pattern reflect the unavoidable fine-scale inhomogeneity of the smoke plume and constant local peculiarities of the observation point (Ippolitov et al., 2013; Donchenko et al., 2015).

The suggested interpretation of the observed effect does not contradict the known results reported for the adiabatic chamber, which are, in turn, consistent with the diffusion-kinetic model of ion charge of aerosol particles (Smirnov, 2010).

CONCLUSION

The anomalous distribution of water vapor in the atmosphere associated with the peculiarities of the evaporation regime and the absence of advective moisture transfer over the southern regions of Western Siberia during the fires.

With an increase in the height of the homogeneous surface smoke layer with an unchanged aerosol optical thickness, the cooling of the earth's surface and heating of the atmosphere was weakened.

The smoke plume spreads predominantly in the middle of the troposphere, creating aerosol layers elevated above the ground, the lower part of which had a negative volume charge.

The effect of diurnal variations in the electrical field in the near-surface layer, differs from the known similar effects.

The fundamental value of the revealed effects is related to the role of the atmospheric electrical field and the smoke plume in both the formation of weather conditions (through evaporation and condensation) and the biosystem processes (osmotic regulation).

Therefore, the atmospheric electrical field variations must be taken into account during comprehensive climatic–ecological monitoring that addresses different meteorological and medical-biological problems.

ACKNOWLEDGMENT

This work was supported by Institute of Monitoring of Climatic and Ecological Systems SB RAS, project Natural features of climate change over the Asian in Russia [N° AAAA-A17-117013050031-8], as part of the Basic Research Program the Siberian Branch of the RAS Natural and climatic changes in Siberia and the Arctic under the influence of global and regional climate-regulating and environment-forming factors.

REFERENCES

Abel, S., Haywood, J. M., Highwood, E. J., Li, J., & Buseck, P. R. (2003). Evolution of biomass burning aerosol properties from an agricultural fire in South Africa. *Geophysical Research Letters*, *30*(15), ASC1–ASC4. doi:10.1029/2003GL017342

Adushkin, V. V., Soloviev, S. P., & Spivak, A. A. (2018). *Electrical fields of man-made and natural processes*. Moscow, Russia: GEOS [in Russian]. *All Russian Research Institute of Hydrometeorological Information – World Data Center*. Retrieved from http://meteo.ru

Anisimov, S. V., & Mareev, E. A. (2008). Geophysical studies of the global electric circuit. *Izv. Phys. Solid Earth*, *44*(10), 760–769. doi:10.1134/S1069351308100030

Azbukin, A. A., Bogushevich, A. Ya., Il'ichevskii, V. S., et al. (2006). AMK-03 Automated Ultrasound Meteorological Complex. *Meteorologiya I Gidrologiya, (11),* 89-97 [in Russian].

Bennett, A. J., & Harrison, R. G. (2007). Atmospheric electricity in different weather conditions. *Weather*, *62*(10), 277–283. doi:10.1002/wea.97

CALIPSO–Products–Lidar Browse Images of Productions Release. Retrieved from http://www-calipso.larc.nasa.gov/products/lidar/browse_images/show_calendar.php

Chalmers, J. A. (1967). *Atmospheric electricity*. Pergamon Press Ltd.

Cherneva, N. V., & Firstov, P. P. (2018). *Formation of the local electric field of the atmosphere in Kamchatka under the influence of natural processes*. Vladivostok, Russia: Dalnauka [in Russian]. *College of DuPage Next Generation Weather Lab (NEXLAB)*. Retrieved from http://weather.cod.edu/digatmos/upa

Donchenko, V. A., Kabanov, M. V., Kaul, B. V., Nagorskiy, P. M., & Samokhvalov, I. V. (2015). *Electro-optical phenomena in the atmosphere*. Tomsk, Russia: NTL Publishing House. [in Russian]

Elanskii, N. F., Mokhov, I. I., Belikov, I. B., & Berezina, E. V. (2011). Gas composition of the surface air in Moscow during the extreme summer of 2010. *Doklady Earth Sciences*, *437*(1), 357–362. doi:10.1134/S1028334X11030020

EOSDIS Worldview. Retrieved from https://worldview.earthdata.nasa.gov

Firstov, P. P., Akbashev, R. R., Zharinov, N. A., Maksimov, A. P., Manevich, T. M., & Mel'nikov, D. V. (2019). Electrification of eruptive plumes discharged by Shiveluch Volcano in relation to the character of the responsible explosion. *Journal of Volcanology and Seismology*, *13*(3), 172–184. doi:10.1134/S0742046319030035

Gorchakov, G. I., Sitnov, S. A., Karpov, A. V., Gorchakova, I. A., Gushchin, R. A., & Datsenko, O. I. (2019). Eurasian large-scale hazes in Summer 2016. *Izvestiya. Atmospheric and Oceanic Physics*, *55*(3), 261–270. doi:10.1134/S0001433819020063

Gorchakova, I. A., & Mokhov, I. I. (2012). The radiative and thermal effects of smoke aerosol over the region of Moscow during the summer fires of 2010. *Izvestiya. Atmospheric and Oceanic Physics*, *48*(5), 496–503. doi:10.1134/S0001433812050039

Gorchakova, I. A., Mokhov, I. I., Anikin, P. P., & Emilenko, A. S. (2018). Radiative and thermal impacts of smoke aerosol longwave absorption during fires in the Moscow Region in Summer 2010. *Izvestiya. Atmospheric and Oceanic Physics*, *54*(2), 154–161. doi:10.1134/S000143381802007X

Gorchakova, I. A., Mokhov, I. I., & Rublev, A. N. (2015). Radiation and temperature effects of the intensive injection of dust aerosol into the atmosphere. *Izvestiya. Atmospheric and Oceanic Physics*, *51*(2), 113–126. doi:10.1134/S0001433815010053

Ippolitov, I. I., Kabanov, M. V., Nagorskiy, P. M., Pkhalagov, Yu. A., & Smirnov, S. V. (2013). Diurnal variations in the electrical field intensity under smoke from forest fires. *Doklady Earth Sciences*, *453*(Part 1), 1137–1140. doi:10.1134/S1028334X1311010X

Ippolitov, I. I., Kabanov, M. V., Nagorskiy, P. M., & Smirnov, S. V. (2012). Dynamics of variability of the electric field in the prestrom atmosphere. [in Russian]. *Opt. Atmos. Okeana*, *25*, 132–136.

Ippolitov, I. I., Kabanov, M. V., & Smirnov, S. V. (2011). Strategy of monitoring network for climatic and environmental processes in Siberia. [in Russian]. *Opt. Atmos. Okeana*, *24*, 7–14.

Katz, S., Yair, Y., Price, C., Yaniv, R., Silber, I., Lynn, B., & Ziv, B. (2018). Electrical properties of the 8–12th September, 2015 massive dust outbreak over the Levant. *Atmospheric Research*, *201*, 218–225. doi:10.1016/j.atmosres.2017.11.004

Kondrat'ev, K. Ya., & Grigor'ev, Al. A. (2004). Forest fires as a component of natural ecodynamics. [in Russian]. *Opt. Atmos. Okeana*, *17*, 279–290.

Konovalov, I. B., Beekmann, M., Berezin, E. V., Petetin, H., Mielonen, T., Kuznetsova, I. N., & Andreae, M. O. (2015). The role of semi-volatile organic compounds in the mesoscale evolution of biomass burning aerosol: A modeling case study of the 2010 mega-fire event in Russia. *Atmospheric Chemistry and Physics*, *15*(23), 13269–13297. doi:10.5194/acp-15-13269-2015

Kupovykh, G. V., Morozov, V. N., & Shvarts, Ya. M. (1998). *Theory of electrode effect in the atmosphere*. Taganrog, Russia: Taganrog Radiotechnical Institute. [in Russian]

Leblanc, F., Aplin, K. L., Yair, Y., Harrison, R. G., Lebreton, J. P., & Blanc, M. (Eds.). (2008). Planetary atmospheric electricity. *Space Science Reviews*, *137*(1-4).

Mareev, E. A., Stasenko, V. N., Bulatov, A. A., Dement'eva, S. O., Evtushenko, A. A., Il'in, N. V., ... Shatalina, M. V. (2016). Russian studies of atmospheric electricity in 2011–2014. *Izvestiya. Atmospheric and Oceanic Physics*, *52*(2), 154–164. doi:10.1134/S0001433816020080

Pkhalagov, Yu. A., Uzhegov, V. N., Panchenko, M. V., & Ippolitov, I. I. (2006). Electro-optical relations in atmosphere under smog conditions. [in Russian]. *Opt. Atmos. Okeana*, *19*(10), 861–864.

Pustovalov, K. N., & Nagorskiy, P. M. (2018). Response in the surface atmospheric electric field to the passage of isolated air mass cumulonimbus clouds. *Journal of Atmospheric and Solar-Terrestrial Physics*, *172*, 33–39. doi:10.1016/j.jastp.2018.03.008

Remote sensing satellite data receiving ground station V. E. Zuev Institute of Atmospheric Optics. Retrieved from http://cosmo.iao.ru/ru

Sedunov, Yu. S. (1991). *Atmosphere: A handbook (Reference data and models)*. Leningrad, Russia: Gidrometeoizdat. [in Russian]

Seinfeld, J. H., & Pandis, S. N. (1998). *Atmospheric chemistry and physics*. New York: Wiley Intersci. Publ.

Semoutnikova, E. G., Gorchakov, G. I., Sitnov, S. A., Kopeikin, V. M., Karpov, A. V., Gorchakova, I. A., ... Kuznetsov, G. A. (2018). Siberian smoke haze over European Territory of Russia in July 2016: Atmospheric pollution and radiative effects. *Atmospheric and Oceanic Optics*, *31*(2), 171–180. doi:10.1134/S1024856018020124

Sitnov, S. A., Mokhov, I. I., Gorchakov, G. I., & Dzhola, A. V. (2017). The smoke haze over the European Part of Russia in Summer of 2016: A link to wildfires in Siberia and atmospheric circulation anomalies. *Meteorologiya I Gidrologiya (8), 50-63.*

Smirnov, V. V. (2010). Electrization of aerosol wetted in bipolarly ionized air. [in Russian]. *Izvestiya. Atmospheric and Oceanic Physics*, *46*(3), 294–303. doi:10.1134/S0001433810030035

Sow, M., Crase, E., Rajot, J. L., Sankaran, R. M., & Lacks, D. J. (2011). Electrification of particles in dust storms: Field measurements during the monsoon period in Niger. *Atmospheric Research*, *102*(3), 343–350. doi:10.1016/j.atmosres.2011.08.010

University of Wyoming. Atmospheric Soundings. Retrieved from http://weather.uwyo.edu/ /upperair/sounding.html

Vakkari, V., Kerminen, V. M., Beukes, J. P., Tiitta, P., van Zyl, P. G., Josipovic, M., ... Laakso, L. (2014). Rapid changes in biomass burning aerosols by atmospheric oxidation. *Geophysical Research Letters*, *41*(7), 2644–2651. doi:10.1002/2014GL059396

Van Donkelaar, A., Martin, R. V., Levy, R. C., da Silva, M. A., Krzyzanowski, M., Chubarova, N. E., ... Cohen, A. J. (2011). Satellite-based estimates of ground-level fine particle matter during extreme events: A case study of the Moscow fires in 2010. *Atmospheric Environment*, *45*(34), 6225–6232. doi:10.1016/j.atmosenv.2011.07.068

Weather in Tomsk. rp5 (Reliable Prognosis). Retrieved from https://rp5.ru/Weather_archive_in_Tomsk

ADDITIONAL READING

Bluestein, H. B. (1992). Synoptic-Dynamic Meteorology in Midlatitudes.: Vol. I. *Principles of Kinematics and Dynamics*. New York: Oxford University Press.

Bluestein, H. B. (1993). Synoptic-Dynamic Meteorology in Midlatitudes.: Vol. II. *Observations and Theory of Weather Systems*. New York: Oxford Univ. Press.

Fischer, E. M., & Schar, C. (2009). Future changes in daily summer temperature variability driving processes and role for temperature extremes. *Climate Dynamics*, *33*(7), 917–935. doi:10.100700382-008-0473-8

Ito A., and Penner J. (2005). Historical emissions of carbonaceous aerosols from biomass and fossil fuel burning for the period 1870-2000. *Global Biochemical Cycles 19, GB2028.*

Mei, L., Xue, Y., de Leeuw, G., Guang, J., Wang, Y., Li, Y., ... Chen, Z. (2011). Integration of remote sensing data and surface observations to estimate the impact of the Russian wildfires over Europe and Asia during August 2010. *Biogeosciences*, *8*(12), 3771–3791. doi:10.5194/bg-8-3771-2011

Samsonov, Y. N., Koutsenogii, K. P., Makarov, V. I., Ivanov, A. V., Ivanov, V. A., McRae, D. J., ... Ivanova, G. A. (2005). Particulate emissions from fires in central Siberian Scots pine forests. *Canadian Journal of Forest Research*, *35*(9), 2207–2217. doi:10.1139/x05-199

Seinfeld, J. H., & Pandis, S. N. (2006). *Atmospheric chemistry and physics: From air pollution to climate change*. New York: Wiley Interscience, J. Wiley & Sons, Inc.

Vicente, A., Alves, C., Calvo, A. I., Fernandes, A. P., Nunes, T., Monteiro, C., ... Casimiro, P. (2013). Emission factors and detailed chemical composition of smoke particles from the 2010 wildfire season. *Atmospheric Environment*, *71*, 295–303. doi:10.1016/j.atmosenv.2013.01.062

Williams, E., & Mareev, E. (2014). Recent progress on the global electrical circuit. *Atmospheric Research*, *135*, 208–227. doi:10.1016/j.atmosres.2013.05.015

KEY TERMS AND DEFINITIONS

Aerological Sounding: Measuring of the atmospheric air characteristics with height based on devices raised by pilot balls, radiosondes, aerostats, airplanes, rockets and satellites.

Cloud Types: Cb - Cumulonimbus; Cu - Cumulus; Ns – Nimbostratus, Frnb - Fractonimbus; St - Stratus; As - Altostratus; Ac - Altocumulus; Cs - Cirrostratus; Cc — Cirrocumulus; Ci - Cirrus.

Electric Field of the Atmosphere: Constantly exists in the atmosphere, it is caused by charges of the Earth and the ionosphere. The field strength is 130 V / m on average and decreases exponentially with height.

Electrode Effect: The formation of a space charge near the earth's surface, since the vertical conduction current in this layer is caused by the upward movement of positive ions at closed to zero negative ion conduction currents.

Global Electrical Circuit: A distributed current circuit that formed by highly conductive layers of the earth's surface and ionosphere, separated by atmosphere, whose conductivity is negligible in the boundary layer, but exponentially increases with height.

Insolation: Irradiation of the earth's surface with solar radiation of all types, which has light, heat and bactericidal effects.

Light Ions: Atmospheric ions consisting of a complex (up to a dozen or more) of molecules carrying one elementary charge. They have the greatest mobility and therefore play a major role in the electrical conductivity of air.

Meteorological Range of Visibility: It characterizes the transparency of the atmosphere and the ability to distinguish distant objects visually.

Optical Thickness: A dimensionless quantity that characterizes the attenuation of light in atmosphere due to its absorption and scattering.

Temperature Inversion: A phenomenon observed when the temperature in the troposphere increases with height instead of its decreasing.

Troposphere: The lowest part of the Earth's atmosphere, which contains about 80% of the mass of the atmosphere and 99% of its water vapor. The upper boundary of the troposphere is located in the polar latitudes at heights of ~ 8-10 km, in the middle latitudes - at heights of ~ 10-12 km, in the equatorial latitudes - at heights of ~ 16-18 km.

Chapter 15
The Influence of Fires on Forest Ecosystems

Olga Anatolievna Pasko
Tomsk Polytechnic University, Russia

Vasily Fedorovich Kovyavin
St. Petersburg Mining University, Russia

Nadezhda Anatolyevna Lebedeva
https://orcid.org/0000-0003-4095-2631
Kherson State Agrarian University, Ukraine

ABSTRACT

The chapter focuses on the role of fire as an environmental factor affecting the components of the forest ecosystem. The classification of forest fires according to the fire location in the forest ecosystem is given. Negative impact of fire on communities and people, components of forest ecosystems: a phytocoenosis (the forest under-growth, understory, ground cover), tocinos (animal world), klimatop (atmosphere and climate) and adatom (topography and soil) are discussed. The role of fire as a factor of organisms and natural ecosystems' evolution in general is studied. The problems of forest restoration on fires (fumes) are considered. Research established that the degree of damage to the ecosystem components and social consequences depend on the forest fire type, natural conditions, and forest conditions of the region.

INTRODUCTION

At the earth Summit in Rio de Janeiro in 1992, the UN Convention on climate change was solemnly adopted. It entered into force on 21 March 1994 and was signed by representatives of more than 180 countries. The Convention specifies the need for cooperation in the development and dissemination of technologies, methods and processes that reduce anthropogenic emissions of greenhouse gases, in particular forestry, as well as the rational use of sinks and reservoirs of all greenhouse gases, including forests (United Nations framework, 1992). In the development of modern public policies, most States,

DOI: 10.4018/978-1-7998-1867-0.ch015

along with the effects of climate change, attach great importance to the conservation and sustainable use of living natural capital based on biodiversity conservation (Johannesburg Declaration, 2002). The global environmental role of forests is also reflected in the UN Convention On biological diversity (The UN Convention, 2002), the statement of principles for a global consensus on the management, conservation and development of all types of forests (Management, 1992), and the national sustainable development strategies of many countries.

The international community recognizes that forests are essential for biodiversity conservation, carbon sequestration and oxygen release. They are the valuable source of renewable energy. One resource of nature use is replaced by an integrated management, taking into account characteristics of the ecosystem.

The forest is considered to be not only as a source of wood, but also as a habitat of various species and some carrier of the most important functions, including soil protection from erosion processes, an active participant of water and gas exchange, recreation area, part of the landscape and natural heritage, "business card" of a territory (Perelet, 2006; Pasko, 2013). The ecosystem approach involves the analysis of the forest as a community of living organisms that interact systematically with each other and with non-living components of the habitat. Biotic and abiotic components are linked together through nutrient cycles and energy flows (Ecosystems and People, 2019) and their disruption to one component leads to changes in the structure and relationships of the entire ecosystem.

Up to 400,000 forest fires occur on earth each year, the area of burning reaches 0.5% of the total forest area and about 1-4 billion tons of coal are released into the atmosphere each year by forest fires around the world (Andreae & Goldammer, 1992).

Extreme and catastrophic forest fires are known. In 1871 the fire "of Peshtigo" has been. It's area of about 1.5 million hectares resulted to the deaths of 1,150 people in Wisconsin and Michigan (USA). In 1915, Siberian forest fires covered an area of 12 million hectares and led to the formation of thick smog, which lasted about 50 days. In 1972 forests and peat bogs were burning in the European territory of Russia in the area of over 100 thousand square kilometers during the summer months. 104 people were killed by their fire only in the Moscow region.

The main causes of forest fires are: human activity, lightning, spontaneous combustion of peat edges and agricultural fires in hot weather. Forest fires destroy trees, undergrowth, underwood, ground cover and harvested wood (Fig. 2) (Pasko @, Baranova, 2014). They also reduce the value of forest land significantly. Water protection, protectiveness and wood together with other useful forest properties are decreasing as a result of fires. The fauna perishes, engineering constructions, and in some cases – the whole settlements are destroyed. The landscape of the district changes (Fig. 3) the atmospheric pollution by combustion products.

A number of scientists regard forest fires as an integral factor of evolution that forms the biodiversity of natural habitats along with the statement of negative economic, environmental and social consequences. They note that organisms in ecosystems have adapted to forest fires and even use them in their life cycles. Fire is regarded as a natural phenomenon similar to flooding, volcanic eruptions, etc., which is an instrument of natural selection of species and contributes to the survival of the fittest individuals.

The new scientific direction "Fire ecology" is actively developing nowadays. The objects of its research are often prairies, savannas, coniferous forests, etc. They are formed with the participation of fire as an essential evolutionary factor that determines the viability of the ecosystem and the renewal of the habitat of living organisms. The subject of the study of fire ecology are natural processes due to the presence of fires in the ecosystem, such as; the interaction of fire, abiotic and biotic components of the ecosystem; ecosystem approach to the evolution of organisms and so on.

Forest science notes that the impact of fires on forest ecosystems depends on the composition and type of plantings, as well as the intensity of burning of forest materials. For example, post-fire water logging may occur in blueberry pine stands with lower fertility rate due to reduction of evaporation from the surface of the earth. On the contrary, in green-mossed pine forests and peaty pine forests, the process of waterlogging due to land drying often develops and the growth of trees in height and diameter increases. Consequently, the fire impact on the place of growth conditions of plantations is ambiguous – it can be both negative and positive.

Forest fires are one of the main reasons for the transformation of vegetation in the boreal zone. They initiate long-term periodic changes in all components of the landscape, microclimatic conditions of the surface layer of the atmosphere (light, humidity, wind), temperature and humidity of the upper soil layers, water regime and chemical properties of the soil; affect on many other processes that determine the quality of growing conditions, as well as the very direction of the soil formation process. Characteristic of post-fire successions are partial or complete change of species composition, stand structure, productivity, etc. In General, fires create specific features of ecological regimes of natural landscapes (Yadav & Devi, 2018). This Chapter discusses the complex relationships within the forest ecosystem that occur in it under the influence of forest fires.

BACKGROUND

Classification of Forest Fires at the Location of the Fire in the Phytocenosis

Forest fire is determined as the spontaneous spread of fire in the forest. As opposed to it the fire which is controlled since ancient times consciously used by man in productive activities: clearings for burning of felling reminders, and the creation of protective strips annealing, agricultural fells – for crop yields increasing. Hiking and on holiday – for cooking and heating in the forest to prepare the soil for planting seedlings.

The degree of negative impact of fire on the forest ecosystem depends on the type-forest fire. In science, "Forest pyrology" (Kalinin, Kovyazin et al., 1993) distinguish three types of forest fires: grassroots, crown and underground (peat), which set at the location of the fire in the phytocenosis, and the intensity of burning of forest materials.

Litter burning in a grassroots fire, moss and shrub cover, undergrowth, grass, fallen trees, protruding roots and lower parts of the trees. Such a fire causes the least damage to the forest ecosystem; the speed of burning does not exceed 3,0 m/min; width of the burning edge – 1.0 m, and the flame height of 1.5 m. ground fire is dangerous for people.

During the crown fire treetops, forest floor and ground cover are burning at the same time. Such kind of fire is formed from the grassroots at a wind speed of 10-15 m/s and at low lowered tree crowns. The wind carries burning branches and sparks that ignite the ground cover tens of meters ahead and create new fires. A runaway crown fire in a strong wind moving through the treetops at a speed of 5-100 m/min the speed of the fire front reaches up to 1 km/hour, the width of the edges of the combustion of 6-8 meters.

Crown fire is a dangerous phenomenon, which is a great danger for people, plants and animals. Thus, in 1972 in the Republic of Mari El in 5 hours coniferous forest area of 14.45 thousand hectares has been burned (Kolesnikova, 2002). In October 1976, in the Khabarovsk region as the result of crown fires an area of the forest in 150 000 ha. per hour burned. Smokiness of the atmosphere was located over an area

Figure 1. Mounted (A) and lower (B) forest fire

Table 1. Ratio of forest fire types in the Russian Federation

Years	The Types of Forest Fires, %		
	Ground	Riding	Underground
1971-1985	73,6	25,4	1,0
1986-2000	83,0	16,7	0,3
2001-2018	82,0	17,7	0,3

of 400-500 sq. km. The fire has brought the hurricane of the Chinese forest burning. The speed of the fire front spread reached 40 km/h, the length – 40 km (Fig. 1).

In case of **an underground (peat)** fire, a layer of peat with a thickness of up to 1.5 m burns. Usually peat accumulates moisture, but in a dry summer peat dries out and becomes easily combustible. Underground fire speed of fire spread is small, up to 5.0 m/ per day, but burning steady and flameless (corruption). Underground fires are extremely dangerous unexpected breakthroughs of fire from the underground hearth and the fact that its edge is not always visible and it's possible for the burnt peat to a depth of one and a half meters.

The ratio of types of fires in Russia by year is given in table 1.

From the above data it can be seen that the forests of the Russian Federation are dominated by grassroots fires, which under certain meteorological conditions pass into more dangerous and destructive riding.

Fires are divided into three categories according to their speed of spread:

Strong – more then100 м/min.;
Average power 3-100 m/min.;
Weak – 0 – 3m/min.

The typical scenario of a large forest fire includes the following stages:

- deviation of meteorological conditions from the average in the direction of increasing the number of days without precipitation,
- reduction of air humidity,
- wind up to 8-30 m /s;
- ignition (self-ignition) of forest combustible materials;
- development of forest fires to be larger (the distribution and the merger of numerous fires in the extensive zone);
- burning of the forest in the removal of the fire front that accompanies by strong smoke and fumes;
- firefighting by fire protection or natural and planting (Methods of assessing ..., 1996; Kalinin, Kovyazin et al., 1993; Pasko, 2013).

Climate and topography play an important role in determining the severity and extent of the fire. Fires are most widespread in dry years, most severe on the upper slopes, and depend on the type of vegetation growing (Emery and Gross, 2005). They follow the path of the least moisture and most combustible material.

The Effect of Fire on Components of the Ecosystem

The forest is the totality of the land, tree, shrub and herbaceous plants, animals, microorganisms and other components of the environment, the biological interrelated and influenced each other in their development. The forest performs crucial biological and ecological functions: regulates and filters of water flow, prevent soil from erosion, maintains and improves its fertility and enriches the atmosphere by oxygen influences the formation of the climate and reduces air pollution.

The main effects of burning on the forest ecological processes can be summarized as follows:

Impact on environmental processes:

- Creation of bare areas that contribute to the invasion of weeds.
- The appearance of plant mosaics containing different succession stages.
- Violation of local ecological balance.
- Reduction of species diversity.
- Increased homogeneity with fewer environmental niches.
- Increase in the number of herbivores in areas with new vegetation
- Distribution of pioneer species such as Euphorbiaceous, Sonneratiaceae and Verbenaceae.
- Suppression of climax species such as Dipterocarpaceae, Annonaceae, Fagaceae, Myristicaceae.

Impact on organic production and decomposition processes:

- Loss of total biomass.
- Reduction of primary production.
- Reduction of secondary production.
- Reduction of organic turnover by decomposition.

Figure 2.

Influence on the circulation of nutrients:

- Loss of nutrients and minerals.
- Destruction and interruption of nutrient cycles.
- Reduced nutrient capital retention in organic matter.
- Excessive loss of elements by surface runoff and lixiviate.
- Change the speed of nitrogen fixation.

Effect on water circulation:

- The increase in permeability and water consumption.
- Reduction of rainfall interception, transpiration and soil moisture.

Impact on soil development:

- Increased pH and soil erosion.
- Loss of organic matter.
- Formation of a base-rich surface layer of soil.
- Increasing soil temperature.
- Increase nutrient loss through leaching and salinity.
- Long-term prospect of depletion of minerals in the soil.
- Death and decay of plant roots (Göltenboth & Widmann, 2006) (Fig. 2).

One of the main terrestrial ecosystems that absorb CO_2 is the Siberian boreal forest. According to experts, they are able to neutralize up to 30-40% of anthropogenic impacts.

Figure 3. The complex effect of forest fire on the ecosystem
Source: Keeley, 2015

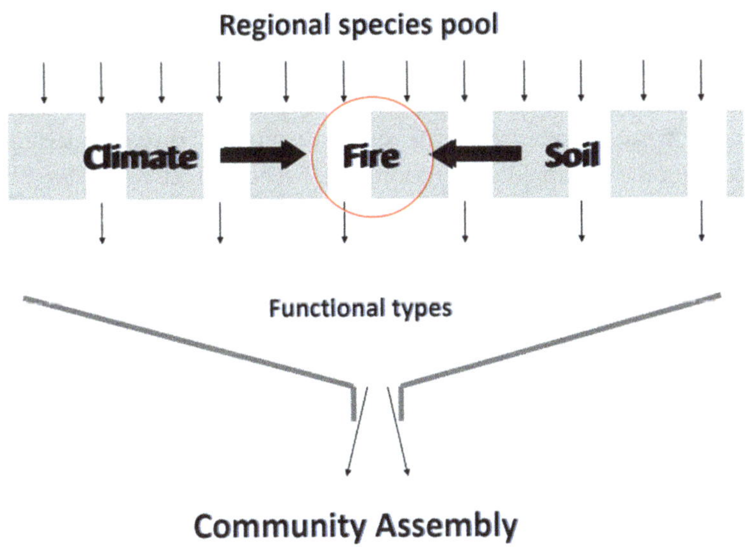

Effect on the Air

As a result of forest fires, the mass of organic matter plant burns annually, a huge amount of carbon compounds and other products of combustion enter the atmosphere (Fig. 3). Direct carbon emission into the atmosphere is complemented by carbon emission in the form of carbon-acid from the decomposition and mineralization of dead vegetation.

Artificial burning of forests is widely used, especially in Central and South America, for deforestation of tropical rainforests and clearing areas for pasture. This is an important way to reduce the catastrophic risk of forest fires. Unlike burning, a forest fire is a natural process of burning in tropical, temperate and boreal forests spreading over the lands of the forest Fund. A natural forest fire involves unplanned burning of the forest due to lightning, while a man-caused forest fire is the result of unauthorized forest burning practices to reach agricultural land.

Natural forest fire destroys an important flow of atmospheric carbon dioxide, increases its emission into the atmosphere from 3 to 150 million tons per year and leads to the spreading of nitric oxide, carbon monoxide (carbon monoxide) and small particles of soot concentration in the air of. At intervals of 6 to 7 years there is a sharp increase. Carbon monoxide is poisonous, colorless and odorless; in small doses, it causes headache, nausea, a feeling of rapid fatigue and lassitude.

Another dangerous product of combustion, soot, the International Agency for research on cancer has classified to the first class of carcinogens, along with asbestos and benzolamide. With an increase in the content of soot in the air by 20%, people feel oxygen starvation. There is a decrease in immunity and epidemics of viral and infectious diseases. The smoke of forest fires in more than 50% consists of tar droplets and water vapour, 25% gases and black carbon, 20% ash. There are 80 – 100 tons of dust particles, 10-12 tons of gas mixture (carbon monoxide, sulfur and nitrogen oxides) enters the atmosphere from one hectare of burning forest.

In the autumn of 1998 in Khabarovsk the content of toxic products of combustion in the air exceeded the permissible norm: carbon monoxide – 5 times, and nitrogen dioxide and sulfur – up to 22 times. Damage from air emissions amounted to 825 million Nine million tons of pollutants has got the air of cities and villages of the region.

Forest fires in Indonesia in 1997 took the form of a global environmental disaster. More than 50 thousand local residents were hospitalized with severe respiratory diseases. Residents of Singapore, Thailand, Brunei and even the Philippines were forced to wear protective masks. Ships in the Strait of Malacca were in conditions of almost zero visibility. There were some cases of collision and death of the crew. In the North of Sumatra Island, an Indonesian Airbus "A-300" crashed with 234 passengers and crew on Board (Riswan & Yusuf, 1986).

The researchers found that the circulation of smoke clouds and smog from forest fires in the atmosphere occurs in different trajectories. Smoke from Indonesian fires moved more slowly, more than light smog, which moved faster and spread over large areas.

Annually in the actively protected area of the Russian forest Fund from 10 to 35 thousand forest fires are registered, covering an area of 0.5 to 2.5 million hectares. Taking into account the burning of forests in the unprotected and occasionally protected areas of the Northern regions of Siberia and the Far East, the total area covered by the fire is from 2.0 to 5.5 million hectares. As a result of combustion of organic materials from these areas, 14.0 to 40.0 MT of carbon is released annually. In recent years, due to the problem of emission of large amounts of greenhouse gases (carbon dioxide, methane, etc.) into the atmosphere, forest and other natural ecosystems have been considered in a completely new aspect for them. The preservation and cultivation of forests has been regarded as the method of binding (the Deposit) of the atmospheric carbon that allows at least partly balancing the powerful emissions of carbon dioxide into the atmosphere by burning fossil fuels. The forests of Russia play a crucial role in absorbing carbon dioxide and maintaining the relative stability of the atmospheric gas composition. They are the huge reservoir of carbon, they absorb 500 million tons of carbon every year. Its total stock in the forests of Russia is equal to 233 billion tons (Vorobiev, Akimov et al., 2004).

Impact of Forest Fires on Soil

Forest fires, depending on temperature, lead to evaporation of water and combustion of organic matter. This produces pyrogenic charcoal, which is a natural fertilizer and temporarily increases the pH of the soil/ microbial activity may increase in the soil (Fire and Grazing, 2000). although there are cases of complete loss of microbes in its upper layer in the Combustion and elimination of vegetation contributes to a better heating of the soil surface by day and cool at night, increasing the availability of water for plants, which in General stimulates the germination of seeds and development of plants.

Long-term studies of burnt areas in different regions of Siberia has allowed to identify two groups of elements with different behavior in fires: 1 – active air migrants (Hg, Cd, Pb, Zn, Mn, As, Sb, Se, U, and artificial radio-nuclides are 90Sr, 137Cs and 239Pu, 240Pu) and 2 – passively accumulating in the fire (Cr, Ni, Co, V, Mg, Si, Fe, Th, Ca, K). Soil and plants are contaminated with migrant elements that lead to a change in the geochemical background of vegetation and soil cover, changing the face and properties of the Siberia landscapes (Vorobyev, Akimov et al., 2004).

Table 2. The value of the complex fire hazard indicator, in which the fire of a forest area is possible

The Name of the Forest	
Pine forests-cowberry	300
Spruce-cowberry	500
Pine forests	550
Mixed	800
Deciduous	900
Birch and blueberry	900
Herbal plantings	5000

(Methods of assessing, 1996)

Table 3. Share of unfit wood by species after wildfires

Fire	Pine	Cedar	Spruce
Riding steady, %	50	30	70
Horse runaway, %	30	20	60

(Methods of assessing, 1996)

Effect of Forest Fires on Plants

The values of the complex indicator of fire danger, in which it is possible to ignite the forest area, that is maximum in grass plantations, the minimum in lingonberry pine (table 2). Comparative range probability of ignition has to be like as follows: herbal planting > birch-blueberry fields, deciduous > mixed > pine > spruce-a regarding red bilberry > pine-a regarding red bilberry.

The calculated indicators of wood unsuitable for implementation revealed that in the upper fire, both fluent and stable the most damaged spruce wood, the least – cedar: spruce > pine, fir > cedar (table 3).

The amount of wood fallout depends on the type and intensity of forest fire, topography, the presence/absence of permafrost, weather conditions, stand composition, growth and diameter of trees, and many other factors. On average, the post-fire drop of wood in the taiga zone is (as a percentage of the stock of stands before the fire): after grassroots fires - 6 – 12%; peat - 70-80% and wildfires - 75-95%. In many types of forests peat and horse fires lead to the complete destruction of the forest stand.

Depending on the degree of damage, the state of the forest stands changes and the amount of wood suitable for harvesting accordingly. For example, at the first degree of damage the condition is maintained 0 – 75% of the stand; it is slightly damaged by fire and almost does not thin out. At the V degree of damage, the condition of the stand is maintained 0 - 39%, it turns into dead wood.

Figure 4.

As a whole:

- Post-fire renewal of boreal forests has significant potential – the area of hares and dead plantations over the past 47 years has decreased by about 2.5 times (from 70 million hectares in 1961 to 30 million hectares in 2018).
- Post-fire reforestation in the far North (forest tundra, Northern and rare taiga) is usually very slow, lasting up to 30-35 years, sometimes longer, due to the lack and poor quality of seeds.
- Productivity of the first post-fire generation of forests in the permafrost is 2-3 times higher than in the unburned areas, due to the increase in the active soil layer after the fire, improving the hydrological regime and better mobilization of nutrients.
- Almost all bioclimatic zones, excluding larch forests of North of areas, the fires that destroy the forest stand and initiate secondary succession change of species.
- Recurrent fires often lead to the degradation of forests, development of the processes water logging and bushclustering ultimately, to an indefinite treeless condition of the original forest landscape and "green desertification».
- Renewal under the forest canopy after grass-roots fires depends on the frequency of recurrent fires.

However, in the process of evolution, plants have developed many adaptations to resist fire, or its use as a necessary factor of growth and development to cope with fire. One of the most well-known devices is the process of releasing seeds from the fruit under the influence of smoke or extremely high temperature. In its absence, any population would die irretrievably, but in reality it is being restored and renewed (Klinkenborg, 2007). Moreover, the fire destroys the debris of dead, old and diseased trees, clearing the area for young growth and providing it with ash – natural fertilizer (Fig. 4).

For example, some plants contain highly flammable oils. A well-known example is the "not burning kupina» (lat. Dictamnus) is an oligotype genus of the Rutaceae family, subfamilies Toddalioideae. In a dry hot year, it releases a large amount of essential oils, which are easily ignited by the flame, but do

Figure 5. Birds hunt small animals fleeing fire

Figure 6a. Map of forests of Canada

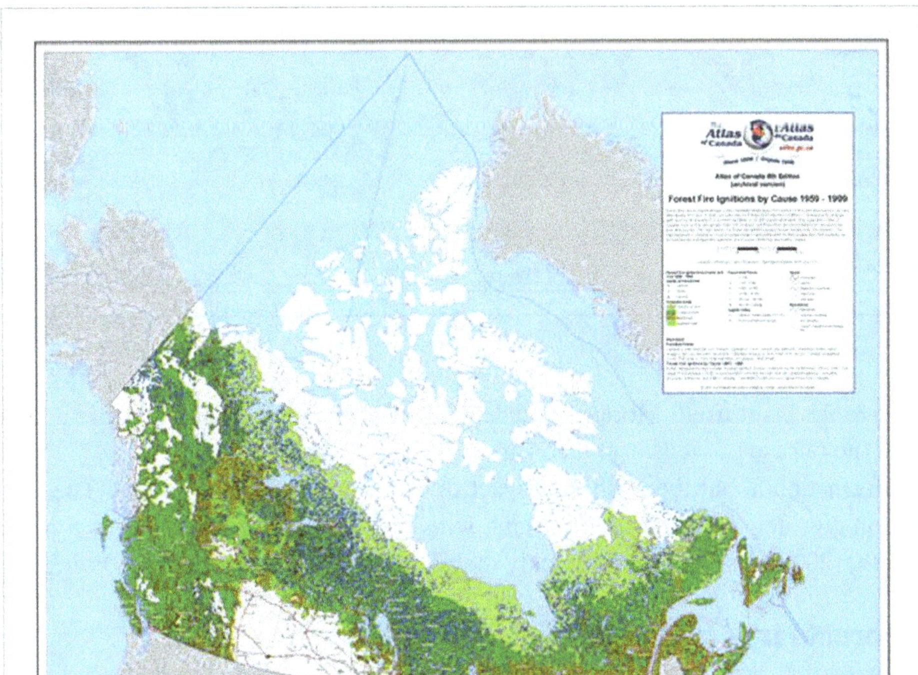

Figure 6b. Forest fire map in Canada

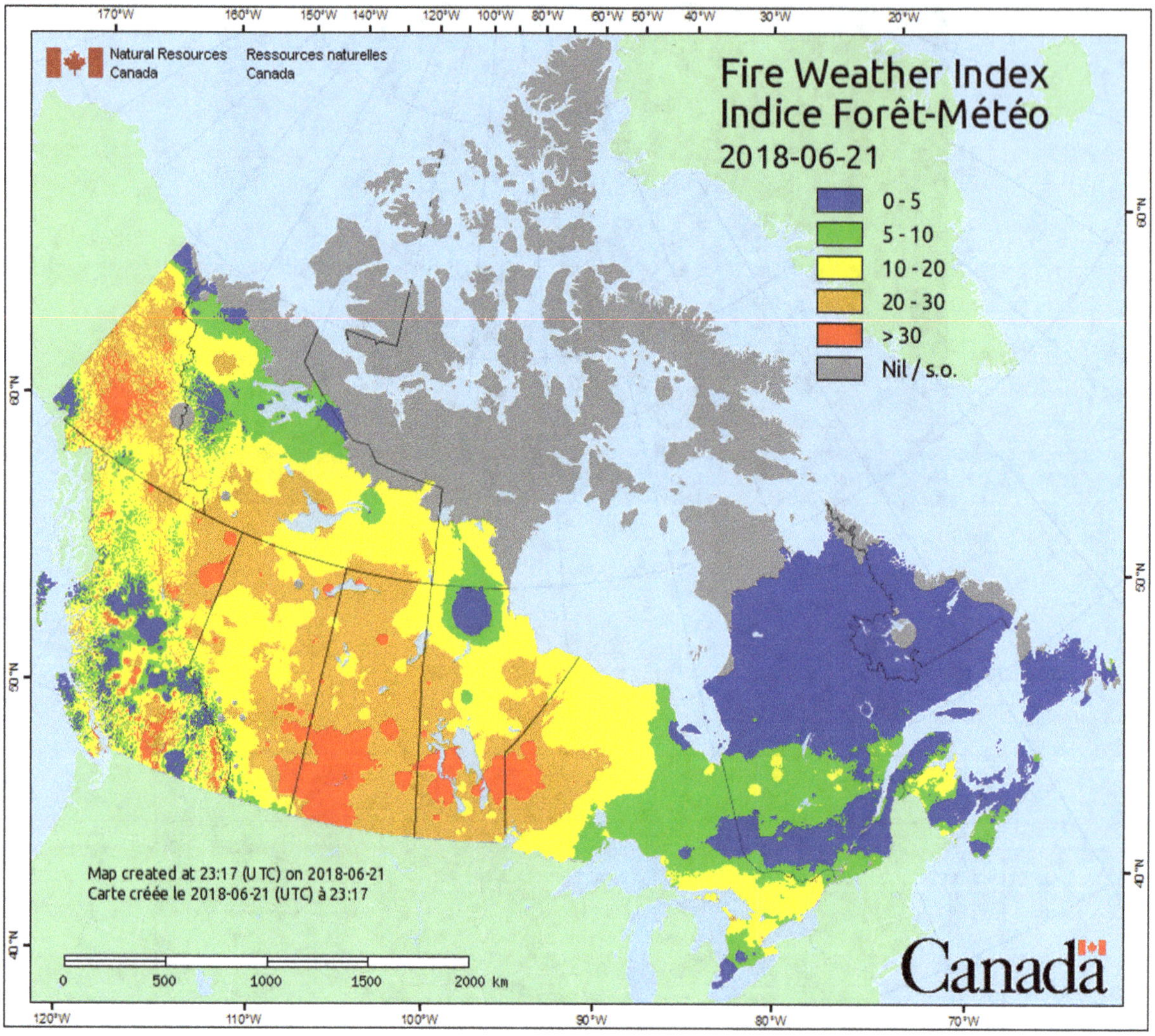

not cause harm to the plant itself. Moreover, under the influence of the flame there is a rapid cracking of the fruit and the release of seeds over a considerable distance.

There is information about the positive impact of fires on light-loving plants. The giant Sequoia (Sequoiadendron giganteum), being fire-resistant, wins the competition for territory from other plants (Brown & Smith, (2000). Its shoots will quickly melt after the fire and capture the living space.

Impact of Forest Fires on Animals

Like plants, animals in certain conditions are able to cope with fire, but this is due to their ability to actively move. Mammals are usually able to escape from fire. Amphibians and reptiles hide in the ground, mud or burrows of other animals. Birds leave a dangerous place, but often attack small animals escaping a fire (Fig. 5) and at its end, they quickly settle post-fire area (Gosford & Bonta, 2015).

Arthropods and microorganisms take refuge in soil fires (Keeley, 2015) and then displace other species. Indians on plains specially burn out a grass to attract of new birds for hunting. They call the forest fire the red bison. Large hoofed animals prefer to graze on pastures faded because of new young grass grows there.

In the South African savannas, the areas covered by forest fires are experiencing rapid growth of herbaceous and shrubby plants, providing animals with a variety of nutritious food compared to the harsher grasses. New pastures attract large herbivores from other areas. Their manure and ash from burned plants stimulate the growth of plants adapted to intensive grazing and allow these species to be preserved in this ecosystem. Most of the old eucalyptus forest in Australia is intended for Oh-wounds. Species such as Eucalyptus grand is survived in fire and forest fires can be beneficial by creating rich soil, reducing competition and promoting seed dispersal.

In the tropics, the long dry seasons of the evergreen trees reset in the movie-charge sheets. Together with the dead wood remaining after felling, this accumulated dry debris contributes to the rapid spread of uncontrolled fire. Many vines become "fire ladders", leading the fire to the tree crowns. Plants located in the lower tier are most vulnerable. 99% of them dies and to 50% – of the middle layer, from a high layer – 20-35%. The areas with low water-holding capacity, peat bog forests, forests on limestone, forests on shallow soils, cut down forests and forests near settlements are most affected by fire. The fire on the island Kalimantan (Borneo) also extends to the underlying peat and underground coal seams and smolders even after the rainy season. Smoldering organic material ignites during the next dry period (Goldammer & Seibert, 1989). Fires destroy more trees in succession forests than in virgin forests (Riswan & Yusuf, 1986).

Since fire plays an important role in nutrient cycling, maintaining diversity and habitat structure, extinguishing it can lead to unforeseen changes in ecosystems. There have been cases of an increase in the number of invasive and non-native plant species that quickly settled the burned areas and violated the ecological balance (Keeley, 2015).

Thus, while planning fire extinguishing it is necessary to take into account the integrity of the ecological system and the consequences caused by possible violations of the ecological balance.

THE METHODOLOGY TO EVALUATE THE EFFECTS OF FOREST FIRES

Post-fire restoration of forests depends on the following factors: geographical location and climate; structure of the landscape and the location of fumes in the landscape; type and peculiarities of relief; the conditions of the habitat, biology and ecology of tree species; specifics of forest types and succession phases; type and fire intensity; the size of fume; the presence of sources of contamination and the quality of seeds, etc.

Ecosystem restoration has regional features, but the sequence of events shown in figure is generally observed. 8.

Forest Fires as a Factor in the Evolution of Ecosystems

In each ecosystem, organisms adapt differently to forest fire. They make use of the heterogeneity of the habitat, starting from the preserved areas, ending with burning, on the one hand, and continuously occurring succession processes, on the other hand. Ecologists usually characterize succession through the

Figure 7. Follow-up of forest restoration after a fire: 1) a fire, 2) the regrowth of aspen one year after fire, 3) burned tree with black-backed woodpecker,one year after fire, 4) a 50-year stand, 5) a 100-year stand, 6) 150 years of growth, 7) old-growth forest (with gap dynamics)

changes in title note that consistently arise. After the fire, the pioneer species will be those whose seeds are already in the soil or can quickly get into it. Typically, these are fast-growing light-loving herbaceous plants that will gradually be suppressed by slower-growing, shade-tolerant woody species.

Conifers are often early succession species, while broadleaved trees replace them in the absence of fire. Consequently, the development of many coniferous forests depends on the repetition of fire.

Weak and moderate fires in the undergrowth burn small trees and grassy ground cover plants. Strong fires capture the crowns of trees, deplete most of the dominant vegetation and contribute to a high level of biodiversity. In case of frequent fires, less plant bushing accumulates the temperature of the soil under the ground rises slightly, which leads to the death of roots and, consequently, trees. Living organisms have adapted to the use of stages of succession, regrowth and habitat changes occurring after each forest fire. Its appearance, speed, size and intensity led to different ways of changing the habitat and influenced the species composition of the population on the fires. In Canada, forests cover about 10% of the land area, they frequent fires, and at the same time, there are 70% of the species of birds and terrestrial mammals (Anderson & Moratto, 1996) (Fig. 7).

As a whole:

- Post-fire renewal of boreal forests has significant potential – the area of hares and dead plantations over the past 47 years has decreased by about 2.5 times (from 70 million hectares in 1961 to 30 million hectares in 2018).
- Post-fire reforestation in the far North (forest tundra, Northern and rare taiga) is usually very slow, lasting up to 30-35 years, sometimes longer, due to the lack and poor quality of seeds.
- Productivity of the first post-fire generation of forests in the permafrost is 2-3 times higher than in the unburned areas, due to the increase in the active soil layer after the fire, improving the hydrological regime and better mobilization of nutrients.
- Almost all bioclimatic zones, excluding larch forests of Northern areas, there are fires that destroy the forest stand and initiate secondary succession change of species.
- Recurrent fires often lead to the degradation of forests, development of the processes water logging and bushing, and, ultimately, to an indefinite treeless condition of the original forest landscape and "green desertification».
- Renewal under the forest canopy after grass-roots fires depends on the frequency of recurrent fires.
- Fires are the main natural disturbance in the forests of the Russian Federation for the following reasons.
- About 95% of the country's forests are in the boreal zone, and most of the forests are represented by coniferous species of high fire hazard.

- A significant part of the forest fund is not managed practically and protected from fires due to the reduction in the number of forest protection and leasing of forests.
- Due to the slow decomposition of plant litter due to harsh climatic conditions, significant reserves of organic material accumulate in forest litter and peat, which is a material for burning combustion.
- The majority of boreal forests are located in areas with limited rainfall and frequent droughts during the fire season. The diversity of native composition, productivity and structure of forests and forest vegetation conditions, land-shaft features, anthropogenic load and so on determine the diversity of forest fire species, their spread, intensity, as well as the nature of the impact on ecosystems and natural landscapes.

All UN member States are recommended to develop coordinated approaches to the conservation of their forests and sustainable forestry to ensure sustainable development of mankind.

According to the results of forest fires study it can be concluded that:

- Fires most often occur because of the human activity and have their distinct landscape basis.
- Quantitative indicators of fire regimes are determined by the conditions of production, the specificity of plantings and the level of anthropogenic impact.
- The impact of forest type on the frequency and intensity of fires is more significant than the impact of the terrain.
- Low-level fires of high frequency (e.g. in pine forests of well-drained habitats) do not usually lead to significant damage to stands or change of species, but significantly affect the process of natural regeneration and the structure of stands.
- Fires in waterlogged habitats are very rare (up to 200-300 years or more), but lead to the complete destruction of stands.
- Accessibility of forests to the population and availability of roads is an important factor affecting the forest cover of the Russian Federation.

The amount of wood fallout depends on the type and intensity of forest fire, relief, the presence of permafrost, weather conditions, the composition of the forest stand, age and diameter of trees, and many other factors. On average, the post-fire drop of wood in the taiga zone is (as a percentage of the stock of forest stands before the fire): after grassroots-6 – 12%; peat - 70-80% and wildfires - 75-95%, although the variability of these values is very high. In many types of forests peat and crown fires lead to the complete destruction of the forest stand.

Post-fire restoration of forests depends on the following factors: geographical location and climate; structure of the landscape and the location of fume in the landscape; type and peculiarities of relief; the conditions of the habitat, biology and ecology of tree species; specifics of forest types and succession phases; type and fire intensity; the size of fume; the presence of sources of contamination and the quality of seeds, etc.

FUTURE RESEARCH DIRECTIONS

The consequences of forest fires are studied by scientists from different countries and continents, but the problems associated with the dynamics of forest ecosystems in the post-pyrogenic period, after 10 or more years of exposure to fire remain unexplored. This problem is particularly relevant for suburban areas of megacities. In the context of urbanization, the role of forests as a source of recreational resources is increasing. The number of tourists traveling to the suburban area of megacities increases annually. Recreants build fires, trample the ground cover, affect the root systems and destroy flowering plants.

In recreational forests often occur and forest fires, through the fault of man. Ecosystems of suburban forests are exposed to the components and campers and fires. This problem remains poorly understood, so further research will be devoted to it.

The main focus will be on the dual impact on ecosystems: human and fire. The main component of the ecosystem under study will be the root systems of forest stands, their dynamics in 5 and 10 years after intensive recreational and pyrogenic impact, as well as the dynamics of plant species diversity in the forest and with varying degrees of anthropogenic impact on the forest ecosystem.

CONCLUSION

The impact of fire on natural ecosystems is determined by the structure and composition of forests, the type of fire and the natural conditions of the region. All types of forest fires occur in the Russian Federation. The cause of forest fires, in most cases, is the anthropogenic factor. Fires damage the national economy and affect the components of natural ecosystems.

The majority of boreal forests are located in areas with limited rainfall and frequent droughts during the fire season. The diversity of species composition, productivity and structure of forests and forest vegetation conditions, landscape features, anthropogenic load and so on determine the variety of forest fire species, their distribution, intensity, as well as the nature of the impact on ecosystems and natural landscapes.

Forest fires affect all components of the natural ecosystem. Primarily the wood stands suffer from the fire. The degree of damage depends on the predominant wood species, type of fire and wind speed. A smoke is formed after the fire; there are wind and burnt trees that do not have marketing sales. Wood-destroying mushrooms and entom pests appear on them. The problem of cleaning deadwood and renewal of burned areas valuable species arises for forestry. Despite the fact that ash elements are formed in the soil after the fire, the process of renewal of conifers is difficult due to the complexity of the use of technical means and the lack of seeds. Preserved in grass-roots fire trees reduce the current gain of the stock by almost 2 times. Fires initiate succession with the change of rocks.

Fire destroys the forest young generation – undergrowth of any height and quality. Therefore, the source of renewal of the forest is only seeds, which also often burn in case of fire. Source of contamination there may be tested – by fire ripe growing coniferous trees that have remained. Fire destroys underwood trees. The ground cover burns as well. Favorable conditions for reforestation are being created, but the lack of sources of insemination will extend the process of reforestation for several decades. Post-fire reforestation is very slow, for a long time, up to 30-35 years, sometimes longer, due to the lack and poor quality of seeds

Fire has significant changes on the soil. The degree of fire impact depends on the type of fire and the type of soil. At the grass-roots and underground in the heat of the burned forest litter or peat. As a result, the morphology of the soil profile changes. The soil is enriched with ash and other nutrients. It is soot-bearing, soil is enriched with nitrogen, which leads to the appearance of nitrophilic plants on the fumes such as wormwood and raspberries. Renewal under the forest canopy after grassroots fires depends on the frequency of recurring fires.

Crown fires are dangerous for humans and human settlements. Every year in places of forest fires people die – employees of forestry and the Ministry of emergency situations and residents of settlements. Often local residents remain without shelter: burn houses, buildings, livestock and other personal courtyard.

SOLUTIONS AND RECOMMENDATIONS

Fire hazards and fire risks depend on forest ecosystems, weather conditions and the availability of fire sources. All forests in Russia are divided into 5 classes of natural fire hazard. Starting from forests of fire hazard class I (the highest risk), where fires are possible throughout the fire season (coniferous young growth, cluttered felling in dry forest conditions), to class V, where fires occur only in adverse weather conditions (prolonged drought in the spruce forest sphagnum).

Fire-fighting arrangement of forests includes the following types of work:

- the device of the mineralized strips
- controlled burning of dry combustible materials
- construction and maintenance of fire-fighting roads
- laying of glades, fire-prevention barriers and gaps
- arrangement of fire reservoirs and entrances to water sources
- construction of landing sites for aircraft, helicopters, used for aerial work on the protection and the protection of forests

ACKNOWLEDGMENT

The research is carried out at Tomsk Polytechnic University within the framework of Tomsk Polytechnic University Competitiveness Enhancement Program.

REFERENCES

Anderson, M. K., & Moratto, M. J. (1996). Native American land-use practices and ecological impacts. Sierra Nevada Ecosystem Project, Final report to Congress, vol. II, Assessments and scientific basis for management options. Davis, CA: University of California, Centers for Water and Wildland Resources, 191-200.

Andreae, M. O., & Goldammer, J. G. (1992). Tropical wildland fires and other biomass burning: Environmental impacts and implications for land use and fire management. *Conservation of West and Central African Rainforests*, 79.

Archibald, S., Bond, V. J., Stock, W. D. & Fairbanks, F. (2005). Shaping the landscape: Fire-grazer interactions in an African savanna. *Ecological Applications,* 1, 15, 96-109. doi:. doi:10.1890/03-5210

Ash is a fertilizer and promotes rapid plant growth. (n.d.). Retrieved from https://phototass4.cdnvideo.ru/width/1200_4ce85301/tass/m2/uploads/i/20160709/4288060.jpg

Begon, M., Harper, J. L., & Townsend, C. R. (1996). *Ecology* (p. 692). Oxford, UK: Blackwell Science.

Birds hunt small animals fleeing fire. (n.d.). Retrieved from https://freethoughtblogs.com/affinity/files/2018/01/Kites.jpg

Brown, J. K., & Smith, J. K. R. (2000). *Wildland fire in ecosystems: effects of fire on flora*. UT: U.S. Department of Agriculture, Forest Service, Rocky Mountain Research Station. doi:10.2737/RMRS-GTR-42-V2

Díaz, S., Demissew, S., Carabias, J., Joly, C., Lonsdale, M., Ash, N., ... Báldi, A. (2015). The IPBES conceptual framework – connecting nature and people. *Current Opinion in Environmental Sustainability*, *14*, 1–10. doi:10.1016/j.cosust.2014.11.002

Emery, S. M., & Gross, K. L. (2005). Effects of timing of prescribed fire on the demography of an invasive plant, spotted knapweed Centaurea maculosa. *Journal of Applied Ecology*, *42*(1), 60–69. doi:10.1111/j.1365-2664.2004.00990.x

Forest fire in Ukraine. (n. d.).

Forest Fire Map in Canada. (n.d.). Retrieved from https://yandex.ru/images/search?text=2019%20map%20forest%20fire%20Canada&pos=3&img_url=http%3A%2F%2Fcwfis.cfs.nrcan.gc.ca%2Fdata%2Fmaps%2Ffwi_fbp%2F2018%2Ffwi20180621.png&rpt=simage

Gill, A. M., & Moore, P. H. R. (n. d.). Fire situation in Australia. Retrieved from http://www.fao.org/3/AD653E/ad653e80.htm

Global biodiversity. (2015). Montréal (Canada): *Convention on Biological Diversity, 155.* Retrieved from https://www.cbd.int/gbo4

Goldammer, J. G., & Seibert, B. (1989). Natural rain forest fires in Eastern Borneo during the Pleistocene and Holocene, *The Science of Nature, 11,* 76. 518-520. doi:. doi:10.1007/BF00374124

Göltenboth, F., & Widmann, P. (2006). Special forest ecosystems. *Ecology of Insular Southeast Asia*, Retrieved from https://www.sciencedirect.com/topics/agricultural-and-biological-sciences/intsia

Gosford, R., & Bonta, M. (2015). Ornithogenic fire: Raptors as propagators of fire in the Australian Savanna). *Raptor Research Foundation Annual Conference,* Nov 4–8, Sacramento, CA. Retrieved from https://www.researchgate.net/publication/324819947_Ornithogenic_Fire_Raptors_as_Propagators_of_Fire_in_the_Australian_Savanna

Johannesburg Declaration on Sustainable Development (2002). (n.d.). Retrieved from http://ec.europa.eu/environment/archives/wssd/documents/wssd_pol_declaration.pdf

Kalinin, L. B., Kovyazin, V. F., & Belov, S. V. (1993). Forest fire science. St. Petersburg State Forestry Academy, St. Petersburg, Russia.

Keane, R. E., & Long, D. G. (1998). A comparison of coarse scale fire effects simulation strategies. *Northwest Science*, *2*, 72, 76–90.

Keeley, J. E. (2015). Chaparral plant diversity: Evolutionary & Ecological Patterns. U.S. Geological Survey Sequoia Field Station & UCLA, Department of Ecology & Evolutionary Biology, 51.

Klinkenborg, V. (2007). *Splendor of the grass: The prairie's grip is unbroken in the Flint Hills of Kansas.* National Geographic.

Kolesnikova, T. G. (2002). Payments for forest resources as a factor of socio-economic development of forestry: On the example of the Republic of Mari El. *Abstract of Dis. thesis....Kand. Econ. Sciences: 08.00.05: Yoshkar-Ola.*

Kovyazin, V. F., Romanchikov, A. N., & Pasko, O. A. (2015). Comparative analysis of forest lands cadastral appraisal estimated with regards to wood and food resources. *IOP Conference Series: Earth and Environmental Science*. Retrieved from 10.1088/1755-1315/27/1/012039

Krug, E. C., Hollinger, S. E. (2003). Identification of factors that aid carbon sequestration in Illinois agricultural systems. Champaign, IL: *Illinois State Water Survey*, 103.

Lower forest fire. (n.d.). Retrieved from http://itd2.mycdn.me/image?id=812140978872&t=20&plc=WEB&tkn=*WaY_waiTdgCqHhBa0HMmywyQMH0

Macdougall, A. C., & Turkington, R. (2004). Relative importance of suppression-based and tolerance-based competition in an invaded oak savanna. *Journal of Ecology*, *3*(92), 422–434. doi:10.1111/j.0022-0477.2004.00886.x

Management, conservation and sustainable development of forests. (1992). (n.d.). Retrieved from http://www.fao.org/3/y5574e/y5574e05.pdf

Map of forests of Canada. (n.d.). Retrieved from http://ftp.geogratis.gc.ca/pub/nrcan_rncan/raster/atlas_6_ed/eng/6432_forest_fire_ignitions_cause_1959_1999_browse.jpg

McCaw, L., Cheney, P., & Sneeuwjagt, R. (2003). *Development of scientific understanding of fire behavior and use in southwest Western Australia. Fire in ecosystems of southwest Western Australia* (pp. 171–187). Leiden, The Netherlands: Backhuys Publishers.

Methods of assessing the consequences of large forest fires. (1996). Moscow, Russia: VNII GOCHS, 13.

Mounted forest fire. (n.d.). Retrieved from https://www.eurekalert.org/multimedia/pub/web/116249_web.jpg

Natural Resources Canada. (n.d.). Retrieved from https://www.nrcan.gc.ca/our-natural-resources/forests-and-forestry/wildland-fires-insects-and-disturbances/forest-fires/fire-ecology/13149

Ostrom, E. (2009). A general framework for analyzing sustainability of social-ecological systems. *Science*, *325*(5939), 419–422. doi:10.1126cience.1172133 PMID:19628857

Paris, L. J., Dietz, T., Carpenter, S. R., Alberti, M., Folke, C., Moran, E., ... Lubchenco, J. (2007). Complexity of coupled human and natural systems. *Science*, *317*(5844), 1513–1516. doi:10.1126cience.1144004 PMID:17872436

Pasko, O. A. (2013). The forest resource of Tomsk Region and its use. *Applied and Fundamental Studies,* 102-110.

Pasko, O. A., & Baranova, A. V. (2015). Forest fire situation analysis over forest reserve land in Tomsk petroleum province. *IOP Conference Series: Earth and Environmental Science*. Retrieved from 10.1088/1755-1315/24/1/012022

Pasko, O. A., Tokareva, O. S., Alshaibi, A. J. A., Chernikova, T. Y., & Cabral, P. (2019). Assessment of state of cedar forests in Tomsk region using remote sensing data of the earth. *Bulletin of the Tomsk Polytechnic University. Geo Assets Engineering*, *330*(1), 89–101. doi:10.18799/24131830/2019/1

Perelet, R. A. (2006). An ecosystem approach to environmental management and environmental activities. *Mehanizm regulyuvannya economy,* 1, 36-53.

Shvidenko, A. (2009). Non-boreal forests of Eastern Europe in a changing world: the role in the Earth systems. *Regional Aspects of Interactions in Non – boreal Eastern Europe, Springer Science: Business Media B.V.*, 123-133.

Smoke and smog from a forest fire in Canada. (n. d.). Retrieved from https://i.pinimg.com/originals/29/31/5c/29315c49b896cb9d51322a4d461c13d2.jpg

Soja, A. J., Tchebakova, N. M., French, N. H. F., Flannigan, M. D., Shugart, H. H., Stocks, B. J., ... Stackhouse, P. W. Jr. (2007). Climate-induced boreal change: Predictions versus current observations. *Global and Planetary Change*, *56*(3-4), 274–296. doi:10.1016/j.gloplacha.2006.07.028

Strategic Research Agenda. (2014). Retrieved from http://www.futureearth.org/media/strategic-research-agenda-2014

The IPBES conceptual framework – connecting nature and people. (2015). *Curr Opin Environ Sustainability, 14,* 1–16.

The Rio declaration on environment and development. (1992). Retrieved from http://www.prometeus.nsc.ru/koptyug/ideas/unrio92/030-031.ssi

The UN Convention. On biological diversity. (2002). Retrieved from https://www.cbd.int/doc/legal/cbd-en.pdf

Tng, D. Y., Goosem, S., Jordan, G. J., & Bowman, D. M. (2014). Letting giants be - Rethinking active fire management of old-growth eucalypt forest in the Australian tropics. *Journal of Applied Ecology, 51*(3), 555–559. doi:. doi:10.1111/1365-2664.12233

United Nations framework Convention on climate change. (1992). Retrieved from https://unfccc.int

Van Oudenhoven, A. P. E., Martín-López, B., Schröter, M., & de Groot, R. (2018). Advancing science on the multiple connections between biodiversity, ecosystems and people. *Journal Biodiversity Science Ecosystem, l,* 14,1, 127–131.

Vinton, M. A., Harnett, D. C., Finck, E. J., & Briggs, J. M. (1993). Interactive effects of fire, bison (Bison bison) grazing and plant community composition in tallgrass prairie. *American Midland Naturalist, 129*(1), 10–18. doi:10.2307/2426430

Vorobyev, Yu. L., Akimov, V. A., & Sokolov, Yu. (2004). *Forest fires in Russia: the state and problems.* Moscow, Russia: Dex Press.

Williams, G. W. (2003). References on the American Indian use of fire in ecosystems. Retrieved from https://www.nrcs.usda.gov/Internet/FSE_DOCUMENTS/nrcs144p2_051334.pdf.

Yadav, I. C., & Devi, N. L. (2018). Biomass burning, regional air quality and climate change, reference module. *Earth Systems and Environmental Sciences.* Retrieved from https://www.sciencedirect.com/science/article/pii/B978012409548911022X?via%3Dihub

ADDITIONAL READING

Belotserkovets, A. V. (2016). The Role of forest fires in changing ecosystems in the regional aspect on the example of the Khabarovsk territory. *The Young scientist.* 3, 381-385. Retrieved September 4, 2019, from https://moluch.ru/archive/107/25524

Bytnerowicz, A., Arbaugh, M. J., Riebau, A. R., & Andersen, C. (2009). Wildland fires and air pollution. Elsevier, 643.

Dangerous fire factors associated with and influencing people. (n. d.). Retrieved September 4, 2019, from https://protivpozhara.com/tipologija/prirodnye/posledstvija-lesnyx-pozharov

Kasischke, E. S., & Turetsky, M. R. (2006). Recent changes in the fire regime across the North American boreal region. Spatial and temporal patterns of burning across Canada and Alaska. *Geophysical Research Letters*, *9*(33), 5.

Kudryavtsev, M., Lukin, V., Malinetskii, G., Mitin, N., Naumenko, S., Podlazov, A., Rumyantsev, A., Toropygina, S. (2008). Forest conflagration extinguishing problems on Russian Federation territory. *Preprint, Institute of applied mathematics of Russian Academy of Science,* 40.

Lloyd, A. H. & Bunn, A. G. (2007). Responses of the circumpolar boreal forest to 20th century climate variability. *Environmental research letters, 2,* 4, 13.

McCaw, L., Cheney, P., Sneeuwjagt, R., McCaw, L., Abbott, I., & Burrows, N. (2003). Development of scientific understanding of fire behavior and use in south-west Western Australia. *Fire in ecosystems of south-west Western Australia. Backhuys Publishers, Leiden*, 171-187. Negative impacts of forest fires to the environment. Retrieved September 4, 2019, from https://protivpozhara.com/tipologija/prirodnye/posledstvija-lesnyx-pozharov

Orozko, S. J. (2008) Forest fire risk model for Michoacan, Mexico. *Thesis submitted to the International Institute for Geo-information Science and Earth Observation in partial fulfillment of the requirements for the degree of Master of Science in Geo-information Science and Earth Observation, Specialisation: Integrated Landscape Management*, 88.

Pasko, O. A. (2012). *Interaction of organism and habitat. Textbook.* Tomsk: Tomsk Polytechnic University.

Sofronov, M. A., Goldammer, J.A., Volokitina, A.V., Sofronova, T. M., & Sofronov, M.A. (2005). Wildland fire danger, *Krasnoyarsk: IL SO RAN, Max Planck Institute for Chemistry, SibGTU,* 330.

KEY WORDS AND DEFINITIONS

Emergency Forest Fire: The situation in the certain territory, developed as a result of the source of natural emergencies – forest fire (forest fires), which can entail or entailed human victims, the damage to human health and the environment, considerable material losses and violation of conditions of activity of people.

Fire Monitoring: A system of observation and control of fire danger in the forest according to weather conditions, the condition of forest combustible materials, fire sources and forest fires in order to timely develop and implement measures to prevent forest fires and (or) reduce damage from them.

Fire Suppression: Actions of participants of fire extinguishing aimed at the final cessation of combustion, as well as to exclude the possibility of its re-occurrence Gorenje.

Forest Fire: A spontaneous, uncontrolled spread of fire over forest areas.

Forest Fund: A set of forest and non-forest land categories of administrative economic territories.

Forest Protection: A special service organized for the protection of forests under the jurisdiction of the state forestry bodies and control over the protection of collective farm forests, urban forests, forest reserves and fixed forests.

Compilation of References

A forest fire in the South of Italy. (n.d.). Retrieved from https://tipolog.livejournal.com/305437.html

A forest fire in the Tomsk region. (n.d.). Retrieved from https://yandex.ru/images/search?text=%D0%BB%D0%B5%D1%81%D0%BD%D1%8B%D0%B5%20%D0%BF%D0%BE%D0%B6%D0%B0%D1%80%D1%8B%20%D0%BA%D0%BE%D1%81%D0%BC%D0%BE%D1%81%20%D0%A2%D0%BE%D0%BC%D1%81%D0%BA%D0%B0%D1%8F%20%D0%BE%D0%B1%D0%BB%D0%B0%D1%81%D1%82%D1%8C&pos=168&p=5&img_url=https%3A%2F%2Fpbs.twimg.com%2Fprofile_banners%2F1254679254%2F1383113680%2F1500x500&rpt=simage

Abatzoglou, J. T., & Kolden, C. A. (2013). Relationships between climate and macroscale area burned in the western United States. *International Journal of Wildland Fire*, *22*(7), 1003–1020. doi:10.1071/WF13019

Abduragimov, I. M. (2012). The problem of extinguishing large forest fires and large-scale fires of solid combustible materials in buildings. *Fire and Explosion Safety*, *21*(2), 69–74.

Abel, S., Haywood, J. M., Highwood, E. J., Li, J., & Buseck, P. R. (2003). Evolution of biomass burning aerosol properties from an agricultural fire in South Africa. *Geophysical Research Letters*, *30*(15), ASC1–ASC4. doi:10.1029/2003GL017342

Abushenko, N. A., Bartalev, S. A., Belyaev, A. I., Ershov, D. V., Zakharov, M. Yu., Korovin, G. N., … Schetinsky, V. E. (1998). Experience and prospects of organizing operational satellite monitoring of the territory of Russia for the purpose of the service of fire protection of forests. Study of Earth from space, 3, 89-95. (in Russia)

Adab, H., Kanniah, K. D., & Solaimani, K. (2013). Modeling forest fire in the northeast of Iran using remote sensing and GIS techniques. *Natural Hazards*, *65*(3), 1723–1743. doi:10.100711069-012-0450-8

Adame, J. A., Lope, L., Hidalgo, P. J., Sorribas, M., Gutiérrez-Álvarez, I., Del Águila, A., ... Yela, M. (2018). Study of the exceptional meteorological conditions, trace gases and particulate matter measured during the 2017 forest fire in Donana Natural Park, Spain. *The Science of the Total Environment*, *645*, 710–720. doi:10.1016/j.scitotenv.2018.07.181 PMID:30031329

Adámek, M., Jankovská, Z., Hadincová, V., Kula, E., & Wild, J. (2018). Drivers of forest fire occurrence in the cultural landscape of Central Europe. *Landscape Ecology*, *33*(11), 2031–2045. doi:10.100710980-018-0712-2

Adushkin, V. V., Soloviev, S. P., & Spivak, A. A. (2018). *Electrical fields of man-made and natural processes*. Moscow, Russia: GEOS [in Russian]. *All Russian Research Institute of Hydrometeorological Information – World Data Center*. Retrieved from http://meteo.ru

Adzhieva, A. A. & Shapovalov, V. A. (2016). Cluster analysis in the automatic detection and support of thunderstorm sites according to the thunderstorming network data. *Engineering Bulletin of the Don. 2*, 8.

Adzhiev, A. H., & Bogachenko, E. M. (2011). *Grozy Severnogo Kavkaza (Thunderstorms of the North Caucasus)*. Nalchik, Russian Federation: Poligrafservis i T.

Adzhieva, A. A., Shapovalov, V. A., Boldyrev, A. S., & Ignatyev, V. V. (2018, November). Characteristics of lightning activity in the North Caucasus according to the LS 8000 from 2009 to 2016. In *Remote sensing of clouds and the atmosphere XXIII* (Vol. 10786, p. 107861A). International Society for Optics and Photonics.

Afonin, S. V., & Belov, V. V. (2001). System of space monitoring of forest fires in the Tomsk region, *Optics of the atmosphere and ocean, 8,* 14, 692-696.

Ahmed, B. M., Hata, T., Tanakamarua, H., Abdelhadi, A., & Tadac, A. (2002). The spatial analysis of surface temperature and evapotranspiration for some land use/ cover types in the Gezira Area, Sudan. Proceedings of Joint Symposium URBAN – VRS 2005 held at Temple, Arizona, March 14-16, 2005. (Eds.) M. Moeller, & E. Wenth, ISPRS Archieves, Volume XXXVI- 8/W27, p. 5.

Al Janabi, S., Al Shourbaji, I., & Salman, M. A. (2017). *Assessing the suitability of soft computing approaches for forest fires prediction.* Applied Computing and Informatics; doi:10.1016/j.aci.2017.09.006

Albini, F. A., & Reinhardt, E. D. (1995). Modeling ignition and burning rate of large woody natural fuels. *International Journal of Wildland Fire*, *5*(2), 81–91. doi:10.1071/WF9950081

Albini, F., Alexander, M., & Cruz, M. (2012). A mathematical model for predicting the maximum potential spotting distance from a crown fire. *International Journal of Wildland Fire*, *21*(5), 609–627. doi:10.1071/WF11020

Alexander, M. E. (1979). Crown fire thresholds in exotic pine C.E. In *Proceedings 10th Conference on Fire and Forest Meteorology* (pp. 207-212). Ottawa, Ontario.

Algorithms and programs for calculating operational systems for detecting and extinguishing forest fires on a computer. (1974). LenNIILH. (in Russia)

Allam, A., Bakeir, M., & Abo-Tabl, E. (2008). Some methods for generating topologies by relations. *Bulletin of the Malaysian Mathematical Sciences Society*, *2*(31), 35–45.

Amarnath, G., Babar, S., & Murthy, M. S. R. (2017). Evaluating MODIS-vegetation continuous field products to assess tree cover change and forest fragmentation in India–A multi-scale satellite remote sensing approach. *The Egyptian Journal of Remote Sensing and Space Sciences*, *20*(2), 157–168. doi:10.1016/j.ejrs.2017.05.004

Amatulli, G., Perez-Cabello, F., & Du la Riva, J. (2007). Mapping lightning / human-caused wildfires occurrence under ignition point location uncertainty. *Ecological Modelling*, *200*(3-4), 321–333. doi:10.1016/j.ecolmodel.2006.08.001

Anderson, M. K., & Moratto, M. J. (1996). Native American land-use practices and ecological impacts. Sierra Nevada Ecosystem Project, Final report to Congress, vol. II, Assessments and scientific basis for management options. Davis, CA: University of California, Centers for Water and Wildland Resources, 191-200.

Andreae, M. O., & Goldammer, J. G. (1992). Tropical wildland fires and other biomass burning: Environmental impacts and implications for land use and fire management. *Conservation of West and Central African Rainforests*, 79.

Andreev, Yu. A. (1986). *Laws of distribution of forest fires. Methods and means of fighting forest fires*. Moscow, Russia: VNIILM. (In Russian)

Andreev, Yu. A., & Larchenko, G. F. (1987). *Socio-psychological aspects of recreational forest visits and the occurrence of fires. Forest fires and the fight against them.* Moscow, Russia: VNIILM. (In Russian)

Andreu, A., & Hermansen-Baez, L. (2008) Fire in the South 2: the southern wildfire risk assessment. A report by Southern Group of State Forester.

Andrews, P. L. (2013). Current status and future needs of the BehavePlus fire modeling system. *International Journal of Wildland Fire*, *23*(1), 21–33. doi:10.1071/WF12167

Anisimov, S. V., & Mareev, E. A. (2008). Geophysical studies of the global electric circuit. *Izv. Phys. Solid Earth*, *44*(10), 760–769. doi:10.1134/S1069351308100030

Anthenien, R. A., Tse, S. D., & Fernandez-Pello, A. C. (2006). On the trajectories of embers initially elevated or lofted by small scale ground fire plumes in high winds. *Fire Safety Journal*, *41*(5), 349–363. doi:10.1016/j.firesaf.2006.01.005

Antropov, V. F., Seredkin, A. D., & Schepin, A. A. (2013). *Forestry of Buryatia*. Ulan-Ude, Russia: Ecos. (In Russian)

Applied reference book on climate of the USSR. Series 3. Long-term data. Parts 1-6. Issue 20. St. Petersburg, Russia: Gidrometeoizdat, 1993.

Apte, A., Hairer, M., Stuart, A. M., & Voss, J. (2007). Sampling the posterior: An approach to non-Gaussian data assimilation. *Physica D. Nonlinear Phenomena*, *230*(1-2), 50–64. doi:10.1016/j.physd.2006.06.009

Archibald, S., Bond, V. J., Stock, W. D. & Fairbanks, F. (2005). Shaping the landscape: Fire-grazer interactions in an African savanna. *Ecological Applications,* 1, 15, 96-109. doi:. doi:10.1890/03-5210

Arroyo, L. A., Pascual, C., & Manzanera, J. A. (2008). Fire models and methods to map fuel types: The role of remote sensing. *Forest Ecology and Management*, *256*(6), 1239–1252. doi:10.1016/j.foreco.2008.06.048

Artsybashev, E. S. (1974). Forest fires and fight with them. Moscow, Russia: Forest industry.

Artsybashev, E. S. (2014). The impact of fires on forest biogeocenoses. *Biosphere*, *6*(1), 53–59.

Ash is a fertilizer and promotes rapid plant growth . (n.d.). Retrieved from https://phototass4.cdnvideo.ru/width/1200_4ce85301/tass/m2/uploads/i/20160709/4288060.jpg

Asylbaev, N. A. (2010). Mathematical modeling of steppe fires. *Computer Research and Modeling*, *4*(4), 377–384. doi:10.20537/2076-7633-2010-2-4-377-384

Asylbaev, N. A. (2013). On the spread of a steppe fire along an inclined underlying surface. *Bulletin of Bashkir University*, *18*(4), 978–983.

Aven, T., Ben-Heim, Y., Andersen, H., Cox, T., Droguett, E., Greenberg, M., . . . Zio, E. (2015). Society for Risk Analysis Glossary.

Aven, T. (2011). On different types of uncertainties in the context of the precautionary principle. *Risk Analysis*, *31*(10), 1515–1525. doi:10.1111/j.1539-6924.2011.01612.x PMID:21477085

Aven, T. (2013). On how to deal with deep uncertainties in a risk assessment and management context. *Risk Analysis*, *33*(12), 2082–2091. doi:10.1111/risa.12067 PMID:23656628

Aven, T. (2015). Comments to the short communication by Jan Erik Vinnem and Stein Haugen titled "Perspectives on risk and the unforeseen". *Reliability Engineering & System Safety*, *137*, 69–75. doi:10.1016/j.ress.2015.01.001

Aven, T. (2015). Implications of black swans to the foundations and practice of risk assessment and management. *Reliability Engineering & System Safety*, *134*, 83–91. doi:10.1016/j.ress.2014.10.004

Aven, T., & Zio, E. (2013). Model output uncertainty in risk assessment. *International Journal of Performability Engineering*, *9*(5), 475–486.

Azbukin, A. A., Bogushevich, A. Ya., Il'ichevskii, V. S., et al. (2006). AMK-03 Automated Ultrasound Meteorological Complex. *Meteorologiya I Gidrologiya, (11),* 89-97 [in Russian].

Azmetov, R. R., Belyaev, A. I., & Moskovenko, V. M. (2000*). Prospects for the creation of the Russian system of electromagnetic monitoring of thunderstorms for the needs of forest protection from fires, energy, aviation, meteorology and disasters forecasting.* Paper presented at Proceedings of International Conference Conjugated Problems of Mechanics and Ecology: (pp. 9-11). Tomsk, Russia: Publishing house of Tomsk University.

Babrauskas, V. (2003). *Ignition handbook.* Issaquah, Washington: Fire Science Publishers.

Bachmann, A., & Allgower, B. (1998) A framework for wildfire risk analysis. In *Proceedings of the 14th Conference on Fire and Forest Meteorology*; Nov. 16-20, Coimbra, Portugal. Vol. 2. Coimbra, Portugal: ADAI. pp. 2177-2190.

Bachmann, A., & Allgower, B. (2001). A consistent wildland fire risk terminology is needed! *Fire Management Today, 61*(4), 28–33.

Badmaev, N. B., Kulikov, A. I., & Mironov, I. A. (2011). Permafrost. In Buryatia. Encyclopedic Reference Book in 2 Volumes (Vol. 1, pp. 54-55). Ulan-Ude, Russia: LLC Publishing House "Ecos". (In Russian).

Badmaev, N., & Bazarov, A. (2019). Monitoring network for atmospheric and soil parameters measurements in permafrost area of Buryatia, Russian Federation. *Geosciences, 9*(1), 6. doi:10.3390/geosciences9010006

Bailey, J. C., Blakeslee, R. J., Buechler, D. E., & Christian, H. J. (2007). *Diurnal lightning distributions as observed by the Optical Transient Detector (OTD) and the Lightning Imaging Sensor (LIS).* Retrieved from https://ntrs.nasa.gov/archive/nasa/casi.ntrs.nasa.gov/20070038367.pdf

Baranovskiy N. V., & Grishin A. M. (2002). Mathematical modeling of drying of a layer of forest combustible materials under the canopy of pine stand, *Computational Technologies, 7,* Part 1. (Joint issue Vestnik KazNU, N 4 (32)), pp. 320-324.

Baranovskiy, N. V. (2004). The influence of anthropogenic load and lightning activity on the probability of forest fires. *Siberian Journal of Ecology, 6.* pp. 835-842.

Baranovskiy, N. V. (2007a). Testing of the system of data assimilation on the level of anthropogenic load for a linear source of anthropogenic load, *Environmental Safety Management System: A Collection of Proceedings of a Correspondence International Scientific and Practical Conference*, Yekaterinburg, Russia: USTU - UPI, 319-322. (In Russian).

Baranovskiy, N. V. (2007b). *Mathematical modeling of the most probably scenarios and conditions of forest fires occurrence.* (Thesis). Tomsk, Russia: Tomsk State University. (In Russian)

Baranovskiy, N. V., Grishin, A. M., & Loskutnikova, T. P. (2003). Information and prognostic system for determining the probability of occurrence of forest fires. *Computational Technologies, 2,* pp. 16-26.

Baranovskiy, N., & Zharikova, M. (2014). A web-oriented geoinformation system application for forest fire danger prediction in typical forests of the Ukraine. *Lecture Notes in Geoinformation and Cartography, (199669),* pp. 13-22.

Baranovskiy, N., & Zharikova, M. (2014). A web-oriented geoinformation system application for forest fire danger prediction in typical forests of the Ukraine. *Lecture Notes in Geoinformation and Cartography.* 13-22.

Baranovskiy, N. V. (2017). Forest fire danger assessment using SPMD-model of computation for massive parallel system. *International Review on Modelling and Simulations., 10*(3), 193–201. doi:10.15866/iremos.v10i3.10570

Baranovskiy, N. V. (2018). Mathematical simulation of anthropogenic load from a linear source using partial differential equations in the context of the forest fires occurrence. *Advances in Differential Equations and Control Processes, 19*(3), 237–262. doi:10.17654/DE019030237

Baranovskiy, N. V., Krechetova, S. Y., Belikova, M. Y., Kocheeva, N. A., & Yankovich, E. P. (2016). Joint processing of RS and WWLLN data for forest fire danger estimation: New concept. *Proceedings of SPIE - The International Society for Optical Engineering, 10001, Article 1000113*, 6.

Baranovskiy, N. V., & Kuznetsov, G. V. (2017). Coniferous tree ignition by cloud-to-ground lightning discharge using approximation of "ideal" crack in bark. *JP Journal of Heat and Mass Transfer*, *14*(1), 173–186. doi:10.17654/HM014010173

Baranovskiy, N. V., & Kuznetsov, G. V. (2017). *Forest fire occurrences and ecological impact prediction: Monograph*. Novosibirsk, Russia: Publishing House of the Siberian Branch of the Russian Academy of Science. doi:10.15372/FOREST2017BNV

Baranovskiy, N. V., & Kuznetsov, G. V. (2017). Mathematical simulation of deciduous tree ignition by cloud-to-ground lightning discharge using large vessels approximation. *JP Journal of Heat and Mass Transfer*, *14*(4), 533–545. doi:10.17654/HM014040533

Baranovskiy, N. V., Kuznetsov, G. V., & Nemova, T. N. (2017). Mathematical computing of coniferous tree ignition by the cloud-to-ground lightning discharge using Joule-Lenz's law. *Iranian Journal of Electrical and Computer Engineering*, *7*(3), 1337–1346.

Baranovskiy, N. V., & Yankovich, E. P. (2015). Geoinformation monitoring of forest fire danger on the basis of remote sensing data of surface by the artificial Earth satellite. *Journal of Automation and Information Sciences*, *47*(8), 11–23. doi:10.1615/JAutomatInfScien.v47.i8.20

Baranovskiy, N., Barakhnin, V., & Yankovich, E. (2017). GIS-Technologies and mathematical simulation as tools for lightning-caused forest fire danger prediction. *CEUR Workshop Proceedings*, *1839*, 2–15.

Baranovskiy, N., Krechetova, S., Belikova, M., & Perelygin, A. (2016, December). WWLLN Data cluster analysis methods for lightning-caused forest fires monitoring. [IJECE]. *Iranian Journal of Electrical and Computer Engineering*, *6*(6), 3112–3120.

Baranovsky, N. V. (2009). Estimation of probability of occurrence of forest fires taking into account meteorological conditions, anthropogenic load and thunderstorm activity. *Fire safety*, 1, 93-99.

Barber, I. R. (1977). Fire preparation as an essential factor in roadside management. *Vic. Resur*, *2*, 18–21.

Barbero, R., Abatzoglou, J. T., Larkin, N. K., Kolden, C. A., & Stocks, B. (2015). Climate change presents increased potential for very large fires in the contiguous United States. *International Journal of Wildland Fire*, *24*(7), 892–899. doi:10.1071/WF15083

Bartalev, S. A., Egorov, V. A., Loupian, E. A., & Uvarov, I. A. (2007). Multi-year circumpolar assessment the area burnt in Boreal ecosystems using SPOT-vegetation. *International Journal of Remote Sensing*, *28*(6), 1397–1404. doi:10.1080/01431160600840978

Bartoli, P., Simeoni, A., Biteau, H., Torero, J. L., & Santoni, P. A. (2011). Determination of the main parameters influencing forest fuel combustion dynamics. *Fire Safety Journal*, *46*(1-2), 27–33. doi:10.1016/j.firesaf.2010.05.002

Battaglia, F., McGrattan, K., Rehm, R., & Baum, H. (2000). Simulating fire whirls. *Combustion Theory and Modelling*, *4*(2), 123–138. doi:10.1088/1364-7830/4/2/303

Bazarov, A. V., Badmaev, N. B., Kurakov, S. A., & Gonchikov, B.-M. N. (2018). A mobile measurement system for the coupled monitoring of atmospheric and soil parameters. *Russian Meteorology and Hydrology*, *43*(4), 271–275. doi:10.3103/S106837391804009X

Bedia, J., Herrera, S., & Gutierrez, J. M. (2014). Assessing the predictability of fire occurrence and area burned across phytoclimatic regions in Spain. *Natural Hazards and Earth System Sciences*, *14*(1), 53–66. doi:10.5194/nhess-14-53-2014

Begon, M., Harper, J. L., & Townsend, C. R. (1996). *Ecology* (p. 692). Oxford, UK: Blackwell Science.

Beltsova, T. G. & Korolchenko, O. N. (2008). Flammability indicators of fire-protected wood. *Fire and explosion safety*, 4, 31-33.

Belyakin, A. A., & Volokitina, A. V. (2010). Forest type pyrological characteristics in the southern Pribaikalye. [In Russian]. *Bulletin of the Krasnoyarsk State Agrarian University*, *7*, 91–96.

Benavent-Corai, J., Rojo, C., Suarez-Torres, J., & Velasco-Garcia, L. (2007). Scaling properties in forest fire sequences: The human role in the order of nature. *Ecological Modelling*, *205*(3-4), 336–342. doi:10.1016/j.ecolmodel.2007.02.028

Benkorichi, S., Fateh, T., Richard, F., Consalvi, J.-L., & Nadjai, A. (2017). Investigation of thermal degradation of pine needles using multi-step reaction mechanisms. *Fire Safety Journal*, *91*, 811–819. doi:10.1016/j.firesaf.2017.03.058

Bennett, A. J., & Harrison, R. G. (2007). Atmospheric electricity in different weather conditions. *Weather*, *62*(10), 277–283. doi:10.1002/wea.97

Berezhnaya, T. V., Golubev, A. D., & Parshina, L. N. (2012). Anomalous hydrometeorological phenomena on the Russian Federation Territory in July 2012. [in Russian]. *Meteorology and Hydrology*, *11*, 107–118.

Bernardi, M., & Ferrari, D. L. (2004). Evaluation of the LLS efficiency, on the ground flash density, using the Italian lightning detection system SIRF. *Journal of Electrostatics*, *60*(2-4), 131–140. doi:10.1016/j.elstat.2004.01.013

Berry, L. E., & Arsenault, A. (2015). Regional case studies. The ecological importance of mixed-severity fires. Retrieved from https://www.sciencedirect.com/book/9780128027493/the-ecological-importance-of-mixed-severity-fires

Betz, H. D., Schumann, U., & Laroche, P. (Eds.). (2008). *Lightning: principles, instruments and applications: review of modern lightning research.* Springer Science & Business Media. doi:10.1007/978-1-4020-9079-0

Birds hunt small animals fleeing fire. (n.d.). Retrieved from https://freethoughtblogs.com/affinity/files/2018/01/Kites.jpg

Bliege Bird, R., Bird, D. W., & Codding, B. F. (2016). People, El Niño southern oscillation and fire in Australia: Fire regimes and climate controls in hummock grasslands. *Philosophical Transactions of the Royal Society*, *371*(1696), 20150343. doi:10.1098/rstb.2015.0343 PMID:27216513

Bogorad, V. I., Litvinskaya, T. V., Shevchenko, I. A., Dybach, A. M., & Slepchenko, A. Yu. (2016). Radiation consequences of a fire in the exclusion zone of the Chernobyl NPP. *Nuclear and Radiation Safety*, *1*(69), 64–68. doi:10.32918/nrs.2016.1(69).10

Bogushov, A. K., & Panyukov, A. V. (2009) Location of interconnected objects under uncertainty. *IV All-Russian Conference Optimization Problems and Economic Applications: Conference Materials, Omsk, Russia, June 29-July 4, 2009, Omsk Branch of the Institute of Mathematics of the SB RAS* (p. 113). Omsk, Russia: Polygraph. Center for CAS.

Boholm, A., & Coevellec, H. (2011). Relational theory of risk. *Journal of Risk Research*, *14*(2), 175–190. doi:10.1080/13669877.2010.515313

Bondur, V. G. (2011). Space monitoring of wildfires in Russia in the conditions of the abnormal heat of 2010. *The study of the earth from space*, 3, 3-13.

Boonmee, N., & Quintiere, J. G. (2002). Glowing and flaming autoignition of wood. *Proceedings of the Combustion Institute*, *29*(1), 289–296. doi:10.1016/S1540-7489(02)80039-6

Borodkin, A. N. (1997). Model of the spread of fire between wooden buildings. *Fire and Explosion Safety*, *1*, 36–39.

Bowman, D. M. J. S., Moreira-Muñoz, A., Kolden, C. A., Chávez, R. O., Muñoz, A. A., Salinas, F., ... Johnston, F. H. (2019). Human–environmental drivers and impacts of the globally extreme 2017 Chilean fires. *Ambio*, *48*(4), 350–362. doi:10.100713280-018-1084-1 PMID:30128860

Branca, C., & Di Blasi, C. (2004). Global intrinsic kinetics of wood oxidation. *Fuel*, *83*(1), 81–87. doi:10.1016/S0016-2361(03)00220-5

Brown, J. K., & Smith, J. K. R. (2000). *Wildland fire in ecosystems: effects of fire on flora*. UT: U.S. Department of Agriculture, Forest Service, Rocky Mountain Research Station. doi:10.2737/RMRS-GTR-42-V2

Brun, C., Artes, T., Cencerrado, A., Margalef, T., & Cortés, A. (2017). A high performance computing framework for continental-scale forest fire spread prediction. *Procedia Computer Science*, *108C*, 1712–1721. doi:10.1016/j.procs.2017.05.258

Budyko, M. I. (1980). *The climate in the past and the future*. Leningrad, Russia: Gidrometeoizdat.

Bufacchi, P., Krieger, G. C., Mell, W., Alvarado, E., Santos, J. C., & Carvalho, J. A. Jr. (2016). Numerical simulation of surface forest fire in Brazilian Amazon. *Fire Safety Journal*, *79*, 44–56. doi:10.1016/j.firesaf.2015.11.014

Bui, D. T., Bui, Q.-T., Nguyen, Q.-P., Pradhan, B., Nampak, H., & Trinh, P. T. (2017). A hybrid artificial intelligence using GIS-based neural-fuzzy inference system and particle swarm optimization for forest fire susceptibility modeling at a tropical area. *Agricultural and Forest Meteorology*, *233*, 32–44. doi:10.1016/j.agrformet.2016.11.002

Building materials. Flammability test method: GOST 30402-96. Enter 03/30/97. Minsk: Ministry of Architecture and Construction of the Republic of Belarus: FSUE TsNIISK them. V. A. Kucherenko, 1996.

Bunn, A. G., Goetz, S. J., Kimball, J. S., & Zhang, K. (2007). Northern high-latitude ecosystems respond to climate change. *Eos (Washington, D.C.)*, *88*(34), 333–335. doi:10.1029/2007EO340001

Burasov, D. M., & Grishin, A. M. (2006). *Mathematical modeling of surface forest and steppe fires*. Kemerovo, Russia: Practica. (in Russian)

Burcev, A. V., Nevretdinov, Yu. M., & Sytina, A. N. (2014). Opyt registracii grozovoj aktivnosti na territorii Kolskogo poluostrova (Experience of registration of storm activity in the territory of the Kola Peninsula). *Vestnik Kol'skogo nauchnogo centra RAN, (2), 130-137.*

Butt, A., Shabbir, R., Ahmad, S. S., & Aziz, N. (2015). Land use change mapping and analysis using Remote Sensing and GIS: A case study of Simly watershed, Islamabad, Pakistan. *The Egyptian Journal of Remote Sensing and Space Sciences*, *18*(2), 251–259. doi:10.1016/j.ejrs.2015.07.003

Byram, G. M., & Martin, R. E. (1970). The modeling of fire whirlwinds. *Forest Science*, *16*(4), 586–398.

CALIPSO–Products–Lidar Browse Images of Productions Release. Retrieved from http://www-calipso.larc.nasa.gov/products/lidar/browse_images/show_calendar.php

Campos, L. Z. S., Saba, M. M. F., Pinto, O. Jr, & Ballarotti, M. G. (2007). Waveshapes of continuing currents and properties of M-components in natural negative cloud-to-ground lightning from high-speed video observations. *Atmospheric Research*, *84*(4), 302–310. doi:10.1016/j.atmosres.2006.09.002

Canadian Wildland Fire Information System. Official site. Retrieved from http://cwfis.cfs.nrcan.gc.ca/home

Cancellieri, D., Innocenti, E., & Leroy-Cancellieri, V. (2013). WinGPYRO: A software platform for kinetic study of forest fuels. *Fire Safety Journal*, *58*, 103–111. doi:10.1016/j.firesaf.2013.01.005

Carrió, D. S., Homar, V., & Wheatley, D. M. (2019). Potential of an EnKF storm-scale data assimilation system over sparse observation regions with complex orography. *Atmospheric Research*, *216*, 186–206. doi:10.1016/j.atmosres.2018.10.004

Catry, F. X., Rego, F. C., Bacao, F. L., & Moreira, F. (2009). Modeling and mapping wildfire ignition risk in Portugal. *International Journal of Wildland Fire*, *18*(8), 921–931. doi:10.1071/WF07123

Chalmers, J. A. (1967). *Atmospheric electricity*. Pergamon Press Ltd.

Chang, J., Chaur, C., & Tzuhn, C. (2015). Discourse on Taiwanese forest fires, *Wildfire Hazards, Risks and Disasters*. 145-166. Retrieved from https://www.sciencedirect.com/science/article/pii/B9780124104341000087

Charrad, M., Ghazzali, N., Boiteau, V., & Niknafs, A. (2014). NbClust: An R Package for Determining the Relevant Number of Clusters in a Data Set. *Journal of Statistical Software*, *61*(6), 1–36. doi:10.18637/jss.v061.i06

Chen, Y., Aanjaneya, K., & Atreya, A. (2017). A study to investigate pyrolysis of wood particles of various shapes and sizes. *Fire Safety Journal*, *91*, 820–827. doi:10.1016/j.firesaf.2017.03.079

Cherneva, N. V., & Firstov, P. P. (2018). *Formation of the local electric field of the atmosphere in Kamchatka under the influence of natural processes*. Vladivostok, Russia: Dalnauka [in Russian]. *College of DuPage Next Generation Weather Lab (NEXLAB)*. Retrieved from http://weather.cod.edu/digatmos/upa

Chowdhury, E. H., & Hassan, Q. K. (2015). Operational perspective of remote sensing-based forest fire danger forecasting systems. *ISPRS Journal of Photogrammetry and Remote Sensing*, *104*, 224–236. doi:10.1016/j.isprsjprs.2014.03.011

Christian, H. J., Blakeslee, R. J., Boccippio, D. J., Boeck, W. L., Buechler, D. E., Driscoll, K. T., ... Stewart, M. F. (2003). Global frequency and distribution of lightning as observed from space by the Optical Transient Detector. *Journal of Geophysical Research, D, Atmospheres*, *108*(D1), ACL-4. doi:10.1029/2002JD002347

Chuah, K. H., & Kushida, G. (2007). The prediction of flame heights and flame shapes of small fire whirls. *Proceedings of the Combustion Institute*, *31*(2), 2599–2606. doi:10.1016/j.proci.2006.07.109

Chuai, X., Huang, X., Wang, W., & Bao, G. (2013). NDVI, temperature and precipitation changes and their relationships with different vegetation types during 1998–2007 in Inner Mongolia, China. *International Journal of Climatology*, *33*(7), 1696–1706. doi:10.1002/joc.3543

Churches, C., Wampler, P., Sun, W., & Smith, A. (2014). Evaluation of forest cover estimates for Haiti using supervised classification of Landsat data. *Int. J. Appl. Earth Obs.*, *30*, 203–216. doi:10.1016/j.jag.2014.01.020

Clarke, H., Gibson, R., Cirulis, B., Bradstock, R. A., & Penman, T. D. (2019). Developing and testing models of the drivers of anthropogenic and lightning-caused wildfire ignitions in south-eastern Australia. *Journal of Environmental Management*, *235*, 34–41. doi:10.1016/j.jenvman.2019.01.055 PMID:30669091

Conedera, M., Cesti, G., Pezzatti, G. B., Zumbrunnen, T., & Spinedi, F. (2006). Lightning-induced fires in the Alpine region: An increasing problem. *Forest Ecology and Management*, *234*(1), S68. doi:10.1016/j.foreco.2006.08.096

Cruz, M. G. (2004). *Ignition of crown fuels above a spreading surface fire*. (PhD Dissertation), University of Montana. Missoula.

Cruz, M. G., Alexander, M. E., & Wakimoto, R. H. (2002). Predicting crown fire behavior to support forest fire management decision-making. In *Proceedings IV International Conference on Forest Fire Research*, 11. Luso-Coimbra, Portugal: Millpress.

Cummins, K. L., Murphy, M. J., Bardo, E. A., Hiscox, W. L., Pyle, R. B., & Pifer, A. E. (1998). A combined TOA / MDF technology upgrade of the U.S. national lightning detection network. *Journal of Geophysical Research*, *103*(D8), 9035–9044. doi:10.1029/98JD00153

Curic, M., & Zivanovic, S. (2013). Dependence between deficit and surplus of precipitation and forest fires. *Disaster Advances*, *6*(6), 64–69.

Data archive on forest fires in the Timiryazevskiy forestry of the Tomsk region, CD. (2010). Department of Forestry of Tomsk Region. (In Russian)

Data User Guide. International Space Station (ISS) Lightning Imaging Sensor (LIS) Provisional datasets. Retrieved from https://ghrc.nsstc.nasa.gov/pub/lis/iss/doc/isslis_dataset.pdf

De la Cueva, A. V., Quintana, J. R., & Canellas, I. (2012). Fire activity projections in the SRES A2 and B2 climatic scenarios in peninsular Spain. *International Journal of Wildland Fire*, *21*(6), 653–665. doi:10.1071/WF11013

De Vicente, J., & Crespo, F. (2012). A new wildland fire danger index for a Mediterranean region and some validation aspects. *International Journal of Wildland Fire*, *21*(8), 1030–1041. doi:10.1071/WF11046

Deeming, J. E., Burgan, R. E., & Cohen, J. D. (1977). *The National Fire Danger Rating System - 1978*. United States Department of Agriculture, Forest Service, General Technical Report INT-39, Intermountain Forest and Range Experiment Station, Odgen, UT.

Deeming, J. E., Burgan, R. E., & Cohen, J. D. (1978). The National Fire-Danger Rating System — 1978. General technical report INT-39. Intermountain Forest and Range Experiment Station, Ogden, UT: USDA Forest Service.

Deeming, J. E., Lancaster, J. W., Fosberg, M. A., Furman, R. W., & Schroeder, M. J. (1972). The national fire-danger rating system. Ft. Collins, CO: USDA Forest Serv. Res. Pap. RM-84 doi:10.5962/bhl.title.98933

Deeming, J. E., Lancaster, J. W., Fosberg, M. A., Furman, W. R., & Schroeder, M. J. (1974). *The National Fire Danger Rating System*. United States Department of Agriculture, Forest Service, Research Paper RM 84, Rocky Mountain Forest and Range Experiment Station, Fort Collins, CO.

Dennison, P. E., Brewer, S. C., Arnold, J. D., & Moritz, M. A. (2014). Large wildfire trends in the western United States, 1984–2011. *Geophysical Research Letters*, *41*(8), 2928–2933. doi:10.1002/2014GL059576

Department of Forestry of Tomsk region. (2010). Data archive on forest fires in the Timiryazevsk forestry of the Tomsk region. CD.

Department of natural resources and environmental protection of Tomsk region. (n.d.). Retrieved from http://green.tsu.ru/fires

Dhanapriya, M., Kumaraperumal, R., Kannan, B., & Hiren Bhatt, P. (2018). Spatio-temporal analysis of vegetation dynamics for Saurashtra region of Gujarat. AgricINTERNATIONAL, 5(2), 43-56.

Di Blasi, C. (2008). Modeling chemical and physical processes of wood and biomass pyrolysis. *Progress in Energy and Combustion Science*, *34*(1), 47–90. doi:10.1016/j.pecs.2006.12.001

Díaz-Delgado, R., Lloret, F., & Pons, X. (2004). Spatial patterns of fire occurrence in Catalonia, NE, Spain. *Landscape Ecology*, *19*(7), 731–745. doi:10.100710980-005-0183-1

Díaz, S., Demissew, S., Carabias, J., Joly, C., Lonsdale, M., Ash, N., ... Báldi, A. (2015). The IPBES conceptual framework – connecting nature and people. *Current Opinion in Environmental Sustainability*, *14*, 1–10. doi:10.1016/j.cosust.2014.11.002

Dibble, A. C., White, R. H., & Lebow, P. K. (2007). Combustion characteristics of northeastern USA vegetation tested in the cone calorimeter: Invasive versus noninvasive plants. *International Journal of Wildland Fire*, *16*(4), 426–443. doi:10.1071/WF05103

Diez-Delgado, R., Lloret, F., & Pons, X. (2004). Statistical analysis of fire-frequency models for Catalonia (NE Spain, 1975-1998) based on fire scar maps from Landsat MSS data. *International Journal of Wildland Fire*, *13*(1), 89–99. doi:10.1071/WF02051

Dmitriev, A. N., Krechetova, S. Yu., & Kocheeva, N. A. (2011). *Grozy i lesnye pozhary ot groz na territorii Respubliki Altaj (Thunderstorms and wildfires from thunderstorms in the territory of Altai Republic)*. Gorno-Altaysk, Russian Federation: Publishing House of the Gorno-Altaysk State University.

Doerr, S. H., & Santín, C. (2016). Global trends in wildfire and its impacts: perceptions versus realities in a changing world, *Philosophical Transactions of the Royal Society B: Biological Sciences, 5,* 371(1696), 20150345. doi:. doi:10.1098/rstb.2015.0345

Donchenko, V. A., Kabanov, M. V., Kaul, B. V., Nagorskiy, P. M., & Samokhvalov, I. V. (2015). *Electro-optical phenomena in the atmosphere*. Tomsk, Russia: NTL Publishing House. [in Russian]

Donovan, G. H., & Brown, T. C. (2007). Be careful what you wish for: The legacy of Smokey Bear. *Front. Ecol.*, *5*(2), 73–79. doi:10.1890/1540-9295(2007)5[73:BCWYWF]2.0.CO;2

Doroshenko, A. M., & Kogan, R. M. (2007). Pyrological vulnerability evaluation of the Middle Amur Region. (By the example of the Jewish Autonomous Region) Fourth International meeting of Regional Northeast Asia wildland fire network, Dec. 16-17, 2007, Harbin, China, Reports. Harbin, 2007. pp. 46-52. (in Russia)

Doroshenko, S. G., & Bondarev, D. A. (2001). Monitoring of forest fires according to the space survey. Remote sensing of the earth's surface and atmosphere by aerospace means, Krasnoyarsk, 221-225.

Dorrer, G. A. (2008). *Dynamics of forest fires*. Novosibirsk, Russia: Publishing House of the Siberian Branch of the Russian Academy of Sciences.

Dorzhiev, T. Z., Yuhai, B., Badmaeva, E. N., Batsaihan, V., Urbazaev, C. B., & Yushan, Y. (2017). Forest fires in republic of Buryatia for 2002–2016. [In Russian]. *Nature of Inner Asia*, *3*, 22–37.

Dowden, R. L., Brundell, J. B., & Rodger, C. J. (2002). VLF lightning location by time of group arrival (TOGA) at multiple sites. *Journal of Atmospheric and Solar-Terrestrial Physics*, *64*(7), 817–830. doi:10.1016/S1364-6826(02)00085-8

Dubovyk, O., Menz, G., & Khamzina, A. (2012). Trend analysis of MODIS time-series using different vegetation indices for monitoring of cropland degradation and abandonment in Central Asia. *Proceedings of IEEE International Geoscience and Remote Sensing Symposium (IGARSS)*, Munich, Germany, July 22-27, 2012, pp. 6589-6592. 10.1109/IGARSS.2012.6352089

Dupuy, J. L., Maréchal, J., Portier, D., & Valette, J. C. (2011). The effects of slope and fuel bed width on laboratory fire behavior. *International Journal of Wildland Fire*, *20*(2), 272–288. doi:10.1071/WF09075

Dupuy, J., & Larini, M. (1999). Fire spread through a porous forest fuel bed: A radiative and convective model including fire-induced flow effects. *International Journal of Wildland Fire*, *9*(3), 155–172. doi:10.1071/WF00006

Dutta, D., Kundu, A., Patel, N. R., Saha, S. K., & Siddiqui, S. R. (2015). Assessment of agricultural drought in Rajasthan (India) using remote sensing derived Vegetation Condition Index (VCI) and Standardized Precipitation Index (SPI). *The Egyptian Journal of Remote Sensing and Space Sciences*, *18*(1), 53–63. doi:10.1016/j.ejrs.2015.03.006

Eckert, S., Hüsler, F., Liniger, H., & Hodel, E. (2015). Trend analysis of MODIS NDVI time-series for detecting land degradation and regeneration in Mongolia. *Journal of Arid Environments*, *113*, 16–28. doi:10.1016/j.jaridenv.2014.09.001

Elanskii, N. F., Mokhov, I. I., Belikov, I. B., & Berezina, E. V. (2011). Gas composition of the surface air in Moscow during the extreme summer of 2010. *Doklady Earth Sciences*, *437*(1), 357–362. doi:10.1134/S1028334X11030020

Elia, M., Giannico, V., Lafortezza, R., & Sanesi, G. (2019). Modeling fire ignition patterns in Mediterranean urban interfaces. *Stochastic Environmental Research and Risk Assessment*, *33*(1), 169–181. doi:10.100700477-018-1558-5

Emery, S. M., & Gross, K. L. (2005). Effects of timing of prescribed fire on the demography of an invasive plant, spotted knapweed Centaurea maculosa. *Journal of Applied Ecology*, *42*(1), 60–69. doi:10.1111/j.1365-2664.2004.00990.x

Environmental movement instrument of legal regulation of forest resources use in Tomsk region: Materials of the round table. (2001). *Tomsk: OSU Oblkompriroda*, 30. European Forest Fire Information System. Retrieved from http://effis.jrc.ec.europa.eu

EOS (Earth Observing System). (n.d.). Retrieved from https://eos.com/platform

EOSDIS Worldview. Retrieved from https://worldview.earthdata.nasa.gov

Ester, M., Kriegel, H.-P., Sander, J., & Xu, X. (1996). A density-based algorithm for discovering clusters in large a spatial database. *Proc. 1996 Intern. Conf. on Knowledge Discovery and Data Mining*, 226-231.

EUCLID (EUropean Cooperation for LIghtning Detection). Retrieved from http://www.euclid.org

EUMETView — EUMETSAT. Retrieved from http://eumetview.eumetsat.int/mapviewer/

Eurocode 1. Impacts on structures. Part 1-2. General effects. Impact for fire resistance determination: TCP EN 1991-1-2-2009. Enter 01/01/10 Minsk, Belarus: Ministry of Architecture and Construction of the Republic of Belarus: Stroytekhnorm RUE, 2009.

European Forest Fire Information System. Official site. Retrieved from http://effis.jrc.ec.europa.eu

Evdokimenko, M. D. (2013). Forest fire causes and distribution in the Baikal Region. [Forestry Journal]. *Lesnoy Zhurnal*, *4*, 33–39.

Everitt, B. S., Landau, S., Leese, M., & Stahl, D. (2011). *Cluster analysis* (5th ed.). Hoboken, NJ: John Wiley & Sons. doi:10.1002/9780470977811

Evseeva, N. C. (2001). Geography of Tomsk region. Natural conditions and resources, *Tomsk: Publishing house TSU*, 223.

Fateh, T., Richard, F., Batiot, B., Rogaume, T., Luche, J., & Zaida, J. (2016). Characterization of the burning behavior and gaseous emissions of pine needles in a cone calorimeter – FTIR apparatus. *Fire Safety Journal*, *82*, 91–100. doi:10.1016/j.firesaf.2016.03.008

Federal Forestry Agency. (2008). *On the determination of the number of forest areas on the territory of the Republic of Buryatia and the establishment of their borders (Order of Oct. 16, 2008 No. 310)*. Moscow, Russia: Federal Forestry Agency. (In Russian)

Federal Hydrometeorology and Environmental Monitoring Service. (2008a). Assessment report on climate change and its effects on the territory of the Russian Federation: Vol. I. *Climate change*. Moscow, Russia: Roshydromet. (In Russian)

Federal Hydrometeorology and Environmental Monitoring Service. (2008b). Assessment report on climate change and its effects on the territory of the Russian Federation: Vol. II. *Implications of climate change*. Moscow, Russia: Roshydromet. (In Russian)

Federal service of statistics. Retrieved from http://www.gks.ru

Federal State Budgetary Institution. West-Siberian Administration for Hydrometeorology and Environmental Monitoring. *The list and the criteria for severe weather events for the service territory of the West Siberian Administration of Hydrometeorology.* Retrieved from http://www.meteo-nso.ru/pages/115 (in Russian)

Feltman, J. A., Straka, T. J., Post, C. J., & Sperry, S. L. (2012). Geospatial analysis application to forecast wildfire occurrences in South Carolina. *Forests, 3*(2), 265–282. doi:10.3390/f3020265

Fernandez-Gomez, I., de Castro, A. J., Guijarro, M., Madrigal, J., Aranda, J. M., Diez, C., ... Lopez, F. (2011). Characterization of forest fuels in a Mass loss calorimeter by short open-path FTIR spectroscopy. *Journal of Quantitative Spectroscopy & Radiative Transfer, 112*(3), 519–530. doi:10.1016/j.jqsrt.2010.10.004

Fernandez-Pello, A. C. (2017). Wildland fire spot ignition by sparks and firebrands. *Fire Safety Journal, 91*, 2–10. doi:10.1016/j.firesaf.2017.04.040

Field, C. B., Barros, V. R., Dokken, D. J., Mach, K. J., Mastrandrea, M. D., Bilir, T. E., ... White, L. L. (Eds.). (2014). *IPCC 2014: Climate Change 2014: Impacts, Adaptation, and Vulnerability. Summaries, frequently asked questions, and cross-chapter boxes. A contribution of Working Group II to the Fifth Assessment Report of the Intergovernmental Panel on Climate Change.* New York, NY: Cambridge University Press.

Finney, M. A. (1993). Modeling the spread and behaviour of prescribed natural fires. *Proceedings of the 12th Conference on Fire and Forest Meteorology.* Society of American Foresters. Jekill Island, Georgia, pp. 138-143.

Finney, M. A. (1998). FARSITE: Fire area simulator — Model development and evaluation. USDA Forest Service, Rocky Mountain Research Station, Research Paper RMRS-RP-4 (Fort Collins, CO).

Finney, M. (2005). The challenge of quantitative risk analysis for wildland fire. *Forest Ecology and Management, 211*(1-2), 97–10. doi:10.1016/j.foreco.2005.02.010

Fire in the Far East. (n.d.). Retrieved from https://www.sibreal.org/a/29352746.html

Firstov, P. P., Akbashev, R. R., Zharinov, N. A., Maksimov, A. P., Manevich, T. M., & Mel'nikov, D. V. (2019). Electrification of eruptive plumes discharged by Shiveluch Volcano in relation to the character of the responsible explosion. *Journal of Volcanology and Seismology, 13*(3), 172–184. doi:10.1134/S0742046319030035

Flannigan, M. D., Krawchuk, M. A., de Groot, W. J., Wotton, B. M., & Gowman, L. M. (2009). Implications of changing climate for global wildland fire. *International Journal of Wildland Fire, 18*(5), 483–507. doi:10.1071/WF08187

Flannigan, M., Cantin, A. S., de Groot, W. J., Wotton, M., Newbery, A., & Gowman, L. M. (2013). Global wildland fire season severity in the 21st century. *Forest Ecology and Management, 294*, 54–61. doi:10.1016/j.foreco.2012.10.022

Flemming, J., Reimer, E., & Stern, R. (2001). Data assimilation for CT-Modelling based on optimum interpolation. *Preprints of 25th NATO/CCMS International Technical Meeting on Air Pollution Modelling and its Application.* Belgium, Louvain-la-Neuve: Catholic University of Louvain.

Fons, W. L. (1946). Analysis of fire spread in light forest fuels. *Journal of Agricultural Research, 72*(3), 93–121.

Font, R., Conesa, J. A., Moltó, J., & Muñoz, M. (2009). Kinetics of pyrolysis and combustion of pine needles and cones. *Journal of Analytical and Applied Pyrolysis, 85*(1-2), 276–286. doi:10.1016/j.jaap.2008.11.015

Food and Agriculture Organization of the United Nations. (2017) Child labour in agriculture in Lebanon. A guide for practitioners. p. 88.

Ford, R. (1995). *Investigation of wildfires. Bend.* Maverick Publications.

Forest complex of Tomsk region. (n.d.). Retrieved from http://tomsk.gov.ru/ru/ekonomika/lesnoy-kompleks/

Forest fire and environmental management. A technical report on forest fire as an ecological factor (2003). *University of Oslo,* 266.

Forest fire in the Far East. (n.d.). Retrieved from https://www.sibreal.org/a/29352746.html

Forest fire in Ukraine. (n. d.).

Forest Fire Map in Canada. (n.d.). Retrieved from https://yandex.ru/images/search?text=2019%20map%20forest%20fire%20Canada&pos=3&img_url=http%3A%2F%2Fcwfis.cfs.nrcan.gc.ca%2Fdata%2Fmaps%2Ffwi_fbp%2F2018%2Ffwi20180621.png&rpt=simage

Forest fires – one of the most serious problems of Russian forests. (n.d.). Retrieved from https://helpiks.org/6-68946.html

Forest Resources Information System. (n.d.). Retrieved from http://www.fao.org/forestry/4239-0f7d-5c1ac66ee0fe944546751e8d85c85.pdf

Frangieh, N., Morvan, D., Meradji, S., Accary, G., & Bessonov, O. (2018). Numerical simulation of grassland fres behavior using an implicit physical multiphase model. *Fire Safety Journal, 102*, 37–47. doi:10.1016/j.firesaf.2018.06.004

Frantz, D., Stellmes, M., Röder, A., & Hill, J. (2016). Fire spread from MODIS burned area data: Obtaining fire dynamics information for every single fire. *International Journal of Wildland Fire, 25*(12), 1228–1237. doi:10.1071/WF16003

Fravolini, A., Hultine, K. R., Brugnoli, E., Gazal, R., English, N. B., & Williams, D. G. (2005). Precipitation pulse use by an invasive woody legume: The role of soil texture and pulse size. *Oecologia, 144*(4), 618–627. doi:10.100700442-005-0078-4 PMID:15891829

Frolov, A. V., Blinov, V. G., Golitsyn, G. S., Dymnikov, V. P., Izrael, Yu. A., & Kattsov, V. M. … Semenov, S. M. (2014). The second Roshydromet assessment report on climate change and its consequences in the Russian Federation. General summary. Moscow, Russia: Federal Service for Hydrometeorology and Environmental Monitoring (Roshydromet)

Gallina, V., Torresan, S., Critto, A., Sperotto, A., Glade, T., & Marcomini, A. (2016). A review of multi-risk methodologies for natural hazards: Consequences and challenges for a climate change impact assessment. *Journal of Environmental Management, 168*, 123–132. doi:10.1016/j.jenvman.2015.11.011 PMID:26704454

Gamon, J. A., Huemmrich, K. F., Stone, R. S., & Tweedie, C. E. (2013). Spatial and temporal variation in primary productivity (NDVI) of coastal Alaskan tundra: Decreased vegetation growth following earlier snowmelt. *Remote Sensing of Environment, 129*, 144–153. doi:10.1016/j.rse.2012.10.030

Ganteaume, A., Camia, A., Jappiot, M., San-Miguel-Ayanz, J., Long-Fournel, M., & Lampin, C. (2013). A review of the main driving factors of forest fire ignition over Europe. *Environmental Management, 9*(3), 651–662. doi:10.100700267-012-9961-z PMID:23086400

Garcia Diez, E. L., Labajo Salazar, J. L., & de Pablo, F. (1993). Some meteorological conditions associated with forest fires in Galicia (Spain). *International Journal of Biometeorology, 37*(4), 194–199. doi:10.1007/BF01387522

Gardener, M. (2012). *Beginning R: The statistical programming language*. Hoboken, NJ: John Wiley & Sons.

Garmaev, E. Z., Tsydypov, B. Z., Dabaeva, D. B., Andreyev S. G., Ayurzhanayev, A. A., & Kulikov A. I. (2017). The lake Baikal level regime: retrospection and current status. *Water Sector of Russia: Problems, Technologies, Management*, 4-18. (In Russian).

Garmaeva, S. D., & Nosyreva, O. V. (2016). Assessment of the normative depth of seasonally frozen soil layer in Western Siberia using GIS. In Proceedings of 16th International Multidisciplinary Scientific GeoConference SGEM 2016. Sofia, Bulgaria: STEF 92 Technology Ltd. doi:10.5593/SGEM2016/B23/S11.015

Gavrilov, K., Accary, G., Morvan, D., Lyubimov, D., Méradji, S., & Bessonov, O. (2011). Numerical simulation of coherent structures over plant canopy. *Flow, Turbulence, and Combustion*, *86*(1), 89–111. doi:10.100710494-010-9294-z

Geographic information systems. Survey information. (1992). – Moscow, Russia: TsNIIGAIK. (in Russia)

Geostationary Lightning Mapper (GLM). Retrieved from https://ghrc.nsstc.nasa.gov/lightning/overview_glm.html

Giglio, L., Randerson, J. T., & Van Der Werf, G. R. (2013). Analysis of daily, monthly, and annual burned area using the fourth-generation global fire emissions database (GFED4). *Journal of Geophysical Research. Biogeosciences*, *118*(1), 317–328. doi:10.1002/jgrg.20042

Gill, A. M., & Moore, P. H. R. (n. d.). Fire situation in Australia. Retrieved from http://www.fao.org/3/AD653E/ad653e80.htm

Girardin, M. P. (2010). Wildfire risk inferred from tree rings in the Central Laurentians of boreal Quebec, Canada. *Dendrochronologia*, *28*(3), 187–206. doi:10.1016/j.dendro.2009.05.006

Glagolev, V. A. (2015). *Estimation and forecast of grassland fire occurrence within Jewish Autonomous Region territory*. Khabarovsk, Russia: Institute for Water and Environmental Problems of Far East Branch of the Russian Academy of Sciences.

Global biodiversity. (2015). Montréal (Canada): *Convention on Biological Diversity, 155.* Retrieved from https://www.cbd.int/gbo4

Goldammer, J. G., & Seibert, B. (1989). Natural rain forest fires in Eastern Borneo during the Pleistocene and Holocene, *The Science of Nature, 11,* 76. 518-520. doi:. doi:10.1007/BF00374124

Goldammer, J., Mitsopoulos, I., Mallinis, G., & Woolf, M. (2017). *Wildfire hazard and risk assessment* (p. 9). UNISDR.

Göltenboth, F., & Widmann, P. (2006). Special forest ecosystems. *Ecology of Insular Southeast Asia*, Retrieved from https://www.sciencedirect.com/topics/agricultural-and-biological-sciences/intsia

Göltenboth, F., & Widmann, P. (2006). Tropical lowland evergreen rainforest, Ecology of Insular Southeast Asia. (n.d.). Retrieved from https://www.sciencedirect.com/science/article/pii/B9780444527394500176

Goman, P. N., Sobol, V. R., Barovik, D. V., & Taranchuk, V. B. (2011). Experimental and numerical simulation of the combustion process and the spread of fire in the conditions of the forest ground fire. *Technosphere Safety Technologies: Internet Journal*, *3*(37), 14. Retrieved from http://ipb.mos.ru/ttb

Goncalves, Z. J., & Lourenco, L. (1990). Meteorological index of forest fire risk in the Portuguese mainland territory. In *Proceedings of the international conference on forest fire research. Coimbra.* 7, p. 1.

Gorbatenko V. P., Gromnitskaya A. A., Konstantinova D. A., Ershova T. V., & Nechepurenko O. E. (2015). Assessment of the role of climatic factors in the occurrence and spread of forest fires in the Tomsk region. *Bulletin of Tomsk State University, 395*, 233–240. (in Russian). doi:10.17223/15617793/395/38

Gorbatenko, V. P., & Ershova, T. V. (2011). *Lightning as a link of the global electrical circuit: monograph.* Tomsk, Russia: Publisher of TSPU.

Gorbatenko, V. P., & Ershova, T. V. (2011). *Molnija kak zveno global'noj jelektricheskoj cepi (Lightning as link of a Global Electrical Circuit).* Tomsk, Russian Federation: Izdatel'stvo TGPU.

Gorbatenko, V. P., & Konstantinova, D. A. (2011). Mesoscale convection and dangerous weather phenomena in southeast of Western Siberia. In *Proceedings of 7th Asia-Pacific International Conference on Lightning* (pp. 160–164). Chengdu, China. Beijing, China: Tsinghua University 10.1109/APL.2011.6111094

Gorbatenko, V. P., Sevastyanov, V. V., Konstantinova, D. A., & Nosyreva, O. V. (2019). Characteristic of the snow cover for the Western Siberia territory. *IOP Conference Series: Earth and Environmental Science*, *232*, 1–5. 10.1088/1755-1315/232/1/012003

Gorchakova, I. A., & Mokhov, I. I. (2012). The radiative and thermal effects of smoke aerosol over the region of Moscow during the summer fires of 2010. *Izvestiya. Atmospheric and Oceanic Physics*, *48*(5), 496–503. doi:10.1134/S0001433812050039

Gorchakova, I. A., Mokhov, I. I., Anikin, P. P., & Emilenko, A. S. (2018). Radiative and thermal impacts of smoke aerosol longwave absorption during fires in the Moscow Region in Summer 2010. *Izvestiya. Atmospheric and Oceanic Physics*, *54*(2), 154–161. doi:10.1134/S000143381802007X

Gorchakova, I. A., Mokhov, I. I., & Rublev, A. N. (2015). Radiation and temperature effects of the intensive injection of dust aerosol into the atmosphere. *Izvestiya. Atmospheric and Oceanic Physics*, *51*(2), 113–126. doi:10.1134/S0001433815010053

Gorchakov, G. I., Sitnov, S. A., Karpov, A. V., Gorchakova, I. A., Gushchin, R. A., & Datsenko, O. I. (2019). Eurasian large-scale hazes in Summer 2016. *Izvestiya. Atmospheric and Oceanic Physics*, *55*(3), 261–270. doi:10.1134/S0001433819020063

Gorev, G. G. (2004). Assessment of the climatic predisposition of the territory to forest fires (on the example of the Tomsk region). (*Thesis abstract), Tomsk, TSU: Publishing house*, 24.

Gorev, G. V. Evaluation of the climate predisposition of the territory to the occurrence of forest fires (on the example of the Tomsk region). (Dissertation). (Tomsk, TSU, 2004) (In Russian)

Gosford, R., & Bonta, M. (2015). Ornithogenic fire: Raptors as propagators of fire in the Australian Savanna). *Raptor Research Foundation Annual Conference,* Nov 4–8, Sacramento, CA. Retrieved from https://www.researchgate.net/publication/324819947_Ornithogenic_Fire_Raptors_as_Propagators_of_Fire_in_the_Australian_Savanna

Gould, J. S., Patriquin, M. N., Wang, S., McFarlane, B. L., & Wotton, B. M. (2013). Economic evaluation of research to improve the Canadian forest fire hazard rating system. *Forestry*, *86*(3), 317–329. doi:10.1093/forestry/cps082

Government of the Republic of Buryatia. (2008). *Annex to the Resolution dated Dec. 17, 2013 No. 665 "Forest Plan of the Republic of Buryatia" (approved by the decree of the Government of the Republic of Buryatia dated Dec. 31, 2008 No. 608).* Ulan-Ude, Russia: Republican Forestry Agency, official site. (In Russian)

Gradstein, I. S., & Redhead, I. M. (1962). Tables of integrals, sums, series and products. Moscow: State. ed. Phys.-Mat. literature.

GRASS GIS manual: i.atcorr. Retrieved from https://grass.osgeo.org/grass76/manuals/i.atcorr.html

GRASS GIS manual: i.atcorr: https://grass.osgeo.org/grass76/manuals/i.atcorr.html. Accessed: 2018-12-05.

Gregory, V., & Vidal, A. (1999). Fuzziness in Chang's fuzzy topological spaces. *Rend. Instit. Mat. Univ. Trieste*, *XXX*(Suppl.), 111–121.

Gringof, I. G., & Kleshchenko, A. D. (2011). Basics of agricultural meteorology. Obninsk, Russia: Federal Service of Russia for Hydrometeorology and Monitoring Environment. All Russian Research Institute of Hydrometeorological Information – World Data Center. (in Russian)

Grishin, A. M. (2001). Prediction and simulation of forest fires. Forest fires: occurrence, distribution, extinguishing and environmental consequences. Tomsk: TSU publishing house, 35-46.

Grishin, A. M. (1981). *Mathematical models of forest fires*. Tomsk, Russia: Tomsk University Press.

Grishin, A. M. (1997). *Mathematical modeling forest fire and new methods fighting them*. Tomsk, Russia: Publishing House of Tomsk University.

Grishin, A. M. (2002). Ignition of forest combustible materials by a flow of radiant energy. *Physics of Combustion and Explosion*, *1*, 30–35.

Grishin, A. M., & Baranovskiy, N. V. (2003). Comparative analysis of simple models of drying of the forest fuel layer, including data from experiments and field observations. *Journal of Engineering Physics and Thermophysics*, *76*(5), 166–169. doi:10.1023/B:JOEP.0000003234.63878.7b

Grishin, A. M., & Filkov, A. I. (2011). A deterministic-probabilistic system for predicting forest fire hazard. *Fire Safety Journal*, *46*(1-2), 56–62. doi:10.1016/j.firesaf.2010.09.002

Grishin, A. M., Golovanov, A. N., Kataeva, L. Y., & Loboda, E. L. (2001). Formulation and solution of the problem of drying of a layer of combustible forest materials. *Combustion, Explosion, and Shock Waves*, *37*(1), 57–66. doi:10.1023/A:1002868709387

Grishin, A. M., Golovanov, A. N., & Sukov, Y. V. (2006). Experimental determination of thermophysical, thermokinetic, and filtration characteristics of peat. *Journal of Engineering Physics and Thermophysics*, *79*(3), 557–562. doi:10.100710891-006-0135-9

Gruza, G. V., & Rankova, E. Y. (2004). Detection of changes in climate state, climate variability, and climate extremity. [In Russian]. *Russian Meteorology and Hydrology*, *4*, 31–43.

Guha-Sapir, R., & Hoyois, P. (2019). EM-DAT: International Disaster Database. Brussels, Belgium: University Cathol. Louvain. Retrieved from http://www.emdat.be

Guo, F., Innes, J. L., Wang, G., Ma, X., Sun, L., Hu, H., & Su, Z. (2015). Historic distribution and driving factors of human-caused fires in the Chinese boreal forest between 1972 and 2005. *Journal of Plant Ecology*, *8*(5), 480–490. doi:10.1093/jpe/rtu041

Gurney, R. J., Foster, J. L., Gurney, R. J., & Parkinson, C. L. (1993). *Atlas of satellite observations related to global change*. Cambridge, UK: Cambridge University Press.

Gusak, A. A., & Gander, G. M. (1991). Handbook of higher mathematics. Minsk, Belarus: Navuki i tehniki.

Gusev V. G., Erican V. A., & Erican, E. V. (2011). Forecasting of parameters of forest fires and resources to deal with them. SPb, *SPbTU*, 217.

Gusev, V. G. (1989). *Assessment of firemen's exposure to firefighters in front of low and high forest fires. Forest fires and the fight against them: Sat. scientific tr. LenNIILH; Editorial* (pp. 133–137). E. S. Artsybashev. [et al.]

Gusev, V. G. (2005). *Physical and mathematical models of fire propagation and fire barriers in pine forests. SPb*. SRI LH.

Haiganoush, K. (2016). Matthew Jolly Near-term probabilistic forecast of significant wildfire events for the Western United States. *International Journal of Wildland Fire*, *25*(11), 1169–1180. doi:10.1071/WF16038

Hansen, J., Ruedy, R., Sato, M., & Lo, K. (2010). Global surface temperature change. *Reviews of Geophysics*, *48*(4), RG4004. doi:10.1029/2010RG000345

Hansen, M., Defries, R., Townshend, J., & Sohlberg, R. (2000). Global land cover classification at 1 km spatial resolution using a classification tree approach. *International Journal of Remote Sensing*, *21*(6-7), 1331–1364. doi:10.1080/014311600210209

Hansson, S. (2010). Risk: objective or subjective, facts or values. *Journal of Risk Research*, *13*(2), 231–238. doi:10.1080/13669870903126226

Hardy, C., & Hardy, C. E. (2007). Fire danger rating in the United States of America: An evolution since 1916. *International Journal of Wildland Fire*, *16*(2), 217–231. doi:10.1071/WF06076

Heikinheimo, M. (1998). Renewing the system for forest fire risk assessment at the Finnish Meteorological Institute. *International Forest Fire News*, *18*, 65–67.

Hernandez-Leal, P. A., Gonzalez-Calvo, A., Arbelo, M., Barreto, A., & Alonso-Benito, A. (2008). Synergy of GIS and remote sensing data in forest fire danger modeling. *IEEE Journal of Selected Topics in Applied Earth Observations and Remote Sensing*, *1*(4), 240–247. doi:10.1109/JSTARS.2008.2009043

Herrmann, S. M., & Anyamba, A. (2005). Exploring relationship between rainfall and vegetation dynamics in the Sahel using coarse resolution satellite data. *Proceedings of the 31st International Symposium on Remote Sensing of Environment*, 118.

Heyer, J. P., Power, M. J., Field, R. D., & van Marle, M. J. E. (2018). The impacts of recent drought on fire, forest loss, and regional smoke emissions in lowland Bolivia. *Biogeosciences*, *15*(14), 4317–4331. doi:10.5194/bg-15-4317-2018

Himoto, K., Shinohara, M., Sekizawa, A., Takanashi, K., & Saiki, H. (2018). A field experiment on fire spread within a group of model houses. *Fire Safety Journal*, *96*, 105–114. doi:10.1016/j.firesaf.2018.01.003

Hohenemser, C., Kates, R., & Slovic, P. (1985). *A casual taxonomy. Perilous progress: managing the hazards of technology* (pp. 67–125). Boulder, CO: Westview Press.

Holgate, C. M., van Dijk, A. I. J. M., Cary, G. J., & Yebra, M. (2017). Using alternative soil moisture estimates in the McArthur Forest Fire Danger Index. *International Journal of Wildland Fire*, *2017*. doi:10.1071/WF16217

Holsten, A., Dominic, A. R., Costa, L., & Kropp, J. P. (2013). Evaluation of the performance of meteorological forest fire indices for German federal states. *Forest Ecology and Management*, *287*, 123–131. doi:10.1016/j.foreco.2012.08.035

Howitt, D. G. (2015). An assessment of hot metal fragments from heavy mechanical equipment as potential ignition source for forest litter. *Journal of Fire Sciences*, *33*(6), 427–444. doi:10.1177/0734904115611820

Hutchins, M. L., Holzworth, R. H., & Brundell, J. B. (2014). Diurnal variation of the global electric circuit from clustered thunderstorms. *Journal of Geophysical Research*, *199*, 620–629.

Hutchins, M. L., Holzworth, R. H., & Brundell, J. B. (2014). Diurnal variation of the global electric circuit from clustered thunderstorms. *Journal of Geophysical Research: Space Physics*, *119*(1), 620–629.

Hutchins, M. L., Holzworth, R. H., Brundell, J. B., & Rodger, C. J. (2012). Relative detection efficiency of the world wide lightning location network. *Radio Science*, *47*(6), n/a. doi:10.1029/2012RS005049

Hyde, K., Dickinson, M., Bohrer, G., Calkin, D., Evers, L., Gilbertson-Day, J., ... Tague, C. (2013). Research and development supporting risk-based wildfire effects prediction for fuels and fire management: Status and needs. *International Journal of Wildland Fire*, *22*(1), 37–50. doi:10.1071/WF11143

ICNF. (2012). (n.d.). Retrieved from http://www2.icnf.pt/portal/florestas/dfci/relat/rel-if/copy_of_2012

Imetkhenov, A. B. (Ed.). (2000). *Atlas of the Republic of Buryatia*. Moscow, Russia: Federal Service of Geodesy and Cartography of Russia.

Information system for remote monitoring of forest fires ISDM-Rosleskhoz. Official site. Retrieved from https://nffc.aviales.ru/main_pages/index.shtml

Ippolitov, I. I., Kabanov, M. V., Nagorskiy, P. M., Pkhalagov, Yu. A., & Smirnov, S. V. (2013). Diurnal variations in the electrical field intensity under smoke from forest fires. *Doklady Earth Sciences*, *453*(Part 1), 1137–1140. doi:10.1134/S1028334X1311010X

Ippolitov, I. I., Kabanov, M. V., Nagorskiy, P. M., & Smirnov, S. V. (2012). Dynamics of variability of the electric field in the prestrom atmosphere. [in Russian]. *Opt. Atmos. Okeana*, *25*, 132–136.

Ippolitov, I. I., Kabanov, M. V., & Smirnov, S. V. (2011). Strategy of monitoring network for climatic and environmental processes in Siberia. [in Russian]. *Opt. Atmos. Okeana*, *24*, 7–14.

Ivanov, V. A. (1996). *Forest fires from thunderstorms on the Yenisei plain.* (Abstract of doctoral dissertation). Sukachev Institute of Forest SB RAS, Krasnoyarsk.

Ivanov, V. A. (1987). *Lightning activity and forest fires. Forest fires and fighting against them* (pp. 208–217). Moscow, Russia: VNIILM.

Ivanov, V. A., & Ivanova, G. A. (2010). *Pozhary ot groz v lesah Sibiri (Wildfires from thunderstorms in Siberia).* Novosibirsk, Russian Federation: Nauka.

Ivanov, V. A., Ivanova, G. A., & Moskalchenko, S. A. (2011). *Handbook for extinguishing natural fires* (2nd ed.). Krasnoyarsk, Russia: Opera Printing Agency LLC.

Jacobson, A. R., Cummins, K. L., Carter, M., Klingner, P., Roussel-Dupré, D., & Knox, S. O. (2000). FORTE observations of lightning radio-frequency signatures: Prompt coincidence with strikes detected by the National Lightning Detection Network. *Journal of Geophysical Research*, *105*, 15653–15662. doi:10.1029/2000JD900103

Jacobson, A. R., Holzworth, R., Harlin, J., Dowden, R., & Lay, E. (2006). Performance assessment of the world wide lightning location network (WWLLN) using the Los Alamos Sferic Array (LASA) as ground truth. *Journal of Atmospheric and Oceanic Technology*, *23*(8), 1082–1092. doi:10.1175/JTECH1902.1

Jain, A. K. (2010). Data clustering: 50 years beyond K-means. *Pattern Recognition Letters*, *31*(8), 651–666. doi:10.1016/j.patrec.2009.09.011

Jiménez-Ruano, A., Rodrigues Mimbrero, M., & de la Riva Fernández, J. (2017). Understanding wildfires in mainland Spain. A comprehensive analysis of the fire conditions in a climate-human context. *Applied Geography (Sevenoaks, England)*, *89*, 100–111. doi:10.1016/j.apgeog.2017.10.007

Johannesburg Declaration on Sustainable Development (2002). (n.d.). Retrieved from http://ec.europa.eu/environment/archives/wssd/documents/wssd_pol_declaration.pdf

Johnson, E. A. (1996). *Fire and the vegetation dynamics: studies from the North American boreal forest.* Cambridge, UK: Cambridge University Press.

Jolly, W. M., Cochrane, M. A., Freeborn, P. H., Holden, Z. A., Brown, T. J., Williamson, G. J., & Bowman, D. M. J. S. (2015). Climate-induced variations in global wildfire danger from 1979 to 2013. *Nature Communications*, *6*(1), 7537. doi:10.1038/ncomms8537 PMID:26172867

Kachurin, L. G., & Divinskij, L. D. (Eds.). (1992). Aktivno-passivnaya radiolokaciya grozovyh i grozoopasnyh ochagov v oblakah (Active-passive radar observation of thunderstorm and thunderstorm foci in the clouds). Izvestiya Akademii nauk SSSR. Fizika atmosfery i okeana, 24(6), 601-612.

Kale, M. P., Ramachandran, R. M., Pardeshi, S., Chavan, M., Joshi, P. K., Pai, D. S., ... Roy, P. S. (2017). Are climate extremities changing forest fire regimes in India? An analysis using MODIS fire locations during 2003-2013 and gridded climate data of India meteorological department. *Proceedings of the National Academy of Sciences India Section A-Physical Sciences*, *87*(4), 827–843. doi:10.100740010-017-0452-8

Kalinin, L. B., Kovyazin V. F., & Belov, S. V. (1993). Forest fire science. St. Petersburg, Russia: St. Petersburg State Forestry Academy, 76.

Kalinin, L. B., Kovyazin, V. F., & Belov, S. V. (1993). Forest fire science. St. Petersburg State Forestry Academy, St. Petersburg, Russia.

Kaplan, S., & Garrick, B. J. (1981). On the quantitative definition of risk. *Risk Analysis*, *1*(1), 11–27. doi:10.1111/j.1539-6924.1981.tb01350.x PMID:11798118

Kappes, M. (2011). *Multi-hazard risk analyses: a concept and its implementation.* (Dissertation). Universitat Wien, Austria.

Karanina, S. Yu., Kocheeva, N. A., & Karanin, A. V. (2017). Prostranstvennoe i vremennoe raspredelenie molnievyh razryadov po territorii Altae-Sayanskogo regiona (Spatial and temporal distribution of lightning discharges on the territory of the Altai-Sayan region). *Izvestiya vysshih uchebnyh zavedenij. Severo-Kavkazskij region. Estestvennye nauki*, (4-1 (196)).

Karouni, A., Daya, B., Bahlak, S., & Chauvet, P. (2014). A simplified mathematical model for fire spread predictions in wildland fires combining between the models of Anderson and Rothermel. *International Journal of Modeling and Optimization*, *4*(3), 197–200. doi:10.7763/IJMO.2014.V4.372

Kashinskii, S. D. (Ed.). (1979). *Opasnye javlenija pogody na territorii Sibiri i Urala. Chast 1. Altajskij kraj, Kemerovskaja, Novosibirskaja i Tomskaja oblasti (The dangerous phenomena of weather in the territory of Siberia and the Urals. Part 1. Altai Krai, Kemerovo, Novosibirsk, and Tomsk regions)*. Leningrad, USSR: Gidrometeoizdat.

Kasperov, G. I., & Goman, P. N. (2010). Investigation of fire hazardous properties of forest combustible materials of pine plantations. *Proceedings of BSTU. Ser. II, Forest and wood. prom.*, Vol. XVIII, 337-340.

Kataeva, L. Yu. (2000). Application of methods and concepts of fluid and gas mechanics for mathematical modeling of some urgent problems of ecology. (Dissertation), Cand. Phys.-Mat. Sciences, Tomsk, Russia: TSU.

Katz, S., Yair, Y., Price, C., Yaniv, R., Silber, I., Lynn, B., & Ziv, B. (2018). Electrical properties of the 8–12th September, 2015 massive dust outbreak over the Levant. *Atmospheric Research*, *201*, 218–225. doi:10.1016/j.atmosres.2017.11.004

Keane, R. E., Burgan, R. E., & van Wagtendonk, J. (2001). Mapping wildland fuel for fire management across multiple scales: Integrating remote sensing, GIS, and biophysical modeling. *International Journal of Wildland Fire*, *10*(4), 301–319. doi:10.1071/WF01028

Keane, R. E., & Long, D. G. (1998). A comparison of coarse scale fire effects simulation strategies. *Northwest Science*, *2*, 72, 76–90.

Keeley, J. E. (2015). Chaparral plant diversity: Evolutionary & Ecological Patterns. U.S. Geological Survey Sequoia Field Station & UCLA, Department of Ecology & Evolutionary Biology, 51.

Keetch, J., & Byram, G. (1988). A drought index for forest fire control. Res. Paper SE-38. Asheville, NC: U.S. Department of Agriculture, Forest Service, Southeastern Forest Experiment Station, p. 32.

Kelly, R., Montgomery, W. I., & Reid, N. (2008). Differences in soil chemistry remain following wildfires on temperate heath and blanket bog sites of conservation concern. *Geoderma*, *315*, 20–26. doi:10.1016/j.geoderma.2017.11.016

Kerkmann, J., Lutz, H. J., König, M., Prieto, J., Pylkko, P., Roesli, H. P., . . . Zwatz-Meise, V. *MSG channels Interpretation Guide Weather, surface conditions and atmospheric constituents*. Retrieved from http://www.eumetrain.org/IntGuide

Khodakov, V., & Zharikova, M. (2011). Forest fires and methods of their research. *Kherson, FLP Green D. S.*, 470.

Khodakov, V. E., & Zharikova, M. V. (2011). *Forest fires: research methods*. Kherson, Ukraine: Grin DS.

Kirkland, M. W., Suszcynsky, D. M., Guillen, J. L. L., & Green, J. L. (2001). Optical observations of terrestrial lightning by the FORTE satellite photodiode detector. *Journal of Geophysical Research, D, Atmospheres*, *106*(D24), 33499–33509. doi:10.1029/2000JD000190

Kislov, A. V., Varentsov, M. I., & Tarasova, L. L. (2015). Role of spring soil moisture in the formation of large-scale droughts in the East European Plain in 2002 and 2010. *Izvestiya. Atmospheric and Oceanic Physics*, *51*(4), 405–411. doi:10.1134/S0001433815020061

Klinkenborg, V. (2007). *Splendor of the grass: The prairie's grip is unbroken in the Flint Hills of Kansas*. National Geographic.

Kocheeva, N. A., Chankibaeva, M. H., Minaev, A. I., Sukhova, M. G., & Modorov, A. A. (2017). Peculiarities of thunderstorms' occurrences on the border of the Western Siberia Plains and Altai Mountains. *Journal of Environmental Management & Tourism*, *8*(5), 21, 1061–1068.

Kolesnikov, B. P. (1995). Sketch of the vegetation of the Far East. Khabarovsk. p. 102 (in Russia)

Kolesnikova, T. G. (2002). Payments for forest resources as a factor of socio-economic development of forestry: On the example of the Republic of Mari El. *Abstract of Dis. thesis....Kand. Econ. Sciences: 08.00.05: Yoshkar-Ola*.

Kondrat'ev, K. Ya., & Grigor'ev, Al. A. (2004). Forest fires as a component of natural ecodynamics. [in Russian]. *Opt. Atmos. Okeana*, *17*, 279–290.

Konev, E. V. (1977). *Physical basis of burning materials*. Novosibirsk, Russia: Science.

Kononov, I. I., Krutoj, D. M., & Yusupov, I. E. (2013). Alternativnoe obespechenie grozobezopasnosti na ob'ektah vodnogo transporta (Alternative providing safety thunderstorm on objects of the water transport). *Vestnik gosudarstvennogo universiteta morskogo i rechnogo flota im. admirala S.O. Makarova, 3*(19), 116-121.

Kononov, A. A., & Ka, M. H. (2008). Model-associated forest parameter retrieval using VHFSAR data at the individual tree level. *IEEE Transactions on Geoscience and Remote Sensing*, *46*(1), 69–84. doi:10.1109/TGRS.2007.907107

Kononov, I. I., & Yusupov, I. E. (2004). Cluster analysis of storm activity. [Radiotekhnika i elektronika]. *Journal of Communications Technology and Electronics*, *49*(3), 260–266.

Kononov, I. I., & Yusupov, I. E. (2004). Cluster analysis of thunderstorm activity. *Radio Engineering and Electronics*, *49*(3), 283–291.

Konovalov, I. B., Beekmann, M., Berezin, E. V., Petetin, H., Mielonen, T., Kuznetsova, I. N., & Andreae, M. O. (2015). The role of semi-volatile organic compounds in the mesoscale evolution of biomass burning aerosol: A modeling case study of the 2010 mega-fire event in Russia. *Atmospheric Chemistry and Physics*, *15*(23), 13269–13297. doi:10.5194/acp-15-13269-2015

Konstantinova, D. A., Gorbatenko, V. P., & Ershova, T. V. (2010). Spatial distribution of ligtning discharges density on Western Siberia. In *Proceedings of 30th International Conference on Lightning Protection (ICLP)*. Cagliari, Italy: Electrical and Electronics Engineers. 10.1109/ICLP.2010.7845826

Konstantinova, D. A., & Gorbatenko, V. P. (2011). Rezultaty registracii molnij nad yugo-vostochnoj territoriej Zapadnoj Sibiri (Results lightnings registration over the southeast territory of Western Siberia). *Izvestiya vysshih uchebnyh zavedenij Fizika*, *54*(11), 156–162.

Konstantinova, D. A., Gorbatenko, V. P., & Polyakov, D. V. (2017). Spatial distribution of the thunderstorm activity characteristics for the territory of Western Siberia. In *Proceedings of SPIE: 23rd International Symposium on Atmospheric and Ocean Optics: Atmospheric Physics*. doi:10.1117/12.2285994

Konstantinova, D. A., Gorbatenko, V. P., & Polyakov, D. V. (2017, November). Spatial distribution of the thunderstorm activity characteristics for the territory of Western Siberia. In *23rd International Symposium on Atmospheric and Ocean Optics: Atmospheric Physics* (Vol. 10466, p. 1046658). International Society for Optics and Photonics.

Korovin, G. N. (1977). Fire hazard assessment in the forest and calculation of parameters of forest fires on a computer: LenNIILH. p. 69 (in Russia)

Korovin, G. N., & Isaev, A. S. (1998). Forest fire protection as the most important element of national security of Russia, *Forest Bulletin,* 8-9. Retrieved from http://old.forest.ru/rus/problems/fires/080908.html ahhh.

Koshmarov, Yu. A., & Bashkir, M. P. (1987). Thermodynamics and heat transfer in fire engineering. Moscow, Russia: Vneshtorgizdat.

Koutsias, N., Xanthopoulos, G., Founda, D., Xystrakis, F., Nioti, F., Pleniou, M., ... Arianoutsou, M. (2013). On the relationships between forest fires and weather conditions in Greece from long-term national observations (1894-2010). *International Journal of Wildland Fire*, *22*(4), 493–507. doi:10.1071/WF12003

Kovács, F., Legány, C., & Babos, A. (2005, November). Cluster validity measurement techniques. In *Proceedings 6th International Symposium of Hungarian Researchers on Computational Intelligence* (p. 35).

Kovacs, K., Ranson, K. J., Sun, G., & Kharuk, V. I. (2004). The relationship of the Terra MODIS fire product and anthropogenic features in the Central Siberian landscape. *Earth Interactions*, *8*(18), 1–25. doi:10.1175/1087-3562(2004)8<1:TROTTM>2.0.CO;2

Kovalev, A. P. (2009). The current state of the forests of the Russian Far East and the prospects for their use. p. 470 (in Russia)

Kovyazin, V. F., Romanchikov, A. N., & Pasko, O. A. (2015). Comparative analysis of forest lands cadastral appraisal estimated with regards to wood and food resources. *IOP Conference Series: Earth and Environmental Science.* doi:10.1088/1755-1315/27/1/012039

Kozlov, V. I., Mullayarov, V. A., & Karimov, R. R. (2011). Prostranstvennoe raspredelenie plotnosti grozovyh razryadov na Vostoke Rossii po dannym distancionnyh nablyudenij (The spatial distribution of the lightning in the east of Russia on remote sensing data). *Sovremennye problemy distancionnogo zondirovaniya Zemli iz kosmosa, 8*(3), 257.

Kozlov, V. I., Mullayarov, V. A., & Solovev, V. S. (2009). Lesnye pozhary v Yakutii ot groz (Wildfires from thunderstorms in Yakutia). *Sovremennye problemy distancionnogo zondirovaniya Zemli iz kosmosa,* *2*(6), 388.

Kozlov, V. I., & Mullayarov, V. A. (2004). *Grozovaya aktivnost v Yakutii (Thunderstorm activity in Yakutia).* Yakutsk, Russian Federation: YaF Izd-va SO RAN.

Kozlov, V. I., Mullayarov, V. A., Grigorev, Y. M., & Tarabukina, L. D. (2014). Parameters of thunderstorm activity and lightning discharges in Central Yakutia from 2009 to 2012. *Izvestiya. Atmospheric and Oceanic Physics*, *50*(3), 323–329. doi:10.1134/S0001433814030086

Kozlov, V. I., & Mulyayarov, V. A. (2004). *Thunderstorm activity in Yakutia.* Yakutsk, Russia: YaP Publishing House of SB RAS.

Krawchuk, M. A., & Moritz, M. A. (2014). Burning issues: Statistical analyses of global fire data to inform assessments of environmental change. *Environmetrics, 25*(6), 472–481. doi:10.1002/env.2287

Krueger, E. S., Ochsner, T. E., Carlson, J. D., Engle, D. M., Twidwell, D., & Fuhlendorf, S. D. (2016). Concurrent and antecedent soil moisture relate positively or negatively to probability of large wildfires depending on season. *International Journal of Wildland Fire, 25*(6), 657–668. doi:10.1071/WF15104

Krug, E. C., Hollinger, S. E. (2003). Identification of factors that aid carbon sequestration in Illinois agricultural systems. Champaign, IL: *Illinois State Water Survey*, 103.

Kudryavtsev, M., Lukin, V., Malinetskii, G., Mitin, N., Naumenko, S., Podlazov, A., Rumyantsev, A., & Toropygina, S. (2008). Forest conflagration extinguishing problems on Russian Federation territory, *Preprint, Inst. Appl. Math., the Russian Academy of Science*, 40.

Kudryavtsev, N. G., Belikova, M. Yu., Krechetova, S. Yu., Geyman, T. N., Kocheeva, N. A. & Kudin, D. V. (2014). Construction of clusters of lightning discharges on the basis of data of the global network of thunderstorm signaling WWLLN and determination of their parameters. *New information technologies in the study of complex structures in the village.* Katun, Russia. June 9-11, 21.

Kulikov, A. I., Kulikov, M. A., & Smirnova, I. I. (2008) Thermal state of the active layer in the permafrost zone of the Baikal region in the context of global warming. In *Proceedings of the International Symposium "Climate Change in Central Asia: Socio-Economic and Environmental Impacts*". Chita, Russia: Institute of Natural Resources, Ecology, and Cryology of the Siberian Branch of the RAS. (In Russian).

Kumar, V., & Dharssi, I. (2017). Evaluation of daily soil moisture deficit used in Australian forest fire danger rating system. *Forest Ecology and Management, 259*, 685–697.

Kundu, A., Denis, D., Patel, N. D., & Dutta, D. (2018). A Geo-spatial study for analysing temporal responses of NDVI to rainfall. *Singapore Journal of Tropical Geography, 39*(1), 107–116. doi:10.1111jtg.12217

Kundu, A., Patel, N., Saha, S., & Patel, N. D. (2015). Monitoring the extent of desertification processes in western Rajasthan (India) using geo-information science. *Arabian Journal of Geosciences, 8*(8), 5727–5737. doi:10.100712517-014-1645-y

Kunkel, S., Teumer, T., Dörnhofer, P., Schlachter, K., Weldeslasie, Y., Kühr, M., ... Repke, J.-U. (2018). Determination of heat transfer coefficients in direct contact latent heat storage systems. *Applied Thermal Engineering, 145*, 71–79. doi:10.1016/j.applthermaleng.2018.09.015

Kupovykh, G. V., Morozov, V. N., & Shvarts, Ya. M. (1998). *Theory of electrode effect in the atmosphere*. Taganrog, Russia: Taganrog Radiotechnical Institute. [in Russian]

Kurbatskiy, N. P. (1964). *Problem of forest fires. Emergence of forest fires.* Moscow, Russia: Nauka. (In Russian)

Kurbatskiy, N. P., & Kostyrina, T. V. (1977). *US National Fire Danger Calculation System / Detection and analysis of forest fires* (pp. 38–90). Krasnoyarsk, Russia: ILID SO AN USSR.

Kurentsova, G. E. (1967). Sketch of the vegetation of the Jewish Autonomous Region. Vladivostok, Russia: Far Eastern Book Publishing House, p. 64 (in Russia)

Kutinov, Yu. G., & Chistova, Z. B. (2012). Kompleksnaya model' protsessov mezhgeosfernogo vzaimodeystviya v tektonicheskikh uzlakh Severa Russkoy plity (Complex model of geospheres interaction processes in the tectonic nodes at the North of Russian Plate). *Prostranstvo i Vremya,* 1(1).

Kuznetsov, G. V. & Baranovsky, N. V. (2010). Spatial formulation and numerical study of the problem of ignition of a layer of forest fuel with a particle heated to high temperatures. *Butlerov readings,*12, 30-37.

Kuznetsov, G. V., & Baranovskiy, N. V. (2013). Focused sun's rays and forest fire danger: new concept / *Proceedings of SPIE. Vol. 8890*, paper 889011; 10.1117/12.2033929

Kuznetsov, G. V., & Baranovsky, N. V. (2009). *Forecast of forest fires and their environmental consequences*. Novosibirsk, Russia: Publishing House of the Siberian Branch of the Russian Academy of Sciences.

Larjavaara, M., Kuuluvainen, T., & Rita, H. (2005). Spatial distribution of lightning-ignited fires in Finland. *Forest Ecology and Management*, *208*(1-3), 177–188. doi:10.1016/j.foreco.2004.12.005

Larjavaara, M., Pennanen, J., & Tuomi, T. J. (2005). Lightning that ignites forest fires in Finland. *Agricultural and Forest Meteorology*, *132*(3-4), 171–180. doi:10.1016/j.agrformet.2005.07.005

Lasaponara, R., & Lanorte, A. (2007a). On the capability of satellite VHR QuickBird data for fuel type characterization in fragmented landscape. *Ecological Modelling*, *204*(1-2), 79–84. doi:10.1016/j.ecolmodel.2006.12.022

Lasuta, G. F., & Goman, P. N. (2019). Simulation of the processes of emergence and spread of forest ground fire with an assessment of the level of heat load from the flame front. *Bulletin of the University of Civil Protection of the Ministry of Emergency Situations of Belarus*, *3*(2), 138–154.

Latham, D. J. (1991). Lightning flashes from a prescribed fire-induced cloud. *Journal of Geophysical Research*, *96*(D9), 17151–17157. doi:10.1029/91JD01808

Latham, D., & Williams, E. (2001). Lightning and forest fires. In E. A. Johnson & K. Miyanishi (Eds.), *Forest fires: Behavior and ecological effects* (pp. 375–418). Academic Press. doi:10.1016/B978-012386660-8/50013-1

Leblanc, F., Aplin, K. L., Yair, Y., Harrison, R. G., Lebreton, J. P., & Blanc, M. (Eds.). (2008). Planetary atmospheric electricity. *Space Science Reviews*, *137*(1-4).

Lee, B. S., Alexander, M. E., Hawkes, B. C., Lynham, T. J., Stocks, B. J., & Englefield, P. (2002). Information systems in support of wildland fire management decision making in Canada. *Computers and Electronics in Agriculture*, *37*(1-3), 185–198. doi:10.1016/S0168-1699(02)00120-5

Lei, J., Liu, N., Zhang, L., Chen, H., Shu, L., Chen, P., ... de Ris, J. L. (2011). Experimental research on combustion dynamics of medium-scale fire whirl. *Proceedings of the Combustion Institute*, *33*(2), 2407–2415. doi:10.1016/j.proci.2010.06.009

Leonelli, L., Barboni, T., Antoine Santoni, P., Quilichini, Y., & Coppalle, A. (2017). Characterization of aerosols emissions from the combustion of dead shrub twigs and leaves using a cone calorimeter. *Fire Safety Journal*, *91*, 800–810. doi:10.1016/j.firesaf.2017.03.048

Leroy-Cancellieri, V., Cancellieri, D., & Leoni, E. (2006). Thermal degradation of lingo-cellulosic fuels: DSC and TGA studies. *Thermochimica Acta*, *451*(1-2), 131–136. doi:10.1016/j.tca.2006.09.017

Li, F., & Bai, B. (2019). A model of heat transfer coefficient for supercritical water considering the effect of heat transfer deterioration. *International Journal of Heat and Mass Transfer*, *133*, 316–329. doi:10.1016/j.ijheatmasstransfer.2018.12.121

Light, T., Davis, S. M., Boeck, W. L., Jacobson, A. R., & Suszcynsky, D. M. (2003). *Global optical lightning flash rates determined with the Forte satellite (No. LA-UR-03-2614)*. Los Alamos, NM: National Laboratory.

LinacreE. (n.d.). Predicting bushfires in Australia. Retrieved from http://www-as.uwyo.edu/~geerts/cwx/notes/chap16/oz_bush.html

Linn, R. (1997). *A transport model for prediction of wildfire behaviour.* (PhD Dissertation), New Mexico State University. Las Cruces.

Liodakis, S., Katsigiannis, G., & Kakali, G. (2005). Ash properties of some dominant Greek forest species. *Thermochimica Acta*, *437*(1-2), 158–167. doi:10.1016/j.tca.2005.06.041

Liu, B., Zhao, W., & Wen, Z. (2012). Photosynthetic response of two shrubs to rainfall pulses in desert regions of northwestern China. *Photosynthetica*, *50*(1), 109–119. doi:10.100711099-012-0015-9

Liu, W., Wang, S., Zhou, Y., Wang, L., Zhu, J., & Wang, F. (2016). Lightning-caused forest fire risk rating assessment based on case-based reasoning: A case study in DaXingAn Mountains of China. *Natural Hazards*, *81*(1), 347–363. doi:10.100711069-015-2083-1

Li, Y. Z., & Ingason, H. (2017). Scaling of wood pallet fires. *Fire Safety Journal*, *88*, 96–103. doi:10.1016/j.firesaf.2017.01.004

Lobodin, T. V. (1986). Karta summarnogo chisla grozovyh razryadov dlya territorii SSSR. (The map of total number of lightning discharges for the territory of the USSR). *Tr. GGO,* (498), 72-76.

Lower forest fire. (n.d.). Retrieved from http://itd2.mycdn.me/image?id=812140978872&t=20&plc=WEB&tkn=*WaY_waiTdgCqHhBa0HMmywyQMH0

Luche, J., Rogaume, T., Richard, F., & Guillaume, E. (2011). Characterization of thermal properties and analysis of combustion behaviour of PMMA in a cone calorimeter. *Fire Safety Journal*, *46*(7), 451–461. doi:10.1016/j.firesaf.2011.07.005

Lundgren, S., Mitchell, W., & Wallace, M. (1995). A status report on NFMAS—an interagency system update project. *Fire Management Notes*, *55*, 11–12.

Lykov, A. V. (1967). *Theory of heat conduction. M*oscow High School.

Macdougall, A. C., & Turkington, R. (2004). Relative importance of suppression-based and tolerance-based competition in an invaded oak savanna. *Journal of Ecology*, *3*(92), 422–434. doi:10.1111/j.0022-0477.2004.00886.x

MacGorman, D. R., Maier, M. W., & Rust, W. D. (1984). *Lightning strike density for the contiguous United States from thunderstorm duration records (No. NUREG/CR--3759).* National Oceanic and Atmospheric Administration.

Makarenko, E. L. (2016). Forest fires and their consequences in the central ecological zone of the Baikal natural territory. *Interaktivnaya nauka (Interactive science)*, 5, 9-12. (In Russian).

Mallek, C. M., Safford, H., Viers, J., & Miller, J. (2013). Modern departures in fire severity and area vary by forest type, Sierra Nevada and Southern Cascades, California. *Ecosphere*, *4*(12), 1–28. doi:10.1890/ES13-00217.1

Management, conservation and sustainable development of forests. (1992). (n.d.). Retrieved from http://www.fao.org/3/y5574e/y5574e05.pdf

Mao, D., Wang, Z., Luo, L., & Ren, C. (2011). Integrating AVHRR and MODIS data to monitor NDVI changes and their relationships with climatic parameters in Northeast China. *International Journal of Applied Earth Observation and Geoinformation, 18*(2012), 528-536.

Map of forests of Canada. (n.d.). Retrieved from http://ftp.geogratis.gc.ca/pub/nrcan_rncan/raster/atlas_6_ed/eng/6432_forest_fire_ignitions_cause_1959_1999_browse.jpg

Marcelli, T., Santoni, P. A., Simeoni, A., Leoni, E., & Porterie, B. (2004). Fire spread across pine needle fuel beds: Characterization of temperature and velocity distributions within the fire plume. *International Journal of Wildland Fire*, *13*(1), 37–48. doi:10.1071/WF02065

Marchenko, N. A. (1993). Fire hazard assessment of forests by the likelihood of fires with a certain speed of propagation based on the analysis of the state of landscapes. Geography and Natural Resources. 4. pp. 131-136. (in Russia)

Mardiana, R., Kawasaki, Z.-I., & Morimoto, T. (2002). Three-dimensional lightning observations of cloud-to-ground flashes using broadband interferometers. *Journal of Atmospheric and Solar-Terrestrial Physics*, *64*(1), 91–103. doi:10.1016/S1364-6826(01)00099-2

Mareev, E. A., Stasenko, V. N., Bulatov, A. A., Dement'eva, S. O., Evtushenko, A. A., Il'in, N. V., ... Shatalina, M. V. (2016). Russian studies of atmospheric electricity in 2011–2014. *Izvestiya. Atmospheric and Oceanic Physics*, *52*(2), 154–164. doi:10.1134/S0001433816020080

Martell, D. L. (2000). A Markov chain model of day to day changes in the Canadian Forest Fire Weather Index. *International Journal of Wildland Fire*, *9*(4), 265–273. doi:10.1071/WF00020

Matlab official web-site. (n.d.). Retrieved from https://matlab.ru/products/matlab

Matlab official web-site. Retrieved from https://matlab.ru/products/matlab

Matsenko, V. V., Sokolov, A. Ya., Kalinin, S. I., Andrianova, F. I., Andreeva, T. A., Ananyin, S. V., ... Kazantseva, L. V. (1999). *Forest fire protection master plan* (Vol. 1). Barnaul, Russia: Rosgiproles. Altay Branch.

Matsoukis, A., Kamoutsis, A., & Chronopoulos, K. (2018). Estimation of the meteorological forest fire risk in a mountainous region by using remote air temperature and relative humidity data. *International Letters of Natural Sciences, 67,* 1-8. doi:10.18052/www.scipress.com/ILNS.67.1.

McArthur, A. G. (1966). Weather and grassland fire behavior / Leaflet 100. Forestry and Timber Bureau. Commonwealth of Australia, 1966.

McArthur, A. G. (1967). Fire behaviour in eucalypt forests. Canberra, Australian Capital Territory: Commonwealth of Australia Forest and Timber Bureau, 25.

McCaffrey, S., Toman, E., Stidham, M., & Shindler, B. (2015). Social science findings in the United States. In D. Paton & J. F. Shroder (Eds.), *Wildfire hazards, risks, and disasters* (pp. 15–34). Amsterdam, The Netherlands: Elsevier; doi:10.1016/B978-0-12-410434-1.00002-6

McCaw, W. (1998). *Predicting fire spread in Western Australia mallee-heath shrubland.* (PhD Thesis), W. McCaw; University of New South Wales. Canberra, Australia.

McCaw, L., Cheney, P., & Sneeuwjagt, R. (2003). *Development of scientific understanding of fire behavior and use in southwest Western Australia. Fire in ecosystems of southwest Western Australia* (pp. 171–187). Leiden, The Netherlands: Backhuys Publishers.

Mehrabian, R., Scharler, R., & Obernberger, I. (2012). Effects of pyrolysis conditions on the heating rate in biomass particles and applicability of TGA kinetic parameters in particle thermal conversion modeling. *Fuel*, *90*, 567–575. doi:10.1016/j.fuel.2011.09.054

Melekhov, I. S. (1965). Forest pyrology and its tasks. Modern issues of forest protection from fires and control, Moscow, Russia: Forest industry, 5-25.

Melekhov, I. S. (1947). *Nature of the forest and forest fires*. Arkhangelsk, Russia. (In Russian)

Melluma, A. Z., Rungule, R. H., & Emsis, I. V. (1982). *Camping as a nature conservation problem*. Riga, Latvia: Zinatne. (In Russian)

Mell, W. E., Manzello, S. L., Maranghides, A., Butry, D., & Rehm, R. G. (2010). The wildland-urban interface fire problem-current approaches and research needs. *International Journal of Wildland Fire*, *19*(2), 238–251. doi:10.1071/WF07131

Mell, W., Jenkins, M. A., Gould, J., & Cheney, P. (2007). A physics-based approach to modelling grassland fires. *International Journal of Wildland Fire*, *16*(1), 1–22. doi:10.1071/WF06002

Meng, Z., & Zhang, F. (2015). Data assimilation and predictability. Ensemble-based data assimilation. Encyclopedia of Atmospheric Sciences (2nd Ed.). 241-247.

Methods of assessing the consequences of large forest fires. (1996). Moscow, Russia: VNII GOCHS, 13.

Mezuman, K., Price, C., & Galanti, E. (2014). On the spatial and temporal distribution of global thunderstorm cells. *Environmental Research Letter, Vol. 9. Paper*, *124023*, 9.

Michele, S., Bachisio, A., & Fermin, A. (2016). Predicting wildfire spread and behaviour in Mediterranean landscapes. *International Journal of Wildland Fire*, *25*(10), 1015–1032. doi:10.1071/WF15081

Miller, C., & Ager, A. (2013). A review of recent advances in risk analysis for wildfire management. *International Journal of Wildland Fire*, *22*(1), 1–14. doi:10.1071/WF11114

Miloua, H. (2015). Numerical prediction of the fire spread in vegetative fuels using NISTWFDS model. *Int. J. Eng. Res. Africa*, *20*, 177–192. doi:10.4028/www.scientific.net/JERA.20.177

Mimbrero, M. (2015). *Review and new methodological approaches in human-caused wildfire modeling and ecological vulnerability: Risk modeling at mainland Spain.* (Doctoral thesis), University of Zaragoza, Spain.

Mindykowski, P., Fuentes, A., Consalvi, J. L., & Porterie, B. (2011). Piloted ignition of wildland fuels. *Fire Safety Journal*, *46*(1-2), 34–40. doi:10.1016/j.firesaf.2010.09.003

Mingjun, D., Yili, Z., Linshan, L., Wei, Z., Zhaofeng, W., & Wanqi, B. (2007). The relationship between NDVI and precipitation on the Tibetan Plateau. *Journal of Geographical Sciences*, *17*(3), 259–268. doi:10.100711442-007-0259-7

Ministry of Natural Resources of the Republic of Buryatia (2008-2018). *State report "On the state and environmental protection of the Republic of Buryatia"* (Vols. 2007-2017). Ulan-Ude, Russia: The Official Portal of the Republic of Buryatia.

Mistry, J., Bilbao, B. A., & Berardi, A. (2016). Community owned solutions for fire management in tropical ecosystems: Case studies from indigenous communities of South America. *Philosophical Transactions of the Royal Society*, *371*(1696), 20150174. doi:10.1098/rstb.2015.0174 PMID:27216507

Mitchell, M., Wilson, R., Twedt, D., Mini, A., & James, J. (2016). Object-based forest classification to facilitate landscape-scale conservation in the Mississippi Alluvial Valley. *Remote Sens. Appl. Soc. Environ.*, *4*, 55–60.

MODIS (Moderate Resolution Imaging Spectroradiometer). (n.d.). Retrieved from https://terra.nasa.gov/about/terra-instruments/modis

Monod, B., Collin, A., Parent, G., & Boulet, P. (2009). Infrared radiative properties of vegetation involved in forest fires. *Fire Safety Journal*, *44*(1), 88–95. doi:10.1016/j.firesaf.2008.03.009

Morandini, F., Santoni, P. A., & Balbi, J. H. (2001). The contribution of radiant heat transfer to laboratory-scale fire spread under the influences of wind and slope. *Fire Safety Journal*, *36*(6), 519–543. doi:10.1016/S0379-7112(00)00064-3

Moritz, M. A., Morais, M. E., Summerell, L. A., Carlson, J. M., & Doyle, J. (2005). Wildfires, complexity, and highly optimized tolerance. *Proceedings of the National Academy of Sciences of the United States of America*, *102*(50), 17912–17917. doi:10.1073/pnas.0508985102 PMID:16332964

Morvan, D. (2011). Physical phenomena and length scales governing the behaviour of wildfires: A case for physical modeling. *Fire Technology*, *47*(2), 437–460. doi:10.100710694-010-0160-2

Morvan, D. (2014). Wind effects, unsteady behaviors, and regimes of propagation of surface fires in open field. *Combustion Science and Technology*, *186*(7), 869–888. doi:10.1080/00102202.2014.885961

Morvan, D. (2015). Numerical study of the behaviour of a surface fire propagating through a firebreak built in a Mediterranean shrub layer. *Fire Safety Journal*, *71*, 34–48. doi:10.1016/j.firesaf.2014.11.012

Morvan, D., Accary, G., Meradji, S., Frangieh, N., & Bessonov, O. (2018). A 3D physical model to study the behavior of vegetation fires at laboratory scale. *Fire Safety Journal*, *101*, 39–52. doi:10.1016/j.firesaf.2018.08.011

Morvan, D., & Dupuy, J. L. (2004). Modeling the propagation of wildfire through a Mediterranean shrub using a multiphase formulation. *Combustion and Flame*, *138*(3), 199–210. doi:10.1016/j.combustflame.2004.05.001

Morvan, D., Meradji, S., & Accary, G. (2008). Wildfire behavior study in a Mediterranean pine stand using a physically based model. *Combustion Science and Technology*, *180*(2), 230–248. doi:10.1080/00102200701600978

Morvan, D., Méradji, S., & Accary, G. (2009). Physical modelling of fire spread in Grasslands. *Fire Safety Journal*, *44*(1), 50–61. doi:10.1016/j.firesaf.2008.03.004

Moskovenko, V. M., Znamenshchikov, B. P., & Zolotarev, S. V. (2012). Primenenie sistemy grozopelengacii Vereya-MR" v interesah elektroenergetiki Rossii (Application of a system of a fixing of lightnings "Vereya-MR" for the benefit of power industry of Russia). *Novoe v rossijskoj elektroenergetike*, (2), 15-23.

Mounted forest fire. (n.d.). Retrieved from https://www.eurekalert.org/multimedia/pub/web/116249_web.jpg

Mullayarov, V. A., Toropov, A. A., Kozlov, V. I., & Karimov, R. R. (2009). Patterns of spatial distribution of positive thunderstorm discharges in Eastern Siberia. *Russian Meteorology and Hydrology*, *34*(6), 364–370. doi:10.3103/S1068373909060041

Müller, M. M., & Vacik, H. (2017). Characteristics of lightnings igniting forest fires in Austria. *Agricultural and Forest Meteorology*, *240*, 26–34. doi:10.1016/j.agrformet.2017.03.020

Nasanbat, E., & Lkhamjav, O. (2016). Wild fire risk map in the eastern steppe of Mongolia using spatial multi-criteria analysis. *The International Archives of the Photogrammetry, Remote Sensing, and Spatial Information Sciences*, Volume XLI-B1, 2016 XXIII ISPRS Congress, July 12–19, 2016, Prague, Czech Republic, 469-473.

National Aeronautics and Space Administration. (NASA). *Fire information for resource management system*. Retrieved from https://firms.modaps.eosdis.nasa.gov/map/

National Aeronautics and Space Administration. (NASA). *Optical Transient Detector (OTD) data sets*. Retrieved from https://ghrc.nsstc.nasa.gov/lightning/data/data_otd.html

National atlas of Russia. Thunderstorms. Retrieved from http://xn--80aaaa1bhnclcci1cl5c4ep.xn--p1ai/cd2/181/181.html

Natural Resources Canada. (n.d.). Retrieved from https://www.nrcan.gc.ca/our-natural-resources/forests-and-forestry/wildland-fires-insects-and-disturbances/forest-fires/fire-ecology/13149

Nelson, R. M. (2015). Re-analysis of wind and slope effects on flame characteristics of Mediterranean shrub fires. *International Journal of Wildland Fire*, *24*(7), 1001–1007. doi:10.1071/WF14155

Nemani, R. R., Keeling, C. D., Hashimoto, H., Jolly, W. M., Piper, S. C., Tucker, C. J., ... Running, S. (2003). Climate-driven increases in global terrestrial net primary production from 1982 to 1999. *Science*, *300*(5625), 1560–1563. doi:10.1126cience.1082750 PMID:12791990

Nesterov, V. G. (1949). Combustibility of forests and methods for its determination. Moscow, Russia: Goslesbumizdat. Retrieved from https://www.originlab.com/

Nesterov, V. G. (1949). *Burnability of the forest and methods for its determination. L.* Goslesbumizdat.

Nieto, H., Aguado, I., García, M., & Chuvieco, E. (2012). Lightning-caused fires in Central Spain. *Agricultural and Forest Meteorology*, *162-163*, 35–43. doi:10.1016/j.agrformet.2012.04.002

Nikishchenko, N. (2011). Comprehensive assessment of natural and anthropogenic factors of fires on the lands of the forest Fund of the Voronezh region thesis abstract, *Voronezh, VSU*, 22.

Nogina, N. A. (1964). *Soils of Transbaikalia.* Moscow, Russia: Nauka. (In Russian)

North, B. M. P., Stephens, S. L., Collins, B. M., Agee, J. K., Aplet, G., Franklin, J. F., & Fulé, P. Z. (2015). Reform forest fire management. *Science*, *349*(6254), 1280–1281. doi:10.1126cience.aab2356 PMID:26383934

Nosyreva, O. V., Barashkova, N. K., & Kizhner, L. I. (2018). Determining the duration of the heating season in Tomsk. *IOP Conference Series: Earth and Environmental Science*, *211,* 1–9. doi:10.1088/1755-1315/211/1/012056

Obyazov, V. A. (2012). Effects of changes in meteorological conditions on the forest fire conditions in Zabaikal'skii krai. *Russian Meteorology and Hydrology*, *37*(6), 373–378. doi:10.3103/S1068373912060039

Occupational safety standards system. Fire safety. General requirements: GOST 12.1.004-91. Enter 07/01/92. Moscow, Russia: State. Standardization Committee: USSR Ministry of the Interior, 1991.

Oliveira, S. L., Pereira, J. M., & Carreiras, J. M. (2012). Fire frequency analysis in Portugal (1975–2005), using Landsat-based burnt area maps. *International Journal of Wildland Fire*, *21*(1), 48–60. doi:10.1071/WF10131

Origin lab official web-site. Retrieved from https://www.originlab.com/

Osipov, D. I. (2006). *Professional programming. Moscow, Russia.* Delphi: Symbol Plus.

Ostrom, E. (2009). A general framework for analyzing sustainability of social-ecological systems. *Science*, *325*(5939), 419–422. doi:10.1126cience.1172133 PMID:19628857

Ovsyannikov, I. V. (1978). Forest fire control system. Moscow, Russia: Forest industry.

Padhi, S., Shotorban, B., & Mahalingam, S. (2016). Computational investigation of flame characteristics of a nonpropagating shrub fire. *Fire Safety Journal*, *81*, 64–73. doi:10.1016/j.firesaf.2016.01.016

Padhi, S., Shotorban, B., & Mahalingam, S. (2017). A computational study of the interactions of three adjacent burning shrubs subjected to wind. *Fire Safety Journal*, *91*, 749–757. doi:10.1016/j.firesaf.2017.03.028

Pakshirajan, R. P. (2013). Probability theory / Texts and readings in mathematics book series (TRIM, volume 63). Springer.

Papadopoulos, A., Paschalidou, A. K., Kassomenos, P. A., & McGregor, G. (2013). Investigating the relationship of meteorological/climatological conditions and wildfires in Greece. *Theoretical and Applied Climatology*, *112*(1-2), 113–126. doi:10.100700704-012-0726-7

Papagiannopoulou, C., Miralles, D. G., Decubber, S., Demuzere, M., Verhoest, N. E., Dorigo, W. A., & Waegeman, W. (2017). A non-linear Granger-causality framework to investigate climate-vegetation dynamics. *Geoscientific Model Development*, *10*(5), 1945–1960. doi:10.5194/gmd-10-1945-2017

Parente, J., Pereira, M. G., Amraoui, M., & Fischer, E. M. (2018). Heat waves in Portugal: Current regime, changes in future climate and impacts on extreme wildfires. *The Science of the Total Environment, 631-632*, 534–549. doi:10.1016/j.scitotenv.2018.03.044 PMID:29529441

Parisien, M. A., Snetsinger, S., Greenberg, J. A., Nelson, C. R., Schoennagel, T., Dobrowski, S. Z., & Moritz, M. A. (2012). Spatial variability in wildfire probability across the western United States. *International Journal of Wildland Fire, 21*(4), 313–327. doi:10.1071/WF11044

Paris, L. J., Dietz, T., Carpenter, S. R., Alberti, M., Folke, C., Moran, E., ... Lubchenco, J. (2007). Complexity of coupled human and natural systems. *Science, 317*(5844), 1513–1516. doi:10.1126cience.1144004 PMID:17872436

Paritsis, J., Landesmann, J., Kitzberger, T., Tiribelli, F., Sasal, Y., Quintero, C., ... Nuñez, M. A. (2018). Pine plantations and invasion alter fuel structure and potential fire behavior in a Patagonian forest-steppe ecotone. *Forests, 117*(9).

Pasko, O. A. (2013). The forest resource of Tomsk Region and its use. *Applied and Fundamental Studies,* 102-110.

Pasko, O. A. (2013). *The forest resource of Tomsk Oblast and its use* (pp. 102–110). St. Louis, MO: Applied and Fundamental Studies, Ltd.

Pasko, O. A., & Baranova, A. V. (2015). Forest fire situation analysis over forest reserve land in Tomsk petroleum province. *IOP Conference Series: Earth and Environmental Science*. Retrieved from 10.1088/1755-1315/24/1/012022

Pasko, O. A., Tokareva, O. S., Alshaibi, A. J. A., Chernikova, T. Y., & Cabral, P. (2019). Assessment of state of cedar forests in Tomsk region using remote sensing data of the earth, *Bulletin of the Tomsk Polytechnic University. Geo Assets Engineering, 330*(1), 89–101. doi:10.18799/24131830/2019/1

Patankar, S. V. (1981). *Numerical heat transfer and fluid flow*. New York: Hemisphere Publishing Corp.

Peng, D., Huang, J., & Wang, X. (2007). Correlation analysis of regional vegetation seasonal fluctuation and climate factors based on MODIS-EVI. *Journal of Applied Ecology, 18*(5), 983–989. PMID:17650845

Perelet, R. A. (2006). An ecosystem approach to environmental management and environmental activities. *Mehanizm regulyuvannya economy,* 1, 36-53.

Pérez, Y., Pastor, E., Àgueda, A., & Planas, E. (2011). Effect of wind and slope when scaling the forest fires rate of spread of laboratory experiments. *Fire Technology, 47*(2), 475–489. doi:10.100710694-010-0168-7

Perminov, V. A., Loboda, E. L., & Reyno, V. V. (2014, November). Mathematical modeling of ignition of woodlands resulted from accident on the pipeline. In *Proceedings 20th International Symposium on Atmospheric and Ocean Optics: Atmospheric Physics (Vol. 9292, p. 929224).* International Society for Optics and Photonics.

Perry, G. L. (1998). Current approaches to modelling the spread of wildland fire: A review. *Progress in Physical Geography, 22*(2), 222–245. doi:10.1177/030913339802200204

Petrov, E. S., Novorotsky, P. N., & Lenshin, V. T. (2000). The climate of the Khabarovsk Territory and the Jewish Autonomous Region. Khabarovsk, Russia: Dal'nauka. pp. 174 (in Russia)

Pew, K. L., & Larsen, C. P. S. (2001). GIS analysis of spatial and temporal patterns of human-caused wildfires in the temperate rain forest of Vancouver Island, Canada. *Forest Ecology and Management, 140*(1), 1–18. doi:10.1016/S0378-1127(00)00271-1

Pimont, F., Dupuy, J. L., Caraglio, Y., & Morvan, D. (2009). Effect of vegetation heterogeneity on radiative transfer in forest fires. *International Journal of Wildland Fire, 18*(5), 536–553. doi:10.1071/WF07115

Pinto, C., Viegas, D., Almeida, M., & Raposo, J. (2017). Fire whirls in forest fires: An experimental analysis. *Fire Safety Journal*, *87*, 37–48. doi:10.1016/j.firesaf.2016.11.004

Pinto, I. R. C. A., & Pinto, O. Jr. (2003). Cloud-to-ground lightning distribution in Brazil. *Journal of Atmospheric and Solar-Terrestrial Physics*, *65*(6), 733–737. doi:10.1016/S1364-6826(03)00076-2

Pinto, O. Jr, Pinto, I. R. C. A., Diniz, J. H., Cherchiglia, L. C. L., Filho, A. C., & Carvalho, A. M. (2003). A seven-year study of the negative cloud-to-ground flash characteristics in Southeastern Brazil. *Journal of Atmospheric and Solar-Terrestrial Physics*, *65*, 739–748. doi:10.1016/S1364-6826(03)00077-4

Pinto, O. Jr, Pinto, I. R. C. A., & Naccarato, K. P. (2007). Maximum cloud-to-ground lightning flash densities observed by lightning location systems in the tropical region: A review. *Atmospheric Research*, *84*(3), 189–200. doi:10.1016/j.atmosres.2006.11.007

Pkhalagov, Yu. A., Uzhegov, V. N., Panchenko, M. V., & Ippolitov, I. I. (2006). Electro-optical relations in atmosphere under smog conditions. [in Russian]. *Opt. Atmos. Okeana*, *19*(10), 861–864.

Podolskaya, A. S., Ershov, D. V., & Shulyak, P. P. (2011). Application of the method for assessing the probability of occurrence of forest fires in ISDM-Rosleskhoz / Modern problems of remote sensing of the Earth from space, 8. pp. 118-126. (in Russian) RAD Studio. Retrieved from https://www.embarcadero.com/ru/products/rad-studio

Podolskaya, A. S., Ershov, D. V., & Shulyak, P. P. (2011). Application of the method for assessing the probability of occurrence of forest fires in ISDM-Rosleskhoz. *Modern problems of remote sensing of the Earth from space. 8,* 118 - 126. (In Russian)

Polyakov, D. V., Barashkova, N. K., & Kuzhevskaya, I. V. (2014). Weather and climate description of anomalous summer 2012 in Tomsk region. *Russian Meteorology and Hydrology*, *1*(1), 22–28. doi:10.3103/S106837391401004X

Ponomarev, E. I., & Kharuk, V. I. (2016). Wildfire occurrence in forests of the Altai-Sayan Region under current climate changes. *Contemporary Problems of Ecology*, *9*(1), 29–36. doi:10.1134/S199542551601011X

Ponomarev, E. I., Kharuk, V. I., & Ranson, K. J. (2016). Wildfires dynamics in Siberian Larch Forests. *Forests*, *7*(6), 125. doi:10.3390/f7060125

Poulos, H. M., Villanueva Díaz, J., Cerano Paredes, J., Camp, A. E., & Gatewood, R. G. (2013). Human influences on fire regimes and forest structure in the Chihuahuan Desert Borderlands. *Forest Ecology and Management*, *298*, 1–11. doi:10.1016/j.foreco.2013.02.014

Prasad, V. K., Badarinath, K. S., & Eaturu, A. (2008). Effects of precipitation, temperature and topographic parameters on evergreen vegetation greenery in the Western Ghats, India. *International Journal of Climatology*, *28*(13), 1807–1819. doi:10.1002/joc.1662

Preisler, H. K., Brillinger, D. R., Burgan, R. E., & Benoit, J. W. (2004). Probability based models for estimation of wildfire risk. *International Journal of Wildland Fire*, *13*(2), 133–142. doi:10.1071/WF02061

Proctor, D. E., Uytenbogaardt, R., & Meredith, B. M. (1988). VHF radio pictures of lightning flashes to ground. *Journal of Geophysical Research, D, Atmospheres*, *93*(D10), 12683–12727. doi:10.1029/JD093iD10p12683

Pummakarnchana, O., Tripathi, N., & Dutta, J. (2005). Air pollution monitoring and GIS modeling: A new use of nanotechnology based solid state gas sensors. *Science and Technology of Advanced Materials*, *6*(3-4), 251–255. doi:10.1016/j.stam.2005.02.003

Pustovalov, K. N., & Nagorskiy, P. M. (2018). Response in the surface atmospheric electric field to the passage of isolated air mass cumulonimbus clouds. *Journal of Atmospheric and Solar-Terrestrial Physics*, *172*, 33–39. doi:10.1016/j.jastp.2018.03.008

RAD Studio. Retrieved from https://www.embarcadero.com/ru/products/rad-studio

Rakov, V. A., & Uman, M. A. (2003). *Lightning: physics and effects*. Cambridge, UK: Cambridge University Press. doi:10.1017/CBO9781107340886

Ramachandra, T. V., Kumar, U., & Dasgupta, A. (2016). *Analysis of land surface temperature and rainfall with landscape dynamics in Western Ghats, India. Technical report* (p. 181). Bangalore, India: Centre for Ecological Sciences, Indian Institute of Science.

Rawlings, J. O., Pantula, S. G., & Dickey, D. A. (2015). *Applied regression analysis. A research tool / Springer texts in statistics book series (STS)*. Springer.

Read, N., Duff, T. J., & Taylor, P. G. (2018). A lightning-induced wildfire ignition forecasting model for operational use. *Agricultural and Forest Meteorology*, *253-254*, 233–246. doi:10.1016/j.agrformet.2018.01.037

Real-Time Lightning Map. Retrieved from http://www.lightningmaps.org]

Regueira, R., & Guaita, M. (2018). Numerical simulation of the fire behaviour of timber dovetail connections. *Fire Safety Journal*, *96*, 1–12. doi:10.1016/j.firesaf.2017.12.005

Reifsnyder, W. E. (1978). *Systems for evaluating and prediction*.

Remote sensing satellite data receiving ground station V. E. Zuev Institute of Atmospheric Optics. Retrieved from http://cosmo.iao.ru/ru

Reports of the Ministry of emergency situations and the media. (n.d.). Retrieved from https://www.mchs.gov.ru/activities/results

Research and Production Center. Map-Maker. Retrieved from http://gismeteo.ru

Restuccia, F., Huang, X., & Rein, G. (2017). Self-ignition of natural fuels: Can wildfires of carbon-rich soil start by self-heating? *Fire Safety Journal*, *91*, 828–834. doi:10.1016/j.firesaf.2017.03.052

Revadekar, J. V., Patwardhan, S. K., & Kumar, K. (2011). Characteristic features of precipitation extremes over India in the warming scenarios. *Advances in Meteorology*, *2011*, 1–11. doi:10.1155/2011/138425

Revadekar, J. V., Tiwari, Y. K., & Kumar, K. (2012). Impact of climate variability on NDVI over the Indian region during 1981–2010. *International Journal of Remote Sensing*, *33*(22), 7132–7150. doi:10.1080/01431161.2012.697642

Richards, G. D., & Bryce, R. W. (1995). A computer algorithm for simulating the spread of wildland fire perimeters for heterogeneous fuel and meteorological conditions. *International Journal of Wildland Fire*, *5*(2), 73–80. doi:10.1071/WF9950073

Roberts, D., Green, R., & Adams, J. (1997). Temporal and spatial patterns in vegetation and atmospheric properties from AVIRIS. *Remote Sensing of Environment*, *62*(3), 223–240. doi:10.1016/S0034-4257(97)00092-8

Rocca, M. E., Miniat, C. F., & Mitchell, R. J. (2014). Introduction to the regional assessments: Climate change, wildfire, and forest ecosystem services in the USA. *Forest Ecology and Management*, *327*, 265–268. doi:10.1016/j.foreco.2014.06.007

Rodger, C. J., Werner, S., Brundell, J. B., Lay, E. H., Thomson, N. R., Holzworth, R. H., & Dowden, R. L. (2006). Detection efficiency of the VLF World-Wide Lightning Network Location Network (WWLLN): Initial case study. *Annales Geophysicae*, *24*(12), 3197–3214. doi:10.5194/angeo-24-3197-2006

Rogeau, M.-P., Flannigan, M. D., Hawkes, B. C., Parisien, M.-A., & Arthur, R. (2016). Spatial and temporal variations of fire regimes in the Canadian Rocky Mountains and foothills of southern Alberta. *International Journal of Wildland Fire*, *25*(11), 1117–1130. doi:10.1071/WF15120

Roitman, M. Y., Komissarov, E. P., & Pchelintsev, V. A. (1978). Fire prevention in construction. Moscow, Russia: stroiizdat.

Rolinski, T., Capps, S. B., Fovell, R. G., Cao, Y., D'Agostino, B. J., & Vanderburg, S. (2016). The Santa Ana wildfire threat index: Methodology and operational implementation. *Weather and Forecasting*, *31*(6), 1881–1897. doi:10.1175/WAF-D-15-0141.1

Rose, E. (1998) Metatheoretical foundations for post-normal risk. *Journal of Risk Research, 1,* pp. 15-41.

Rothermal, R. C. (1972). *Predicting behaviour and size of crown fires in the Northern Rocky Mountains*, USDA Forest Service, Intermountain Forest and Range Experiment Station: Research Paper: INT-438. (Ogden, UT).

Rothermel, R. C. (1972). *A mathematical model for fire spread predictions in wildland fuels.* USDA Forest Service Research Paper INT-115, Ogden, 1972. (Intermountain Forest and Range Exp. Stn.).

Rousseeuw, P. J. (1987). Silhouettes: A graphical aid to the interpretation and validation of cluster analysis. *Journal of Computational and Applied Mathematics*, *20*, 53–65. doi:10.1016/0377-0427(87)90125-7

Roy, P. S. (2003) Forest fire and degradation assessment using satellite remote sensing and geographic information system. Available at http://www.wamis.org/agm/pubs/agm8/Paper-18.pdf

Royal Society. (1992). *Risk: Analysis, perception and management.* London, UK: Royal Society. p. 201.

Ruffault, J., & Mouillot, F. (2017). Contribution of human and biophysical factors to the spatial distribution of forest fire ignitions and large wildfires in a French Mediterranean region. *International Journal of Wildland Fire*, *26*(6), 498–508. doi:10.1071/WF16181

Rules for fire prevention of forests in the Republic of Belarus: TCP 193-2009. Enter 11/01/09. Minsk, Belarus: Ministry of Forestry of the Republic of Belarus: Forestry Institute of the National Academy of Sciences of Belarus, 2009.

Russian Federation. (2006). *Article 81 of the Forest Code of the Russian Federation. The powers of the state authorities of the Russian Federation in the field of forest relations* (Forest Code of the Russian Federation of 04.12.2006 N 200-ФЗ). Russian Federation. (In Russian).

Sable, V. R., Goman, P. N., & Yan, V. I. (2011). On the question of the spatial distribution of the intensity of the radiant energy flux from a large area source. *Emergencies: Education and Science, 2*(6), 116-124.

Sable, V. R., Goman, P. N., & Yan, V. I. (2010). On the influence of the form of an extended heat source on the density of the radiant energy flux. *All BDPU. Ser. III, Physics. Matematyka Infarmatika. Bialogiya. Geography (Sheffield, England)*, *4*, 3–9.

Saikia, A. (2009). NDVI variability in North East India. *Scottish Geographical Journal*, *125*(2), 195–213. doi:10.1080/14702540903071113

Sakellariou, S., Tampekis, S., Samara, F., Sfougaris, A., & Christopoulou, O. (2017). Review of state-of-the-art decision support systems (DSSs) for prevention and suppression of forest fires. *Journal of Forestry Research*, *28*(6), 1107–1117. doi:10.100711676-017-0452-1

Samarskii, A. A., & Vabishchevich, P. N. (1995). Computational heat transfer: Vol. 1. *Mathematical modelling*. Chichester, UK: Wiley.

Samarskii, A. A., & Vabishchevich, P. N. (1995). Computational heat transfer: Vol. 2. *The finite difference method*. Chichester, UK: Wiley.

San-Miguel-Ayanz, J., Moreno, J. M., & Camia, A. (2013). Analysis of large fires in European Mediterranean landscapes: Lessons learned and perspectives. *Forest Ecology and Management*, *294*, 11–22. doi:10.1016/j.foreco.2012.10.050

Sapkota, A., Symons, J. M., Kleissl, J., Wang, L., Parlange, M. B., Ondov, J., Breysse, P. N., Diette, G. B., Eggleston, P. A., & Buckley, T. J. (2005). Impact of the 2002 Canadian forest fires on particulate matter air quality in Baltimore city. *Environmental Science & Technology*, 1, 39(1), 24-32.

Satellite monitoring of forest fires in Russia. (2003). Results problems. Perspectives. Novosibirsk, Russia. p. 135. (in Russia)

ScanEx. (n.d.). Retrieved from http://new.scanex.ru

Scheller, R., Kretchun, A., Hawbaker, T. J., & Henne, P. D. (2019). A landscape model of variable social-ecological fire regimes. *Ecological Modelling*, *401*, 85–93. doi:10.1016/j.ecolmodel.2019.03.022

Schemel, C. F., Simeoni, A., Biteau, H., Rivera, J. D., & Torero, J. L. (2008). A calorimetric study of wildland fuels. *Experimental Thermal and Fluid Science*, *32*(7), 1381–1389. doi:10.1016/j.expthermflusci.2007.11.011

Scherbov, B. L., Lazareva, E. V., & Zhurkova, I. S. (2015). Forest fires and their consequences on the example of Siberian objects, Novosibirsk, Russia: Geo, 154.

Schunk, C., Wastl, C., Leuchner, M., Schuster, C., & Menzel, A. (2013). Forest fire danger rating in complex topography - results from a case study in the Bavarian Alps in autumn 2011. *Natural Hazards and Earth System Sciences*, *13*(9), 2157–2167. doi:10.5194/nhess-13-2157-2013

Scientific-applied reference book on the climate of the USSR. Parts 1-6. Issue. 20. (Tomsk, Novosibirsk, Kemerovo regions and Altai Territory). (1993). St. Petersburg, Russia: Gidrometeoizdat.

Scott, J. H., & Reinhardt, E. D. (2001). *Assessing crown fire potential by linking models of surface and crown fire behavior*, USDA Forest Service, Rocky Mountain Research Station. Fort Collins, CO: Research Paper RMRS-RP-29.

Seager, R., Hooks, A., Williams, A. P., Cook, B. I., Nakamura, J., & Henderson, N. (2015). Climatology, variability and trends in United States vapor pressure deficit, an important fire-related meteorological quantity. *Journal of Applied Meteorology and Climatology*, *54*(6), 1121–1141. doi:10.1175/JAMC-D-14-0321.1

Sedunov, Yu. S. (1991). *Atmosphere: A handbook (Reference data and models)*. Leningrad, Russia: Gidrometeoizdat. [in Russian]

Seinfeld, J. H., & Pandis, S. N. (1998). *Atmospheric chemistry and physics*. New York: Wiley Intersci. Publ.

Seity, Y., Soula, S., & Sauvageot, H. (2000). Radar observation and lightning detection in coastal thunderstorms. *Physics and Chemistry of the Earth. Part B: Hydrology, Oceans and Atmosphere*, *25*(10-12), 1107–1110. doi:10.1016/S1464-1909(00)00161-1

Semoutnikova, E. G., Gorchakov, G. I., Sitnov, S. A., Kopeikin, V. M., Karpov, A. V., Gorchakova, I. A., ... Kuznetsov, G. A. (2018). Siberian smoke haze over European Territory of Russia in July 2016: Atmospheric pollution and radiative effects. *Atmospheric and Oceanic Optics*, *31*(2), 171–180. doi:10.1134/S1024856018020124

Shabaganova, S. N., Karimov, R. R., Kozlov, V. I., & Mullayarov, V. A. (2012). Characteristics of storm cells from observations in Yakutia. *Russian Meteorology and Hydrology*, *37*(11-12), 746–751. doi:10.3103/S1068373912110088

Shabbar, A., Skinner, W., & Flannigan, M. D. (2011). Prediction of seasonal forest fire severity in Canada from large-scale climate patterns. *Journal of Applied Meteorology and Climatology*, *50*(4), 785–799. doi:10.1175/2010JAMC2547.1

Sheriza, M. (2010, May 7). Razali, Ainuddin Nuruddin, Ismail A. Malek, and Norizan A. Patah Forest fire hazard rating assessment in peat swamp forest using Landsat thematic mapper image. *Journal of Applied Remote Sensing*, *4*(1), 043531. doi:10.1117/1.3430040

Sherstjuk, V., Zharikova, M., & Sokol, I. (2018). Forest fire-fighting monitoring system based on UAV team and remote sensing. In *Proc. of the 38th Int. Conf. "Electronics and Nanotechnology"*, pp. 99-106. Kyiv, Ukraine. 10.1109/ELNANO.2018.8477527

Shindo, T., & Yokoyama, S. (1998). Lightning occurrence data observed with lightning location systems in Japan: 1992-1995. *IEEE Transactions on Power Delivery*, *13*(4), 1468–1474. doi:10.1109/61.714797

Shvidenko, A. (2009). Non-boreal forests of Eastern Europe in a changing world: the role in the Earth systems. *Regional Aspects of Interactions in Non – boreal Eastern Europe, Springer Science: Business Media B.V.*, 123-133.

Sidorov, A. A., & Khankhunov, Yu. M. (2015). On the problems with forest fires in the Republic of Buryatia. In *Proceedings of the Scientific-Practical Conference with International Participation*. Ulan-Ude, Russia: East Siberia State University of Technology and Management. (In Russian).

Sidorov, A. A., & Khankhunov, Yu. M. (2016). Forest fires in the Republic of Buryatia: liquidation, prevention and consequences. In *Proceedings of the Scientific-Practical Conference with International Participation*. Ulan-Ude, Russia: East Siberia State University of Technology and Management. (In Russian).

Sidorov, A. A., & Sanzhieva, S. E. (2018). Chronology of forest fires in the Republic of Buryatia. [In Russian]. *Bulletin of the Krasnoyarsk State Agrarian University*, *4*, 204–208.

Simeoni, A., Thomas, J. C., Bartoli, P., Borowieck, P., Reszka, P., Colella, F., ... Torero, J. L. (2012). Flammability studies for wildland and wildland–urban interface fires applied to pine needles and solid polymers. *Fire Safety Journal*, *54*, 203–217. doi:10.1016/j.firesaf.2012.08.005

Sitnov, S. A., Mokhov, I. I., Gorchakov, G. I., & Dzhola, A. V. (2017). The smoke haze over the European Part of Russia in Summer of 2016: A link to wildfires in Siberia and atmospheric circulation anomalies. *Meteorologiya I Gidrologiya (8), 50-63.*

Slayback, D. A., Pinzon, J. E., Los, S. O., & Tucker, C. J. (2003). Northern hemisphere photosynthetic trends 1982–1999. *Global Change Biology*, *9*(1), 1–15. doi:10.1046/j.1365-2486.2003.00507.x

Smelyansky, I. E., Buffalos, Yu. A., Bazhenov, Yu. A., & ... . (2015). *Steppe fires and fire management in the steppe protected areas: environmental and ecological aspects. Analytical Review*. Moscow, Russia: Wildlife Conservation Center. (in Russian)

Smirnov, V. V. (2010). Electrization of aerosol wetted in bipolarly ionized air. [in Russian]. *Izvestiya. Atmospheric and Oceanic Physics*, *46*(3), 294–303. doi:10.1134/S0001433810030035

Smith, C. B. (2008). *Lightning: Fire from the Sky*. Dockside Consultants Inc.

Smith, D. A., Heavner, M. J., Jacobson, A. R., Shao, X. M., Massey, R. S., Sheldon, R. J., & Wiens, K. C. (2004). A method for determining intracloud lightning and ionospheric heights from VLF/LF electric field records. *Radio Science*, *39*(1), n/a. doi:10.1029/2002RS002790

Smoke and smog from a forest fire in Canada. (n. d.). Retrieved from https://i.pinimg.com/originals/29/31/5c/29315c49b896cb9d51322a4d461c13d2.jpg

Sochava, V. B. (Ed.). (1967). *Atlas of Transbaikalia. (Buryat ASSR and Chita region)* [*Maps*]. Moscow, Irkutsk, Russia: Main Department of Geodesy and Cartography. (In Russian)

Soja, A. J., Tchebakova, N. M., French, N. H. F., Flannigan, M. D., Shugart, H. H., Stocks, B. J., ... Stackhouse, P. W. Jr. (2007). Climate-induced boreal change: Predictions versus current observations. *Global and Planetary Change*, *56*(3-4), 274–296. doi:10.1016/j.gloplacha.2006.07.028

Solovyev V. A. (1969). Prostranstvenno-vremennye harakteristiki grozovoj aktivnosti po dannym pelengovaniya atmosferikov (Space-time characteristics of storm activity by results of a bearing of atmospherics). *Tr. GGO* (242), 52-61.

Solovyev, V. S., Kozlov, V. I., Karimov, R. R., & Vasiliev, M. S. (2010). Kompleksnyj monitoring grozovoj aktivnosti i lesnyh pozharov po dannym nazemnyh i sputnikovyh nablyudenij (Simultaneous monitoring of thunderstorm activity and forest fires on data of ground-base and remote sensing observations). *Sovremennye problemy distantsionnogo zondirovaniya Zemli iz kosmosa. 7*(4), 218.

Sommers, W. T., Coloff, S. G., & Conard, S. G. (2012). Synthesis of knowledge: fire history and climate change. *Final report to the Joint Fire Science Program*, JFSP Project, 09-02-1-09.

Sonnenschein, R., Kuemmerle, T., Udelhoven, T., Stellmes, M., & Hostert, P. (2011). Differences in Landsat-based trend analyses in drylands due to the choice of vegetation estimate. *Remote Sensing of Environment*, *115*(6), 1408–1420. doi:10.1016/j.rse.2011.01.021

Soriano, L. R., De Pablo, F., & Tomas, C. (2005). Ten-year study of cloud-to-ground lightning activity in the Iberian Peninsula. *Journal of Atmospheric and Solar-Terrestrial Physics*, *67*(16), 1632–1639. doi:10.1016/j.jastp.2005.08.019

Sow, M., Crase, E., Rajot, J. L., Sankaran, R. M., & Lacks, D. J. (2011). Electrification of particles in dust storms: Field measurements during the monsoon period in Niger. *Atmospheric Research*, *102*(3), 343–350. doi:10.1016/j.atmosres.2011.08.010

Space Research Institute of the Russian Academy of Sciences. Retrieved from http://meteo.infospace.ru

Srivas, T., de Callafon, R. A., Crawl, D., & Altintas, I. (2017). Data assimilation of wildfires with fuel adjustment factors in farsite using ensemble kalman filtering. *Procedia Computer Science*, *108C*, 1572–1581. doi:10.1016/j.procs.2017.05.197

Starodumov, A. M. (1964). Scale of fire hazard weather for the conditions of the Khabarovsk Territory, Primorye and Amur Region. Collected Works DalNIIHL. 6. pp. 176-185. (in Russia)

State hydrometeorological center of the Russian Federation. Retrieved from http://meteoinfo.ru

State Standard GOST R 22.1.09-99. Monitoring and prediction of forest fires. General requirements. (Moscow, Gosstandart of the Russian Federation, 1999) (In Russian)

Stepanenko, V. D., & Galperin, S. M. (1983). *Radiotekhnicheskie metody issledovaniya groz (Radio Means for Thunderstorm Detection)*. Leningrad, USSR: Gidrometeoizdat.

Stepanov, K. L., Stanchits, L. K., & Stankevich, Yu. A. (2011). Heat radiation of an explosion: Risk of thermal injury to people and the occurrence of fires. *Vestn. Team Engineer Inst. Ministry of Emergency Situations of the Republic of Belarus*, *1*(13), 31–45.

Stephens, S. L., Burrows, N., Buyantuyev, A., Gray, R. W., Keane, R. E., Kubian, R., ... van Wagtendonk, J. W. (2014). Temperate and boreal forest mega-fires: Characteristics and challenges. *Frontiers in Ecology and the Environment*, *12*(2), 115–122. doi:10.1890/120332

Stocker, T. F., Qin, D., Plattner, G.-K., Tignor, M., Allen, S. K., Boschung, J., ... Midgley, P. M. (Eds.). (2013). *IPCC 2013: Climate change 2013: The physical science basis. Contribution of Working Group I to the Fifth Assessment Report of the Intergovernmental Panel on Climate Change*. New York, NY: Cambridge University Press.

Strategic Research Agenda. (2014). Retrieved from http://www.futureearth.org/media/strategic-research-agenda-2014

Strydom, S., & Savage, M. J. (2017). Potential impacts of climate change on wildfire dynamics in the midlands of KwaZulu-Natal, South Africa. *Climatic Change, 143*(3-4), 385–397. doi:10.100710584-017-2019-8

Sukhinin, A. I., Kashkin, V. B., & Ponomarev, E. I. (1999). Monitoring forest fire in eastern Siberia from space. *Proceedings of the Society for Photo-Instrumentation Engineers, 3983*, 206–214. doi:10.1117/12.370494

Sullivan, A. L. (2009). Wildland surface fire spread modelling, 1990-2007. 1: Physical and quasi-physical models. *International Journal of Wildland Fire, 18*(4), 349–368. doi:10.1071/WF06143

Sun, D., & Kafatos, M. (2007). Note on the NDVI-LST relationship and the use of temperature-related drought indices over North America. *Geophysical Research Letters, 34*(24), 4. doi:10.1029/2007GL031485

Sun, L., Zhou, X., Mahalingam, S., & Weise, D. R. (2006). Comparison of burning characteristics of live and dead chaparral fuels. *Combustion and Flame, 144*(1-2), 349–359. doi:10.1016/j.combustflame.2005.08.008

Suszcynsky, D. M., Kirkland, M. W., Jacobson, A. R., Franz, R. C., Knox, S. O., Guillen, J. L. L., & Green, J. L. (2000). FORTE observations of simultaneous VHF and optical emissions from lightning: Basic phenomenology. *Journal of Geophysical Research, D, Atmospheres, 105*(D2), 2191–2201. doi:10.1029/1999JD900993

Suzuki, S., & Manzello, S. L. (2017). Experimental investigation of firebrand accumulation zones in front of obstacles. *Fire Safety Journal, 94*, 1–7. doi:10.1016/j.firesaf.2017.08.007 PMID:31092969

Svedeniya o neblagopriyatnyh usloviyah pogody i opasnyh gidrometeorologicheskih yavleniyah, nanesshih social'nye i ekonomicheskie poteri na territorii Rossii (Information on adverse weather conditions and hydrometeorological hazards that caused social and economic losses in Russia). Retrieved from http://meteo.ru/data/310-neblagopriyatnye-usloviya-pogody-nanjosshie-ekonomicheskie-poteri

Taflove, A., & Hagness, S. (2005). *Computational electrodynamics: The finite-difference time domain method*. Boston, MA: Arthech House Publishers.

Tantisattayakul, T., Masugata, K., Kitamura, I., & Kontani, K. (2005). Broadband VHF sources locating system using arrival-time differences for mapping of lighting discharge process. *Journal of Atmospheric and Solar-Terrestrial Physics, 67*(11), 1039–1039. doi:10.1016/j.jastp.2005.04.002

Tarabukina, L. D., & Kozlov, V. I. (2018). Sravnenie izmerenij neskolkih sistem registracii molnievyh radioimpulsov (Comparison of measurements of several systems of registration lightning of radio impulses). *Vestnik Severo-Vostochnogo federal'nogo universiteta im. M.K. Ammosova,* 2(64), 77-86.

Tarabukina, L. D., & Kononova, N. K. (2018). Cirkulyaciya atmosfery, grozy i lesnye pozhary v Severnoj Azii v 2009-2016 gg (Circulation of the atmosphere, thunderstorm, and wildfires in Northern Asia in 2009-2016). *Zhizn Zemli, 40*(1), 22–30.

Tarabukina, L. D., Kozlov, V. I., Karimov, R. R., & Mullayarov, V. A. (2017). Spatial pattern of lightning strikes in North Asia. *Russian Meteorology and Hydrology, 42*(2), 88–94. doi:10.3103/S1068373917020029

Taufik, M., Setiawan, B. I., & van Lanen, H. A. J. (2015). Modification of a fire drought index for tropical wetland ecosystems by including water table depth. *Agricultural and Forest Meteorology, 203*, 1–10. doi:10.1016/j.agrformet.2014.12.006

Teakles, A. D., So, R., Ainslie, B., Nissen, R., Schiller, C., Vingarzan, R., ... Huang, L. (2017). Impacts of the July 2012 Siberian fire plume on air quality in the Pacific Northwest. *Atmospheric Chemistry and Physics*, *17*(4), 2593–2611. doi:10.5194/acp-17-2593-2017

Tedim, F., Xanthopoulos, G., & Leone, V. (2015). Forest fires in Europe: Facts and challenges wildfire. Hazards, risks, and disasters, 1, 5. Elsevier: Douglas Paton, 77-99.

Teferi, E., Uhlenbrook, S., & Bewket, W. (2015). Inter-annual and seasonal trends of vegetation condition in the Upper Blue Nile (Abay) Basin: Dual-scale time-series analysis. *Earth System Dynamics*, *6*(2), 617–636. doi:10.5194/esd-6-617-2015

Telitsin, G. P. (1984). Study of the relationship of forest attendance and the emergence of fires. [In Russian]. *Forest Science*, *1*, 59–63.

Telitsin, G. P. (1987). *Method of determining the fire danger of forest territory. Forest fires and the fight against them.* Moscow, Russia: VNIILM. (In Russian)

The book of account of forest fires. (n.d.). Retrieved from https://lektsii.com/1-157112.html

The combustibility of the forests of Russia. Retrieved from https://vuzlit.ru/1100313/gorimost_lesov_rossii

The Comprehensive R Archive Network | Submitting to CRAN Retrieved from https://cran.r-project.org

The fire in Oklahoma. A photo from space . (n.d.). Retrieved from https://yandex.ru/images/search?text=forest%20 fires%20%D0%BA%D0%BE%D1%81%D0%BC%D0%BE%D1%81&pos=12&img_url=https%3A%2F%2Fstatic2. businessinsider.com%2Fimage%2F5ad8e5e238e2fcb0628b4818-2048%2F26623968147a3e3b45f4ak-2.jpg&rpt=simage

The IPBES conceptual framework – connecting nature and people. (2015). *Curr Opin Environ Sustainability, 14,* 1–16.

The Rio declaration on environment and development. (1992). Retrieved from http://www.prometeus.nsc.ru/koptyug/ideas/unrio92/030-031.ssi

The satellite images forest fires in the Tomsk region. (n.d.). Retrieved from https://i.mr 7.ru/photos/2012/07/640x480_84nZNFB0S4QswU4Nis0X.jpg

The system of fire safety standards. Fire safety of technological processes. Methods for assessing and analyzing fire hazard. General requirements: STB 11.05.03-2010. Enter 01/01/11. Minsk, Belarus: State. Standardization Commitee: Research Institute of Emergency Situations and Emergencies Ministry of Emergency Situations of the Republic of Belarus, 2010.

The UN Convention. On biological diversity. (2002). Retrieved from https://www.cbd.int/doc/legal/cbd-en.pdf

Thi, V. D., Khelifa, M., El Ganaoui, M., & Rogaume, Y. (2016). Finite element modelling of the pyrolysis of wet wood subjected to fire. *Fire Safety Journal*, *81*, 85–96. doi:10.1016/j.firesaf.2016.02.001

Thomas, J. C., Mueller, E. V., Santamaria, S., Gallagher, M., El Houssami, M., Filkov, A., ... Simeoni, A. (2017). Investigation of firebrand generation from an experimental fire: Development of a reliable data collection methodology. *Fire Safety Journal*, *91*, 864–871. doi:10.1016/j.firesaf.2017.04.002

Thompson, M., Zimmerman, T., Mindar, D., & Taber, M. (2016) Risk terminology primer: Basic principles and glossary for the wildland fire management community. *Gen. Tech. Rep. RMRS-GTR-349. Fort Collins, CO: U.S. Department of Agriculture, Forest Service, Rocky Mountain Research Station.*

Thompson, M., Calkin, D., Finney, M., Ager, A., & Gilbertson-Day, J. (2011). Integrated national-scale assessment of wildfire risk to human and ecological values. *Stochastic Environmental Research and Risk Assessment*, *25*(6), 761–780. doi:10.100700477-011-0461-0

Thompson, M., Haas, J., Gilbertson-Day, J., Scott, J., Langowski, P., Bowne, E., & Calkin, D. (2015). Development and application of a geospatial wildfire exposure and risk calculation tool. *Environmental Modelling & Software*, *63*, 61–72. doi:10.1016/j.envsoft.2014.09.018

Tian, X., Zhao, F., Shu, L., & Wang, M. (2013). Distribution characteristics and the influence factors of forest fires in China. *Forest Ecology and Management*, *310*, 460–467. doi:10.1016/j.foreco.2013.08.025

Tihay, V., Simeoni, A., Santoni, P. A., Rossi, L., Garo, J. P., & Vantelon, J. P. (2009). Experimental study of laminar flames obtained by the homogenization of three forest fuels. *International Journal of Thermal Sciences*, *48*(3), 488–501. doi:10.1016/j.ijthermalsci.2008.03.018

Tng, D. Y., Goosem, S., Jordan, G. J., & Bowman, D. M. (2014). Letting giants be - Rethinking active fire management of old-growth eucalypt forest in the Australian tropics. *Journal of Applied Ecology,* 51(3), 555–559. doi:. doi:10.1111/1365-2664.12233

Tohid, A., & Kaye, N. B. (2017). Stochastic modeling of firebrand shower scenarios. *Fire Safety Journal*, *91*, 91–102. doi:10.1016/j.firesaf.2017.04.039

Tolstykh, M. A. (2001). High resolution semi-lagrangian atmospheric model for numerical weather prediction, *Meteorology and Hydrology, 4*. pp. 5-15.

Tork, H. F. (2012). Spatio-temporal clustering methods classification. *Doctoral Symposium on Informatics Engineering*. 199-209.

Tulokhonov, A. K., & Puntsukova, S. D. (2016) Forest fires in the Republic of Buryatia in a changing climate. *Society: Politics, economics, law,* (3), 72-78.

Tulokhonov, A. K., & Puntsukova, S. D. (2016). Forest fires in the Republic of Buryatia in the context of climate change. [In Russian] Society: Politics, Economics, *Law*, *3*, 72–78.

Tuomi, T. J., & Mäkelä, A. (2008). Thunderstorm climate of Finland 1998–2007. *Geophysica*, *44*(1-2), 67–80.

Ukraintsev, A. V., & Plyusnin, A. M. (2015). Forest fires in Zaigraevsky district of the Republic of Buryatia in 2010–2012: Causes of fire and damage. [In Russian]. *Geography and Natural Resources*, *2*, 60–65.

Uman, M. N. (1984). *Lightning*. New York, NY: Dover Publications.

Unique Vaisala Global Lightning Dataset GLD360. Retrieved from https://www.vaisala.com/sites/default/files/documents/WEA-MET-GLD360%20Brochure-B211271EN.pdf

United Nations framework Convention on climate change. (1992). Retrieved from https://unfccc.int

University of Wyoming. Atmospheric Soundings. Retrieved from http://weather.uwyo.edu/ /upperair/sounding.html

Usenya, V. V. (2002). *Forest fires, consequences and fight against them.* Gomel, Belarus: Forestry Institute of the National Academy of Sciences of Belarus.

Ustin, S. L., Roberts, D. A., Gamon, J. A., Asner, G. P., & Green, R. O. (2004). Using imaging spectroscopy to study ecosystem processes and properties. *Bioscience*, *54*(6), 523–534. doi:10.1641/0006-3568(2004)054[0523:UISTSE]2.0.CO;2

Vakkari, V., Kerminen, V. M., Beukes, J. P., Tiitta, P., van Zyl, P. G., Josipovic, M., ... Laakso, L. (2014). Rapid changes in biomass burning aerosols by atmospheric oxidation. *Geophysical Research Letters*, *41*(7), 2644–2651. doi:10.1002/2014GL059396

Valendik, E. N., Matveev, P. M., & Sofronov, M. A. (1979). *Large forest fires*. Moscow, Russia: Science.

Van Dijk, A. I. J. M., & Bruijnzeel, L. A. (2001). Modelling rainfall interception by vegetation of variable density using an adapted analytical model. Part 1. model description. *Journal of Hydrology (Amsterdam), 247*(3-4), 230–238. doi:10.1016/S0022-1694(01)00392-4

Van Donkelaar, A., Martin, R. V., Levy, R. C., da Silva, M. A., Krzyzanowski, M., Chubarova, N. E., ... Cohen, A. J. (2011). Satellite-based estimates of ground-level fine particle matter during extreme events: A case study of the Moscow fires in 2010. *Atmospheric Environment, 45*(34), 6225–6232. doi:10.1016/j.atmosenv.2011.07.068

Van Oudenhoven, A. P. E., Martín-López, B., Schröter, M., & de Groot, R. (2018). Advancing science on the multiple connections between biodiversity, ecosystems and people. *Journal Biodiversity Science Ecosystem, l,* 14,1, 127–131.

Van Wagner, C. E. (1987). *Development and structure of the Canadian Forest Fire Weather Index System.* Can. For. Serv., Ottawa, Canada: For. Tech. Rep. 35.

Van Wagner, C. E. (1977). Conditions for the start and spread of crown fire. *Canadian Journal of Forest Research, 7*(1), 23–34. doi:10.1139/x77-004

van Wagner, C. E. (1987). *Development and structure of the Canadian forest fire weather index system. Rep. No. Forest Technology Report 35.* Ottawa, Canada: Canadian Forestry Service.

Van Wagner, C. E. (1999). Prediction of crown fire behavior in two stands of jack pine. *Canadian Journal of Forest Research, 23*, 445–449.

Varol, T., & Ertugˇrul, M. (2016). Analysis of the forest fires in the Antalya region of Turkey using the Keetch–Byram drought index. *Journal of Forest Research, 27*(4), 811–819. doi:10.100711676-016-0235-0

Vashchuk, L. N. (2014). *Irkutsk forest inventory 120 years: the stages of development of forest accounting.* Irkutsk, Russia: Forward LLC.

Verhulst, N., & Govaerts, B. (2010). The normalized difference vegetation index (NDVI) GreenSeekerTM handheld sensor: Toward the integrated evaluation of crop management. Part A: Concepts and case studies. Mexico, D.F.; CIMMYT. 2010.

Vermesi, I., DiDomizio, M. J., Richter, F., Weckman, E. J., & Rein, G. (2017). Pyrolysis and spontaneous ignition of wood under transient irradiation: Experiments and a-priori predictions. *Fire Safety Journal, 91*, 218–225. doi:10.1016/j.firesaf.2017.03.081

Viedma, O., Urbieta, I. R., & Moreno, J. M. (2018). Wildfires and the role of their drivers are changing over time in a large rural area of west-central Spain. *Scientific Reports, 8*(1), 17797. doi:10.103841598-018-36134-4 PMID:30542114

Viegas, D. X., Bovio, G., Ferreira, A., Nosenzo, A., & Sol, B. (2000). Comparative study of various methods of fire danger evaluation in southern Europe. *International Journal of Wildland Fire, 9*(4), 235–246. doi:10.1071/WF00015

Vinton, M. A., Harnett, D. C., Finck, E. J., & Briggs, J. M. (1993). Interactive effects of fire, bison (Bison bison) grazing and plant community composition in tallgrass prairie. *American Midland Naturalist, 129*(1), 10–18. doi:10.2307/2426430

Virts, K. S., Wallace, J. M., Hutchins, M. L., & Holzworth, R. H. (2013). Highlights of a new ground-based, hourly global lightning climatology. *Bulletin of the American Meteorological Society, 94*(9), 1381–1391. doi:10.1175/BAMS-D-12-00082.1

Volkova, M. A., Cheredko, N. N., & Titovskaya, A. A. (2019). Spatio-temporal distribution of periods with low and high temperatures in Western Siberia for 1961-2016. In *Proceedings International Young Scientists School and Conference on Computational Information Technologies for Environmental Sciences CITES 2019.* Moscow, Russia: IMCES.

Vorobyev, Yu. L., Akimov, V. A., & Sokolov, Yu. (2004). *Forest fires in Russia: the state and problems*. Moscow, Russia: Dex Press.

Walsh, S. F., Nyman, P., Sheridan, G. J., Baillie, C. C., Tolhurst, K. G., & Duff, T. J. (2017). Hillslope-scale prediction of terrain and forest canopy effects on temperature and near-surface soil moisture deficit. *International Journal of Wildland Fire*, *26*(3), 191–208. doi:10.1071/WF16106

Wang, C., Zheng, D., Zhang, Y., & Liu, L. (2017). Relationship between lightning activity and vertical airflow characteristics in thunderstorms. *Atmospheric Research*, *191*, 12–19. doi:10.1016/j.atmosres.2017.03.003

Wang, H. (1979). *Basic formulas and data on heat transfer for engineers: Per. from English Handbook. Moscow, Russia: Atomizdat.*

Wang, J., Rich, P. M., & Price, K. (2003). Temporal responses of NDVI to precipitation and temperature in the central Great Plains, USA. *International Journal of Remote Sensing*, *24*(11), 2345–2364. doi:10.1080/01431160210154812

Wang, L., Sousa, W. P., Gong, P., & Biging, G. (2004). Comparison of IKONOS and QuickBird images for mapping mangrove species on the Caribbean coast of Panama. *Remote Sensing of Environment*, *91*(3-4), 432–440. doi:10.1016/j.rse.2004.04.005

Wastl, C., Schunk, C., Leuchner, M., Pezzatti, G. B., & Menzel, A. (2012). Recent climate change: Long-term trends in meteorological forest fire danger in the Alps. *Agricultural and Forest Meteorology*, *162*, 1–13. doi:10.1016/j.agrformet.2012.04.001

Wastl, C., Schunk, C., Lüpke, M., Cocca, G., Conedera, M., Valese, E., & Menzel, A. (2013). Large-scale weather types, forest fire danger, and wildfire occurrence in the Alps. *Agricultural and Forest Meteorology*, *168*, 15–25. doi:10.1016/j.agrformet.2012.08.011

Weather in Tomsk. rp5 (Reliable Prognosis). Retrieved from https://rp5.ru/Weather_archive_in_Tomsk

Weber, R. (1991). Modelling fire spread through fuel beds. *Progress in Energy and Combustion Science*, *17*(1), 62–82. doi:10.1016/0360-1285(91)90003-6

Weise, D. R., White, R. H., Beall, F. C., & Etlinger, M. (2005). Use of the cone calorimeter to detect seasonal differences in selected combustion characteristics of ornamental vegetation. *International Journal of Wildland Fire*, *14*(3), 321–338. doi:10.1071/WF04035

WFAS Wildland Fire Assessment System. Official site. Retrieved from http://www.wfas.net

Wierzchowski, J., Heathcott, M., & Flannigan, M. D. (2002). Lightning and lightning fire, central cordillera, Canada. *International Journal of Wildland Fire*, *11*(1), 41–51. doi:10.1071/WF01048

Williams, G. W. (2003). References on the American Indian use of fire in ecosystems. Retrieved from https://www.nrcs.usda.gov/Internet/FSE_DOCUMENTS/nrcs144p2_051334.pdf.

Williams, A. P., Seager, R., Macalady, A. K., Berkelhammer, M., Crimmins, M. A., Swetnam, T. W., ... Rahn, T. (2015). Correlations between components of the water balance and burned area reveal new insights for predicting forest fire area in the southwest United States. *International Journal of Wildland Fire*, *24*(1), 14–26. doi:10.1071/WF14023

Williams, E. R. (1989). The triple structure of thunderstorms. *Journal of Geophysical Research*, *94*(D11), 13151–13167. doi:10.1029/JD094iD11p13151

Wilson, H. L., & Connarro, A. M. (1969). Fire breaks of many uses. *Fire Control Notes. Winter*, *3*(1), 11-13.

Won, M. S., Lee, S. Y., Lee, M. B., & Ohga, S. (2010). Development and application of a forest fire danger rating system in South Korea. *Journal of the Faculty of Agriculture, Kyushu University*, *55*(2), 221–229.

World fire map. (2019). (n.d.). Retrieved from https://fires.globalforestwatch.org/home

World fire map. (n.d.). Retrieved from https://fires.globalforestwatch.org/map/#activeLayers=viirsFires%2CactiveFires&activeBasemap=topo&activeImagery=&planetCategory=PLANET-MONTHLY&planetPeriod=Jan%200000&x=-163.652344&y=50.196019&z=2

World Meteorological Organization (WMO). (1956). World distribution of thunderstorms days [Geneva, Switzerland.]. *Publ.*, *21*, TP96.

World Wide Lightning Location Network (WWLLN). Retrieved from wwlln.net.

Wotton, B. M. (2009). Interpreting and using outputs from the Canadian forest fire danger rating system in research applications. *Environmental and Ecological Statistics*, *16*(2), 107–131. doi:10.100710651-007-0084-2

Wu, D., Zhao, X., Zhao, W., Tang, B., & Xu, W. (2014). Response of vegetation to temperature, precipitation and solar radiation time scales: A case study over mainland Australia. *Proceedings of 2014 IEEE International Geoscience and Remote Sensing Symposium (IGARSS)*, Quebec city, QC. pp. 855-858.

Xanthopoulos, G. (1990). *Development of a wildland crown fire initiation model.* (PhD thesis), University of Montana, Montana.

Xu, R., & Wunsch, D. II. (2005). Survey of clustering algorithms. *IEEE Transactions on Neural Networks*, *16*(3), 645–678. doi:10.1109/TNN.2005.845141 PMID:15940994

Yadav, I. C., & Devi, N. L. (2018). Biomass burning, regional air quality and climate change, reference module. *Earth Systems and Environmental Sciences*. Retrieved from https://www.sciencedirect.com/science/article/pii/B978012409548911022X?via%3Dihub

Yadav, I. C., & Devi, N. L. (2018). Biomass burning, regional air quality, and climate change Earth Systems and Environmental Sciences. USA: Encyclopedia of Environmental Science and Engineering.

Yagoub, Y. E., Li, Z., Musa, O. S., Anjum, M. N., Wang, F., Xu, C., & Bo, Z. (2017). Investigation of vegetation cover change in Sudan by using MODIS data. *Journal of Geographic Information System*, *9*(3), 279–292. doi:10.4236/jgis.2017.93017

Yan, E., Wang, G., Lin, H., Xia, C., & Sun, H. (2015). Phenology-based classification of vegetation cover types in Northeast China using MODIS NDVI and EVI time series. *International Journal of Remote Sensing*, *36*(2), 489–512. doi:10.1080/01431161.2014.999167

Yao, J., Raffuse, S. M., Brauer, M., Williamson, G. J., Bowman, D. M. J. S., Johnston, F., & Henderson, S. B. (2018). Predicting the minimum height of forest fire smoke within the atmosphere using machine learning and data from the CALIPSO satellite. *Remote Sensing of Environment*, *206*, 98–106. doi:10.1016/j.rse.2017.12.027

Yashwanth, B. L., Shotorban, B., Mahalingam, S., & Weise, D. R. (2015). An investigation of the influence of the heating modes on ignition and pyrolysis of woody wildland fuel. *Combustion Science and Technology*, *187*(5), 780–796. doi:10.1080/00102202.2014.973948

Ye, J., Wu, M., Deng, Z., Xu, S., Zhou, R., & Clarke, K. C. (2017). Modeling the spatial patterns of human wildfire ignition in Yunnan province, China. *Applied Geography (Sevenoaks, England)*, *89*, 150–162. doi:10.1016/j.apgeog.2017.09.012

Yi, K., Bao, Y., & Zhang, J. (2016). Spatial distribution and temporal variability of open fire in China. *International Journal of Wildland Fire*, *26*(2), 122–135. doi:10.1071/WF15213

You, W., Lin, L., Wu, L., Ji, Z., Yu, J., Zhu, J., ... He, D. (2017). Geographical information system-based forest fire risk assessment its spatiotemporal variability. *Ecological Indicators*, *77*, 176–184. doi:10.1016/j.ecolind.2017.01.042

Zagoruiko, N. G., Borisova, I. A., Dyubanov, V. V., & Kutnenko, O. A. (2008). Methods of recognition based on the function of rival similarity. *Pattern Recognition and Image Analysis*, *18*(1), 1–6. doi:10.1134/S105466180801001X

Zdereva, M. Y., & Vinogradova, M. V. (2009). Medium-range forecasting of the extent of forest fire hazard from meteorological conditions. *Russian Meteorology and Hydrology*, *34*(1), 8–15. doi:10.3103/S1068373909010026

Zhang, Y., Song, C., Band, L. E., Sun, G., & Li, J. (2017). Reanalysis of global terrestrial vegetation trends from MODIS products: Browning or Greening. *Remote Sensing of Environment*, *191*, 145–155. doi:10.1016/j.rse.2016.12.018

Zhang, Y., Zhang, X., & Liu, S. (2011). Correlation analysis on Normalized Difference Vegetation Index (NDVI) of different vegetation and climatic factors in Southwest China. *Journal of Applied Ecology*, *22*(2), 323–330. PMID:21608242

Zhao, Z., Kooperman, G. J., Pritchard, M. S., Russell, L. M., & Somerville, R. C. J. (2014). Investigating impacts of forest fires in Alaska and western Canada on regional weather over the northeastern United States using CAM5 global simulations to constrain transport to a WRF-Chem regional domain. *Journal of Geophysical Research, D, Atmospheres*, *119*(12), 7515–7536. doi:10.1002/2013JD020973

Zharikova, M. (2018) *Methodological basis of geoinformation technology of decision support in combined natural and man-made systems in destructive processes conditions.* (Doctoral Thesis), KNTU, Ukraine. p. 503.

Zharikova, M., Sherstjuk, V., & Baranovskij, N. (2015). The plausible wildfire model in geoinformation decision support system for wildfire response. In Proc. of the 15th Int. Multidisciplinary Sc. Geoconference SGEM-2015, 2(3), pp. 575–583. Albena, Bulgaria.

Zharikova, M., & Sherstjuk, V. (2015). Development of the model of natural emergencies in decision support system. *Eastern European Journal of Enterprise Technologies*, *4*(1), 62–69. doi:10.15587/1729-4061.2015.37801

Zharikova, M., & Sherstjuk, V. (2017). Threat assessment method for intelligent disaster decision support system. *Advances in Int. Systems and Computing*, *512*, 81–99.

Zhdanko, V. A., & Gritsenko, M. V. (1980). Analysis method for forest fire seasons, Practical recommendations. Leningrad.

Zhou, L., Tucker, C. J., Kaufmann, R. K., Slayback, D., Shabanov, N. V., & Myneni, R. (2001). Variations in northern vegetation activity inferred from satellite data of vegetation index during 1981 to 1999. *Journal of Geophysical Research, D, Atmospheres*, *106*(17), 20069–20083. doi:10.1029/2000JD000115

Zhurnist, V. I., Kogan, R. M., Kodyakova, T. E., Rubtsova, T. A., ... (2004). Far East Branch of the Russian Academy of Sciences. Birobidzhan, Russia: ICARP. p. 111 (in Russia)

Zink, M., Samaniego, L., Kumar, R., Thober, S., Mai, J., Schäfer, D., & Marx, A. (2016). The German drought monitor. *Environmental Research Letters*, *11*(7), Article number 074002. doi:. doi:10.1088/1748-9326/11/7/074002

About the Contributors

Nikolay Baranovskiy received Specialist Degree in Mechanics from Tomsk State University in 2000. He received PhD Degree in Physics and Mathematics from Tomsk State University in 2007. His current position is Associate Professor at Tomsk Polytechnic University. His current interests include forest fire danger and ecological impact prediction, parallel computing and spatial data analysis using GIS and RS technologies. Dr. Baranovskiy is a Member of International Association of Wildland Fire and International Association of Engineers.

* * *

Nimazhap Badmaev, Doctor of Biological Sciences, Head of the Laboratory of Soil Geography and Ecology of the Institute of General and Experimental Biology of the Siberian Branch of the Russian Academy of Sciences, Professor of the Department of Land Use and Land Cadastre of Buryat State University. The area of his scientific interests is geography, physics, and quantitative soil ecology. He developed an information-cartometric approach to assess the diversity of soils, which is a universal tool for describing, summarizing, streamlining and formalizing the available cartographic material on soils. This approach allows one to determine the quantitative parameters of the ecological niches of the soil, on the basis of which an information-recognition system has been created for factor diagnosis of the soil.

Marina Belikova graduated from Gorno-Altaisk State University in 2001 with a degree in Mathematics. Since 2001 Marina Belikova has been working in Gorno-Altaisk State University at the Department of Mathematics, Physics and Computer Science at the position of senior lecturer. Area of scientific interests: data mining, geoinformation systems.

Hiren Bhatt has more than 32 years of experience in the field of remote sensing doing microwave applications for agriculture, calibration and validation (CALVAL) of satellite data, crop acreage and crop condition assessment and conducting training and research in the field of Earth ecosystem. He has worked as Project Manager training under RISAT Utilization programme (RISAT-UP) and Forecasting Agricutural output using Space Agrometeorology and Land based observations (FASAL) programmes.

Vladimir Glagolev is a researcher from Russia, that lives in the Jewish Autonomous Region in the city of Birobidzhan. He graduated with honors from Birobidzhan State Pedagogical Institute (now Sholom-Aleichem Priamursky State University) by Specialist of Mathematics and computer science in 2002 and Master of Information systems and technologies in 2013. Since 2015, PhD., Geoecology Science, Institute of Water and Ecology problem FEB RAS, Khabarovsk, Russia, specializing in geoecology. Currently, he works at a research institute in the laboratory of regional geoecology as a researcher. The main areas of scientific activity are development the information system of fire hazard assessment of the territory, create database and electronic map forecast prediction of vegetation fires. He has 32 publications in peer-reviewed foreign and Russian journals.

Alyona Glebova graduated from Gorno-Altaisk State University in 2013 with a degree in Mathematics. Since 2012 she has been working for Gorno-Altaisk State University at the Department of Mathematics, Physics and Computer science at the position of senior lecturer. Area of scientific interests: data mining.

Pavel Goman works for the University of Civil Protection of the Ministry for Emergency Situations of the Republic of Belarus at the position of assistant professor. He has a Ph.D. degree and an associate professor certificate. The main areas of scientific activity are related to limiting the spread of forest fires, research of the process of thermal radiation effects on forest fuel, evaluating the parameters of fire breaks, protecting the population and the territory from natural and man-made disasters. Scientific results are exposed in 80 scientific publications.

Valentina Gorbatenko, head of the Department of Meteorology and Climatology of TSU (National Research Tomsk State University), was born in 1955 in the Tomsk region. She graduated from Tomsk State University and received a diploma of engineer-meteorologist in 1979. V. Gorbatenko received the Ph.D. degree in meteorology from the Tomsk Polytechnic University, Tomsk, Russia in 1993. In the year 2003 she became a Full Professor in meteorology at High Voltage Research Institute of Tomsk Polytechnic University. For over 30 years she is carrying out the research of thunderstorm activity based on the data from radar, from the meteorological stations, from the networks of lighting flash counters, and as derived from the lighting related power line outage statistics. She is author or coauthor of more than 250 scientific papers and technical reports on various aspects of lightning, geoecology, climatology.

Mikhail Kabanov is a specialist in the field of atmospheric physics, corresponding member of USSR Academy of Sciences (1987), RAS (1991), doctor of physical and mathematical sciences, professor, was born in 1937. He graduated from the Faculty of Physics of TSU in 1959, the specialty - physics (optics and spectroscopy). He worked: from 1962 to 1975 – at SPTI of TSU (senior researcher, head of the laboratory); since 1975 - the head of laboratory, the head of department, the deputy director of IOA SB AS USSR; from 1984 to 1992 - director at SPTI of TSU; from 1997 to 2008 - director of IMCES SB RAS (Tomsk).

Andrey Karanin was born on August 6, 1978. He graduated at the Gorno-Altaysk state university from 1995 to 2000. He defended the dissertation in 2004.He work at the Gorno-Altaysk State University, department of geography and environmental management.

Svetlana Karanina was born on February 11, 1980. I graduated at the Gorno-Altaisk state university from 1996 to 2001. She defended the dissertation in 2007. She works at the Gorno-Altaisk state university, depatment of mathematics, physics and informatics.

Nina Kocheeva was born on January 31, 1962. She graduated at the Saint-Petersburg (Leningrad) mining university from 1979 to 1984. She defended the dissertation in 2002. She works at the Gorno-Altaisk state university, department of geography and environmental management.

Vasily Kovyazin, Doctor of Biological Sciences, Professor, Deputy Head of the Department of Engineering Geodesy at St. Petersburg Mining University, Academician of the International Academy of Sciences for Ecology and Security (MANEB) and the Russian Academy of Natural Sciences (RANS), a member of three doctoral dissertation councils. He was awarded by the Diploma of the Ministry of Education and Science of the Russian Federation. The range of scientific interests includes the formation of sustainable ecosystems and the rational use of natural resources of megalopolises, the development of an inventory of forest and urban plantings for recreation and environmental management. Research interests: Formation of sustainable ecosystems and rational use of soil and plant resources of megacities, development of forest and city cadastres for the purposes of recreational and rational nature management.

Irina Kuzhevskaia, Associate Professor of Department of Meteorology and Climatology, National Research Tomsk State University, Tomsk, Russia. Kuzhevskaia Irina was born in Tomsk, Russia on 15 February 1974. She graduated from Tomsk State University, with a qualification in Meteorology on May 1996. She remained at the University of Tomsk to complete her PhD in Meteorology, Clymatology and Agrometeorology in 1996. After receiving her education, Kuzhevskaia Irina V. worked as a Engineer in Department of Meteorology and Climatology. Since 2007 she worked as an Associate Professor at TSU. Certificate of World Meteorological Organization by Training Manager and Trainers with competencies established in the WMO Technical Requirements for Education and Training for Meteorological, Hidrologycal and Climat Services in 2016. She has over 60 scientific papers in the field of applied meteorology.

Nadezhda Lebedeva works in Kherson State Agrarian University, but she is also a Doctor of Philosophy in the field of cultural studies, Professor of Philosophy of the International Personnel Academy, full member of the Eurasian Academy of Television and Radio. She is an author of more than 100 scientific articles. The focus of her research are state, cultural and social aspects of regional development.

Dhanapriya M. is a Research Scholar, has a two years of exemplary expertise encompassing Research, Analysis, Geographical Library maintenance, Crop Classification, Crop Area Estimation, Crop Cutting Experiment, Crop Simulation modeling, Crop Health, Accuracy Assessment, Yield Estimation, Agricultural Resource Management, and Precision Agriculture. Her articles have appeared in Society for Advancement of Human Nature (SADHNA) and won Best Paper and Oral Presentation Award in National Seminar conducted at Dr. Yashwant Singh Parmar University of Horticulture and Forestry, Solan, India. Her articles focus on qualitative analysis, interpreting results, developing agriculture and forestry and performing various image processing techniques.

Petr Nagorskiy, Professor, Doctor of Phys.-Mat. Sc., born in 1943, in 1960 he entered and in 1966 graduated from the Faculty of Radio Physics of Tomsk State University (TSU). From 1966 to 1972 - Researcher at TSU. From 1972 to 2006 - Senior Researcher at the Siberian Physical and Technical Institute at the TSU. From 2006 - Leading Researcher at the Institute for Monitoring Climate and Ecological Problems of the SB RAS. Currently – the chief researcher of this institute. His research interests are electricity from the surface layer of the atmosphere, physics of the upper atmosphere and ionosphere, and the propagation of short radio waves.

Olga V. Nosyreva associate Professor of Department of Meteorology and Climatology, National Research Tomsk State University, Tomsk, Russia. Nosyreva Olga was born on 27 February 1985 in Kiseljovsk, Kemerovo region. She studied at Tomsk State University, at the Faculty of Geology and Geography, at the Department of Meteorology and Climatology from 2002 to 2007. She remained at the University of Tomsk to complete her PhD in Meteorology, Clymatology and Agrometeorology in 2012. After graduating from graduate school, Olga Nosyreva works at the Department of Meteorology and Climatology. Since 2014, she is an Assistant Professor at the Department of Meteorology and Climatology of Tomsk State University. She has about 40 scientific papers in the field of climatology and agrometeorology.

Olga Pasko's main scientific interest is the stimulation of yields and the improvement of the quality of crops by effective and environmentally friendly methods. Author (co-author) of more than one hundred scientific papers, including 25 patents and inventions, five monographs. Chief editor of six scientific collections of articles. Awards She was awarded the Jubilee Medal "400 Years to the City of Tomsk", Diplomas of the State Duma of the Tomsk Region, the Administrations of Tomsk and Novosibirsk Regions, and the City Hall of Tomsk "For significant contribution to the development of the scientific and educational complex." Master of Sports of the USSR in rhythmic gymnastics, full member of the Eurasian Academy of Television and Radio, winner of international and Russian film and television festivals (screenwriter and director of documentaries). In 2012, she twice became the winner of the XIII Interregional Tomsk Festival of Travelers "56th parallel". She won the video competition and the literary competition in the nomination "Professional work". Research interests Natural resources, monitoring, assessment, forestry, agriculture, dendrology, land management, landscape science.

Valeriy Perminov, born in Gureevsk, Kemerovo region, studied for his master degree in mathematics at the Kemerovo State University, Russia. Then he graduated at the Tomsk State University and was awarded a PhD for research on modelling of forest fires. Worked at associate professor in mechanics of fluid and gas at the Tomsk State University. Was Deputy Director on educational work and associate professor at the Belovo Branch of Kemerovo State University from 1996 to 2011. Then he was awarded a degree of doctor of physical and mathematical sciences. He moved to the Tomsk Polytechnic University (Department of Ecology and Basic Safety) in 2012 to a position as Associate Professor. His career began at the Theory of forest fires Labs (Tomsk State University), where he gained valuable experience of mathematical modeling of forest fires and environmental pollution. He worked on numerical solution of differential equations using method of finite volume. This work led to the development of the new mathematical models and numerical methods for getting of solution. The next phase of his career was a series of research projects on the applications of these methods for mathematical modeling of forest fires and environmental pollution: the investigation of the new conjugate problems of continuum reactive multiphase media mechanics, mathematical modeling of environment pollution from the motor transport, investigation of the mechanism and limiting conditions of forest fire spread, mathematical modeling and heat and mass transfer of gas dynamics at the natural and technogenic catastrophes(mathematical prediction of the ecological consequences of the natural and technogenic catastrophes (initiation and spread of the mass forest fires, ignition of forest massifs by radiant energy as a result of nuclear bursts and Tunguska celestial body fall, the recurring radioactive contamination from forest fires). This research theme has been very fruitful and highly collaborative, involving extensive interactions with Russian Foundation for Basic Research, International Science Foundation, Moscow State University, Ministry of Forestry and etc. As part of this research, he has managed the Russian Foundation for Basic Research projects, which had the goals of investigation of the mechanism and limiting conditions of forest fires initiation and spread. He is currently a professor in the Department of Control and Diagnostics. He has published over 100 academic papers and have written two text books.

Konstantin Pustovalov is candidate of physical and mathematical sciences. Pustovalov K. was born on 28 October 1989 in Gorelovka, Tomsk region. In 2014 he graduated from National Research Tomsk State University (TSU). After graduation, he has been working at the Institute for Monitoring Climate and Ecological Systems of the Siberian Branch of the Russian Academy of Sciences (IMCES SB RAS), and since 2018 he also has been working at the Department of Meteorology and Climatology of TSU. At present time – he is researcher of IMCES SB RAS and associate professor of TSU. His research interests deals with electricity from the atmospheric surface layer, physics of the clouds and precipitation, and the mesometeorology processes.

Roman Sychev was born June 21, 1994. EDUCATION: 1) 2012-2016: Buryat State Academy of Agriculture, Institute of Land Management, Cadastre and Soil Reclamation, Bachelor's degree: "Land Management and Cadastre"; 2) 2016-2018: Buryat State Academy of Agriculture, Institute of Land Management, Cadastre and Soil Reclamation, Master's degree: "Environment and Water Use"; 3) 2018-Present: Institute of Physical Materials Science of the Siberian Branch, Russian Academy of Sciences, Postgraduate education: "Physics and Astronomy". WORK EXPERIENCE: 1) 2017, January - 2017, March: Limited Liability Company (LLC) "Geostandart" (assistant Cadastre engineer); 2) 2018, March - Present: Federal forestry Agency «Roslesinforg» (engineer of the Department for Monitoring the Use of Forests).

Vyas S. P. has more than 30 years of experience towards his scientific contribution for Agriculture and has published more than 50 papers in International Journal and Symposium with scientific reports. His support leads to receiving of ISRO team excellence Award for his contribution towards crop acreage estimation and production using satellite technology in 2007 from Dr. Abdul Kalam (Former President of India).

Marina Volkova is an Associate Professor of Department of Meteorology and Climatology, National Research Tomsk State University, Tomsk, Russia. Volkova Marina was born in Tomsk, Russia on 21 January 1974. She graduated from Tomsk State University, with a qualification in Meteorology on May, 1996. She remained at the University of Tomsk to complete her PhD in Meteorology, Clymatology and Agrometeorology in 1996. After receiving her education, Volkova Marina worked as a assistent in Department of Meteorology and Climatology. Since 2007 she worked as an Associate Professor at TSU. She has over 45 scientific papers in the field of applied meteorology.

Elena Yankovich obtained her PhD from Tomsk Polytechnic University. The focus of her research is to use geoinformation technologies in the study of the environment. Highlights of recent work include prediction of forest fire danger based on GIS and satellite imagery, methods for assessing the quality of groundwater based on the theory of the interaction of water rock and geoinformation technologies. Elena is also very active in teaching undergraduate and masters students on several courses.

Ksenia Yankovich obtained her Master's degree from ITMO University. The focus of her research is to use geoinformation technologies and remote sensing in the study of the environment.

Maryna Zharikova is currently Associate Professor of Program Tools and Technologies department at Kherson National Technical University, Ukraine. She earned her PhD in Mathematical Modeling and Computational Methods from Kharkiv National University of Radioelectronics in 2004, and the degree of Doctor of Technical Science in Information Technologies from Academy of Printing, Lviv, in 2018. A subject of her doctoral thesis is "Methodological basis of geoinformation technology of decision support in combined natural and man-made systems in destructive processes conditions". After completing her PhD, Maryna Zharikova accepted a position as Senior Lecturer of the same department. In 2008, she accepted an Associate Professor position at Information Technologies department. Maryna Zharikova has over 70 refereed publications.

George Zhuravlev is Associate Professor, Department of Meteorology and Climatology, TSU, Tomsk. In 1974 he graduated from Tomsk Polytechnic Institute. After graduation, he worked at the Research Institute of Applied Mathematics and Mechanics at TSU, since 1976 he has been working at the Department of Meteorology and Climatology of TSU. In 1998 he defended his thesis. Currently engaged in scientific and teaching activities, works as an assistant professor at the Department of Meteorology and Climatology of TSU.

Anna Zubareva is a researcher from Russia, that lives in the Jewish Autonomous Region in the city of Birobidzhan. In 2004 she graduated with honors from the Birobidzhan State Pedagogical Institute, Department of Geography and Nature Management with a degree in Geography. Since 2013, PhD., Geo-ecology Science, Institute of Water and Ecology problem FEB RAS, Khabarovsk, Russia with a degree in Geoecology. Currently she is working at a research institute in a laboratory of regional geoecology as a researcher. From 2014 till now she is a member of the International Association of Wildland Fire. The main areas of scientific activity are fire hazard assessment of the territory, vegetation fires, analysis of natural and anthropogenic factors affecting the occurrence and spread of vegetation fires. She has 25 publications in peer-reviewed foreign and Russian journals.

Index

G

H

I

J

K

L

M

N

O

P

R

S

T

U

V

W

www.ingramcontent.com/pod-product-compliance
Lightning Source LLC
LaVergne TN
LVHW080534300926
849162LV00007B/1601
9781799818670